Undergraduate Texts in Physics

Undergraduate Texts in Physics (UTP) publishes authoritative texts covering topics encountered in a physics undergraduate syllabus. Each title in the series is suitable as an adopted text for undergraduate courses, typically containing practice problems, worked examples, chapter summaries, and suggestions for further reading. UTP titles should provide an exceptionally clear and concise treatment of a subject at undergraduate level, usually based on a successful lecture course. Core and elective subjects are considered for inclusion in UTP.

UTP books will be ideal candidates for course adoption, providing lecturers with a firm basis for development of lecture series, and students with an essential reference for their studies and beyond.

More information about this series at http://www.springer.com/series/15593

Kyriakos Tamvakis

Basic Quantum Mechanics

Kyriakos Tamvakis
Department of Physics
University of Ioannina
Ioannina, Greece

ISSN 2510-411X   ISSN 2510-4128   (electronic)
Undergraduate Texts in Physics
ISBN 978-3-030-22776-0   ISBN 978-3-030-22777-7   (eBook)
https://doi.org/10.1007/978-3-030-22777-7

This Springer imprint is published by the registered company Springer Nature Switzerland AG
The registered company address is: Gewerbestrasse 11, 6330 Cham, Switzerland

To my daughters Anna and Maria

Preface

This textbook on Quantum Mechanics has been designed for use in two-semester undergraduate courses and as a supplemental textbook in graduate courses. It is true that there are many excellent books on the subject, like the classic examples of Baym, Gottfried, Messiah, or Schiff, which, although written half a century ago, are still fit to cover most of the basic material. Nevertheless, new topics have entered into the realm of interest of the present-day student of physics that have to be included in the basic material. In addition to that, each new book represents a challenge in the selection of topics, in the emphasis on each of them and the overall organization of the material, which is to a large extent subjective. The present book has come out as a result of teaching Quantum Mechanics since 1982 at the University of Ioannina. It aims to describe the basic concepts of Quantum Mechanics, to explain the use of the mathematical formalism and to provide illustrative examples of both concepts and methods. In that sense, its purpose is quite conventional as to the training of physics students. Although it is intended to provide a mastery of the use of quantum mechanics as a tool, it also provides some discussion on the meaning of quantum concepts, despite the fact that no general consensus as to what its fundamental principles mean has been reached. After a brief discussion of the basic features of Quantum Mechanics in the framework of a simplified version of the two-slit experiment, the reader is introduced to the Schroedinger equation and is familiarized with its application on a number of simple one-dimensional systems. The following chapters introduce the full mathematical formalism and basic postulates of Quantum Mechanics. Further applications follow on a number of simple systems among which two-state systems, the one-dimensional lattice, periodic potentials, etc. The following chapters are devoted to angular momentum and spin. Next, three-dimensional systems are considered with a number of applications in central potentials. The following chapters deal with many-particle systems, atoms, and molecules. A chapter on particle interactions with an external electromagnetic field is devoted to Landau levels and the Bohm–Aharonov phenomena. The next chapter on approximation methods includes the WKB method, the adiabatic approximation, variational methods, and perturbation theory with a number of applications. A chapter on symmetries deals with rotations, tensor operators, and the Wigner–Eckart theorem, discrete symmetries as well as dynamical symmetries. Scattering theory is considered next with a number of applications. A chapter on quantum behavior deals with the measurement process, the concepts of mixed states and density matrices, entanglement, the EPR issue, and Bell's theorem. Next, a chapter is devoted to the quantization of the electromagnetic field and its interaction with matter. Finally, the last chapter consists of an introduction on the path integral formulation of Quantum Mechanics.

It is important to stress that the material of this book should not be approached passively, a very important factor in the learning process being the initiative exercised by the prospecting student. To this end, apart from about 60 worked out examples within each chapter, there are around 200 problems and exercises at the end of each chapter, which can be quite helpful, if not necessary, towards achieving a command of the subject.

Ioannina, Greece   Kyriakos Tamvakis

Contents

Chapter 1
Introduction

1.1 A Thought Experiment

In order to illustrate some of the basic principles of quantum physics, we shall employ
the ideal experimental setup [1] shown in Fig. 1.1. A tungsten wire, heated by electric
current, plays the role of an *electron gun*. A cylindrical metal box with a hole in it,
kept at a suitable opposite voltage with respect to the wire, can accelerate the emitted
electrons so that some of the electrons will pass through the hole. Thus, we obtain
a beam of electrons of, more or less, the same energy. A thin metal plate with two
holes (or slits) is placed in front of the electron beam. Finally, beyond this plate there
is another plate which stops all electrons that pass through the holes. On that final
plate, we place a movable electron detector (e.g., a Geiger counter). This detector
could very well be a photomultiplier connected to a loudspeaker. The whole system
is in a vacuum in order to avoid collisions of the electrons to air molecules.[1]

The electrons that reach the detector at the final plate are all detected through
almost identical detection signals. In case that we have incorporated a loudspeaker
to the detector, these signals materialize as sharp *"click"* noises. We can count the
number of electrons that arrive in the detector in a given time interval. This number
will be a fraction of the total number of electrons that arrive on the final wall in that
time interval. Provided that the electric current that flows through the heated wire
is constant, we shall have a constant total electron flux. The electron rate varies as
we move the detector at different locations. Nevertheless, the electrons are always

---

[1]Note that this is not an easy experiment to do. In order to see the phenomena, we are about to describe
the apparatus should be microscopic. Since a number of similar realistic microscopic experiments
have been done, it makes us confident to use the above oversimplified "ideal" experiment in order
to sketch the behavior of microscopic particles. In a realistic experimental setup, the size of the slits
should be of the order of $10^{-7} - 10^{-8}$ cm, while the accelerating voltage can be of the order of
100 V. The distance of the two slits should not be more than a few per cents of a micron. Of course,
the role of the two-slit diaphragm is played by a microscopic system, such as a metallic crystal
layer.

© Springer Nature Switzerland AG 2019
K. Tamvakis, *Basic Quantum Mechanics*, Undergraduate Texts in Physics,
https://doi.org/10.1007/978-3-030-22777-7_1

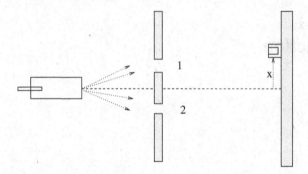

Fig. 1.1 The two-slit experiment

detected through identical sharp signals. If we were to decrease the current in the emitting wire, we would get a decreased average electron flux and a decreased rate of detection signals but the individual detection signals will stay exactly the same.

Common sense seems to dictate that the electrons arriving at the detector pass either through slit 1 or slit 2. One would expect that the plot of the average fraction of electrons detected at point x of the final plate $N(x)/N$ would be the direct sum of the corresponding fractions $N_1(x)/N$, $N_2(x)/N$ with either of the two slits closed. Note that, in the limit of large N, this fractional numbers correspond to *the probabilities* $\mathcal{P}(x)$, $\mathcal{P}_{1,2}(x)$ of finding an electron at x with both slits open, or with either slit closed. In contrast to common sense, the experimental plot is the one shown in Fig. 1.2.

This plot has the characteristic shape of the *interference pattern* encountered in the superposition of classical waves. Note that there are points x where the number of electrons with both slits open is *less* than the number with one slit closed. This cannot be interpreted as a catastrophic interference between electrons because it persists even when the flux has been decreased so that the whole apparatus is traversed by one electron at a time. Even if we were to assign to each electron some kind of wave, we would have to consider interference of this wave with itself, in contrast to the classical wave picture, where we have interference of two different waves. Nevertheless, we may adopt a pragmatic attitude for the above probability that displays the wave-like interference pattern and treat it mathematically as the intensity of a classical wave.

Fig. 1.2 The electron interference pattern

Since intensities are related to amplitudes as $I \propto |A|^2$, we introduce *a probability amplitude* $\psi(x)$ related to the probability as

$$P(x) = |\psi(x)|^2. \tag{1.1}$$

Next, we may assume that a *Superposition Principle*, analogous to the *Huygens Principle* for classical waves, holds and state that the probability amplitude for a process, such as the detection of an electron at a point x, is the superposition (linear combination) of the corresponding amplitudes for each alternative version of the process. Thus, the amplitude to detect an electron at point x will be the superposition of the corresponding amplitudes for the detection of electrons that have arrived at x through slit 1 or 2, namely,

$$\psi(x) = \psi_1(x) + \psi_2(x). \tag{1.2}$$

For the corresponding probability, we have

$$P = |\psi_1(x)|^2 + |\psi_2(x)|^2 + \psi_1(x)\psi_2^*(x) + \psi_1^*(x)\psi_2(x) \tag{1.3}$$

or

$$P(x) = |\psi_1(x)|^2 + |\psi_2(x)|^2 + 2\,|\psi_1(x)|\,|\psi_2(x)|\,\cos(\phi_1(x) - \phi_2(x)), \tag{1.4}$$

where $\phi_1(x)$ and $\phi_2(x)$ are the phases of the complex functions $\psi_{1,2}(x) = |\psi_{1,2}(x)|\,e^{i\phi_{1,2}(x)}$. Note the existence of extrema at $\phi_1(x_n) - \phi_2(x_n) = n\pi$, with n being an integer. This is sufficient to show that, in general, an interference pattern will be present. Note however that, in contrast to classical waves, here the amplitude ψ has to be a complex number, since the phases are necessary for the occurrence of an interference pattern.

The above assumption explains the experimental curve but, despite its mathematical simplicity, it seems to be in conflict with common logic since

$$P(x) \neq P_1(x) + P_2(x). \tag{1.5}$$

Since our commonsense notion that each individual electron goes through one particular slit would imply the equality sign in (1.5), we may check if indeed the observed electrons can be divided into these two classes. In order to do that, we modify our apparatus by introducing a strong light source between the two slits. A sketch of the modified apparatus is given in Fig. 1.3.

Each time that an electron passes through a particular slit, light coming from this direction is observed. Thus, each detection signal or sharp sound from the loudspeaker is accompanied by a flash of light coming from the neighborhood of a particular slit. The electrons are clearly separated in two classes, namely, those accompanied by a flash near slit 1 and those accompanied by a flash near slit 2. We, therefore, conclude that electrons pass either from slit 1 or from slit 2. No electron passes simultaneously

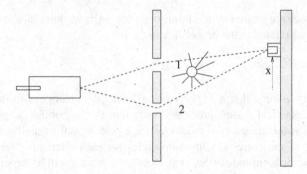

Fig. 1.3 The modified two-slit experimental apparatus

through both slits. Nevertheless, if we plot the associated probabilities we obtain the plots shown in Fig. 1.4., the dotted plot corresponding to the some of the other two.

According to these plots, the probability with both slits open is the sum of probabilities with either slit closed, namely,

$$P'(x) = P'_1(x) + P'_2(x), \qquad (1.6)$$

in agreement with common logic. No interference pattern is observed. We have no alternative but to conclude that our modification of the experimental apparatus, designed so that it can distinguish the electrons according to which slit they are coming from, has changed the final distribution of electrons and removed the interference. This is not entirely unexpected, since the light scattered on electrons transfers momentum which can modify the final distribution. What is new here is that this phenomenon cannot be entirely removed. If we decrease the intensity of the light source, we observe that each flash does not change in size. When we decrease substantially the intensity of the source, the only thing that happens is that some

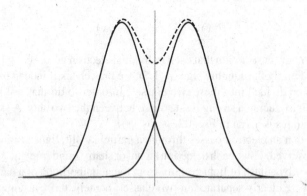

Fig. 1.4 Classical behavior

electrons are detected without an associated flash of light.[2] If we plot this last set of measurements, corresponding to electrons for which we do not know from which slit they are coming from, we recover the interference pattern. Thus, we are led to the conclusion that, whenever we are in a position to determine the path of electrons, we do not obtain an interference pattern. Equivalently, we may conclude that knowledge of the path is complementary to interference. This is a particular manifestation of a general principle, according to which the particle aspects of a system (e.g., well-defined path) are complementary to its wave aspects (e.g., interference).

Summarizing, we may generalize our conclusions as follows:

A quantum system is characterized by *probability* $\mathcal{P}$ to be in any given state. This probability is the absolute square of a complex *probability amplitude*

$$\mathcal{P} = |\psi|^2. \tag{1.7}$$

If a quantum system can be in either of two states characterized by probability amplitudes ψ_1 and ψ_2, then it can also be in any state corresponding to linear combination

$$\psi = a\,\psi_1 + b\,\psi_2. \tag{1.8}$$

This is the so-called *Superposition Principle*.

Complete knowledge of the particle aspects of a system (e.g., path) is not compatible with complete knowledge of its wave aspects (interference pattern). This is the so-called *Complementarity Principle* [2]. This principle expresses the fact that for a quantum system its particle aspects are complementary to its wave aspects.

Related to the Principle of Complementarity is the *Uncertainty Principle* [3]. In its simplest version it states that, if we make a (ideal) measurement on any object and determine its position along a given direction with uncertainty Δx, its momentum along this direction will necessarily be known with uncertainty no less than $\frac{h}{4\pi \Delta x}$ or

$$\Delta x \, \Delta p_x \geq \frac{h}{4\pi}, \tag{1.9}$$

where h is Planck's constant equal to 6.63×10^{-34} J s. Thus, viewed from a classical viewpoint, in a microscopic system, complete knowledge of x and p_x is impossible. This is an inherent physical indeterminacy of nature not related to common experimental uncertainties. The Uncertainty or Heisenberg's Principle is irrelevant at macroscopic scales due to the smallness of Planck's constant.

---

[2]Light consists of *photons* and displays particle behavior. Decreasing the intensity of the source amounts to decreasing the number of photons and not their momentum. As a result some electrons pass without encountering a photon. If we attempt to decrease not the number of photons but their frequency, then, their wavelength will increase and when it becomes of the order of magnitude of the slit distance, the associated flashes will be so fuzzy that we shall not be in a position to distinguish among flashes from different slits. In that case, we shall get an experimental plot with an interference pattern.

In order to complete the list of general principles obeyed by quantum systems, we must also include the *Correspondence Principle* [4], according to which any quantum physical quantity should tend to its classical counterpart in the macroscopic limit. Thus, Classical Mechanics should in principle be recovered in that limit. Of course, there are physical quantities with no classical counterpart (e.g., spin), which will necessarily scale with Planck's constant.

1.2   The de Broglie Relations

The analysis of the Compton effect, i.e., the inelastic scattering of photons by electrons, made clear that light, classically thought to consist of electromagnetic waves, should be associated with a particle, the *photon*. Not very long after, the reverse question was asked of whether a particle like the electron could be associated with a wave. Experimental support for this idea came from electron diffraction experiments, the analysis of which, as well as other analogous experiments, leads us to a quantitative relation between physical quantities that describe the particle aspects of the system and those that describe its wave aspects. For a free particle of definite momentum $\mathbf{p}$ and, therefore, of energy $E = p^2/2m$, the corresponding wave with amplitude ψ (probability amplitude) will be a monochromatic wave of wavelength λ and frequency ν, related to the momentum and energy of the particle through the simple relations [5]

$$\lambda = \frac{h}{p}, \quad \nu = \frac{E}{h}, \tag{1.10}$$

where h is Planck's constant. These are the so-called *de Broglie relations*. The *de Broglie wavelength* characterizes the spatial order of magnitude of quantum phenomena. For a macroscopic particle (g.e. $m = 10\,gr$, $v = 10\,m\,s^{-1}$), this length is entirely negligible. In contrast, for a microscopic particle this length and the related wave phenomena are important. For instance, for a proton ($m \sim 10^{-24}\,gr$) with velocity $v \approx 3 \times 10^5\,m\,s^{-1}$, the de Broglie wavelength is $\lambda \approx 10^{-12}\,m$. This length is of the order of magnitude of the effective proton radius $r_p = 1.2 \times 10^{-12}\,m$.

The de Broglie relations can also be expressed in terms of the *wave number* $k \equiv \frac{2\pi}{\lambda}$ and the *angular frequency* $\omega \equiv 2\pi\nu$ as

$$\mathbf{p} = \hbar\mathbf{k}, \quad E = \hbar\omega, \tag{1.11}$$

where $\hbar \equiv \frac{h}{2\pi}$. This implies the *dispersion relation*

$$\omega(k) = \frac{\hbar k^2}{2m}. \tag{1.12}$$

1.3   The Wave Function

Let $\psi(x, t)$ be the *amplitude* of the wave associated with a point particle moving in one dimension. The function $\psi(x, t)$ is called the *wave function* of the particle. The *probability*[3] of finding the particle at the point x or its vicinity is proportional to the absolute square of the wave function [6]. In fact, since space is continuous, the probability of finding the particle in an infinitesimal space interval $(x, x + dx)$ will be infinitesimal and related to the wave function as

$$dP(x, t) = |\psi(x, t)|^2 \, dx. \tag{1.13}$$

This means that the *probability density*

$$\mathcal{P}(x, t) = \frac{dP(x, t)}{dx} \tag{1.14}$$

to find the particle at some point x will be

$$\mathcal{P}(x, t) = |\psi(x, t)|^2. \tag{1.15}$$

Furthermore, the wave function of the particle (or any quantum system) characterizes fully the *state* of the system at each instant of time. This is in sharp contrast to the case of a classical particle where the state of the system is characterized by its instantaneous position $x(t)$ and velocity $v(t)$. The classical concept of a well-defined trajectory of the particle ceases to be meaningful for a quantum system of particles.

If $|\psi(x, t)|^2$ is to be a meaningful probability density, its integral over all space should give the probability of finding the particle *anywhere*. This is equal to 1, since it corresponds to certainty. This is expressed mathematically through the *wave function normalization condition*[4]

$$\int_{-\infty}^{+\infty} dx \, |\psi(x, t)|^2 = 1. \tag{1.16}$$

The concept of the wave function and that of the probability density is readily generalizable to a particle moving in three-dimensional space as

$$\psi(\mathbf{r}, t) \quad \Longrightarrow \quad \mathcal{P}(\mathbf{r}, t) = |\psi(\mathbf{r}, t)|^2 \tag{1.17}$$

and

$$\int d^3r \, |\psi(\mathbf{r}, t)|^2 = 1. \tag{1.18}$$

---

[3]For some of the introductory probability notions see the Mathematical Appendix.

[4]Note that for an acceptable wave function it is sufficient that it is *normalizable*, i.e., it is *square-integrable* or $\int dx \, |\psi(x)|^2 = C < \infty$. Then, we can always define a *normalized* wave function $\psi_N(x) = \psi(x)/\sqrt{C} \implies \int_{-\infty}^{+\infty} dx \, |\psi_N(x)|^2 = 1$.

A final remark before we close this section is that the wave function is in general a complex number, although its absolute square is always a nonnegative real number as a probability distribution should be. A second remark is that the normalization condition should hold for all times or, equivalently, probability should be conserved at all times, meaning that it is not possible for a particle to disappear or appear from nothingness.

1.4   A Free Particle of Definite Momentum

Consider now a free particle of definite momentum p moving in one dimension. Its energy will be $E = \frac{p^2}{2m}$. Let $\psi(x, t)$ be the amplitude of the probability wave associated with the particle, i.e., its wave function. Since the definite momentum p through the de Broglie relations corresponds to a definite wave number $k = p/\hbar$, this wave function will correspond to a *monochromatic* wave. Such a wave traveling to the right[5] will have an amplitude of the form

$$\psi_p(x, t) = A \cos(kx - \omega t) + B \sin(kx - \omega t). \tag{1.19}$$

Note however that for a free particle which is not subject to any kind of force, each point in space should appear identical and the probability density to find the particle should be independent of x. This amounts to

$$|\psi_p(x, t)|^2 = |A|^2 \cos^2(kx - \omega t) + |B|^2 \sin^2(kx - \omega t)$$

$$+ \left(AB^* + A^*B\right) \cos(kx - \omega t) \sin(kx - \omega t)$$

being space independent. This can happen only if the coefficients are related by $A = \pm i B$. In that case, we have $|\psi_p(x, t)|^2 = |A|^2$ and

$$\psi_p(x, t) = A e^{i(kx - \omega t)}. \tag{1.20}$$

The case $A = -i B$ corresponds to the same situation replacing ψ_p with its complex conjugate. The plane wave (1.20) has an infinite spatial extension and gives the constant probability density $|A|^2$. The particle can be observed at any point in space with the same likelihood. This corresponds to an infinite uncertainty for the position of the particle ($\Delta x = \infty$). Note that this is found to be in complete agreement with the so-called *Uncertainty Principle or Heisenberg's Inequality*, which will be elaborated later on in this book. According to it the product of the uncertainties in position and momentum has to be larger than Planck's constant divided by 4π. Since the particle has definite momentum p and the corresponding uncertainty in momentum vanishes ($\Delta p = 0$), the product $(\Delta x)(\Delta p)$ can be finite and does not violate Heisenberg's

---

[5]Alternatively, a monochromatic wave traveling to the left will have a phase $kx + \omega t$.

inequality. It should be stressed however that a particle of definite momentum is an extreme abstraction and idealization. Realistic particles always have an uncertainty δp in their momentum value, having a momentum in an interval of size δp. This limits the extent of the corresponding wave function to a finite spatial interval of size $\sim \hbar/\delta p$ instead of the full real line.

An immediate consequence of the constancy of the probability density $|\psi_p(x, t)|^2$ is that this wave function *is not square-integrable* and cannot satisfy the normalization condition

$$\int_{-\infty}^{+\infty} dx \, |\psi_p(x, t)|^2 = |A|^2 \int_{-\infty}^{+\infty} dx = \infty. \tag{1.21}$$

Nevertheless, as we remarked above this is a consequence of the fact that this state, i.e., a particle with a definite momentum, is an extreme idealization. Square-integrable wave functions describing realistic particles can always be constructed as linear superpositions of plane waves of definite momentum.[6]

Closing this section we may return to (1.20) and fix by convention[7] the factor A (which otherwise should have been fixed by the normalization condition) to $A = \frac{1}{\sqrt{2\pi\hbar}}$. Thus, finally, the *"plane wave"* wave function corresponding to a free particle of momentum p moving in one or three dimensions will be

$$\psi_p(x, t) = \frac{e^{\frac{i}{\hbar}(px-Et)}}{\sqrt{2\pi\hbar}}, \qquad \psi_{\mathbf{p}}(\mathbf{r}, t) = \frac{e^{\frac{i}{\hbar}(\mathbf{p}\cdot\mathbf{r}-Et)}}{(2\pi\hbar)^{3/2}}. \tag{1.22}$$

Example 1.1 (*Free particle motion in the half-line*) Consider a particle of given energy E that can move freely in one dimension for $x > 0$ but experiences an infinite repulsive force for $x \leq 0$. Are there any other points at which the probability density to find the particle vanishes? (Fig. 1.5)

As a result of the infinite repulsive force the particle cannot penetrate in the region $x \leq 0$ and its probability density there will have to vanish or, equivalently, its wave function will be vanishing there

$$\psi(x \leq 0) = 0. \tag{1.23}$$

Nevertheless, in the region $x > 0$ the particle can move freely. Its wave function can be either

$$\psi_p(x) = \frac{e^{\frac{i}{\hbar}(px-Et)}}{(2\pi\hbar)^{1/2}} \quad \text{or} \quad \psi_{-p}(x) = \frac{e^{-\frac{i}{\hbar}(px+Et)}}{(2\pi\hbar)^{1/2}}. \tag{1.24}$$

---

[6]The *Superposition Principle*—i.e., the fact that any linear combination of wave functions is an acceptable wave function—enables us to construct arbitrary normalizable wave functions out of these non-normalizable wave functions of definite momentum. For example, the wave functions $\psi_{p,\delta}(x, t) = \frac{1}{\sqrt{\delta}} \int_p^{p+\delta} dp' \psi_{p'}(x, t)$. It can be easily shown that these are normalizable. See Messiah [7].

[7]This corresponds to *"delta function normalization"* to be employed later on.

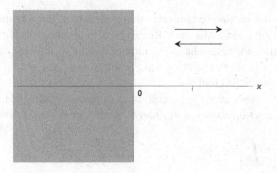

Fig. 1.5 Free motion in the half-plane

Thus, the wave function of the particle in this region must be a linear combination of the above. The overall wave function will be

$$\psi_E(x, t) = \psi_E(x)e^{-\frac{i}{\hbar}Et} \quad \text{with} \quad \psi_E(x) = \begin{cases} 0 & (x \leq 0) \\ A e^{\frac{i}{\hbar}px} + B e^{-\frac{i}{\hbar}px} & (x > 0), \end{cases} \tag{1.25}$$

where $E = p^2/2m$. The probability, as a measurable physical quantity, has to be continuous everywhere. As a result, the wave function also has to be continuous and, therefore, we must have $A + B = 0$ at $x = 0$. Thus, we are led to

$$\psi_E(x, t) = \begin{cases} 0 & (x \leq 0) \\ 2iA \, \sin(px/\hbar) \, e^{-\frac{i}{\hbar}Et} & (x \geq 0). \end{cases} \tag{1.26}$$

The corresponding probability density will be

$$\mathcal{P}_E(x, t) = \begin{cases} 0 & (x \leq 0) \\ |A|^2 \, \sin^2(px/\hbar) & (x \geq 0). \end{cases} \tag{1.27}$$

The points of vanishing probability—at which the particle *cannot be*—are

$$\forall x \leq 0 \quad \text{and} \quad x_n = \frac{n\pi\hbar}{\sqrt{2mE}} \quad (n = 1, 2, \ldots). \tag{1.28}$$

Example 1.2 (Particle in a circle) Consider a particle of definite energy that can move freely in a circle of radius R. Find the allowed values of the energy.

The position of the particle can be expressed in terms of an angle θ according to

$$x = \theta R \quad \text{with} \quad 0 \leq \theta < 2\pi . \tag{1.29}$$

The values

$$x \quad and \quad x + 2\pi R \tag{1.30}$$

correspond to the same point. The wave function will be

$$\psi_p(x, t) = A e^{\frac{i}{\hbar}(xp - Et)} . \tag{1.31}$$

In order to have a single-valued wave function, we must have

$$e^{\frac{i}{\hbar} 2\pi R p} = 1 \implies p_n = n \frac{\hbar}{R} \quad for \quad n = \pm 1, \pm 2, \ldots \tag{1.32}$$

Thus, the allowed energies are the discrete values

$$E_n = \frac{\hbar^2 n^2}{2m R^2} . \tag{1.33}$$

As a result of the compact nature of the space in which the particle moves, the values of the energy are *"quantized"*. This is in contrast to a particle moving in infinite space where the energy values span a continuum. The corresponding wave functions are $\psi_n(\theta, t) = A e^{in\theta} e^{-i E_n t / \hbar}$. The coefficient A can be determined from the normalization condition, demanding that the total probability is unity, i.e.,

$$\int_0^{2\pi R} dx \, |\psi(x, t)|^2 = 1 \implies 2\pi R |A|^2 = 1$$

or

$$\psi_n(\theta, t) = \frac{e^{in\theta}}{\sqrt{2\pi R}} . \tag{1.34}$$

Note the *"degeneracy"* $E_n = E_{-n}$, meaning that $\psi_n(x, t)$, i.e., a right-moving particle and $\psi_{-n}(x, t)$, i.e., a left moving one, have the same energy.

1.5 The Schroedinger Equation for a Free Particle

The free particle wave function (1.20) is manifestly a differentiable function of space and time. Specifically, we have

$$\frac{\partial \psi_p}{\partial t} = -\frac{i}{\hbar} E \psi_p \quad \textbf{and} \quad \frac{\partial^2 \psi_p}{\partial x^2} = -\frac{p^2}{\hbar^2} \psi_p . \tag{1.35}$$

However, in view of the energy–momentum relation $E = p^2/2m$, the above derivatives satisfy

$$-\frac{\hbar^2}{2m} \frac{\partial^2 \psi_p}{\partial x^2} = i\hbar \frac{\partial \psi_p}{\partial t} . \tag{1.36}$$

This is the *Schroedinger equation* satisfied by the plane wave (1.20).

Notice that (1.36) is a linear partial differential equation. An immediate consequence of that is that any linear combination (*superposition*) of solutions will also be a solution. Acting with the *"Schroedinger differential operator"* on a superposition

$$\psi(x, t) = \sum_p g_p \psi_p(x), \tag{1.37}$$

we get

$$\left(-\frac{\hbar^2}{2m}\frac{\partial^2}{\partial x^2} - i\hbar\frac{\partial}{\partial t} \right) \sum_p g_p \psi_p = \sum_p g_p \left(-\frac{\hbar^2}{2m}\frac{\partial^2}{\partial x^2} - i\hbar\frac{\partial}{\partial t} \right) \psi_p = 0$$

or

$$-\frac{\hbar^2}{2m}\frac{\partial^2 \psi}{\partial x^2} = i\hbar\frac{\partial\psi}{\partial t}. \tag{1.38}$$

Thus, the superposition of plane waves considered in (1.37) satisfies also the same Schroedinger equation.

For a particle of any energy, moving freely in infinite space, there is no restriction on its momentum values and, therefore, p should be a continuous variable taking values over the full real line. Therefore, the symbolic sum in (1.37) should be replaced with an integral

$$\psi(x, t) = \int_{-\infty}^{+\infty} \frac{dp}{\sqrt{2\pi\hbar}}\, g(p)\, e^{\frac{i}{\hbar}(px - Et)}. \tag{1.39}$$

From a mathematical point of view, (1.39) is a Fourier transform.[8] Thus, any free particle of a general wave function (1.39) satisfies the Schroedinger equation.

The Schroedinger equation is a *time-evolution equation* of first order. Therefore, $\psi(x, t)$ can be determined for any time t in terms of $\psi(x, 0)$ alone. Since the second spatial derivative appears in the equation, not only $\psi(x, t)$ should be continuous (and, of course, differentiable) but the first derivative $\partial_x\psi$ should also be continuous.

Notice that the Schroedinger equation is a *complex* equation, since i is manifestly present in it. As a result, even if the superposition (1.39) were to be chosen real, in the course of time the wave function would necessarily develop an imaginary part. This is in contrast to classical waves which can always be chosen to be real.

Closing this section we may write down the three-dimensional version of (1.38) and (1.39) by replacing $\frac{\partial^2}{\partial x^2}$ with $\frac{\partial^2}{\partial x^2} + \frac{\partial^2}{\partial y^2} + \frac{\partial^2}{\partial z^2} = \nabla^2$. We obtain

$$-\frac{\hbar^2}{2m}\nabla^2\psi(\mathbf{r}, t) = i\hbar\frac{\partial}{\partial t}\psi(\mathbf{r}, t), \tag{1.40}$$

where

---

[8] See the Mathematical Appendix.

$$\psi(\mathbf{r}, t) = \int \frac{d^3 p}{(2\pi\hbar)^{3/2}} \, g(\mathbf{p}) \, e^{\frac{i}{\hbar}(\mathbf{p}\cdot\mathbf{r} - Et)}. \tag{1.41}$$

1.6 Wave Packets

In contrast to a particle of sharply defined momentum, which has a wave function of infinite extension, a realistic particle will have a wave function of a finite spatial size, meaning that the probability to find the particle is nonzero only in a finite spatial region.[9] A characteristic example is a *Gaussian wave packet*, corresponding to a wave function having the form of the Gauss distribution. To be specific, let's consider a free particle described by the wave function

$$\psi(x) = \left(\frac{2\alpha}{\pi}\right)^{1/4} e^{-\alpha x^2}. \tag{1.42}$$

It can be checked that (1.42) satisfies the correct normalization condition[10]

$$\int_{-\infty}^{+\infty} dx \, |\psi(x)|^2 = 1. \tag{1.43}$$

The parameter $\alpha > 0$ determines the extent of the localization of the particle. For points beyond $\pm 1/\sqrt{2\alpha}$, the probability density to find the particle diminishes rapidly, being less than $e^{-1} \approx 37\%$ of its maximum value. Thus, we may roughly designate

$$\Delta x \sim \frac{1}{\sqrt{2\alpha}} \tag{1.44}$$

as the *uncertainty* in the position of the particle. Note that in the limit $\alpha \to 0$ the wave function becomes approximately constant and the uncertainty becomes very large. In the opposite limit $\alpha \to \infty$, the wave function is almost zero everywhere except at the point $x = 0$, where it becomes very large. In this limit, the particle is localized at the origin. For a value of the parameter α between these extremes, the wave packet $\psi(x)$ has a finite size.

The Gaussian wave packet $\psi(x)$ is manifestly a *square-integrable* function (g.e. condition (1.43)) and, therefore, can be Fourier transformed according to

$$\psi(x) = \int_{-\infty}^{+\infty} \frac{dp}{\sqrt{2\pi\hbar}} \, g(p) \, e^{\frac{i}{\hbar}px}. \tag{1.45}$$

---

[9]For a more general treatment of wave packets see Merzbacher [8].

[10]See the section on Gaussian Integrals in the Mathematical Appendix.

This is exactly the kind of expression employed before in (1.39) when we discussed the general solution of the Schroedinger equation as a superposition of plane waves.[11]

We may take a moment to discuss the physical interpretation of the function $g(p)$. Since $\psi_p(x) = e^{\frac{i}{\hbar}px}/\sqrt{2\pi\hbar}$ stands for the probability amplitude to find the particle at the position x having momentum p, in order to obtain the probability amplitude to find the particle at the position x whatever its momentum is, we would have to multiply[12] $\psi_p(x)$ by the probability amplitude to find the particle having momentum p whatever its position, and, then, sum over all possible momenta. Thus, we may read from (1.45) that $g(p)$ should be interpreted as the *probability amplitude to find the particle with momentum p*. This justifies the fact that the name *momentum wave function* is often used for $g(p)$. As in the case of the standard wave function, the probability density to find the particle with momentum p is equal to the absolute square $\Pi(p) = |g(p)|^2$. It is interesting that by *Plancherel's Theorem*, we have

$$1 = \int_{-\infty}^{+\infty} dx |\psi(x)|^2 = \int_{-\infty}^{+\infty} dp\, |g(p)|^2 \tag{1.46}$$

and the probability that the particle has any momentum equals one as it should.

By virtue of the Fourier theorem (1.45) implies the *inverse Fourier transform*

$$g(p) = \int_{-\infty}^{+\infty} \frac{dx}{\sqrt{2\pi\hbar}} e^{-\frac{i}{\hbar}px} \psi(x). \tag{1.47}$$

Substituting $\psi(x) = (2\alpha/\pi)^{1/4} e^{-\alpha x^2}$ and carrying out the integration we obtain

$$g(p) = \frac{e^{-\frac{p^2}{4\hbar^2\alpha}}}{(2\pi\alpha\hbar^2)^{1/4}}. \tag{1.48}$$

In Fig. 1.6., we have plotted the position and momentum (dotted line) probability densities. From the Gaussian form of this amplitude, it is clear that the probability to find the particle with momentum beyond $\pm\hbar\sqrt{2\alpha}$ is negligible. Thus, we may conclude that the momentum uncertainty is roughly

$$\Delta p \sim \hbar\sqrt{2\alpha}. \tag{1.49}$$

Therefore, the product of uncertainties

$$(\Delta x)(\Delta p) \sim \hbar \tag{1.50}$$

is bounded and independent of α, in accordance with the anticipated Heisenberg's inequality.

---

[11] In this section the whole analysis refers to a particular instant of time taken to be $t = 0$. Thus, $\psi(x)$ stands for $\psi(x, 0)$.

[12] See the note on conditional probability in the section on Probability of the Mathematical Appendix.

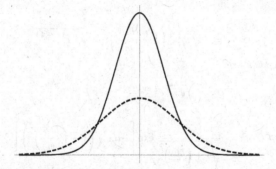

Fig. 1.6 Gaussian position/momentum distributions

Example 1.3 (Time-evolved wave function from the Schroedinger equation) A free particle is at $t = 0$ described by the Gaussian wave packet

$$\psi(x, 0) = (\alpha/\pi)^{1/4} e^{-\frac{\alpha}{2}x^2} . \tag{1.51}$$

Use the Schroedinger equation to obtain $\psi(x, t)$ at $t > 0$. Hint: Introduce the Ansatz

$$\psi(x, t) = N(t)\, e^{-\frac{1}{2}A(t)x^2} \tag{1.52}$$

into the free Schroedinger equation and solve for $N(t)$ and $A(t)$.

Introducing this Ansatz into the Schroedinger equation,

$$-\frac{\hbar}{2m}\psi'' = i\hbar\dot\psi \implies -\frac{\hbar}{2m}\frac{\partial^2}{\partial x^2}\left(N(t)e^{-\frac{1}{2}Ax^2}\right) = i\hbar\frac{\partial}{\partial t}\left(Ne^{-\frac{1}{2}Ax^2}\right)$$

we obtain

$$\frac{\hbar^2}{2m}NA - i\hbar\dot N + x^2\left(-i\frac{\hbar}{2}N\dot A + \frac{\hbar^2}{2m}NA^2\right) = 0$$

or

$$\frac{\dot A}{A^2} = -\frac{i\hbar}{m} \implies A(t) = \frac{\alpha}{1 - \frac{i\hbar t\alpha}{m}} \tag{1.53}$$

and

$$\frac{\dot N}{N} = -\frac{i\hbar}{2m}A \implies N(t) = (\alpha/\pi)^{1/4}e^{-\frac{i\hbar\alpha}{2m}\int_0^t \frac{dt}{1-\frac{i\hbar t\alpha}{m}}} = (\alpha/\pi)^{1/4}\left(1 - \frac{i\hbar t\alpha}{m}\right)^{-1/2} . \tag{1.54}$$

Finally,

$$\psi(x,t) = (\alpha/\pi)^{1/4} \left(1 - \frac{i\hbar t\alpha}{m}\right)^{-1/2} e^{-\frac{\alpha x^2}{2\left(1 - \frac{i\hbar t\alpha}{m}\right)}} .$$

This can be set in a *polar form*

$$\psi = \sqrt{\psi\psi^*} \, e^{\frac{1}{2}\ln(\psi/\psi^*)} = \sqrt{\mathcal{P}} \, e^{i\phi}$$

with

$$\mathcal{P}(x,t) = \frac{\sqrt{\alpha/\pi}}{\sqrt{1 + \left(\frac{\hbar t\alpha}{m}\right)^2}} \exp\left[-\alpha x^2 / \left(1 + \left(\frac{\hbar t\alpha}{m}\right)^2\right)\right] \tag{1.55}$$

and

$$\phi(x,t) = \frac{1}{2}\arctan(\hbar t\alpha/m) - \frac{\hbar t\alpha^2 x^2/2m}{\left(1 + \left(\frac{\hbar t\alpha}{m}\right)^2\right)} . \tag{1.56}$$

Problems and Exercises

1.1 The wave function of a free particle at a particular instant (taken to be $t = 0$) is real, i.e.,

$$\psi(x,0) = \psi^*(x,0) .$$

Show that the evolved wave function at time t has the property

$$\psi(x,t) = \psi^*(x,-t) .$$

1.2 A particle constrained to move in a circle of radius R has initially ($t = 0$) a wave function

$$\psi(\theta,0) = \frac{1}{2\sqrt{\pi R}} \left(e^{i\pi N\theta} + e^{i(N+1)\pi\theta}\right) ,$$

where N is a given integer. Calculate the evolved probability density $P(\theta, t)$ at times $t > 0$ and show that it is periodic in time.

1.3 Consider the wave functions

$$\psi_p(x) = N \, e^{\frac{i}{\hbar}xp - \epsilon x^2/2} \text{ with } \epsilon \to 0 .$$

Show that, with a suitably chosen coefficient N, they satisfy the following *orthonormality relation*:

$$\int_{-\infty}^{+\infty} dx \, \psi_p^*(x)\psi_{p'}(x) = \begin{cases} 1 & (p = p') \\ 0 & (p \neq p'). \end{cases}$$

Show also the following *completeness relation*:

$$\int_{-\infty}^{+\infty} dp \, \psi_p(x)\psi_p^*(x') = 0 \ \ for \ \ x \neq x'.$$

1.4 Consider the Gaussian wave packet of *Example* 1.3 Show that the Heisenberg inequality (uncertainty relation) holds at any time $t > 0$.

1.5 A free particle has at a particular instant t_0 an *even* wave function, i.e., one satisfying

$$\psi(x, t_0) = \psi(-x, t_0).$$

Show that this property holds at all times, i.e.,

$$\psi(x, t) = \psi(-x, t) \ \ \forall t.$$

Show that this property holds also for the case of an *odd* wave function.

1.6 Consider a particle of energy $E = p^2/2m$ whose wave function is the following superposition of two plane waves of opposite direction:

$$\psi(x, t) = \frac{A}{\sqrt{2\pi\hbar}} e^{ipx/\hbar - iEt/\hbar} + \frac{B}{\sqrt{2\pi\hbar}} e^{-ipx/\hbar - iEt/\hbar},$$

where $A = |A|e^{i\phi_a}$ and $B = |B|e^{i\phi_b}$ are complex coefficients. Find the points in space characterized by vanishing probability density.

1.7 Any wave function can be written in a *polar form*, i.e., as $\psi(x, t) = (\mathcal{P}(x, t))^{1/2} e^{i\phi(x,t)}$. Prove the relation

$$\frac{\hbar}{m} \, \phi'(x, t)\mathcal{P}(x, t)\Big|_{-a}^{a} = -\frac{dP_\Delta}{dt},$$

where $P_\Delta(t) = \int_{-a}^{a} dx \, \mathcal{P}(x, t)$ is the probability to find the particle in a spatial region $\Delta = [-a, a]$. Verify this relation for the wave function of *Example* 1.3.

1.8 The Schroedinger equation for a free particle is invariant under *spatial translations* ($x \to x + \alpha$) as well as *time translations* ($t \to t + \beta$). Show that the corresponding momentum probability density is not affected by spatial translations of the wave function.

1.9 Consider the general wave function for a free particle

$$\psi(x, t) = \int_{-\infty}^{+\infty} \frac{dp}{(2\pi\hbar)^{1/2}} g(p) e^{ipx/\hbar - iEt/\hbar}$$

and by virtue of the inverse Fourier transform for $g(p)$ rewrite it as

$$\psi(x, t) = \int dx' \mathcal{K}(x, x'; t)\, \psi(x', 0) \quad \text{where} \quad \mathcal{K}(x, x'; t) = \int \frac{dp}{2\pi\hbar} e^{ip(x-x')/\hbar - iEt/\hbar}.$$

Calculate the amplitude $\mathcal{K}(x, x'; t)$. What is its physical interpretation?

1.10 Consider the system of a particle constrained to move (otherwise freely) on a circle of radius R. Periodicity leads to momentum quantization according to $p_n = n\pi\hbar/R$ with $n = \pm 1, \pm 2, \ldots$ (*Example* 1.2.). The wave functions corresponding to definite momentum and energy $E_n = \hbar^2 n^2 \pi^2 / 2mR^2$ are $\psi_n(\theta) = e^{in\theta}/\sqrt{2\pi R}$. The general wave function of the system, in terms of the unknown coefficients ψ_n, can be written as[13]

$$\psi(\theta, t) = \sum_{n=-\infty}^{+\infty} \psi_n\, e^{in\theta}\, e^{-iE_n t/\hbar}.$$

Assuming that the initial $(t = 0)$ wave function of the system is the function $\psi(\theta, 0) = 2\cos^2\theta/\sqrt{3\pi}$, determine the exact evolved wave function at times $t > 0$.

References

1. M. Sands, R. Feynman, R. Leighton, Lectures on Physics, vol. III. (Pearson, 1970)
2. N. Bohr, The quantum postulate and the recent development of atomic theory. Nature **121**, 580–590 (1928)
3. W. Heisenberg, The Physical Contents of Quantum Kinematics and Mechanics, in *Quantum Theory and Measurement*, ed. by J.A. Wheeler, W.H. Zurek (Princeton University Press, Princeton, 1983), p. 62
4. N. Bohr, in *Niels Bohr, Collected Works; The Correspondence Principle*, vol. 3, ed. by L. Rosenfeld, J. Nielsen (Amsterdam, North-Holland, 1976)
5. L. De Broglie, Radiation-waves and quanta. Comptes Rendus (in French) **177**, 507–510, 548, 630; Physical aspects of quantum mechanics. Nature **119**(2992), 354–357 (1923)
6. M. Born, Physical aspects of quantum mechanics. Nature **119**(2992), 354–357 (1927)
7. A. Messiah, Quantum Mechanics. (Dover publications, Mineola, 1958). Single-volume reprint of the Wiley, New York, two-volume 1958 edn
8. E. Merzbacher, *Quantum Mechanics*, 3rd edn. (Wiley, New York, 1998), pp. 18–22

---

[13]Note that $\int_0^{2\pi} d\theta\, e^{i(n-n')\theta} = 2\pi\delta_{n,n'}$.

Chapter 2
The Schroedinger Equation

2.1 The Fundamental Time-Evolution Equation

In the previous chapter, we saw that starting from the particular form $e^{i(kx-\omega t)}$ for the wave function of a particle of definite momentum $p = \hbar k$ and energy $E = \hbar \omega$, we were led to a wave equation satisfied by a general free particle wave function. Alternatively, we could have started by postulating the wave equation, namely, the *Schroedinger equation* (1.38), and subsequently derive particular solutions like the above plane wave.

It is not difficult to see that for the case of a particle moving in a constant potential V_0 we may write down the appropriate Schroedinger equation with a minimal modification. Indeed, the equation

$$-\frac{\hbar^2}{2m}\nabla^2 \psi(\mathbf{r},\, t) + V_0 \psi(\mathbf{r},\, t) = i\hbar \frac{\partial}{\partial t}\psi(\mathbf{r},\, t)$$

has plane wave solutions $e^{\frac{i}{\hbar}(\mathbf{p}\cdot\mathbf{r} - Et)}$ with the correct energy–momentum relation $E = \frac{p^2}{2m} + V_0$.

Since any smooth potential function is approximately constant in the vicinity of a particular point, it is a reasonable generalization to assume that the general wave equation for a particle of mass m moving under the influence of a potential $V(\mathbf{r})$ is

$$i\hbar\frac{\partial}{\partial t}\psi(\mathbf{r},\, t) = \left(-\frac{\hbar^2}{2m}\nabla^2 + V(\mathbf{r})\right)\psi(\mathbf{r},\, t). \tag{2.1}$$

This is the Schroedinger equation for an interacting particle.[1]

Again we note that this is a complex equation and if we consider its complex conjugate, this will be a different equation

---

[1] For an unorthodox *"derivation"* of the Schroedinger equation, see [1].

© Springer Nature Switzerland AG 2019
K. Tamvakis, *Basic Quantum Mechanics*, Undergraduate Texts in Physics,
https://doi.org/10.1007/978-3-030-22777-7_2

$$-i\hbar\frac{\partial}{\partial t}\psi^*(\mathbf{r},\,t) = \left(-\frac{\hbar^2}{2m}\nabla^2 + V(\mathbf{r})\right)\psi^*(\mathbf{r},\,t),\tag{2.2}$$

and, even if we were to start with an initially real wave function, time evolution will generate a nonzero imaginary part. We shall always assume that the potential function is a real function. It should also be stressed that in the presence of forces the general solution of the Schroedinger equation is not anymore a superposition of plane waves as in the free case. Nevertheless, we can always write the wave function as a Fourier transform

$$\psi(\mathbf{r},\,t) = \int \frac{d^3p}{(2\pi\hbar)^{3/2}}\,g(\mathbf{p},\,t)\,e^{\frac{i}{\hbar}\mathbf{p}\cdot\mathbf{r}},\tag{2.3}$$

where the momentum space wave function $g(\mathbf{p},\,t)$ does not have the simple exponential time evolution as in the free particle case.[2]

Example 2.1 (*Time-evolved wave function of a particle subject to a harmonic force*) Consider a particle moving in one dimension under the influence of a so-called harmonic oscillator potential $V(x) = \frac{1}{2}m\omega^2 x^2$. Assume that the particle is initially ($t = 0$) in a state described by a Gaussian wave function

$$\psi(x, 0) = (\alpha/\pi)^{1/4}\,e^{-\alpha x^2/2}\tag{2.4}$$

with $\alpha > 0$. Use the Schroedinger equation to show that the evolved wave function at time $t > 0$ will also be Gaussian.

We may consider the ansatz $\psi(x, t) = N(t)\,e^{-\frac{1}{2}A(t)x^2}$ with $N(0) = (\alpha/\pi)^{1/4}$ and $A(0) = \alpha$. Substituting this ansatz into the Schroedinger equation, we obtain

$$\frac{\dot{N}}{N} = -\frac{i\hbar}{2m}A \quad \text{and} \quad \frac{\dot{A}}{A^2 - \left(\frac{m\omega}{\hbar}\right)^2} = -\frac{i\hbar}{m}.\tag{2.5}$$

Integrating with respect to $A(t)$, we obtain

$$A(t) = \alpha\left(\frac{1 + i\beta\tan(\omega t)}{1 + i\beta^{-1}\tan(\omega t)}\right),\tag{2.6}$$

where we have set $\beta = m\omega/\alpha\hbar$. The normalization factor is

$$N(t) = (\alpha/\pi)^{1/4}\,e^{-\frac{i\hbar}{2m}\int_0^t dt\,A(t)},$$

---

[2]For a discussion on the general properties of the solutions of the Schroedinger equation see [2].

postponing the integration in the exponent for later. The probability density is

$$P(x, t) = \sqrt{\frac{\alpha}{\pi}} \, e^{\frac{\hbar}{m} \int_0^t dt' \, Im(A(t'))} \, e^{-\frac{1}{2} Re(A(t)) x^2} \,,$$

the real and imaginary parts of A being

$$Re(A) = \alpha \left(\frac{1 + \tan^2(\omega t)}{1 + \beta^{-2} \tan^2(\omega t)} \right) \quad and \quad Im(A) = \alpha(\beta - \beta^{-1}) \frac{\tan(\omega t)}{1 + \beta^{-2} \tan^2(\omega t)} \,.$$

After a simple integration in the exponent we obtain

$$P(x, t) = \left(\frac{\alpha \beta^2 / \pi}{(\beta^2 - 1) \cos^2(\omega t) + 1} \right)^{1/2} e^{-\frac{\alpha}{2} x^2 \left(\frac{1 + \tan^2(\omega t)}{1 + \beta^{-2} \tan^2(\omega t)} \right)} \,. \tag{2.7}$$

This is a periodic function with an oscillating spread Δx in the interval defined by $\sqrt{2/\alpha}$ and $\beta^{-1} \sqrt{2/\alpha}$, the upper bound being determined by whether β is smaller or larger than 1.

Example 2.2 (Motion in the half-plane) A particle of definite energy E moves in two dimensions. Its motion is restricted in the half-plane $(x < 0, \, y)$ where it moves freely, while in the other half it experiences an infinite repulsive force. Determine its wave function (Fig. 2.1).

Since the particle cannot enter the $(x \geq 0, \, y)$ region its wave function there must vanish. In the $(x < 0, \, y)$ half-plane, the solution of the (free) Schroedinger equation will be a superposition of standard plane wave $e^{i\mathbf{k} \cdot \mathbf{r}}$ with different wave numbers. Two arbitrary wave number vectors are sufficient to cover all possible directions in the plane (x, y). We may write

$$\psi(\mathbf{r}, t) = \begin{cases} 0 & (x \geq 0, \, y) \\ \left(A \, e^{i\mathbf{k} \cdot \mathbf{r}} + B \, e^{i\mathbf{k}' \cdot \mathbf{r}} \right) e^{-iEt/\hbar} & (x < 0, \, y). \end{cases} \tag{2.8}$$

Since

$$E = \frac{\hbar^2 k^2}{2m} = \frac{\hbar^2 k'^2}{2m} \implies k' = k \,.$$

Continuity at $x = 0$, $\forall y$ implies

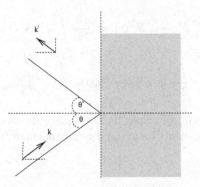

Fig. 2.1 Motion in the half-plane

$$\left(A\,e^{ik(\cos\theta x + \sin\theta y)} + B\,e^{ik(-\cos\theta' x + \sin\theta' y)} \right)\Big|_{x=0} = 0$$

or

$$A\,e^{ik\sin\theta y} + B\,e^{ik\sin\theta' y} = 0 \implies \theta' = \theta \quad \text{and} \quad B = -A\,.$$

The two terms of the wave function have the obvious interpretation as incident and reflected wave. We see that the standard optical law of reflection is valid here. Thus, finally the solution is

$$\psi(\mathbf{r},\, t) = \begin{cases} 0 & (x \geq 0,\, y) \\ 2i\,A\,e^{iky\sin\theta - iEt/\hbar}\,\sin(kx\cos\theta) & (x < 0,\, y). \end{cases} \tag{2.9}$$

2.2 Conservation of Probability

The probability to find the particle anywhere in space must be equal to one, i.e., certainty. This has to be true at any instant in time. Therefore, the normalization condition on the wave function must hold for any t, namely,

$$\int d^3r\, |\psi(\mathbf{r},\, t)|^2 = 1 \implies \frac{d}{dt} \int d^3r\, |\psi(\mathbf{r},\, t)|^2 = 0\,. \tag{2.10}$$

The integral is over all space. Therefore, proceeding with the left-hand side, we have

$$\int d^3r\, \frac{\partial}{\partial t}\left(|\psi(\mathbf{r},\, t)|^2 \right) = \int d^3r\, \left(\dot{\psi}^*\psi + \psi^*\dot{\psi} \right),$$

where the dot indicates time derivative. Substituting Eqs. (2.1), (2.2), this becomes

$$\frac{i}{\hbar} \int d^3r \left(\left(-\frac{\hbar^2}{2m}\nabla^2\psi^* + V\psi* \right)\psi - \psi^*\left(-\frac{\hbar^2}{2m}\nabla^2\psi + V\psi \right) \right)$$

or

$$-\frac{i\hbar}{2m} \int d^3r \left(\psi\nabla^2\psi^* - \psi^*\nabla^2\psi \right) = \frac{i\hbar}{2m} \int d^3r\, \nabla \cdot \left(\psi^*\nabla\psi - \psi\nabla\psi^* \right)$$

or

$$\frac{i\hbar}{2m} \oint d\mathbf{S} \cdot \left(\psi^*\nabla\psi - \psi\nabla\psi^* \right)\,.$$

The last step was taken with the help of the *Divergence Theorem*. The final expression is a surface integral performed over a closed surface at infinity that encloses all space.

On this surface, the wave function and its derivative vanish.[3] Thus, this is indeed equal to zero and probability is conserved at all times. Note that the above proof depends on the reality of the potential. An imaginary part of the potential corresponds to an instability of the system and, as a result, to a loss of probability.

The conservation of probability can be also expressed locally as follows:

$$\frac{\partial}{\partial t}|\psi|^2 = (\psi^*\dot\psi + \dot\psi^*\psi) = \cdots = \frac{i\hbar}{2m} \nabla \cdot (\psi^*\nabla\psi - \psi\nabla\psi^*), \qquad (2.11)$$

where the dots signify the use of the Schroedinger equation and its complex conjugate. Here, we can introduce the *Probability Current Density* **J** as

$$\mathbf{J}(\mathbf{r},\,t) \equiv \frac{\hbar}{2mi}(\psi^*\nabla\psi - \psi\nabla\psi^*). \qquad (2.12)$$

Then, (2.11) can be written as

$$\frac{\partial}{\partial t}\mathcal{P}(\mathbf{r},\,t) + \nabla \cdot \mathbf{J}(\mathbf{r},\,t) = 0, \qquad (2.13)$$

where $\mathcal{P} = |\psi|^2$ is the probability density. This equation is called the *Continuity Equation* and describes the local conservation of probability. It has the familiar form of the analogous equation of Classical Electrodynamics expressing locally the conservation of electric charge.

Example 2.3 A particle of mass m and energy E is moving in one dimension. The wave function of the particle is

$$\psi_E(x,t) = \frac{1}{\sqrt{2\pi}}\left(e^{ikx} + f(k)\,e^{ik|x|}\right)e^{-iEt/\hbar}.$$

Show that local conservation of probability (continuity equation) implies the following relation $f^{-1} + f^{*-1} + 2 = 0$.

The current density corresponding to each branch of the wave function is

$$\psi_E(x,t) = \begin{cases} (1+f(k))e^{ikx}\frac{e^{-iEt/\hbar}}{\sqrt{2\pi}} \implies J_+ = \frac{\hbar k}{m(2\pi)}|1+f(k)|^2 & (x \ge 0) \\[2mm] \left(e^{ikx} + f(k)e^{-ikx}\right)\frac{e^{-iEt/\hbar}}{\sqrt{2\pi}} \implies J_- = \frac{\hbar k}{m(2\pi)}\left(1-|f(k)|^2\right) & (x \le 0). \end{cases}$$

The probability density $P = |\psi|^2$ is time-independent and, therefore, the continuity equation is just

---

[3] A normalizable wave function should be square-integrable, i.e., $\int d^3r\,|\psi|^2 < \infty$. This requires that ψ vanishes at spatial infinity. For example, for $\psi \propto \frac{1}{r^\alpha}$ with $\alpha > 3/2$, the above integral converges.

$\frac{\partial J(x)}{\partial x} = 0 \implies J(x) = const. \implies J_+ = J_-$, leading to the above relation

$$|1 + f|^2 = 1 - |f|^2 \implies 1 + |f|^2 + f + f^* = 1 - |f|^2 \implies \frac{1}{f} + \frac{1}{f^*} + 2 = 0.$$

2.3 The Hamiltonian

Consider a classical system described by N generalized coordinates $q_j(t)$ and an equal number of *canonical momenta* $p_j(t)$. The equations of motion are expressed in terms of the *Hamiltonian* $\mathcal{H}(q, p)$. They have the form (*Hamilton's equations of motion*) [3]

$$\frac{dq_j}{dt} = \frac{\partial \mathcal{H}}{\partial q_j}, \quad \frac{dp_j}{dt} = -\frac{\partial \mathcal{H}}{\partial q_j}. \tag{2.14}$$

The physical meaning of the Hamiltonian is that of the *energy of the system as a function of coordinates and momenta*.

A single particle of mass m, described in terms of its position $\mathbf{r}(t)$ and momentum $\mathbf{p}(t)$, subject to the force $\mathbf{F} = -\nabla V(\mathbf{r})$ that results from a potential $V(\mathbf{r})$, has a Hamiltonian

$$\mathcal{H} = \frac{p^2}{2m} + V(\mathbf{r}). \tag{2.15}$$

On the other hand, the quantum version of this system is set up through the wave function $\psi(\mathbf{r}, t)$ and the Schroedinger equation

$$i\hbar \frac{\partial \psi}{\partial t} = \left(-\frac{\hbar^2}{2m} \nabla^2 + V(\mathbf{r}) \right) \psi. \tag{2.16}$$

It is evident that the operator in parenthesis, appearing in the right-hand side, has exactly the structure of the Hamiltonian (2.15). In fact the *Schroedinger operator* in (2.16) can be obtained considering the Hamiltonian and replacing the momentum of the particle with a differential *operator*

$$p_j \rightarrow \hat{p}_j \equiv -i\hbar \frac{\partial}{\partial x_j} = -i\hbar \nabla_j. \tag{2.17}$$

Note that this is a highly nontrivial replacement since we substitute a set of real numbers with a differential operator. This is a particular case of a general correspondence recipe (quantization): For any quantum system, we consider the Hamiltonian of its classical analogue and we construct a *Hamiltonian operator* corresponding to it[4] $\hat{\mathcal{H}}(\hat{p}_j, \hat{x}_j)$ by the substitution (2.17). Then, the Schroedinger equation, describing

---

[4]The position operator coincides with the real number position variables.

the time evolution of the system is

$$i\hbar\frac{\partial\psi}{\partial t} = \hat{\mathcal{H}}\psi,\tag{2.18}$$

where *hats* denote operators in contrast to real or complex numbers. We are going to come back to this correspondence which is the most important cornerstone in the quantum mechanical description of physical systems.

2.4 Stationary States

Consider the Schroedinger equation

$$i\hbar\frac{\partial\psi}{\partial t} = \left(-\frac{\hbar^2}{2m}\nabla^2 + V(\mathbf{r})\right)\psi.\tag{2.19}$$

We have remarked earlier that this is the time-evolution equation for the system. As we have learnt in the simple case of the free particle of definite energy and momentum, this time evolution for the wave function (plane wave) can be particularly simple consisting of an exponential factor

$$\psi_\mathbf{p}(\mathbf{r},\,t) = \frac{e^{\frac{i}{\hbar}(\mathbf{p}\cdot\mathbf{r}-Et)}}{(2\pi\hbar)^{3/2}} = \psi_\mathbf{p}(\mathbf{r},\,0)\,e^{-\frac{i}{\hbar}Et}.\tag{2.20}$$

As a result, the corresponding probability density of these states is time independent. These *plane-wave* wave functions are simple special solutions of the free Schroedinger equation.

Nevertheless, in the general case of an interacting particle we may still consider special solutions of (2.19) of the form

$$\psi_E(\mathbf{r},\,t) = \psi_E(\mathbf{r})\,e^{-\frac{i}{\hbar}Et},\tag{2.21}$$

where, of course, $\psi_E(\mathbf{r})$ is not any more a plane wave. Inserting (2.21) in (2.19), we obtain that $\psi_E(\mathbf{r})$ must satisfy the equation

$$\left(-\frac{\hbar^2}{2m}\nabla^2 + V(\mathbf{r})\right)\psi_E(\mathbf{r}) = E\,\psi_E(\mathbf{r}).\tag{2.22}$$

This is the so-called *Time-Independent Schroedinger equation*. The wave functions (2.21) correspond to a time-independent probability density

$$\mathcal{P}_E(\mathbf{r}) = |\psi_E(\mathbf{r},\,t)|^2 = |\psi_E(\mathbf{r})|^2\tag{2.23}$$

and are called *Stationary States*.

There is an alternative way to look at the time-independent Schroedinger equation (2.22), namely, in terms of the Hamiltonian operator as

$$\hat{\mathcal{H}}\psi_E(\mathbf{r}) = E\,\psi_E(\mathbf{r}). \tag{2.24}$$

According to this equation, the wave functions $\psi_E(\mathbf{r})$ are such that the Hamiltonian operator acts on them just as if it were a number E. This is a particular case of an important concept on which we shall elaborate later, namely, the concept of *eigenfunctions* and *eigenvalues* of an operator. For each operator, there is a special set of functions, characteristic of the operator, on which the operator acts as a number. These are the *eigenfunctions* of the operator and the resulting numbers are its *eigenvalues*. Therefore, the stationary states are the eigenfunctions of the Hamiltonian or, equivalently, of the energy of the system.

There is a crucial property of the energy eigenfunctions which will be proven in a subsequent chapter but it is important to underline now and make use of it. Stationary states represent *mutually exclusive* physical situations. For example, the probability for the system to be in a state of energy E_1 *or* E_2 should be the sum of the probabilities for each of the situations, namely,

$$\int d^3r\,|\psi_{E_1}(\mathbf{r},\,t) + \psi_{E_2}(\mathbf{r},\,t)|^2 = \int d^3r\,|\psi_{E_1}(\mathbf{r})|^2 + \int d^3r\,|\psi_{E_2}(\mathbf{r})|^2.$$

The immediate mathematical consequence of this is that no interference terms arise or

$$\int d^3r\,\psi_{E_1}^*(\mathbf{r})\psi_{E_2}(\mathbf{r}) = 0 \quad (E_1 \neq E_2). \tag{2.25}$$

This property is called *orthogonality* of the energy eigenfunctions.

2.5 General Solution of the Schroedinger Equation

It is clear that any linear combination of solutions of the Schroedinger equation will also be a solution. In particular, a superposition of stationary wave functions[5]

$$\psi(\mathbf{r},\,t) = \sum_E C_E\,\psi_E(\mathbf{r})\,e^{-\frac{i}{\hbar}Et} \tag{2.26}$$

---

[5]The summation symbol in this superposition is purely symbolic. It could very well correspond to an integral as, for example, in the case of a free particle where the energy eigenvalues are continuous. In general, the energy *spectrum* (i.e., the set of eigenvalues) can consist of a discrete and/or continuous part.

will automatically satisfy (2.19). As we have remarked earlier, since the Schroedinger equation is a differential equation first order in time, the time-evolved $\psi(\mathbf{r}, t)$ will be fully determined in terms of the initial wave function $\psi(\mathbf{r}, 0)$. The general solution expressed by the above superposition (2.26) involves the eigenfunctions $\psi_E(\mathbf{r})$, which are characteristic of the system at hand, and the coefficients C_E, which should be expressible in terms of the initial wave function.

Let's consider (2.26) at an initial time chosen to be $t = 0$, namely,

$$\psi(\mathbf{r},\, 0) = \sum_E C_E\, \psi_E(\mathbf{r}), \tag{2.27}$$

multiply both sides with a particular-but arbitrary-eigenfunction $\psi_{E'}^*(\mathbf{r})$ and then integrate. We get

$$\int d^3r\, \psi_{E'}^*(\mathbf{r})\, \psi(\mathbf{r},\, 0) = \sum_E C_E \int d^3r\, \psi_{E'}^*(\mathbf{r})\, \psi_E(\mathbf{r}).$$

The sum in the right-hand side, thanks to the orthogonality property of the energy eigenfunctions (2.25), collapses to just the single term with $E = E'$, namely, $C_{E'} \int d^3r\, |\psi_{E'}(\mathbf{r})|^2$, which, if the eigenfunctions are taken to be normalized, is just $C_{E'}$. Thus, finally, for any eigenvalue E, we can express the corresponding coefficient in terms of the initial wave function as

$$C_E = \int d^3r\, \psi_E^*(\mathbf{r})\psi(\mathbf{r},\, 0). \tag{2.28}$$

The time-independent Schroedinger equation and the pair of (2.26), (2.28) comprise the general solution to the Schroedinger equation, consisting of first obtaining the energy eigenfunctions and eigenvalues characteristic of the system, and then writing down the expansion with coefficients determined by the initial state of the system.

What is the physical meaning of the expansion coefficients C_E? It is clear that each number C_E measures the degree of participation of the eigenfunction $\psi_E(\mathbf{r})$ in the wave function $\psi(\mathbf{r}, t)$. Therefore, the appropriate interpretation of the coefficient C_E is that it is the *probability amplitude* to find the particle in the stationary state $\psi_E(\mathbf{r}, t)$.

Example 2.4 (Time evolution of a Gaussian wave packet) A free particle of mass m is moving in one dimension. The wave function of the particle at an initial time $t = 0$ is

$$\psi(x, 0) = \left(\frac{m}{\hbar\pi a}\right)^{1/4} e^{-\frac{mx^2}{2\hbar a}}, \tag{2.29}$$

where $a > 0$ is a real parameter. Find the solution to the Schroedinger equation $\psi(x, t)$ at all times $t > 0$.

Expanding $\psi(x, t)$ in stationary states (plane waves), we have

$$\psi(x, t) = \int_{-\infty}^{+\infty} dk \, \frac{e^{ikx}}{\sqrt{2\pi}} \, \tilde{\psi}(k) \, e^{-\frac{i\hbar k^2 t}{2m}} \text{ with } \tilde{\psi}(k) = \int_{-\infty}^{+\infty} dx \frac{e^{-ikx}}{\sqrt{2\pi}} \, \psi(x, 0) \,,$$

in which, substituting (2.29) in the integral for $\tilde{\psi}(k)$, we calculate

$$\left(\frac{m}{\hbar\pi a}\right)^{1/4} \frac{1}{\sqrt{2\pi}} \int_{-\infty}^{+\infty} dx \, e^{-ikx} e^{-\frac{mx^2}{2\hbar a}} = \left(\frac{m}{\hbar\pi a}\right)^{1/4} \frac{1}{\sqrt{2\pi}} \int_{-\infty}^{+\infty} dx \, e^{-\frac{m}{2\hbar a}\left(x+\frac{ik\hbar a}{m}\right)^2} e^{-\frac{\hbar a k^2}{2m}}$$

and obtain

$$\tilde{\psi}(k) = \left(\frac{\hbar a}{m\pi}\right)^{1/4} e^{-\frac{\hbar a k^2}{2m}} \,. \tag{2.30}$$

Substituting $\tilde{\psi}(k)$ into the expression for $\psi(x, t)$, we obtain

$$\psi(x, t) = \left(\frac{\hbar a}{m\pi}\right)^{1/4} \frac{1}{\sqrt{2\pi}} \int_{-\infty}^{+\infty} dx \, e^{ikx} e^{-\frac{\hbar a}{2m}(1+it/a)k^2}$$

or

$$\psi(x, t) = \left(\frac{ma}{\hbar\pi}\right)^{1/4} e^{-i\pi/4} \frac{e^{i\frac{max^2}{2\hbar(t-ia)}}}{\sqrt{t - ia}} \,, \tag{2.31}$$

where we have completed the square in the exponent and performed the Gaussian integration (or used the related formulae of the Mathematical Appendix). It may be checked explicitly that the $\psi(x, t)$ found is indeed a solution of the free Schroedinger equation. Note that it starts as a standard Gaussian of a spread $\sim \sqrt{a}$ at $t = 0$ and at times $t \gg a$ it evolves into a density $|\psi(x, t)|^2 \propto t^{-1}$ that is spatially constant and decreasing in time with an infinite spread.

Problems and Exercises

2.1 Show that the time-independent Schroedinger equation cannot have acceptable solutions corresponding to energies smaller than the minimum of the potential. As an example, consider the potential step $V(x) = V_0 \Theta(x)$ and show that there are no acceptable solutions with $E < 0$.

2.2 The wave function of a particle moving in one dimension is $\psi_E(x) = \frac{1}{\sqrt{2\pi}}$ $\left(e^{ikx} - \frac{e^{ik|x|}}{1-i\alpha k}\right)$, corresponding to energy $E = \hbar^2 k^2 / 2m$. Find the corresponding potential $V(x)$.

2.3 Consider the system of a particle described by a wave function $\psi(\mathbf{r}, t)$. The rate of change of the probability to find the particle in a spatial region of volume $\mathcal{V}$ should be equal to minus the probability current through the surface $\mathcal{S}$ enclosing this volume. Verify this for the wave function

$$\psi(\mathbf{r}, t) = (\alpha/\pi)^{3/4} \left(1 - \frac{i\hbar\alpha t}{m}\right)^{-3/2} e^{-\frac{\alpha r^2}{2\left(1-\frac{i\hbar\alpha t}{m}\right)}} \,,$$

taking this region to be a sphere of radius R.

2.4 Derive the equation giving the rate of probability nonconservation in the case of a system with a potential having an imaginary part ($V^* \neq V$).

2.5 A particle moves in one dimension under the influence of a potential of the form

$$V(x) = \begin{cases} 0 & (x \leq 0) \\ \\ U(x) & (x \geq 0), \end{cases}$$

where $U(x)$ is a function with the property $U(+\infty) > 0$. Show that the general form of the wave function corresponding to energy $E = \hbar^2 k^2/2m > 0$ in the region $x < 0$ will be

$$\psi_E(x) = N\left(e^{ikx} + e^{i\alpha}e^{-ikx}\right),$$

where N and $\alpha^* = \alpha$ parameters.

2.6 Consider the general solution to the Schroedinger equation and, substituting the expression of the coefficients in terms of the initial wave function, rewrite it in the form

$$\psi(\mathbf{r}, t) = \sum_E \left(\int d^3r'\, \psi_E^*(\mathbf{r}')\psi(\mathbf{r}', 0)\right)\psi_e(\mathbf{r})\,e^{-iEt} = \int d^3r'\, \mathcal{K}(\mathbf{r}, \mathbf{r}'; t)\,\psi(\mathbf{r}', 0).$$

Verify that

$$\mathcal{K}(\mathbf{r}, \mathbf{r}', 0) = \sum_E \psi_E(\mathbf{r})\psi_E(\mathbf{r}') = \delta(\mathbf{r} - \mathbf{r}')$$

in the case of the particle moving in a circle.

2.7 Show that the momentum operator has the property

$$\int_{-\infty}^{+\infty} d^3r\, \psi_1^*(\mathbf{r}, t)\, \hat{p}_j\, \psi_2(\mathbf{r}, t) = \left(\int_{-\infty}^{+\infty} d^3r\, \psi_2^*(\mathbf{r}, t)\, \hat{p}_j\, \psi_1(\mathbf{r}, t)\right)^*,$$

where ψ_1 and ψ_2 arbitrary wave functions are describing possible states of the system.

2.8 Show that the Fourier transform of the wave function, defined by $g(p, t) = \int_{-\infty}^{+\infty} \frac{dx}{\sqrt{2\pi\hbar}}\psi(x', t)e^{-ipx/\hbar}$, satisfies the following *integro-differential equation*:

$$\frac{p^2}{2m}g(p, t) + \int_{-\infty}^{+\infty} \frac{dq}{\sqrt{2\pi\hbar}}\tilde{V}(p - q)g(q, t) = i\hbar\dot{g}(p, t),$$

where $\tilde{V}(q)$ is the Fourier transform of the potential.

References

1. G. Baym, *Lectures in Quantum Mechanics*, Lecture Notes and Supplements in Physics (ABP, 1969)
2. A. Messiah, *Quantum Mechanics*. (Dover publications, Mineola, 1958). Single-volume reprint of the Wiley, New York, two-volume 1958 edn
3. H. Goldstein, C. Poole, J. Safko, *Classical Mechanics*, 3rd edn. (Pearson International, 2014)

Chapter 3
Some Simple Systems

3.1 The Infinite Square Well

Consider a particle moving in one dimension but confined only in a particular spatial interval taken to be $(-L, L)$. The particle moves freely at $-L < x < L$ but is subject to an infinite repulsive force at $\pm L$. As a result, the particle cannot propagate in either of the regions $x \geq L$, $x \leq -L$. Such a situtation can be expressed in terms of a potential function

$$V(x) = \begin{cases} 0 & (-L < x < L) \\ +\infty & (|x| \geq L) \end{cases} \tag{3.1}$$

In what follows we shall determine the energy eigenvalues and energy eigenfunctions (stationary states) for this system (Fig. 3.1).

The time-independent Schroedinger equation is

$$\left(-\frac{\hbar^2}{2m}\frac{d^2}{dx^2} + V(x) \right) \psi_E(x) = E\,\psi_E(x). \tag{3.2}$$

It is clear that in the $|x| \geq L$ *outside* region, where $V = \infty$, (3.2) can only be satisfied with

$$\forall\, |x| \geq L \implies \psi_E(x) = 0. \tag{3.3}$$

In the *inside* region, the potential vanishes and the particle moves freely with its wave function satisfying the free Schroedinger equation

$$-\frac{\hbar^2}{2m}\frac{d^2\psi_E}{dx^2} = E\,\psi_E. \tag{3.4}$$

This equation can be set in the form

© Springer Nature Switzerland AG 2019
K. Tamvakis, *Basic Quantum Mechanics*, Undergraduate Texts in Physics,
https://doi.org/10.1007/978-3-030-22777-7_3

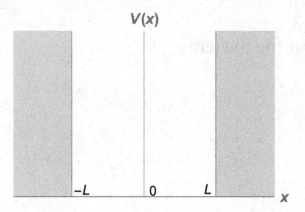

Fig. 3.1 The infinite square well

$$\psi_E''(x) = -k^2 \psi_E(x), \tag{3.5}$$

in terms of the wave number k related to the energy via $E = \frac{\hbar^2 k^2}{2m}$. Note that the energy is positive while the wave number takes up positive and negative values.

The general solution of (3.5) is

$$\psi_E(x) = A e^{ikx} + B e^{-ikx} \quad (|x| < L). \tag{3.6}$$

Continuity of the wave function[1] at $x = \pm L$ gives the conditions $\psi_E(\pm L) = 0$ or

$$A e^{-ikL} + B e^{ikL} = 0 \quad \textbf{and} \quad A e^{ikL} + B e^{-ikL} = 0 \tag{3.7}$$

or

$$B = -A e^{-2ikL} = -A e^{2ikL}. \tag{3.8}$$

This can only be true if

$$4ikL = 2in\pi \implies k = n\frac{\pi}{2L} \quad (n = \pm 1, \pm 2, \ldots). \tag{3.9}$$

The resulting eigenvalues of the energy are

$$E_n = \frac{\hbar^2 n^2 \pi^2}{8mL^2} \quad (n = \pm 1, \pm 2, \ldots). \tag{3.10}$$

---

[1]Continuity of any measurable physical quantity—e.g., the probability density—is always a requirement. Therefore, the wave function must always be continuous. The same is true for the first spatial derivative of the wave function provided the potential is non-singular. For the above case, at the points where the potential is infinite, it is obvious that the second spatial derivative of the wave function does not exist and, therefore, its first derivative is neither differentiable nor continuous. Thus, here there is no continuity condition for the first derivative.

Thus, the wave number and the energy are *quantized*. The corresponding eigenfunctions are[2] (symbolized now as ψ_n instead of ψ_E)

$$\psi_n(x) = A\left(e^{in\frac{\pi x}{2L}} + (-1)^{n+1}e^{-in\frac{\pi x}{2L}}\right).$$

Note that since ψ_n and ψ_{-n} are related by a just a phase factor it is sufficient to restrict the quantum number n to positive integer values $n = 1, 2, \ldots$. The coefficient A is determined from the normalization condition

$$\int_{-L}^{L} dx\,|\psi_n(x)|^2 = 1$$

or

$$1 = |A|^2\int_{-L}^{L} dx\left(1 + 1 + (-1)^{n+1}\left(e^{in\frac{\pi x}{L}} + e^{-in\frac{\pi x}{L}}\right)\right) =$$

$$2|A|^2\int_{-L}^{L} dx\left(1 + (-1)^{n+1}\cos(n\pi x/L)\right) = 2|A|^2\left(2L + \frac{L(-1)^{n+1}}{n\pi}\sin(n\pi x/L)|_{-L}^{L}\right)$$

giving

$$A = \frac{e^{i\phi_n}}{2\sqrt{L}}, \tag{3.11}$$

where ϕ_n is an arbitrary phase that can be chosen at will. Thus, the eigenfunctions are

$$\psi_n(x) = \frac{e^{i\phi_n}}{2\sqrt{L}}\left(e^{in\frac{\pi x}{2L}} + (-1)^{n+1}e^{-in\frac{\pi x}{2L}}\right). \tag{3.12}$$

The first few eigenfunctions are

$$\psi_1(x) = \frac{e^{i\phi_1}}{\sqrt{L}}\cos(\pi x/2L)$$

$$\psi_2(x) = \frac{ie^{i\phi_2}}{\sqrt{L}}\sin(\pi x/L)$$

$$\psi_3(x) = \frac{e^{i\phi_3}}{\sqrt{L}}\cos(3\pi x/2L) \tag{3.13}$$

$$\psi_4(x) = \frac{ie^{i\phi_4}}{\sqrt{L}}\sin(2\pi x/L)$$

$$\cdots$$
$$\cdots$$

The choice of the phase factor $\phi_n = -(1 + (-1)^n)\pi/2$ renders all eigenfunctions real.

---

[2]Note that $e^{\pm 2ik_n L} = e^{\pm in\pi} = (-1)^n$.

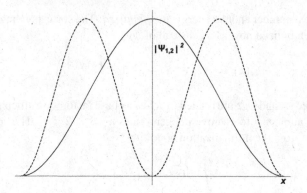

Fig. 3.2 Probability distribution for $n = 1, 2$

Thus, the eigenfunctions are grouped in the infinite sets

$$\psi_n = \begin{cases} \frac{1}{\sqrt{L}} \cos(n\pi x/2L) & (n = 1, 3, 5, \ldots) \\[2mm] \frac{1}{\sqrt{L}} \sin(n\pi x/L) & (n = 2, 4, 6, \ldots) \end{cases} \tag{3.14}$$

The corresponding probability distribution for the ground state and the first excited state (dotted line) is shown in Fig. 3.2.

Let us take a pause and compare this quantum system with the corresponding classical system of a classical particle trapped in such an infinite potential square well. For the classical system the allowed energies span the continuum $0 \le E < \infty$. In contrast, in our case, the energy spectrum is discrete, expressible in terms of an integer $n = 1, 2, \ldots$ as $E_n = E_1 n^2$. Although the gap between two successive energy levels increases with n as $E_{n+1} - E_n = (2n + 1)E_1$, the relative distance of successive energy levels $(E_{n+1} - E_n)/E_n$ decreases as $1/n$ for very large values of n. This means that at very large quantum numbers ($n \to \infty$) the spectrum approaches the continuum. Another point that is in sharp contrast with the classical system is that for the quantum system there are points in the interval $[-L, L]$ where the particle cannot be present or, equivalently, points where the probability density vanishes. For example, if the particle is in the first excited state ($n = 2$), the probability density vanishes at the midpoint $x = 0$. Similarly for the second excited state ($n = 3$) the probability density vanishes at the points $x = \pm L/3$. No such behavior is possible for the system of the trapped classical particle.

Parity. Observe that all eigenfunctions are grouped in two sets of even and odd functions

$$\psi_{2\nu+1}(-x) = \psi_{2\nu+1}(x)$$

$$\psi_{2\nu}(-x) = -\psi_{2\nu}(x) \tag{3.15}$$

or just

Table 3.1 Energy eigenfunctions of the infinite square well

n	$\psi_n(x)$	$|\psi_n(x)|^2$	E_n	$\mathcal{P}$
1	$\frac{1}{\sqrt{L}}\cos(\pi x/2L)$	$\frac{1}{L}\cos^2(\pi x/2L)$	$\frac{\hbar^2}{8mL^2}$	$+1$
2	$\frac{1}{\sqrt{L}}\sin(\pi x/L)$	$\frac{1}{L}\sin^2(\pi x/L)$	$\frac{\hbar^2\pi^2}{2mL^2}$	-1
3	$\frac{1}{\sqrt{L}}\cos(3\pi x/2L)$	$\frac{1}{L}\cos^2(3\pi x/L)$	$\frac{9\hbar^2\pi^2}{8mL^2}$	$+1$
4	$\frac{1}{\sqrt{L}}\sin(2\pi x/L)$	$\frac{1}{L}\sin^2(2\pi x/L)$	$\frac{2\hbar^2\pi^2}{mL^2}$	-1
.	.	.	.	
.	.	.	.	
.	.	.	.	

$$\psi_n(-x) = (-1)^{n+1}\psi_n(x). \tag{3.16}$$

We may talk about the *"evenness"* or *"oddness"* of a wave function in terms of the concept of *Parity* $\mathcal{P}$, defined as having the value $+1$ for even functions and the value -1 for odd ones. Later on we shall introduce a *Parity operator* as a spatial reflection operator

$$\hat{\mathcal{P}}\Psi(x) = \Psi(-x). \tag{3.17}$$

For the time being, we write its action on the energy eigenfunctions found above and obtain

$$\hat{\mathcal{P}}\psi_n(x) = \psi_n(-x) = (-1)^{n+1}\psi_n(x). \tag{3.18}$$

Thus, we may say that the *partity of the n-th eigenfunction is* $(-1)^{n+1}$.

The energy eigenfunctions found have alternating parity. Starting from the *ground state*, i.e., the state of lowest energy (here $E_1 = \frac{\hbar^2\pi^2}{8mL^2}$), which has parity $+1$, we move next to the first excited level ($E_2 = \frac{\hbar^2\pi^2}{2mL^2}$), which has parity -1, and so on (Table 3.1).

The fact that the eigenfunctions came out to have a definite parity automatically is just the particular manifestation of a general property relating to the symmetries of the Hamiltonian. In our case, the Hamiltonian of the system is even under spatial reflections

$$x \to -x \quad \Longrightarrow \mathcal{H}(-x) = \mathcal{H}(x). \tag{3.19}$$

As we will demonstrate in a subsequent chapter this symmetry reflects on the energy eigenfunctions, which will have automatically definite parity, or, in a more mathematical language, will also be eigenfunctions of the parity operator.

Orthogonality and Completeness. The eigenfunctions (3.14) are well known to be a set of mutually orthogonal functions. It can be easily shown that

$$\frac{1}{L} \int_{-L}^{L} dx \cos((2m + 1)\pi x/2L) \cos((2m' + 1)\pi x/2L) = \delta_{mm'}$$

$$\frac{1}{L} \int_{-L}^{L} dx \sin(2m\pi x/L) \sin(2m'\pi x/L) = \delta_{mm'} \qquad (3.20)$$

$$\frac{1}{L} \int_{-L}^{L} dx \sin(2m\pi x/L) \cos((2m' + 1)\pi x/2L) = 0$$

or collectively,

$$\int_{-L}^{L} dx \psi_n^*(x)\psi_n(x) = \delta_{nn'} . \qquad (3.21)$$

Since, they have also been constructed to be normalized, they are not only orthogonal but *orthonormal*.

Furthermore, the set of eigenfunctions (3.14) are well known from Fourier Series Analysis to be a *complete set of functions*, in the sense that any smooth function $\Psi(x)$, defined in the interval $[-L, L]$, can be expanded in terms of them as

$$\Psi(x) = \sum_{n=1}^{\infty} C_n \psi_n(x) . \qquad (3.22)$$

The expansion coefficients C_n can be readily obtained multiplying both sides of (3.22) by one of the ψ_n's, integrating and making use of the orthonormality relation

$$\int_{L}^{L} dx \psi_{n'}(x)\Psi(x) = \sum_{n=0}^{\infty} C_n \int_{L}^{L} dx \psi_{n'}(x)\psi_n(x) = \sum_{n=0}^{\infty} C_n \delta_{nn'}$$

or

$$C_n = \int_{L}^{L} dx \, \psi_n(x)\Psi(x) . \qquad (3.23)$$

Example 3.1 (Particle trapped in the infinite square well) The wave function of a particle trapped in the infinite square well is at a time $t = 0$

$$\Psi(x, 0) = \frac{2}{\sqrt{3L}} \cos^2(\pi x/2L) . \qquad (3.24)$$

Write down the evolved wave function $\Psi(x, t)$ for times $t > 0$ and calculate the probability to find the particle at its ground state.

The wave function at any time can be expanded in terms of the energy eigenfunctions (3.14) as in (2.26). The coefficients are

$$C_n = \int_{-L}^{L} dx \, \psi(x, 0)\psi_n^*(x) = \frac{2}{\sqrt{3L}} \int_{-L}^{L} dx \cos^2(\pi x/2L) \, \psi_n^*(x).$$

Note however that, since $\Psi(x, 0)$ is an even function (i.e., $\Psi(-x, 0) = \Psi(x, 0)$), only the even members of the set $\psi_n(x)$, i.e., the cosines, will give a nonvanishing integral and contribute to the sum. In fact the coefficients corresponding to the sines will exactly vanish. Indeed, $\int_{-L}^{L} dx \, \Psi(x, 0) \sin(n\pi x/L)$ exactly vanish due to the oddness of the intergrand. Thus, we have $\Psi(x, t) = \frac{1}{\sqrt{L}} \sum_{n=1,3,\ldots} C_n \, e^{-\frac{i}{\hbar} E_n t} \cos(n\pi x/2L)$ and

$$C_n = \frac{2}{L\sqrt{3}} \int_{-L}^{L} dx \, \cos^2(\pi x/2L) \cos(n\pi x/2L)$$

with $n = 1, 3, 5, \ldots$. Proceeding to compute C_1 we obtain the required probability to be ($C_1 = \frac{4}{\pi\sqrt{3}} \int_{-\pi/2}^{\pi/2} d\xi \, \cos^3 \xi = \frac{4}{\pi\sqrt{3}} \int_{-1}^{1} d\eta \, (1 - \eta^2) = \frac{16}{3\pi\sqrt{3}}$)

$$|C_1|^2 = \frac{(16)^2}{27\pi^2} \approx 0.9607.$$

The corresponding probabilities for excited states decrease rapidly. For example, for the next excited level, we obtain the probability to be $|C_3|^2 = |C_1|^2/25 \approx 0.036$.

3.2 Piecewise Constant Potentials

Consider a particle moving in one dimension subject to a constant potential V_0. Classically, there is absolutely no difference than the case of vanishing potential since the force is zero in both cases. Nevertheless, in the quantum case, since the value of the potential enters in the Schroedinger equation, there is a slight difference. The time-independent Schroedinger equation reads

$$-\frac{\hbar^2}{2m} \frac{d^2\psi_E(x)}{dx^2} + V_0\psi_E(x) = E\psi_E(x). \tag{3.25}$$

Let's assume that $E \geq V_0$ as in the classical case. Introducing the shifted wave number

$$q^2 = \frac{2m}{\hbar^2}(E - V_0), \tag{3.26}$$

in terms of which our equation becomes

$$\psi_E''(x) = -q^2\psi_E(x), \tag{3.27}$$

we obtain a general solution in the form of plane waves analogous to the zero potential case

$$\psi_E(x) = A e^{iqx} + B e^{-iqx}. \tag{3.28}$$

Next, we may consider the classically forbidden case of $E < V_0$ and see whether acceptable quantum solutions exist. The Schroedinger equation now takes the form

$$\psi_E''(x) = +s^2\psi_E(x) \tag{3.29}$$

in terms of

$$s^2 = \frac{2m}{\hbar^2}(V_0 - E).$$ (3.30)

The general solution of this equation is

$$\psi_E(x) = C\,e^{sx} + D\,e^{-sx}.$$ (3.31)

Nevertheless, if D is nonvanishing, this solution blows up as $x \to -\infty$, giving an infinite probability density. Therefore, we must have $D = 0$. With an analogous argument for $x \to +\infty$, the other coefficient C has to vanish as well. Thus, there is no physically acceptable solution having $E < V_0$, just as in the classical case. In what will follow shortly we shall see that this depends crucially on the fact that the classically disallowed region with $E < V$ extends to infinity. We shall see shortly that classically disallowed potential barriers of finite extend can be penetrated by quantum particles.

Of course, the realistic potentials encountered by microscopic particles are not constant.[3] There are situations though in which the full range of a potential function can be subdivided in a number of smaller parts over which the potential is approximately constant. Thus, the potential is approximated by a *piecewise constant potential*. Apart from their applicability in realistic physical systems, piecewise constant potentials thanks to their simplicity have a great pedagogical value in the study of quantum behavior.

For a piecewise constant potential, if x_0 is a point of discontinuity of the potential, the Schroedinger equation dictates that $\psi_E''(x)$ will also be discontinuous at this point. Nevertheless, existence of the second derivative presupposes the continuity of the first derivative $\psi_E'(x)$. Therefore, continuity of both the wave function $\psi_E(x)$ and its first derivative $\psi_E'(x)$ has to be imposed at each point of discontinuity of the potential.

3.2.1   The One-Dimensional Potential Step

Consider a particle of energy E moving in one dimension under the influence of the potential

$$V(x) = \begin{cases} 0 & (x < 0) \\ V_0 & (x > 0) \end{cases}$$ (3.32)

shown in Fig. 3.3.

The Schroedinger equation can be written as

---

[3]For an analysis of the general properties of the solutions of the one-dimensional Schroedinger equation see [1].

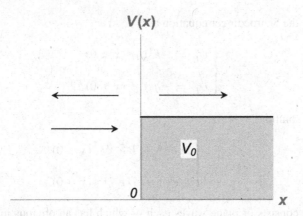

Fig. 3.3 Potential step

$$\psi_E''(x) = -\frac{2m}{\hbar^2}\left(E - V(x)\right)\psi_E(x) = \begin{cases} -\frac{2mE}{\hbar^2}\psi_E(x) & (x < 0) \\[2mm] -\frac{2m}{\hbar^2}(E - V_0)\psi_E(x) & (x > 0) \end{cases} \tag{3.33}$$

We assume that $V_0 > 0$. Classically of course E can only be positive. Nevertheless, there is no harm in investigating whether acceptable solutions of negative energy exist. For $E = -|E| < 0$, our equation in both regions has exponential solutions

$$\psi_E(x) = \begin{cases} C_1\, e^{q_- x} + C_2\, e^{-q_- x} & (x < 0) \\[2mm] C_3\, e^{q_+ x} + C_4\, e^{-q_+ x} & (x > 0), \end{cases} \tag{3.34}$$

where we introduced $q_- \equiv \sqrt{\frac{2m|E|}{\hbar^2}}$, $q_+ \equiv \sqrt{\frac{2m(|E|+V_0)}{\hbar^2}}$. It is obvious that in order to have a bounded probability density we must set $C_2 = C_3 = 0$. Furthermore, continuity of the wave function and its first derivative at the point $x = 0$ gives

$$C_1 = C_4, \qquad C_1\, q_- = C_4\, q_+, \tag{3.35}$$

which are incompatible for any $V_0 \neq 0$. Therefore, we must conclude that there are no physically acceptable solutions of negative energy. This is a particular case of a general property, namely, that *there are no acceptable solutions of the Schroedinger equation of energy smaller than the minimum of the potential.*

For $E > 0$ we still have two possibilities, namely, $E > V_0$ or $0 < E < V_0$. We proceed to analyze each of these cases.

The case $E > V_0$. Introducing the wave numbers

$$k^2 = \frac{2m}{\hbar^2}E, \qquad q^2 = \frac{2m}{\hbar^2}(E - V_0) \tag{3.36}$$

we can write the Schroedinger equation as

$$\begin{cases} \psi_E'' = -k^2\psi_E \ (x < 0) \\[2mm] \psi_E'' = -q^2\psi_E \ (x > 0) \end{cases} \tag{3.37}$$

with general solution[4]

$$\psi_E(x) = \begin{cases} A\,e^{ikx} + B\,e^{-ikx} \ (x < 0) \\[2mm] C\,e^{iqx} + D\,e^{-iqx} \ (x > 0) \end{cases} \tag{3.38}$$

The solution consists of plane waves each of which has an obvious interpretation. $A\,e^{ikx}$ stands for an *incoming* particle moving from the left towards the potential step. $B\,e^{-ikx}$ stands for a particle moving towards the left away from the potential step and can only be interpreted as a *reflected* particle. Similarly, $C\,e^{iqx}$ propagates towards the right away from the potential step and $D\,e^{-iqx}$ is an incoming particle from the right. We may simplify our solution, without any loss of generality, by assuming that we are interested only in particles incident from the left. Thus, we may take $D = 0$ and designate $A\,e^{ikx}$ as the *incoming* wave, $B\,e^{-ikx}$ as the *reflected* wave and $C\,e^{iqx}$ the *transmitted* wave.

Continuity of $\psi_E(x)$ and $\psi_E'(x)$ at $x = 0$ gives

$$\begin{cases} A + B = C \\[2mm] k\,(A - B) = q\,C \end{cases} \implies \begin{cases} \frac{B}{A} = \frac{k-q}{k+q} \\[2mm] \frac{C}{A} = \frac{2k}{k+q}. \end{cases} \tag{3.39}$$

Notice that there is always a reflected wave ($V_0 \neq 0 \implies B \neq 0$). In the classical case for $E > V_0$ there is no reflection and the particle propagates to the right.

In a previous chapter, discussing the local conservation of probability, we introduced the *Probability Current Density* (2.12) and showed that it satisfies the *Continuity Equation* (2.13). The one-dimensional version of this probability current density and the corresponding continuity equation are

$$J(x, t) = \frac{\hbar}{2mi}\left(\psi^*(x, t)\psi'(x, t) - \psi'^*(x, t)\psi(x, t)\right) \tag{3.40}$$

$$\frac{\partial J}{\partial x} + \frac{\partial P}{\partial t} = 0, \tag{3.41}$$

---

[4]Mathematically, the general solution could also be written in terms of $\sin(kx)$ and $\cos(kx)$. Nevertheless, in situations with propagating particles it is always preferable to use plane wave exponentials, since they are associated with a sense of direction. In bound state problems however, in cases where the potential has definite parity, it will be more profitable to write the solution in terms of cosines and sines which are even and odd.

where $\mathcal{P} = |\psi|^2$. Carrying this concept over to our study of stationary states, where time dependence drops out, the above equations simplify to

$$J_E(x) = \frac{\hbar}{2mi} \left(\psi'_E(x)\psi_E(x) - \psi'^*_E(x)\psi_E(x) \right) \quad and \quad J_E = constant. \quad (3.42)$$

Thus, for stationary one-dimensional states the probability current density is constant.

Applying the above in our problem, we obtain that

$$J_E(x < 0) = J_E(x > 0) \quad \Longrightarrow \quad \frac{\hbar k}{m}|A|^2 + \frac{(-\hbar k)}{m}|B|^2 = \frac{\hbar q}{m}|C|^2. \quad (3.43)$$

Each of the appearing terms has a straightforward interpretation, namely, $\frac{\hbar k}{m}|A|^2$ being the velocity of the incident particle times its probability density (*incident current density*), $\frac{(-\hbar k)}{m}|B|^2$ being the velocity of the reflected particle times its probability density (*reflected current density*) and $\frac{\hbar q}{m}|C|^2$ being the velocity of the transmitted particle times its probability density (*transmitted current density*). Denoting these partial current densities with

$$J_i = \frac{\hbar k}{m}|A|^2, \quad J_r = -\frac{\hbar k}{m}|B|^2, \quad J_t = \frac{\hbar q}{m}|C|^2, \quad (3.44)$$

we can write the current conservation as

$$J_i + J_r = J_t. \quad (3.45)$$

A nice way to express the reflection probability is through the quantity

$$\mathcal{R} = \frac{|J_r|}{J_i} \quad (3.46)$$

which is called *Reflection Coefficient*. Similarly, we may define the *Transmission Coefficient*

$$\mathcal{T} = \frac{J_t}{J_i}. \quad (3.47)$$

Note that the current conservation equation, written in terms of these coefficients, reads

$$\mathcal{R} + \mathcal{T} = 1, \quad (3.48)$$

simply expressing the fact that the incident beam can only be reflected and transmitted. In our particular problem of the potential step these coefficients take the values

$$\mathcal{R} = \frac{|B|^2}{|A|^2} = \left(\frac{k-q}{k+q}\right)^2$$

$$\mathcal{T} = \frac{q|C|^2}{k|A|^2} = \frac{4qk}{(k+q)^2}.$$

(3.49)

The case $E < V_0$. This is not possible for a classical particle, since it would mean that the particle has negative kinetic energy. Nevertheless, in the quantum case it is the Schroedinger equation that will decide whether there exist physically acceptable solutions. The Schroedinger equation takes the form

$$\psi_E''(x) = \frac{2m}{\hbar^2}\left(V(x) - E\right)\psi_E(x) \implies \begin{cases} \psi_E''(x) = -k^2\psi_E(x) \ \ (x < 0) \\ \\ \psi_E''(x) = s^2\psi_E(x) \ \ (x > 0), \end{cases}$$

(3.50)

where

$$k^2 = \frac{2m}{\hbar^2}E, \quad s^2 = \frac{2m}{\hbar^2}(V_0 - E).$$

(3.51)

s and k are taken to be positive. The general solution of (3.50) is

$$\psi_E(x) = \begin{cases} A\,e^{ikx} + B\,e^{-ikx} \ \ (x \le 0) \\ \\ C\,e^{-sx} \quad\quad (x \ge 0) \end{cases}$$

(3.52)

No e^{+sx} part has been included in the $x > 0$ branch of ψ_E, since it would blow up at $x \to +\infty$.

Continuity at $x = 0$ for (3.52) and its derivative gives

$$\begin{matrix} A + B = C \\ \\ A - B = \frac{is}{k}\,C \end{matrix} \implies \begin{cases} \frac{B}{A} = \frac{k-is}{k+is} \\ \\ \frac{C}{A} = \frac{2k}{k+is}. \end{cases}$$

(3.53)

From these coefficients we can immediately compute the corresponding current densities

$$J_i = \frac{\hbar k}{m}|A|^2, \quad J_r = \frac{-(\hbar k)}{m}|B|^2 = \frac{-(\hbar k)}{m}|A|^2\left|\frac{k-is}{k+is}\right|^2 = -J_i$$

and

$$J_t = \frac{\hbar}{2mi}\left(C^*e^{-sx}(-s)Ce^{-sx} - Ce^{-sx}(-s)C^*e^{-sx}\right) = 0.$$

Thus, we have

$$\mathcal{R} = 1, \qquad \mathcal{T} = 0 \tag{3.54}$$

and there is no transmission but only total reflection.

Nevertheless, although there is no transmission, the fact remains that the wave function is nonzero inside the classically forbidden region $x > 0$. Note however that the x dependence of the wave function is such that it is appreciable only for a range $[0, 1/s]$. Beyond $1/s$ the wave function and the corresponding probability density dies out exponentially. If we were interested in locating the particle inside the forbidden region we would at most face an uncertainty of $\Delta x \sim 1/s$. On the other hand, the *Uncertainty Principle* dictates that there will necessarily be a corresponding uncertainty in the particle's momentum determination bounded from below through the *Heisenberg Inequality* as

$$(\Delta p) \sim \hbar/(\Delta x) \sim \hbar s .$$

This implies a minimum uncertainty in the energy of the particle

$$\Delta E \sim \frac{(\Delta p)^2}{2m} \sim \frac{\hbar^2 s^2}{2m} = V_0 - E .$$

Thus, even if E was assumed to be smaller than V_0, such an uncertainty could very well lift the particle to the classically allowed region $(E + \Delta E \geq V_0)$. In other words, due to the Uncertainty Principle, any attempt to locate the particle in the classically forbidden region is accompanied with an uncertainty in energy that is larger than the gap that separates it from allowed energies.

In the above analysis of the energy eigenfunctions the allowed values of the energy E span a continuum $0 \leq E < \infty$. In the standard terminology used we say that the *energy spectrum is continuous*. As in the case of the free particle, where the spectrum was also continuous, the eigenfunctions are not normalizable. Note however that they are orthogonal as in the case of the infinite square well energy eigenfunctions. Later on, we shall see that they are normalizable in a generalized sense and their orthonormality is expressible in terms of a *delta function*. This is a general property of the continuous part of the spectrum, namely, that the continuum eigenfunctions are *"delta function-normalizable"*.

Example 3.2 (Particle incident on an increasing potential) A particle of mass m and energy $E > 0$ is moving in one dimension under the influence of the potential $V(x) = \Theta(x) g^2 x^3$. What is the reflection coefficient in this potential?

The wave function of the particle is

$$\psi(x) = \begin{cases} A e^{ikx} + B e^{-ikx} & (x \leq 0) \\ \psi_+(x) & (x \geq 0), \end{cases}$$

where the function $\psi_+(x)$ is unknown. Nevertheless, since the potential is infinite at $x \to +\infty$, the wave function has to vanish there, i.e., $\psi_+(+\infty) = 0$. For a stationary state the probability current density is a constant $(J'(x) = 0)$ and we have *total reflection*

$$J(x < 0) = J(x > 0) = J(+\infty) = 0 \implies J_{inc} + J_{ref} = 0 \implies \mathcal{R} = \frac{-J_{ref}}{J_{inc}} = 1.$$

3.2.2 The Square Barrier

In the case of a particle that encounters a potential step of height $V_0 > E$ we saw that the particle does not propagate in the classically forbidden region. What if this region does not extend to infinity but it is of finite extent? Consider a particle moving in the potential shown in Fig. 3.4. $(V_0 > 0)$

$$V(x) = \begin{cases} 0 & (|x| > L) \\ V_0 & (|x| < L). \end{cases} \tag{3.55}$$

We are interested in the $E < V_0$ case in which a classical particle incident from the left would not be able to penetrate the barrier and move to the $x > L$ region but would be reflected back to $x < -L$. The Schroedinger equation has the form

$$\psi_E''(x) = \begin{cases} -k^2 \psi_E(x) & (|x| > L) \\ s^2 \psi_E(x) & (|x| < L), \end{cases} \tag{3.56}$$

where

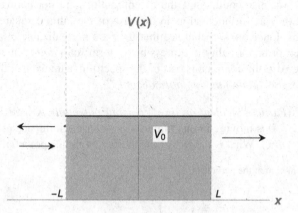

Fig. 3.4 Square barrier

$$k^2 = \frac{2m}{\hbar^2} E, \quad s^2 = \frac{2m}{\hbar^2} (V_0 - E) . \tag{3.57}$$

The corresponding solutions are

$$\psi_E(x) = \begin{cases} A\, e^{ikx} + B\, e^{-ikx} & (x < -L) \\[2mm] C\, e^{-sx} + D\, e^{sx} & (-L < x < L) \\[2mm] F\, e^{ikx} + G\, e^{-ikx} & (x > L) \end{cases} \tag{3.58}$$

Without loss of generality we may assume incidence only from the left and set $G = 0$. Continuity at $x = -L$ and $x = L$ gives us

$$Ae^{-ikL} + Be^{ikL} = Ce^{sL} + De^{-sL}$$

$$ik\left(Ae^{-ikL} - Be^{ikL}\right) = -s\left(Ce^{sL} - De^{-sL}\right)$$

$$Ce^{-sL} + De^{sL} = Fe^{ikL} \tag{3.59}$$

$$-s\left(Ce^{-sL} - De^{sL}\right) = ikFe^{ikL}.$$

This system of four equations can be solved to determine the four ratios B/A, C/A, D/A, F/A. Of particular interest is the last ratio, since it gives as the *transmission coefficient*

$$T = \frac{J_t}{J_i} = \frac{|F|^2}{|A|^2} . \tag{3.60}$$

Any nonzero value for T means that the barrier is penetrated. This is in sharp contrast to the classical case where such a penetration is not possible. This is the so-called *quantum tunneling* effect.

Proceeding to solve the system (3.59) we subtract the two last equations and obtain $D = -C\, e^{-2sL}(ik + s)/(ik - s)$. Then, canceling the B-term between the first two equations and inserting D we obtain $C/A = 2ik(ik - s)e^{-ikL}e^{sL}/[e^{2sL}(ik - s)^2 - e^{-2sL}(ik - s)^2]$. Inserting these to one of the equations for F, we obtain after a little algebra

$$\frac{F}{A} = \frac{-2ikse^{-2iks}}{\left[(s^2 - k^2)\sinh(2sL) - 2iks\cosh(2sL)\right]} . \tag{3.61}$$

The resulting transmission coefficient is

$$T = \left|\frac{F}{A}\right|^2 = \frac{4s^2 k^2}{\left[(s^2 - k^2)^2 \sinh^2(2sL) + 4k^2 s^2 \cosh^2(2sL)\right]}$$

or

$$T = \left[1 + \frac{\sinh^2 \left(2\beta\sqrt{1 - E/V_0} \right)}{4 \left(1 - \frac{E}{V_0} \right) \frac{E}{V_0}} \right]^{-1} \quad \text{where} \quad \beta \equiv \sqrt{\frac{2mV_0L^2}{\hbar^2}}. \quad (3.62)$$

β is a parameter characterizing the height and the width of the barrier. For a very large $\beta \gg$ (i.e., high and/or wide barrier) we have an exponential dependence

$$T \approx 4\frac{E}{V_0} \left(1 - \frac{E}{V_0} \right) e^{-4\beta\sqrt{1-E/V_0}}. \quad (3.63)$$

Note that according to the expression (3.62), in the limit of very slow particles, i.e., $E \to 0$, transmission tends to zero, while reflection is total, as it would be expected $(E \to 0 \implies T \to 0, \quad \mathcal{R} \to 1)$.

Example 3.3 (*Infinite square well with a central square barrier*) A particle of mass m is trapped in a symmetric infinite square well of width $2L$ that has a symmetric central square barrier of width L and height V_0. Investigate the existence of even energy eigenstates of energy $E > V_0$.

Setting $k^2 = 2mE/\hbar^2$ and $q^2 = 2m(E - V_0)/\hbar^2$ we have the following solution

$$V(x) = \begin{cases} +\infty & (x \le -L) \\ 0 & (-L < x < -L/2) \\ V_0 & (-L/2 < x < L/2) \\ 0 & (L/2 < x < L) \\ +\infty & (x \ge L) \end{cases} \implies \psi_E(x) = \begin{cases} 0 \\ A\cos(kx) + B\sin(kx) \\ C\cos(qx) \\ A\cos(kx) - B\sin(kx) \\ 0 \end{cases}$$

Continuity of $\psi_E(x)$ and $\psi'_E(x)$ at $\pm L$ and $\pm L/2$ gives

$$A\cos(kL/2) - B\sin(kL/2) - C\cos(qL/2) = 0$$

$$Ak\sin(kL/2) + Bk\cos(kL/2) - Cq\sin(qL/2) = 0$$

$$A\cos(kL) - B\sin(kL) = 0.$$

In order to have a solution for this system of three linear equations, the corresponding determinant of coefficients has to vanish, i.e.,

$$\begin{vmatrix} \cos(kL/2) & \sin(kL/2) & -\cos(qL/2) \\ k\sin(kL/2) & k\cos(kL/2) & -Cq\sin(qL/2) \\ \cos(kL) & -\sin(kL) & 0. \end{vmatrix} = 0 \implies tan(kL/2)\,tan(qL/2) = k/q$$

Setting $\xi = kL/2$ and $\beta^2 = mV_0L^2/2\hbar^2$, the condition reads

$$\tan \xi \tan \sqrt{\xi^2 - \beta^2} = \frac{\xi}{\sqrt{\xi^2 - \beta^2}}.$$

Plotting both sides of this equation for $\xi > \beta$ we see that there is always at least one point of intersection and, therefore, at least one even solution.

3.2.3 The Square Well

What if instead of the repulsion by the barrier of the previous section, the particle is subject to an attractive force? This is the case of a *square well* potential which is shown in Fig. 3.5.

$$V(x) = \begin{cases} 0 & (|x| > L) \\ -V_0 & (|x| < L) \end{cases} \qquad (V_0 > 0). \qquad (3.64)$$

The case of positive energy for this potential has no major surprises apart from the fact that even with an attractive force there is still nonzero reflection. In fact the equations are identical to those of the square barrier, the only necessary replacement being that of $s \to is$. The resulting square well transmission coefficient is

$$\mathcal{T} = \left| \frac{F}{A} \right|^2 = \frac{4s^2 k^2}{\left[-(s^2 - k^2)^2 \sin^2(2sL) + 4k^2 s^2 \cos^2(2sL) \right]}. \qquad (3.65)$$

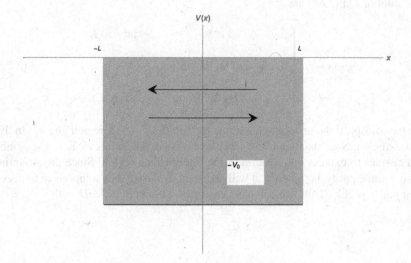

Fig. 3.5 Square well

Note now that there are very specific values of the incoming momentum for which the well is entirely transparent with total transmission $\mathcal{T} = 1$ and no reflection $\mathcal{R} = 0$. These are the special values for which $\sin(2sL) = 0$ or

$$E_n = -V_0 + \frac{n^2\pi^2\hbar^2}{8mL^2} \quad (n = \pm 1, \pm 2, \ldots). \tag{3.66}$$

Such *resonance* phenomena are familiar in optics.

What about the case of negative energies $E < 0$? We have remarked earlier that in general there exist no physically acceptable solutions of the Schroedinger equation for energies smaller than the minimum of the potential.[5] Here the potential is either zero or $-V_0$. Therefore, negative energies should be bounded from below as

$$. - V_0 \leq E \leq 0. \tag{3.67}$$

The Schroedinger equation takes the form

$$\psi_E''(x) = \begin{cases} \kappa^2 \psi_E(x) & (|x| > L) \\[2mm] -q^2 \psi_E(x) & (|x| < L) \end{cases} \tag{3.68}$$

in terms the parameters

$$\kappa^2 = -\frac{2m}{\hbar^2}E = \frac{2m}{\hbar^2}|E|, \quad q^2 = \frac{2m}{\hbar^2}(E + V_0). \tag{3.69}$$

The solutions of (3.68) are

$$\psi_E(x) = \begin{cases} A\,e^{\kappa x} & (x < -L) \\[2mm] C\cos(qx) + D\sin(qx) & (-L < x < L) \\[2mm] B\,e^{-\kappa x} & (x > L). \end{cases} \tag{3.70}$$

We have dropped the unacceptable terms $e^{-\kappa x}$ in the $x < -L$ region and $e^{\kappa x}$ in the $x > L$ region. Note also that instead of plane wave exponentials we have used sines and cosines to express the solution in the intermediate region. Since the potential has a definite parity, being even, it will be useful to classify our solutions in terms of their parity profile. Thus, we can have an even solution ($B = A$, $D = 0$)

---

[5]For an analysis of the issue of bound states in general one-dimensional potentials see [2].

$$\psi_E^{(+)}(x) = \begin{cases} A\,e^{\kappa x} & (x < -L) \\ C\cos(qx) & (|x| < L) \\ A\,e^{-\kappa x} & (x > L) \end{cases} = \begin{cases} A\,e^{-\kappa|x|} & (|x| > L) \\ C\cos(qx) & (|x| < L) \end{cases} \tag{3.71}$$

and an odd solution ($B = -A$, $C = 0$)

$$\psi_E^{(-)}(x) = \begin{cases} A\,e^{\kappa x} & (x < -L) \\ D\sin(qx) & (|x| < L) \\ -A\,e^{-\kappa x} & (x > L). \end{cases} \tag{3.72}$$

We proceed to study first the even solution. Continuity at $x = L$ gives

$$\begin{aligned} Ae^{-\kappa L} &= C\cos(qL) \\ -\kappa A\,e^{-\kappa L} &= -q\,C\sin(qL) \end{aligned} \implies \frac{C}{A} = \frac{e^{-\kappa L}}{\cos(qL)} = \frac{\kappa e^{-\kappa L}}{q\sin(qL)}. \tag{3.73}$$

We observe that in addition to the determination of the coefficient ratio we obtain also a condition on the energy, namely

$$\tan(qL) = \frac{\kappa}{q}. \tag{3.74}$$

The meaning of this condition is that solutions in the interval $-V_0 < E < 0$ can exist only for special values of the energy that will be extracted from this equation. From a mathematical point of view (3.74) is a *transcendental equation* combining trigonometric functions with irrational ones. One of the ways to solve such an equation is to employ a graphical method. This the line we shall follow here.

Let's introduce the variable $\xi \equiv qL$ and the parameter $\beta \equiv \sqrt{\frac{2mV_0}{\hbar^2}L^2}$, the latter representing the strength of the attractive potential, depending on its depth and extent. In terms of them, we can rewrite (3.74) as

$$\tan\xi = \sqrt{\frac{\beta^2}{\xi^2} - 1}. \tag{3.75}$$

In Fig. 3.6. We have plotted both the right hand side and the left hand side (dotted line) of (3.75), the existing solutions corresponding to their intersection points.

We see that there will always be at least one solution. The number of intersection points and, therefore, the number of solutions depends on the parameter β. For $\beta < \pi$ there is only one solution. For $\pi < \beta < 3\pi/2$ there are three and so on. As a result, we obtain a set of increasing discrete energy eigenvalues

Fig. 3.6 Even bound state
solutions

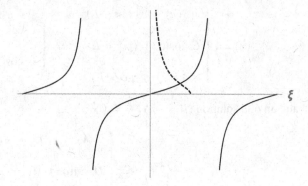

$$E_1^{(+)}, \; E_2^{(+)}, \; E_3^{(+)}, \; \ldots \tag{3.76}$$

The corresponding eigenstates are called *bound states* due to the fact that the wave
functions are localized in the potential well, decreasing exponentially out of it and,
thus, representing particles essentially bound inside the well.

Next, we proceed to study the odd branch of solutions. Continuity at $x = L$ gives

$$\frac{D}{A} = -\frac{e^{-\kappa L}}{\sin(qL)} = \frac{\kappa e^{-\kappa L}}{q \cos(qL)} \tag{3.77}$$

and the odd-eigenvalues condition is

$$\tan(qL) = -\frac{q}{\kappa}. \tag{3.78}$$

In terms of the variable ξ and parameter β this is written as

$$\tan \xi = -\frac{1}{\sqrt{\dfrac{\beta^2}{\xi^2} - 1}}. \tag{3.79}$$

Both sides of this equation are plotted in Fig. 3.7, the dotted lines corresponding to the
right hand side for two different values of the parameter β. The existing intersection
points correspond to the solutions.

We can see that for $\beta < \pi/2$ there is no intersection and therefore no solution.
For $\beta > \pi/2$ we can have at least one solution or more, depending on how large is
the potential strength parameter β. Thus, we obtain a hierarchy of odd bound state
eigenvalues

$$E_1^{(-)}, \; E_2^{(-)}, \; E_3^{(-)}, \; \ldots \tag{3.80}$$

Note that as the parity of adjacent eigenstates changes so does the hierarchy of the

Fig. 3.7 Odd bound state solutions

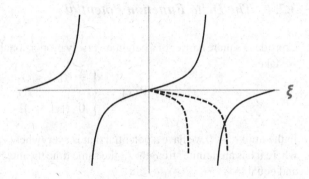

corresponding eigenvalues, namely

$$\psi_1^{(+)} \qquad \psi_1^{(-)} \qquad \psi_2^{(+)} \qquad \psi_2^{(-)} \qquad \cdots$$
$$\downarrow \qquad \downarrow \qquad \downarrow \qquad \downarrow \qquad \cdots \qquad (3.81)$$
$$E_1^{(+)} < E_1^{(-)} < E_2^{(+)} < E_2^{(-)} < \cdots$$

The state of lowest energy (ground state) is always even. There is always at least one bound state independently of how small is β. Note that the square well is an example of an energy eigenvalue spectrum that is mixed, consisting of a continuum of *scattering states* of energy $0 \leq E < \infty$ and a finite number of discrete *bound states* of energies (3.81).

Example 3.4 (Square well with an infinite wall) Consider a particle in the presence of the potential

$$V(x) = \begin{cases} +\infty & (x \leq 0) \\ -V_0 & (0 < x < L) \\ 0 & (x > L). \end{cases} \qquad (3.82)$$

Investigate the existence of bound states.

For energies $-V_0 < E < 0$ the solution to the Schroedinger equation is

$$\psi_E(x) = \begin{cases} 0 & (x \leq 0) \\ A\cos(qx) + B\sin(qx) & (0 \leq x \leq L) \\ Ce^{-\kappa x} & (x \geq L). \end{cases} \qquad (3.83)$$

Continuity at $x = 0$ implies $A = 0$. Then, these solutions coincide with the odd bound state solutions of the standard square well. Solutions exist provided $2mV_0L^2/\hbar^2 > \pi^2$.

3.2.4   The Delta Function Potential

Consider a square barrier of a height inversely proportional to its width. To be specific we take

$$V(x) = \begin{cases} \frac{\lambda}{2\epsilon} & (|x| < \epsilon) \\ 0 & (|x| > \epsilon). \end{cases} \tag{3.84}$$

In the limit $\epsilon \to 0$ we have a potential that is everywhere zero except the point $x = 0$ where it has an infinite strength. At the same time the integral of the potential is finite and equal to λ

$$\int_{-\infty}^{+\infty} dx \, V(x) = \int_{-\epsilon}^{+\epsilon} dx \, \lambda \left(\frac{1}{2\epsilon}\right) = \lambda.$$

Therefore, the potential in that limit can be modeled in terms of the *delta function*[6]

$$V(x) = \lambda \, \delta(x). \tag{3.85}$$

Let's consider the time-independent Schroedinger equation for a particle of mass m and energy E moving under the influence of the delta function potential (3.85)

$$-\frac{\hbar^2}{2m} \psi''(x) + \lambda \, \delta(x) \, \psi(x) = E \, \psi(x). \tag{3.86}$$

The fact that the potential is singular at $x = 0$ means that the second derivative of the wave function, appearing in the Schroedinger equation, will not exist at that point and therefore its first derivative will not be continuous. Of course, the wave function itself must be continuous everywhere. Integrating the Schroedinger equation in the vicinity of the origin we obtain

$$-\frac{\hbar^2}{2m} \int_{-\epsilon}^{+\epsilon} dx \, \psi''(x) + \lambda \int_{-\epsilon}^{+\epsilon} dx \, \delta(x) \, \psi(x) = E \int_{-\epsilon}^{+\epsilon} dx \, \psi(x)$$

or

$$-\frac{\hbar^2}{2m} \left(\psi'(\epsilon) - \psi'(-\epsilon)\right) + \lambda \psi(0) = E \, (2\epsilon) \, \psi(0) = 0$$

or, for $\epsilon \to 0$,

$$\psi'(\epsilon) - \psi'(-\epsilon) = \frac{2m\lambda}{\hbar^2} \psi(0). \tag{3.87}$$

This equation determines the discontinuity of the derivative of the wave function at the point $x = 0$.

---

[6]See the section on Generalized Functions in the Mathematical Appendix.

Scattering in the Delta Function Potential $(E > 0)$. The solution of the Schroedinger equation is

$$\psi_k(x) = \begin{cases} A\,e^{ikx} + B\,e^{-ikx} & (x \leq 0) \\ \\ C\,e^{ikx} & (x \geq 0) \end{cases} \tag{3.88}$$

where $E = \hbar^2 k^2/2m$. Without loss of generality, we have chosen incidence from the left. Continuity of the wave function at $x = 0$ gives

$$A + B = C. \tag{3.89}$$

The discontinuity at $x = 0$ according to (3.87) gives

$$ikC - ik(A - B) = \frac{2m\lambda}{\hbar^2} C. \tag{3.90}$$

From (3.89) and (3.90) we obtain

$$\frac{B}{A} = \frac{1}{-1 + \frac{ik\hbar^2}{m\lambda}}$$
$$\frac{C}{A} = \frac{\frac{ik\hbar^2}{m\lambda}}{-1 + \frac{ik\hbar^2}{m\lambda}} \tag{3.91}$$

Note that the above expressions are equally valid for repulsive $(\lambda > 0)$ or attractive $(\lambda < 0)$ potential. It is straightforward to obtain the reflection and transmission coefficients

$$\mathcal{R} = \frac{J_r}{J_i} = \frac{|B|^2}{|A|^2} = \frac{1}{1 + \left(\frac{\hbar^2 k}{m\lambda}\right)^2}$$
$$\mathcal{T} = \frac{J_t}{J_i} = \frac{|C|^2}{|A|^2} = \frac{\left(\frac{\hbar^2 k}{m\lambda}\right)^2}{1 + \left(\frac{\hbar^2 k}{m\lambda}\right)^2}. \tag{3.92}$$

Before we move to the study of bound states it is not difficult to see that the solution $\psi_k(x)$ that we have obtained can also be written in the form

$$\psi_k(x) = A\,e^{ikx} + \frac{A}{-1 + \frac{ik\hbar^2}{m\lambda}}\,e^{ik|x|}. \tag{3.93}$$

In this form, the wave function is the superposition of an incident plane wave e^{ikx} and a *scattered wave* $e^{ik|x|}$ propagating in both available directions. The coefficient of the scattered wave corresponds to the so-called *scattering amplitude*

$$f(k) = \frac{1}{-1 + \frac{ik\hbar^2}{m\lambda}}. \tag{3.94}$$

Bound States of the Delta Function Potential ($E < 0$). Solutions with $E < 0$ exist only in the case of an attractive potential, namely for $\lambda < 0$. These are

$$\psi_b(x) = \begin{cases} A\,e^{\kappa x} & (x < 0) \\[2mm] A\,e^{-\kappa x} & (x > 0), \end{cases} \tag{3.95}$$

where

$$E = -\frac{\hbar^2 \kappa^2}{2m} < 0. \tag{3.96}$$

Equation (3.95) is written compactly as

$$\psi_b(x) = A\,e^{-\kappa|x|}. \tag{3.97}$$

The discontinuity at $x = 0$ reads

$$-\kappa\,A - \kappa\,A = \frac{2m\lambda}{\hbar^2}\,A$$

or

$$\kappa = \frac{m|\lambda|}{\hbar^2}. \tag{3.98}$$

Thus, there is one bound state of energy

$$E_b = -\frac{m\lambda^2}{2\hbar^2}. \tag{3.99}$$

The coefficient of the wave function is determined by its normalization to be

$$\int_{-\infty}^{+\infty} dx\, A^2\, e^{-2\kappa|x|} = 1 \quad \Longrightarrow \quad A = \sqrt{\kappa}$$

and, finally the bound state wave function is

$$\psi_b(x) = \sqrt{\kappa}\, e^{-\kappa|x|} = \sqrt{\frac{m|\lambda|}{\hbar^2}}\, e^{-\sqrt{\frac{m|\lambda|}{\hbar^2}}|x|}. \tag{3.100}$$

Let's return briefly to the scattering amplitude previously considered to note a peculiar property. The scattering amplitude, considered as a function of the wave number taking values in the *complex k-plane*, is written as

$$f(k) = \frac{m\lambda}{i\hbar^2\,(k - k_p)}, \tag{3.101}$$

where

$$k_p = i\frac{m|\lambda|}{\hbar^2}.$$

Therefore, it has a pole at k_p along the imaginary axis or, equivalently, at the negative energy

$$E_p = \frac{\hbar^2 k_p^2}{2m} = -\frac{m\lambda^2}{2\hbar^2},$$

which is exactly the value of the bound state energy. This is a general property and can be stated as follows: "The bound states correspond to poles of the scattering amplitude considered in the complex k-plane."

Closing this subsection, we briefly comment on the orthonormality and completeness of the energy eigenfunctions, having repeatedly emphasized that the energy eigenfunctions should be automatically *orthogonal* and *complete*. The orthogonality of the scattering eigenfunctions is promoted to *orthonormality* by adjusting the coefficient $A(k)$ through the orthonormality and completeness relations

$$\begin{cases} \int_{-\infty}^{+\infty} dx\, \psi_k^*(x)\, \psi_{k'}(x) = \delta(k - k') \\[2mm] \int_{-\infty}^{+\infty} dx\, \psi_k^*(x)\, \psi_b(x) = 0 \end{cases} \qquad (orthonormality) \qquad (3.102)$$

$$\int_{-\infty}^{+\infty} dk\, \psi_k(x)\, \psi_k^*(x') + \psi_b(x)\psi_b(x') = \delta(x - x') \qquad (completeness). \qquad (3.103)$$

These can be verified explicitly.

Example 3.5 (Orthogonality of bound state and scattering eigenfunctions) Verify explicitly that the bound state wave function is orthogonal to the scattering wave functions.

We need to prove that $\int dx\, \psi_k(x)\psi_b(x) = 0$. We have

$$\int_{-\infty}^{+\infty} dx\, \left(e^{ikx} + f(k)e^{ik|x|}\right) e^{-\kappa|x|} = \int_0^{+\infty} dx\, (1 + f(k))\, e^{-(\kappa - ik)x}$$

$$+ \int_{-\infty}^0 dx\, \left(e^{(ik+\kappa)x} + f(k)\, e^{(-ik+\kappa)x}\right) = \frac{1 + 2f(k)}{\kappa - ik} + \frac{1}{\kappa + ik}.$$

Noting that $f(k) = \dfrac{1}{-1 - i\frac{k\hbar^2}{m|\lambda|}} = -\dfrac{\kappa}{\kappa + ik}$, we obtain

$$\int dx\, \psi_k(x)\psi_b(x) = A\sqrt{\kappa}\left(\frac{1 - \frac{2\kappa}{\kappa + ik}}{\kappa - ik} + \frac{1}{\kappa + ik}\right) = A\sqrt{\kappa}\left(\frac{ik - \kappa}{\kappa^2 + k^2} + \frac{1}{\kappa + ik}\right) = 0.$$

Example 3.6 (*Potential step with delta function spike at the edge*) Consider a potential step with a delta function at its edge, i.e., $V(x) = V_0 \Theta(x) + \lambda \delta(x)$ and calculate the reflection coefficient.

We have

$$\psi(x) = \begin{cases} e^{ikx} + Be^{-ikx} & (x < 0) \\ C e^{iqx} & (x > 0) \end{cases} \tag{3.104}$$

with $E = \frac{\hbar^2 k^2}{2m}$, $E - V_0 = \frac{\hbar^2 q^2}{2m}$. The discontinuity of the derivative at $x = 0$ gives

$$\psi'(+0) - \psi'(-0) - \frac{2m\lambda}{\hbar^2}\psi(0) = 0 \implies iqC - ik(1 - B) - \frac{2m\lambda}{\hbar^2}C = 0$$

while the continuity of the wave function gives just $1 + B = C$. Solving for B we obtain

$$= -\left(\frac{\frac{2m\lambda}{\hbar^2} - i(q - k)}{\frac{2m\lambda}{\hbar^2} - i(q + k)} \right) \implies \mathcal{R} = |B|^2 = \frac{\left(\frac{2m\lambda}{\hbar^2}\right)^2 + (q - k)^2}{\left(\frac{2m\lambda}{\hbar^2}\right)^2 + (q + k)^2}. \tag{3.105}$$

Example 3.7 (*Delta function potential next to an infinite wall*) A particle is subject to the potential

$$V(x) = \begin{cases} +\infty & (x < 0) \\ -\frac{\hbar^2 g^2}{2m}\delta(x - a) & (x > 0). \end{cases} \tag{3.106}$$

Investigate whether there are bound states $(E < 0)$.

The wave function of energy $E = -\frac{\hbar^2 \kappa^2}{2m} < 0$ is

$$\psi(x) = \begin{cases} 0 & (x \le 0) \\ A \sinh(\kappa x) & (0 \le x \le a) \\ B e^{-\kappa x} & (x \ge a) \end{cases} \tag{3.107}$$

The discontinuity/continuity relations at $x = a$ are

$$\kappa B e^{-\kappa a} + A\kappa \cosh(\kappa a) = g^2 B e^{-\kappa a}$$
$$A \sinh(\kappa a) = B e^{-\kappa a}$$
$$\implies \frac{B}{A} = \frac{\cosh(\kappa a)}{\frac{g^2}{\kappa} - 1}e^{\kappa a} = \sinh(\kappa a)e^{\kappa a}.$$

The energy eigenvalue condition can be written as $(\xi = \kappa a)$

$$\tanh \xi = \frac{1}{\frac{ag^2}{\xi} - 1}. \tag{3.108}$$

It is not difficult to see that the right and left hand sides plotted together intersect if $g^2 a > 1$, i.e., for a sufficiently attractive delta function potential there is one bound state.

Example 3.8 (*Infinite square well with an attractive delta function at its center*) A particle of mass m is trapped in a symmetric infinite square well of width $2L$ that has an attractive delta function potential at its center $V(x) = -\frac{\hbar^2 g^2}{2m} \delta(x)$. Investigate the existence of even energy eigenstates.

We have

$$
\psi_E(x) = \begin{cases}
0 & (x \leq -L) \\
A \cos(kx) + B \sin(kx) & (-L \leq x \leq 0) \\
C \cos(kx) - D \sin(kx) & (0 \leq x \leq L) \\
0 & (x \geq L).
\end{cases}
\tag{3.109}
$$

Evenness $\psi_E(-x) = \psi_E(x)$ and continuity of $\psi_E(x)$ at $x = 0$ implies $C = A'$, $D = B$. The discontinuity of $\psi'_E(x)$ at $x = 0$ gives $B = \frac{g^2}{2k} A$. Therefore, we have

$$
\psi_E(x) = A \left(\cos(kx) - \frac{g^2}{2k} \sin(k|x|) \right).
\tag{3.110}
$$

The condition $\psi_E(\pm L) = 0$ at the edges gives

$$
\tan(kL) = \frac{2ka}{g^2 a},
\tag{3.111}
$$

which always has solutions.

3.2.5 Scattering by Two Delta Functions and Resonances

Consider a particle moving in a potential of two repulsive delta function barriers at $x = -a$ and $x = +a$

$$
V(x) = \frac{\hbar^2 g}{2m} \left(\delta(x + a) + \delta(x - a) \right).
\tag{3.112}
$$

The coupling strength of the potential has been suitably parametrized as $\hbar^2 g / 2m$. The resulting Schroedinger equation is

$$
-\psi''_k(x) + g\, \delta(x + a)\, \psi_k(-a) + g\, \delta(x - a)\, \psi_k(a) = k^2\, \psi_k(x)
\tag{3.113}
$$

with $E = \hbar^2 k^2 / 2m$. The discontinuity at $x = -a$ is obtained by integrating the Schroedinger equation in the interval $[-a - \epsilon, -a + \epsilon]$ with $\epsilon \to 0$, where, apart from the second derivative term, only the delta function contributes in this range. Analogously, for the discontinuity at $x = a$, we integrate in the interval $[a - \epsilon, a + \epsilon]$. We obtain the conditions

$$\psi'_k(-a+\epsilon) - \psi'_k(-a-\epsilon) - g\psi_k(-a) = 0$$
$$\psi'_k(a+\epsilon) - \psi'_k(a-\epsilon) - g\psi_k(a) = 0. \tag{3.114}$$

The solution is everywhere a combination of plane waves $e^{\pm ikx}$

$$\begin{cases} A\,e^{ikx} + B\,e^{-ikx} & (x < -a) \\ C\,e^{ikx} + D\,e^{-ikx} & (-a < x < a) \\ F\,e^{ikx} & (x > a). \end{cases} \tag{3.115}$$

Without loss of generality, we have assumed incidence only from the left. The continuity/discontinuity equations are

$$A\xi^{-1} + B\xi = C\xi^{-1} + D\xi$$

$$ik(C\xi^{-1} - D\xi) - ik(A\xi^{-1} - B\xi)$$
$$= g(C\xi^{-1} + D\xi)$$

$$F\xi = C\xi + D\xi^{-1}$$

$$ikF\xi - ik(C\xi - D\xi^{-1})$$
$$= g(C\xi + D\xi^{-1})$$

$$\implies$$

$$\begin{cases} A\xi^{-1} + B\xi = C\xi^{-1} + D\xi \\ A\xi^{-1} - B\xi = \left(1 + i\frac{g}{k}\right)C\xi^{-1} - \left(1 - i\frac{g}{k}\right)D\xi \\ F\xi = C\xi + D\xi^{-1} \\ F\xi = \left(1 - i\frac{g}{k}\right)C\xi + \left(-1 - i\frac{g}{k}\right)D\xi^{-1}, \end{cases}$$

where we have set $\xi = e^{ika}$. Subtracting the last two equations we obtain

$$D = -\frac{ig}{k}\frac{C\xi^2}{\left(2 + \frac{ig}{k}\right)}.$$

Canceling B among the first two equations and substituting D we get

$$\frac{C}{A} = \frac{2\left(2 + i\frac{g}{k}\right)}{\left[\left(2 + i\frac{g}{k}\right)^2 + \frac{g^2}{k^2}\xi^4\right]} \implies \frac{F}{A} = \frac{1}{\left[\left(1 + i\frac{g}{2k}\right)^2 + \frac{g^2}{4k^2}e^{4ika}\right]}. \tag{3.116}$$

From this, we get the transmission coefficient

$$T = \left[1 + \frac{g^2}{k^2}\left(\cos(2ka) + \frac{g}{2k}\sin(2ka)\right)^2\right]^{-1}. \tag{3.117}$$

Total transmission (resonances), corresponding to the condition $T = 1$, is satisfied if

$$\tan(2ka) = -\frac{2ka}{ga}. \tag{3.118}$$

Fig. 3.8 Graphical solution
of two delta function
resonances

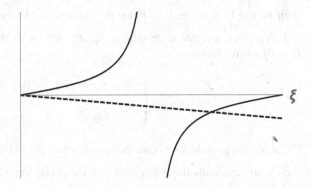

In Fig. 3.8. we have a common plot of the right and left hand sides of this equation
showing that there is always an intersection point and therefore an energy value
corresponding to a resonance.

Note that the amplitude F/A for very large $g \to \infty$, corresponding to impenetrable delta functions, has the form

$$\frac{F}{A} \approx \frac{4k^2/g^2}{e^{4ika} - 1},\qquad(3.119)$$

exhibiting poles at $k_n = n\pi/2a$ associated with energies which are exactly the energy
levels of the infinite square well. Note also that for large coupling g the condition
(3.118) is satisfied for

$$k \approx \frac{\pi}{a} - \frac{\pi}{ga^2} + O(1/g^2).\qquad(3.120)$$

Problems and Exercises

3.1 Particles of mass m and energy E are incident from the left on the potential step
$V(x) = -\Theta(x)V_0$ with $V_0 > 0$. Calculate the reflection coefficient.

3.2 Consider a particle of mass m and energy $E > V_0 > 0$ incident on the potential
step $V(x) = \Theta(x)V_0$. Calculate the reflection coefficient for incidence from the right
and for incidence from the left and compare the results.

3.3 Consider the asymmetric square well

$$V(x) = \begin{cases} V_1 & (x < 0) \\ -V_0 & (0 < x < L) \\ 0 & (x > L) \end{cases}$$

with V_1 and V_0 positive. Investigate the existence of bound states.

3.4 A particle trapped in an infinite square well of width $[-2L, 2L]$ has initially $(t = 0)$ a wave function[7]

$$\psi(x, 0) = \begin{cases} \frac{1}{\sqrt{L}} \cos(\pi x/2L) & (|x| \le L) \\ \\ 0 & (L \le |x| \le 2L). \end{cases}$$

Calculate the probability to find the system in its ground state.

3.5 Verify explicitly the orthogonality of the energy eigenfunctions[8] of the potential step $V(x) = \Theta(x)V_0$

$$\int_{-\infty}^{+\infty} dx \, \psi_E^*(x)\psi_{E'}(x) = 0 \qquad (E \ne E').$$

3.6 A particle of mass m is trapped in an infinite square well of width $[-L, L]$. The particle is initially $(t = 0)$ in a state with wave function

$$\psi(x, 0) = \frac{1}{\sqrt{2}} (\psi_N(x) + i \psi_{N+1}(x)),$$

where $\psi_n(x)$ is the energy eigenfunction corresponding to energy E_n. Find the wave function of the particle at times $t > 0$. Calculate the probability to find the particle in the initial state at time $t > 0$. What is the minimum time that this probability becomes one? Calculate the probability to find the particle at time $t > 0$ in the orthogonal state

$$\psi^{(\perp)}(x) = \frac{1}{\sqrt{2}} (\psi_N(x) - i\psi_{N+1}(x)).$$

3.7 A particle of mass m and energy $E > 0$ is incident from the right on a potential step $V(x) = -V_0\Theta(-x)$ with $V_0 > 0$. What is the value of the ratio E/V_0 if the reflection coefficient is $1/4$?

3.8 A particle of mass m is bound in a potential

$$V(x) = \begin{cases} +\infty & (x \le 0) \\ \\ -\frac{\hbar^2 g^2}{2m}\delta(x - a) & (x > 0). \end{cases}$$

Calculate the probability to find the particle in the region $a \le x < \infty$.

---

[7] Such a situation can occur if a particle occupies the ground state of an infinite square well of width $[-L, L]$ and *suddenly* (in a time interval τ much smaller than the characteristic time of the system, i.e., $\tau << 8mL^2/\hbar\pi^2$) the walls of the well move so that the width gets doubled.

[8] You may assume the regularization $\lim_{x \to \infty} e^{ikx} = \lim_{\epsilon \to 0} \lim_{x \to \infty} e^{i(x+i\epsilon)k} = 0$.

3.9 A particle moves in the potential $V(x) = g^2 |x|$. If the particle is in a state with wave function

$$\psi(x) = N x e^{-\alpha x^2},$$

where N, $\alpha > 0$ are known parameters, calculate the probability to find the particle in the ground state $\psi_0(x)$ of the energy. The exact form of $\psi_0(x)$ is not necessary.

3.10 Consider a particle moving in the potential of two attractive delta functions

$$V(x) = -\frac{\hbar^2 g^2}{2m} \left(\delta(x+a) + \delta(x-a) \right).$$

Investigate the existence of bound states ($E < 0$).

References

1. A. Messiah, *Quantum Mechanics* (Dover publications, Mineola, 1958). Single-volume reprint of the John Wiley & Sons, New York, two-volume 1958 edition
2. F. Levin, *An Introduction to Quantum Theory* (Cambridge University Press, Cambridge, 2002)

Chapter 4
Physical Observables as Operators

4.1 Physical Quantities and Operators

In our discussion of the system of a quantum particle, we have mostly concentrated on its wave function, i.e., its probability distribution in space. This is not the only quantity of interest that may arise in the description of a physical system. Other physical observables are also important as for example the energy of the system or its momenta. The correspondence of each of these observables to measurements on the system is not as straightforward as in classical physics. Nevertheless, since the description of any quantum system is bound to be probabilistic, general concepts like the *"average"* or *"expectation values"* of the various physical observables of the system, central to probabilities, are of particular importance.

Consider such a physical quantity A and assume that it is subject to measurements with outcome a set of values $\{\alpha_1, \alpha_2, \ldots, \alpha_i, \ldots, \alpha_N\}$. Its *"average value"* is defined as

$$\langle A \rangle = \sum_i \alpha_i P_i, \tag{4.1}$$

where P_i is the probability of each value.[1] Let us now apply the above definition in the simple system of a particle moving in one dimension. Since $|\psi(x, t)|^2$ is the probability density to find the particle at a point x, the *"average value"* or *"expectation value"* of the position will be given by the expression

$$\langle x \rangle = \int_{-\infty}^{+\infty} dx \, x \, |\psi(x, t)|^2, \tag{4.2}$$

which is the analogue of (4.1) in the case of a variable that takes continuous values.

---

[1] See the section on *Probability* in the Mathematical Appendix.

© Springer Nature Switzerland AG 2019
K. Tamvakis, *Basic Quantum Mechanics*, Undergraduate Texts in Physics,
https://doi.org/10.1007/978-3-030-22777-7_4

Assume now that the particle is free. Recall that the probability amplitude to find the particle at x having momentum p is $e^{i(xp-Et)/\hbar}/\sqrt{2\pi\hbar}$. Therefore, as we have discussed earlier, the function $g(p,t) = g(p)e^{-iEt/\hbar}$ in the Fourier transformation

$$\psi(x,t) = \int_{-\infty}^{+\infty} \frac{dp}{\sqrt{2\pi\hbar}} e^{\frac{i}{\hbar}(xp-Et)} g(p)$$

stands for the probability amplitude density to find the particle with momentum p, or in other words its *"momentum space wave function"*. We shall assume that the Fourier transform $g(p,t)$ given by

$$\psi(x,t) = \int_{-\infty}^{+\infty} \frac{dp}{\sqrt{2\pi\hbar}} e^{\frac{i}{\hbar}xp} g(p,t) \quad and \quad g(p,t) = \int_{-\infty}^{+\infty} \frac{dx}{\sqrt{2\pi\hbar}} e^{-\frac{i}{\hbar}xp} \psi(x,t)$$
$$(4.3)$$

has the same interpretation in the general case of an interacting particle as well. In terms of this, the corresponding *expectation value of the momentum* will be supplied by the analogous expression in terms of the momentum probability density $|g(p,t)|^2$ as

$$\langle p \rangle = \int_{-\infty}^{+\infty} dp\, p\, |g(p,t)|^2 .$$
$$(4.4)$$

Substituting the Fourier expression for $g(p,t)$ and making use of the integral expression and the properties of the *delta function*,[2] we obtain

$$\langle p \rangle = \int dx \int dx'\, \psi^*(x,t)\, \psi(x',t) \int \frac{dp}{2\pi\hbar}\, p\, e^{\frac{i}{\hbar}(x-x')p}$$

$$= \int dx \int dx'\, \psi^*(x,t)\, \psi(x',t) \int \frac{dp}{2\pi\hbar} \left(-i\hbar\frac{\partial}{\partial x}\right) e^{\frac{i}{\hbar}(x-x')p}$$

$$= \int dx\, \psi^*(x,t) \left(-i\hbar\frac{\partial}{\partial x}\right) \int dx'\, \psi(x',t)\, \delta(x-x')$$

or

$$\langle p \rangle = \int_{-\infty}^{+\infty} dx\, \psi^*(x,t) \left(-i\hbar\frac{\partial}{\partial x}\right) \psi(x,t) .$$
$$(4.5)$$

This is the same type of expression like the one we encountered for the position expectation value with the place of the momentum taken by the operator $-i\hbar\frac{\partial}{\partial x}$ or

$$p \rightarrow \hat{p} = -i\hbar\frac{\partial}{\partial x} .$$
$$(4.6)$$

Recall that we made an analogous correspondence in order to obtain the Schroedinger equation from the classical Hamiltonian. In contrast to momentum, the position operator acts as an ordinary number on the position wave function, the position expectation value being just

---

[2]See the section on Generalized Functions in the Mathematical Appendix.

$$\langle x \rangle = \int_{-\infty}^{+\infty} dx \, \psi^*(x, t) \, x \, \psi(x, t). \tag{4.7}$$

Note that the most important difference of operators to ordinary numbers is the fact that they do not commute. Since operators do not commute in general, an important quantity for operators is the *commutator* of two operators, defined as

$$\left[\hat{A}, \hat{B} \right] \equiv \hat{A}\hat{B} - \hat{B}\hat{A}. \tag{4.8}$$

Of fundamental importance is the commutator of position and momentum, which turns out to be

$$\left[\hat{x}, \hat{p} \right] = i\hbar. \tag{4.9}$$

All these can be carried over to three dimensional motion as

$$\mathbf{r}, \, \mathbf{p} \rightarrow \mathbf{r}, \, -i\hbar\nabla. \tag{4.10}$$

The corresponding commutation relations are

$$\left[\hat{x}_i, \hat{x}_j \right] = \left[\hat{p}_i, \hat{p}_j \right] = 0 \tag{4.11}$$

$$\left[\hat{x}_i, \hat{p}_j \right] = i\hbar\delta_{ij}.$$

These commutation relations are fundamental and go by the name *Canonical Commutation Relations*.

The mathematical concept of operators is not restricted to position and momentum but will cover all other physical observables, as for example the energy, angular momenum and others. Thus, the *expectation value* of a physical quantity represented by an operator $\hat{A}$ will be defined as in the case of (4.5) and (4.7) according to

$$\langle \mathcal{A} \rangle = \int dx \, \psi^*(x) \, \hat{A} \, \psi(x). \tag{4.12}$$

In case the wave function is not normalized to unity, the above definition generalizes to

$$\langle \mathcal{A} \rangle = \frac{\int dx \, \psi^*(x) \, \hat{A} \, \psi(x)}{\int dx \, |\psi(x)|^2}. \tag{4.13}$$

4.2 Eigenvalues and Eigenfunctions

Consider the wave function of a free particle moving in one dimension with a definite momentum p

$$\psi_p(x) = \frac{e^{\frac{i}{\hbar}px}}{\sqrt{2\pi\hbar}}. \tag{4.14}$$

We have dropped the time variable, assuming that our considerations take place at a given instant in time, set to be $t = 0$. Acting on (4.14) with the momentum operator we obtain

$$\hat{p}\,\psi_p(x) = -i\hbar\frac{\partial\psi_p}{\partial x} = p\,\psi_p(x)$$

or

$$\hat{p}\,\psi_p(x) = p\,\psi_p(x). \tag{4.15}$$

Thus, it is evident that the momentum operator, acting on wave functions of definite momentum, behaves just as a number. The wave functions of definite momentum (plane waves) are referred to as the *eigenfunctions* of the momentum operator and the corresponding momentum values that are produced upon acting on them by it are called the momentum *eigenvalues*.[3]

In fact, this is a particular example of a general situation. For every operator $\hat{A}$ there is always a particular set of functions upon which the operator acts as an ordinary number. These are the *eigenfunctions*[4] $\psi_a(x)$ of the operator

$$\hat{A}\,\psi_a(x) = a\,\psi_a(x), \tag{4.16}$$

while a are the *eigenvalues* of the operator. The full set $\{\,a\,\}$ of eigenvalues is called the *spectrum* of the operator. The spectrum of an operator can be continuous or discrete (or mixed). For example, the spectrum of the energy operator for a particle trapped in the infinite square well is discrete, labeled by an integer. In contrast, the energy spectrum of a particle moving in the presence of a finite square well consists of a discrete part of negative energy eigenvalues (*bound states*) and a continuum of positive energy eigenvalues (*scattering states*).

The eigenvalues of an operator can be *degenerate* if more than one eigenfunctions correspond to the same eigenvalue. As an example, we may consider the energy operator (Hamiltonian) of a free particle

$$\hat{\mathcal{H}} = \frac{\hat{p}^2}{2m}. \tag{4.17}$$

Note that the momentum eigenfunctions (4.14) $\psi_p(x)$ and $\psi_{-p}(x)$ correspond to one and the same energy eigenvalue $E = \frac{p^2}{2m}$

$$\hat{\mathcal{H}}\psi_p = E\psi_p, \quad \hat{\mathcal{H}}\psi_{-p} = E\psi_{-p}. \tag{4.18}$$

---

[3] General considerations on the eigenvalue problem of operators acting on square-integral functions can be found in books on *Mathematical Methods*, like Jackson [1], Mathews and Walker [2], Arfken et al. [3], Dennery and Krzywicki [4] as well as in standard QM textbooks like Messiah [5], Merzbacher [6], Levin [7].

[4] Sometimes we will be using also the terms *eigenvector* or *eigenstate* instead of *eigenfunction*. The three terms are entirely equivalent at this level of discussion. A precise meaning to the terms *vector* or *state vector* will be given when we shall be discussing the Hilbert space of states.

In fact, any linear combination $\alpha\psi_p + \beta\psi_{-p}$ is an eigenfunction of the energy with eigenvalue E. Note that this degeneracy is associated with the physical equivalence of motion toward the right and motion toward the left in free space.

4.3 Hermitian Conjugation and Hermitian Operators

Given an operator $\hat{A}$ and any two (square-integrable) wave functions, we may construct the complex number

$$\int dx\, \psi_1^*(x)\, \hat{A}\, \psi_2(x)\,,$$

which is refered to as a *matrix element* of the operator between these wave functions. Consider now an operator $\hat{A}^\dagger$, in general different than $\hat{A}$, defined by the relation

$$\int dx\, \psi_1^*(x)\, \hat{A}^\dagger\, \psi_2(x) = \left(\int dx\, \psi_2^*\, \hat{A}\, \psi_1(x) \right)^*. \tag{4.19}$$

The operator $\hat{A}^\dagger$ is called the *Hermitian conjugate or Hermitian adjoint of* $\hat{A}$. For example, the Hermitian conjugate of the derivative operator $\frac{d}{dx}$ is the operator $\left(\frac{d}{dx}\right)^\dagger = -\frac{d}{dx}$. This can be easily shown. Applying the definition (4.19), we obtain

$$\int dx\, \psi_1^* \left(\frac{d}{dx}\right)^\dagger \psi_2 = \left(\int dx\, \psi_2^* \psi_1' \right)^* = \int dx\, \psi_2 \psi_1'^*$$

$$= \int dx\, \left((\psi_2 \psi_1^*)' - \psi_2' \psi_1 \right) = -\int dx\, \psi_1^* \psi_2' = \int dx\, \psi_1^* \left(-\frac{d}{dx}\right) \psi_2\,,$$

using the fact that $\int_{-\infty}^{\infty} dx\, \left(\psi_2 \psi_1^* \right)' = \psi_2 \psi_1^* \big|_{-\infty}^{\infty} = 0$, since $\psi_{1,2}(\pm\infty) = 0$.

A special subset of operators are those which are equal to their Hermitian adjoints, i.e., those that have the property

$$\hat{A}^\dagger = \hat{A} \implies \int dx\, \psi_1^*(x)\, \hat{A}\, \psi_2(x) = \left(\int dx\, \psi_2^*\, \hat{A}\, \psi_1(x) \right)^*. \tag{4.20}$$

Such operators are called *self-conjugate or self-adjoint* or, simply, just, *Hermitian*. An obvious example of a nontrivial Hermitian operator is the momentum operator

$$\left(-i\hbar \frac{\partial}{\partial x}\right)^\dagger = -i\hbar \frac{\partial}{\partial x}\,. \tag{4.21}$$

Applying the definition of Hermitian conjugation (4.19) on the expectation value of an operator $\mathcal{A}$, we deduce that the expectation value of the Hermitian conjugate $\mathcal{A}^\dagger$ equals the complex conjugate of $\mathcal{A}$, namely

$$\langle \mathcal{A}^\dagger \rangle = \frac{\int dx\, \psi^*(x)\hat{\mathcal{A}}^\dagger \psi(x)}{\int dx |\psi(x)|^2} = \frac{\left(\int dx\, \psi^*(x)\hat{\mathcal{A}}\psi(x)\right)^*}{\int dx |\psi(x)|^2} \tag{4.22}$$

or

$$\langle \mathcal{A}^\dagger \rangle = \langle \mathcal{A} \rangle^*. \tag{4.23}$$

Thus, on account of (4.23) Hermitian operators have real expectation values

$$\mathcal{A}^\dagger = \mathcal{A} \implies \langle \mathcal{A} \rangle^* = \langle \mathcal{A} \rangle. \tag{4.24}$$

It should be noted that the reality of expectation values, singles out Hermitian operators for the representation of physical observables, since only real numbers can meaningfully result from experiment and real measurements will necessarily have real averages.

Apart from the reality of expectation values, Hermitian operators have two very important properties, namely [1–4]

(1) Hermitian operators have real eigenvalues

(2) Eigenfunctions of a Hermitian operator, corresponding to different eigenvalues, are *orthogonal* to each other, in the sense

$$\int dx\, \psi_\alpha^*(x)\, \psi_{\alpha'}(x) = 0 \quad (\alpha \neq \alpha').$$

We proceed to prove each of these properties.

Starting with the eigenvalue conditions for two arbitrary eigenfunctions

$$\hat{\mathcal{A}}\psi_\alpha = \alpha\psi_\alpha, \quad \hat{\mathcal{A}}\psi_{\alpha'} = \alpha'\psi_{\alpha'}$$

and the hermiticity definition

$$\int dx\, \psi_{\alpha'}^* \hat{\mathcal{A}}\psi_\alpha = \left(\int dx\, \psi_\alpha^* \hat{\mathcal{A}}\psi_{\alpha'}\right)^*$$

we obtain

$$\alpha \int dx\, \psi_{\alpha'}^*{}^* \psi_\alpha = \alpha'^* \int dx\, \psi_{\alpha'}^* \psi_\alpha \implies (\alpha - \alpha'^*)\int dx\, \psi_{\alpha'}^* \psi_\alpha = 0. \tag{4.25}$$

In case that the two eigenvalues coincide,[5] the last relation reads

---

[5]For simplicity, we assume that there is no degeneracy.

$$(\alpha - \alpha^*) \int dx \, |\psi_\alpha(x)|^2 = 0.$$

Since $\int_{-\infty}^{+\infty} dx \, |\psi_\alpha(x)|^2 > 0$, necessarily[6] the eigenvalues must be real

$$\alpha = \alpha^* \quad \forall \alpha. \tag{4.26}$$

Returning to (4.25) we have, for $\alpha \neq \alpha'$

$$(\alpha - \alpha') \int dx \, \psi_{\alpha'}^* \psi_\alpha = 0.$$

This necessarily implies that

$$\int_{-\infty}^{+\infty} dx \, \psi_\alpha^*(x) \psi_{\alpha'}(x) = 0 \quad \forall \alpha \neq \alpha'. \tag{4.27}$$

This is the so-called *orthogonality condition*. The underlying physical meaning of the orthonality property is that for a wave function that is a superposition of orthogonal eigenfunctions

$$\psi(x) = \psi_\alpha(x) + \psi_{\alpha'}(x),$$

the corresponding probability

$$\mathcal{P}_{\alpha \cup \alpha'} = \int dx \, |\psi_\alpha(x)|^2 + \int dx \, |\psi_{\alpha'}(x)|^2 + \int dx \, \psi_\alpha^*(x) \psi_{\alpha'}(x) + \int dx \, \psi_{\alpha'}^*(x) \psi_\alpha(x)$$

will be just the sum of the probabilities for each eigenfunction

$$\mathcal{P}_{\alpha \cup \alpha'} = \mathcal{P}_\alpha + \mathcal{P}_{\alpha'},$$

signifying that the two alternatives are independent. Thus, no interference pattern can arise between the alternatives that are represented by each of the orthogonal eigenfunctions.

If the eigenfunctions $\psi_\alpha(x)$ are normalizable, i.e., if we can write

$$\int_{-\infty}^{+\infty} dx \, \psi_\alpha^*(x) \psi_\alpha(x) = 1,$$

we may promote (4.27) into an *orthonormality condition*

$$\int_{-\infty}^{+\infty} dx \, \psi_\alpha^*(x) \, \psi_{\alpha'}(x) = \delta_{\alpha \alpha'}. \tag{4.28}$$

---

[6]The integral $\int dx \, |\psi_\alpha|^2$ cannot be zero, since this would correspond to a wave function that vanishes everywhere, i.e., an empty system.

In the case of a continuous spectrum, for which the eigenfunctions are not normalizable, the orthonormality condition is expressed in terms of the delta function as

$$\int_{-\infty}^{+\infty} dx\, \psi_\alpha^*(x)\psi_{\alpha'}(x) = \delta(\alpha - \alpha'). \tag{4.29}$$

In addition to the above proven properties of orthonormality of the eigenfunctions and of the reality of eigenvalues of a Hermitian operator, there is a third important property stated as follows [1–4]:

(3) The eigenfunctions of a Hermitian operator form a *complete set* of functions in terms of which any square-integrable function can be expanded.

Thus, any function $\Psi(x)$ can be expanded as

$$\Psi(x) = \sum_\alpha C_\alpha \psi_\alpha(x), \tag{4.30}$$

in terms of calculable coefficients C_α. As a substitute for a rigorous proof we may proceed to determine the expansion coefficients in (4.30) in the following fashion. We multiply (4.30) with the complex conjugate of an arbitrary eigenfunction, integrate

$$\int dx\, \psi_\beta^*(x)\, \Psi(x) = \sum_\alpha C_\alpha \int dx\, \psi_\beta^*(x)\psi_\alpha(x) = \sum_\alpha C_\alpha\, \delta_{\beta\alpha} = C_\beta$$

and, using orthonormality, we conclude

$$C_\beta = \int dx\, \psi_\beta^*(x)\, \Psi(x). \tag{4.31}$$

This property is referred to as *completeness* and stated as

$$\Psi(x) = \sum_\alpha C_\alpha \psi_\alpha(x) \quad \text{with} \quad C_\alpha = \int dx\, \psi_\alpha^*(x)\Psi(x). \tag{4.32}$$

In the case of a continuous spectrum, the completeness of the set of eigenfunctions is expressed as

$$\Psi(x) = \int d\alpha\, C(\alpha)\, \psi_\alpha(x), \quad \text{with} \quad C(\alpha) = \int dx\, \psi_\alpha^*(x)\Psi(x). \tag{4.33}$$

An alternative mathematical statement of the completeness can be obtained as follows:

$$\Psi(x) = \sum_\alpha C_\alpha \psi_\alpha = \sum_\alpha \left(\int dx'\, \psi_\alpha^*(x')\Psi(x') \right) \psi_\alpha(x) = \int dx' \left(\sum_\alpha \psi_\alpha(x)\psi_\alpha^*(x') \right) \Psi(x')$$

or

$$\sum_\alpha \psi_\alpha(x)\psi_\alpha^*(x') = \delta(x - x').$$ (4.34)

For a continuous spectrum this property is written as

$$\int d\alpha \, \psi_\alpha(x)\psi_\alpha^*(x') = \delta(x - x').$$ (4.35)

4.4 Physical Observables

Let us consider a Hermitian operator $\hat{A}$ and its orthonormal and complete set of eigenfunctions ψ_α and real eigenvalues α. The *expectation value* of $\hat{A}$ for an arbitrary (square-integrable) wave function $\psi(x)$ will be

$$\langle A \rangle = \frac{\int dx \, \psi^*(x)\hat{A}\psi(x)}{\int dx \, |\psi(x)|^2},$$ (4.36)

allowing for the case that $\psi(x)$ is not normalized. As we showed earlier the expectation value of a Hermitian operator is a real number.

Let us now expand the wave function $\psi(x)$ in terms of the complete set of eigenfunctions of $\hat{A}$

$$\psi(x) = \sum_\alpha C_\alpha \psi_\alpha(x).$$ (4.37)

Substituting (4.37) in (4.36) we obtain

$$\langle A \rangle = \frac{\int dx \sum_{\alpha,\alpha'} C_\alpha^* C_{\alpha'} \psi_\alpha^*(x)\hat{A}\psi_{\alpha'}(x)}{\int dx \sum_{\alpha,\alpha'} C_\alpha^* C_{\alpha'} \psi_\alpha^*(x)\psi_{\alpha'}(x)} = \frac{\sum_{\alpha,\alpha'} C_\alpha^* C_{\alpha'}\alpha' \int dx \, \psi_\alpha^*(x)\psi_{\alpha'}(x)}{\sum_{\alpha,\alpha'} C_\alpha^* C_{\alpha'} \int dx \, \psi_\alpha^*(x)\psi_{\alpha'}(x)}$$

or, using the orthonormality of ψ_α,

$$\langle A \rangle = \frac{\sum_\alpha |C_\alpha|^2 \alpha}{\sum_{\alpha'} |C_{\alpha'}|^2}.$$ (4.38)

The expectation value $\langle A \rangle$ is manifestly real since the eigenvalues α are real. The expansion coefficients are determined in terms of the wave function as (see (4.33))

$$C_\alpha = \int dx \, \psi_\alpha^*(x)\psi(x).$$ (4.39)

For a normalized wave function the sum $\sum_\alpha |C_\alpha|^2$ equals unity. In any case the number $|C_\alpha|^2 / \sum_{\alpha'} |C_{\alpha'}|^2$ corresponds to the probability of each eigenvalue α.

Measurements of various physical quantities like the position or the momentum of a particle come out as real numbers. The operators that represent these quantities at the quantum level should necessarily lead to real eigenvalues and, therefore, real expectation values. This is always satisfied if these operators are Hermitian. Thus, *physical observables* should correspond to Hermitian operators. Furthermore, as the expression (4.38) clearly indicates, the expectation values of these observables are entirely determined by the eigenvalues and the relative frequency (probability) with which these eigenvalues show up. In fact, the eigenvalues are the only possible outcomes of measurement. This will be further clarified in our discussion on quantum measurements. From now on the term *"physical observable"* will rigorously refer to a quantity represented by a Hermitian operator [5–7].

4.5 Uncertainty

It is an outcome of Statistics that for a large number of measurements of a quantity A the *mean* or *average value* $\langle A \rangle$ coincides with the *most probable value*. Nevertheless, very often two different series of measurements with the same average value differ in their spread of values around the mean value. This difference is depicted by higher *statistical moments*. Of particular importance in the case of quantum probability distributions is the so-called *Standard Deviation* ΔA defined as[7]

$$(\Delta A)^2 \equiv \langle (A - \langle A \rangle)^2 \rangle . \tag{4.40}$$

Larger values of (ΔA) correspond to a greater spread of individual measurements and, thus, greater uncertainty as to where any subsequent measurement would lie. Thus, (ΔA) is a measure of the *Uncertainty* in the distribution of a physical observable and is often referred to by this name.

Expanding the square in the average value we have

$$(\Delta A)^2 = \langle \left(A^2 + \langle A \rangle^2 - 2A\langle A \rangle \right) \rangle = \langle A^2 \rangle + \langle A \rangle^2 - 2\langle A \rangle \langle A \rangle ,$$

giving the alternative expression for (ΔA)

$$(\Delta A)^2 = \langle A^2 \rangle - \langle A \rangle^2 . \tag{4.41}$$

For a quantum system characterized by a normalized wave function $\psi(x)$ and a physical quantity A represented by a Hermitian operator $\hat{A}$, the above definition reads

$$(\Delta A)^2 = \int dx\, \psi^*(x)\hat{A}^2\psi(x) - \left(\int dx\, \psi^*(x)\hat{A}\psi(x) \right)^2 . \tag{4.42}$$

---

[7]See the section on *Probability* in the Mathematical Appendix.

Note that if the system occupies an eigenstate ψ_α of the physical quantity at hand, the corresponding uncertainty vanishes, namely

$$(\Delta A)^2 = \int dx \psi_\alpha^* \alpha^2 \psi_\alpha - \left(\int dx \psi_\alpha^* \alpha \psi_\alpha \right)^2 = 0.$$

Thus, in a way, when a system is in an eigenstate of a particular physical observable A, a precise value can be assigned to this quantity, namely, the corresponding eigenvalue α, in contrast to the case of a general wave function for which no particular value can be assigned to it but only a distribution of values.

Example 4.1 (*A Gaussian wave packet*) As an example, we may consider the case of a free particle with a Gaussian wave function (1.42) discussed previously

$$\psi(x) = \left(\frac{2\alpha}{\pi} \right)^{1/4} e^{-\alpha x^2} \tag{4.43}$$

and compute the uncertainty in the position x using the definition (4.40) or (4.41).

First, note that the expectation value of the position vanishes, namely,

$$\langle x \rangle = \sqrt{\frac{2\alpha}{\pi}} \int_{-\infty}^{+\infty} dx \, x \, e^{-2\alpha x^2} = 0, \tag{4.44}$$

since the integrand is odd and integrated over all space gives zero. Then, we have

$$(\Delta x)^2 = \langle x^2 \rangle = \sqrt{\frac{2\alpha}{\pi}} \int_{-\infty}^{+\infty} dx \, x^2 \, e^{-2\alpha x^2} = \cdots = \frac{1}{4\alpha}. \tag{4.45}$$

The corresponding uncertainty for the momentum can also be computed in a straightforward fashion. Note that the average momentum vanishes. This is easily seen either as a result of oddness of the integrand or as a result of the fact that the wave function is real

$$\langle p \rangle = -i\hbar \int_{-\infty}^{+\infty} dx \, \psi(x) \psi'(x) = -\frac{i\hbar}{2} \int_{-\infty}^{+\infty} dx \frac{d\psi^2}{dx} = -\frac{i\hbar}{2} \psi^2(x) \Big|_{-\infty}^{+\infty} = 0,$$

since a square-integrable wave function should vanish at infinity. Next, we proceed to compute $\langle p^2 \rangle$ and have

$$(\Delta p)^2 = \langle p^2 \rangle = -\hbar^2 \int_{-\infty}^{+\infty} dx \, \psi(x) \frac{\partial^2}{\partial x^2} \psi(x) = -\hbar^2 \sqrt{\frac{2\alpha}{\pi}} \int_{-\infty}^{+\infty} dx (4\alpha^2 x^2 - 2\alpha) e^{-2\alpha x^2}$$

or

$$(\Delta p)^2 = -4\alpha^2 \hbar^2 \langle x^2 \rangle + 2\alpha \hbar^2 = \hbar^2 \alpha. \tag{4.46}$$

The product of the uncertainties is

$$(\Delta x)^2 (\Delta p)^2 = \hbar^2/4 \tag{4.47}$$

and it is independent of α. This result is in agreement with the *Heisenberg Uncertainty Relation* between position and momentum

$$(\Delta x)(\Delta p) \geq \frac{\hbar}{2}. \tag{4.48}$$

The fact that in our particular case we got an equality instead of inequality, simply signifies that the wave function chosen in our example corresponds to a *minimal uncertainty wave packet*.

4.6 The Heisenberg Inequality

In this section, we shall give a general proof of the *Heisenberg Inequality* between the position and momentum. For an arbitrary normalized wave function $\psi(x)$ we can always introduce the shifted operators

$$\hat{p}' = \hat{p} - \langle p \rangle, \quad \hat{x}' = \hat{x} - \langle x \rangle, \tag{4.49}$$

which, since the mean values are real, are Hermitian aswell. In addition, we introduce also the functions

$$\chi = \hat{x}'\psi, \quad \phi = \hat{p}'\psi, \tag{4.50}$$

simply corresponding to $\chi(x) = (x - \langle x \rangle)\psi(x)$ and $\phi(x) = -i\hbar\psi'(x) - \langle p \rangle\psi(x)$. These functions will also be square integrable.

Consider now the uncertainties $(\Delta x)^2$ and $(\Delta p)^2$. We have

$$(\Delta x)^2 = \langle (x - \langle x \rangle)^2 \rangle = \langle x'^2 \rangle = \int dx\, \psi^* \hat{x}'^2 \psi$$

$$= \int dx\, \psi^* \hat{x}' \chi = \left(\int dx\, \chi^* \hat{x}' \psi \right)^* = \int dx\, |\chi|^2$$

and

$$(\Delta p)^2 = \langle (p - \langle p \rangle)^2 \rangle = \langle p'^2 \rangle = \int dx\, \psi^* \hat{p}'^2 \psi$$

$$= \int dx\, \psi^* \hat{p}' \phi = \left(\int dx\, \phi^* \hat{p}' \psi \right)^* = \int dx\, |\phi|^2$$

Next, we consider the *Schwartz inequality* that holds true for any two square-integrable functions

$$\left(\int dx\, |\chi(x)|^2 \right) \left(\int dx\, |\phi(x)|^2 \right) \geq \left| \int dx\, \chi^*(x)\phi(x) \right|^2. \tag{4.51}$$

The left-hand side of this inequality is just the product $(\Delta x)^2(\Delta p)^2$. The right-hand side is the absolute square of the complex number[8]

---

[8] We make use of the operator rules $(AB)^\dagger = B^\dagger A^\dagger$. For more, see the section on *Operators* in the Mathematical Appendix.

$$\int dx \left(\hat{x}'\psi\right)^* \hat{p}'\psi = \left(\int dx \, \psi^* \hat{p}'\hat{x}'\psi\right)^* = \int dx \, \psi^* \left(\hat{p}'\hat{x}'\right)^\dagger \psi$$

$$= \int dx \, \psi^* \hat{x}'\hat{p}'\psi = \frac{1}{2}\int dx \, \psi^* \left(\hat{x}'\hat{p}' + \hat{p}'\hat{x}' + \hat{x}'\hat{p}' - \hat{p}'\hat{x}'\right)\psi$$

$$= \frac{1}{2}\int dx \, \psi^* \left((\hat{x}'\hat{p}') + \left(\hat{x}'\hat{p}'\right)^\dagger\right)\psi + \frac{1}{2}\int dx \, \psi^* \left[\hat{x}', \, \hat{p}'\right]\psi.$$

Since,

$$\left[\hat{x}', \, \hat{p}'\right] = \left[\hat{x}, \, \hat{p}\right] = i\hbar,$$

the second term is just $i\hbar/2$. The first term is a real number, since

$$\int dx \, \psi^* \left(\hat{x}'\hat{p}'\right)^\dagger \psi = \left(\int dx \, \psi^* \hat{x}'\hat{p}'\psi\right)^*.$$

Thus, the right-hand side has the form $\left|\frac{i\hbar}{2} + R\right|^2$, R being a real number depending on ψ. Therefore, we have the inequality

$$(\Delta x)^2(\Delta p)^2 \geq \left|\frac{i\hbar}{2} + R\right|^2 \geq \frac{\hbar^2}{4}$$

or

$$(\Delta x)^2(\Delta p)^2 \geq \frac{\hbar^2}{4}. \tag{4.52}$$

An analogous general inequality can be proven for any two observables $\hat{A}$, $\hat{B}$ following the same steps. The result takes the form

$$(\Delta A)^2(\Delta B)^2 \geq \frac{1}{4}\left|\left\langle\left[\hat{A}, \, \hat{B}\right]\right\rangle\right|^2. \tag{4.53}$$

Example 4.2 Verify the validity of the Heisenberg inequality between position and momentum for the energy eigenfunctions of the infinite square well.

Since $|\psi_n(x)|^2$ is always even, the expectation value $\langle\hat{x}\rangle$ vanishes due to the oddness of the integrant. Similarly, $\langle\hat{p}\rangle = -i\hbar\int_{-\infty}^{+\infty} dx \, \psi_n^*(x)\psi_n'(x) = 0$ due to the oddness of $\psi_n\psi_n'$. Thus, the corresponding uncertainties will be $(\Delta x)_n^2 = \langle\hat{x}^2\rangle_n$, $(\Delta p)^2 = \langle\hat{p}^2\rangle_n$. We proceed to compute $(\Delta x)^2$ and have

$$\langle\hat{x}^2\rangle_n = \frac{1}{2L}\int_{-L}^{L} dx \, x^2 \left(1 + (-1)^{n+1}\cos(n\pi x/L)\right).$$

The first term is integrated trivially and gives $L^2/3$. The second term requires the integral

$$\int_{-L}^{L} dx \, x^2 \cos(\alpha x) = -\frac{\partial^2}{\partial\alpha^2}\int_{-L}^{L} dx \, \cos(\alpha x) = -2\frac{\partial^2}{\partial\alpha^2}\left(\frac{\sin(\alpha L)}{\alpha}\right)$$

which for $\alpha = n\pi/L$ gives just $(-1)^n 4L^3/(n\pi)^2$. Therefore, we finally obtain $(\Delta x)_n^2$ $= \frac{L^2}{3}\left(1 - \frac{6}{(n\pi)^2}\right)$. The momentum uncertainty is easier to obtain. We have $(\Delta p)_n^2 = 2m\langle H\rangle_n =$ $2m E_n = \frac{\hbar^2(n\pi)^2}{4L^2}$. Thus, the uncertainty product is

$$(\Delta x)_n^2 \,(\Delta p)_n^2 = \frac{\hbar^2}{4}\left(\frac{n^2\pi^2}{3} - 2\right), \tag{4.54}$$

which is in full agreement with Heisenberg's inequality, since $n^2\pi^2 > 9$.

4.7 Commuting Observables

According to the Heisenberg Inequality proven in the previous section pairs of *commuting* physical observables can have simultaneously vanishing uncertainties. This ultimately means that these observables can be measured simultaneously with complete accuracy. An obvious example is given by a particle moving in three-dimensional space. Among its variables any of the triplets $(\hat{x},\ \hat{p}_y,\ \hat{p}_z), (\hat{y},\ \hat{p}_x,\ \hat{p}_z),$ $(\hat{z},\ \hat{p}_x,\ \hat{p}_y), (\hat{x},\ \hat{y},\ \hat{p}_z), (\hat{x},\ \hat{z},\ \hat{p}_y), (\hat{y},\ \hat{z},\ \hat{p}_x)$ and, of course $(\hat{x},\ \hat{y},\ \hat{z}), (\hat{p}_x,\ \hat{p}_y,\ \hat{p}_z)$ makeup sets of mutually commuting observables that are simultaneously measurable with complete accuracy. In what follows we shall show that, if two such quantities commute, it follows that they possess common eigenfunctions. Also the inverse will be shown to be true, namely that if two observables have common eigenfunctions, they will necessarily commute.

Consider the observables A and B and assume that

$$\left[\hat{A},\ \hat{B}\right] = 0. \tag{4.55}$$

Let the eigenvalue equations for each of then be

$$\hat{A}\,\psi_\alpha = \alpha\,\psi_\alpha, \quad \hat{B}\,\psi_\beta = \beta\,\psi_\beta. \tag{4.56}$$

With ψ_α and α we have symbolized the eigenfunctions and the eigenvalues of A, while ψ_β and β are those of B. For simplicity we shall assume that there is no degeneracy in their spectra.

From the vanishing commutator of the two observables we obtain

$$\hat{B}\hat{A} = \hat{A}\hat{B}.$$

Acting with this operator relation on one of the eigenstates of $\hat{B}$, say ψ_β, we get

$$\hat{B}\hat{A}\psi_\beta = \hat{A}\hat{B}\psi_\beta \implies \hat{B}\left(\hat{A}\psi_\beta\right) = \beta\left(\hat{A}\psi_\beta\right),$$

which means that $\hat{A}\psi_\beta$ is an eigenfunction of $\hat{B}$ with eigenvalue β. Since, by assumption, there is no degeneracy, this state has to be proportional to the designated eigenfunction corresponding to the eigenvalue β, namely, ψ_β. This is written as

$$\hat{A}\psi_\beta \propto \psi_\beta \quad \text{or} \quad \hat{A}\psi_\beta = \alpha\psi_\beta$$

and simply means that ψ_β is a common eigenstate of both $\hat{B}$ and $\hat{A}$. A more appropriate symbolization would be $\psi_{\alpha,\beta}$. Summarizing, we have

$$\left[\hat{A}, \hat{B}\right] = 0 \implies \begin{cases} \hat{A}\psi_{\alpha,\beta} = \alpha\psi_{\alpha,\beta} \\ \hat{B}\psi_{\alpha,\beta} = \beta\psi_{\alpha,\beta}. \end{cases} \tag{4.57}$$

The inverse is also true. Assume that the following hold

$$\hat{A}\psi_{\alpha,\beta} = \alpha\psi_{\alpha,\beta}, \quad \hat{B}\psi_{\alpha,\beta} = \beta\psi_{\alpha,\beta}. \tag{4.58}$$

Multiplying the first with $\hat{B}$ and the second with $\hat{A}$ gives

$$\hat{B}\hat{A}\psi_{\alpha,\beta} = \alpha\hat{B}\psi_{\alpha,\beta} = \alpha\beta\psi_{\alpha,\beta}, \quad \hat{A}\hat{B}\psi_{\alpha,\beta} = \beta\hat{A}\psi_{\alpha,\beta} = \alpha\beta\psi_{\alpha,\beta} \tag{4.59}$$

and subtracting, gives us

$$\left[\hat{B}, \hat{A}\right]\psi_{\alpha,\beta} = 0. \tag{4.60}$$

Since any wave function ψ can be expanded in terms of the complete set $\psi_{\alpha,\beta}$ as $\psi = \sum_{\alpha,\beta} C_{\alpha,\beta}\psi_{\alpha,\beta}$, the relation (4.60) is equivalent to $[\hat{A}, \hat{B}]\psi = 0$ for any ψ. Therefore, we conclude with the operator relation

$$\left[\hat{A}, \hat{B}\right] = 0. \tag{4.61}$$

4.8 Time Evolution of Expectation Values

The expectation values of physical observables depend on time through the time dependence of the wave function. Consider such an observable and the Hermitian operator corresponding to it. Assuming a normalized wave function for simplicity, we have

$$\langle A \rangle_t = \int d^3r \, \psi^*(\mathbf{r}, t) \hat{A}\psi(\mathbf{r}, t). \tag{4.62}$$

Differentiating this expression with time we obtain

$$\frac{d\langle \mathcal{A} \rangle}{dt} = \int d^3r \left(\dot{\psi}^*(\mathbf{r}, t) \, \hat{A} \, \psi(\mathbf{r}, t) + \psi^*(\mathbf{r}, t) \, \hat{A} \, \dot{\psi}(\mathbf{r}, t) \right) \tag{4.63}$$

and, using the Schroedinger equation,

$$\frac{d\langle \mathcal{A} \rangle}{dt} = \frac{i}{\hbar} \int d^3r \left(\left(\hat{H} \psi \right)^* \hat{A} \psi - \psi^* \hat{A} \hat{H} \psi \right) . \tag{4.64}$$

Recalling the definition of hermitian conjugation, this becomes

$$\frac{d\langle \mathcal{A} \rangle}{dt} = \frac{i}{\hbar} \left(\int d^3r \, \psi^* \hat{A}^\dagger \hat{H} \psi \right)^* - \frac{i}{\hbar} \int d^3r \, \psi^* \hat{A} \hat{H} \psi \tag{4.65}$$

and, using $\hat{A}\hat{H} = \hat{A}^\dagger \hat{H}^\dagger = \left(\hat{H} \hat{A} \right)^\dagger$,

$$\frac{d\langle \mathcal{A} \rangle}{dt} = \frac{i}{\hbar} \left(\int d^3r \, \psi^* \left(\hat{H} \hat{A} \right)^\dagger \psi \right)^* - \frac{i}{\hbar} \int d^3r \, \psi^* \hat{A} \hat{H} \psi$$

$$= \frac{i}{\hbar} \int d^3r \, \psi^* \hat{H} \hat{A} \psi - \frac{i}{\hbar} \int d^3r \, \psi^* \hat{A} \hat{H} \psi \tag{4.66}$$

or, finally

$$\frac{d\langle \mathcal{A} \rangle}{dt} = \frac{i}{\hbar} \left\langle \left[\hat{A}, \, \hat{H} \right] \right\rangle . \tag{4.67}$$

This is a differential equation giving the rate of temporal change of the expectation value of any given physical observable. Note that if the operator $\hat{A}$ corresponding to this physical quantity commutes with the Hamiltonian of the system, the expectation value of this quantity in any state $\psi(\mathbf{r}, t)$ will not change with time. Such physical quantities are said to be *conserved* and often go by the name *constants of the motion*. As a very obvious example, consider the momentum of a free particle. Since its Hamiltonian is just $\hat{H} = \hat{p}^2/2m$ and $[\hat{H}, \hat{p}] = [\hat{p}^2, \hat{p}] = 0$, according to (4.67), $\langle \hat{p} \rangle$ is conserved for any state.

4.9   Ehrenfest's Theorem

Consider the simple system of a particle of mass m moving in a potential $V(\mathbf{r})$. Applying the expectation value evolution law (4.67) for the position of the particle, we obtain

$$\frac{d\langle \hat{x}_j \rangle}{dt} = \frac{i}{\hbar} \left\langle \left[\hat{H}, \, \hat{x}_j \right] \right\rangle = \frac{i}{2m\hbar} \left\langle \left[\hat{p}^2, \, \hat{x}_j \right] \right\rangle$$

$$= \frac{i}{2m\hbar} \left\langle \sum_k \hat{p}_k [\hat{p}_k, \hat{x}_j] \right\rangle + \frac{i}{2m\hbar} \left\langle \sum_k [\hat{p}_k, \hat{x}_j] \hat{p}_k \right\rangle = \frac{\langle \hat{p}_j \rangle}{m}$$

or

$$\frac{d\langle \hat{\mathbf{r}} \rangle}{dt} = \frac{\langle \hat{\mathbf{p}} \rangle}{m}. \qquad (4.68)$$

Doing the same for the momentum of the particle, we obtain

$$\frac{d\langle \hat{p}_j \rangle}{dt} = \frac{i}{\hbar} \left\langle \left[\hat{H}, \hat{p}_j \right] \right\rangle = \frac{i}{\hbar} \left\langle \left[V(\hat{\mathbf{r}}), \hat{p}_j \right] \right\rangle = -\left\langle \frac{\partial V}{\partial \hat{x}_j} \right\rangle$$

or

$$\frac{d\langle \hat{\mathbf{p}} \rangle}{dt} = -\langle \nabla V(\hat{\mathbf{r}}) \rangle. \qquad (4.69)$$

The Eqs. (4.68) and (4.69) can be combined to the equation

$$m \frac{d^2 \langle \hat{\mathbf{r}} \rangle}{dt^2} = -\langle \nabla V(\hat{\mathbf{r}}) \rangle. \qquad (4.70)$$

Equations (4.68), (4.69) or their combination (4.70) are known as *Ehrenfest's Theorem*. Note that in the case that the potential function is quadratic in the position (harmonic forces), the expectation values satisfy the classical equations of motion.

Example 4.3 A particle of mass m and electric charge q is moving in three-dimensional space under the influence of a homogeneous electric field $\mathcal{E}$. At an initial time $t = 0$ the particle occupies a state that corresponds to the given position and momentum expectation values $\langle \mathbf{r} \rangle_0$ and $\langle \mathbf{p} \rangle_0$. Find the corresponding expectation values at later times $t > 0$.

The potential corresponding to the electric force is $V = -q\mathcal{E} \cdot \mathbf{r}$. Ehrenfest's theorem gives

$$\frac{d\langle \mathbf{p} \rangle}{dt} = -\langle \nabla V \rangle = q\mathcal{E} \quad \text{and} \quad \frac{d\langle \mathbf{r} \rangle}{dt} = \frac{\langle \mathbf{p} \rangle}{m}. \qquad (4.71)$$

Integrating the first of these equations we obtain

$$\langle \mathbf{p} \rangle_t = \langle \mathbf{p} \rangle_0 + q\mathcal{E}t. \qquad (4.72)$$

Substituting and integrating the second equation we obtain

$$\langle \mathbf{r} \rangle_t = \langle \mathbf{r} \rangle_0 + \frac{\langle \mathbf{p} \rangle_0}{m} t + \frac{q}{2m} \mathcal{E} t^2. \qquad (4.73)$$

4.10   The Virial Theorem

Consider the system of an interacting particle and the operator

$$\hat{G} = \mathbf{r} \cdot \mathbf{p} + \mathbf{p} \cdot \mathbf{r} \tag{4.74}$$

The time evolution Eq. (4.67) of its expectation value is

$$\frac{d\langle \hat{G} \rangle}{dt} = \frac{i}{\hbar} \left\langle \left[\hat{H}, \, \hat{G} \right] \right\rangle . \tag{4.75}$$

Computing the commutator

$$\left[\frac{\hat{p}^2}{2m} + V(\mathbf{r}), \, \hat{x}_j \hat{p}_j + \hat{p}_j \hat{x}_j \right] = \frac{1}{m} [\hat{p}_i \hat{p}_i, \, \hat{p}_j \hat{x}_j] + 2[V, \, \hat{p}_j \hat{x}_j]$$

$$= \frac{\hat{p}_j}{m} [\hat{p}_i \hat{p}_i, \, \hat{x}_j] + 2i\hbar x_j \nabla_j V = -2i\hbar \frac{p^2}{m} + 2i\hbar x_j \nabla_j V ,$$

we obtain

$$\frac{d\langle \hat{G} \rangle}{dt} = -4i\hbar \langle T \rangle + 2i\hbar \langle \mathbf{r} \cdot \nabla V \rangle . \tag{4.76}$$

If the state of the system is a stationary state, we have

$$\langle \hat{G} \rangle_t = \langle e^{iEt/\hbar} \hat{G} e^{-iEt/\hbar} \rangle_0 = \langle \hat{G} \rangle_0 ,$$

which is time-independent. Therefore, the left-hand-side of (4.75) vanishes and we have

$$\langle T \rangle = \frac{1}{2} \langle \mathbf{r} \cdot \nabla V \rangle . \tag{4.77}$$

This is the so-called *Virial Theorem*.

In the case that potential is of the form $V = gr^\nu$, we have $\mathbf{r} \cdot \nabla V = \nu V$ and the Virial Theorem takes the form

$$\langle T \rangle = \frac{\nu}{2} \langle V \rangle . \tag{4.78}$$

Thus, in the case of the attractive Coulomb potential $-r^{-1}$, we have $\langle V \rangle = -2\langle T \rangle$ and $\langle E \rangle = -\langle T \rangle$. In the case of the isotropic harmonic oscillator potential r^2, we have $\langle T \rangle = \langle V \rangle$.

Example 4.4 Consider the operator $\hat{G} = \mathbf{r} \cdot \mathbf{p} + \mathbf{p} \cdot \mathbf{r}$. Show that its expectation value in a state with real wave function ($\psi^*(\mathbf{r}) = \psi(\mathbf{r})$) vanishes.

We have

$$\langle \hat{G} \rangle = -i\hbar \int d^3r \, (\psi \mathbf{r} \cdot \nabla \psi + \psi \nabla \cdot (\mathbf{r} \psi)) = \ldots =$$

$$-i\hbar \int d^3r \left(\mathbf{r} \cdot \nabla(\psi^2) + \psi^2 \nabla \cdot \mathbf{r}\right) = -i\hbar \int d^3r \, \nabla \cdot \left(\mathbf{r}\psi^2\right) = -i\hbar \oint d\mathbf{S} \cdot \left(\mathbf{r}\psi^2\right) = 0.$$

Problems and Exercises

4.1 A particle of mass m is trapped in an infinite square well of width $2L$ having a wave function

$$\psi(x) = \sqrt{\frac{8}{5L}} \cos^3(x\pi/2L).$$

Calculate the probability to find the particle in an eigenstate of energy $\frac{\hbar^2 \pi^2}{8mL^2}$.

4.2 Show that the expectation value of the momentum $\langle p \rangle$ of a particle having a *real* wave function ($\psi^*(x) = \psi(x)$) vanishes. Show also that the expectation value of the position $\langle x \rangle$ or the momentum $\langle p \rangle$ in a wave function of a definite parity (even or odd) vanishes.

4.3 Consider a quantum system with a purely discrete energy spectrum $\{E_n\}$. Assume that initially ($t = 0$) the system is in the state corresponding to the wave function

$$\psi(x, 0) = \frac{1}{\sqrt{2}} \left(\psi_N(x) + e^{i\alpha}\psi_{N+1}(x)\right),$$

where $\alpha = \alpha^*$ known parameter. Calculate $(\Delta E)^2/\langle E \rangle^2$ at any time t, where $\langle E \rangle$ is the expectation value of the energy. Assume that $E_N = E_1/N^2$ and comment on the limit of this ratio at large quantum numbers ($N \to \infty$). Do the same for the cases $E_N = E_1 N^2$ and $E_N = E_0(2N + 1)$.

4.4 The state of a particle of mass m and of definite energy $E = \frac{4\hbar^2 a^2}{m}$ is described by the wave function

$$\psi(x) = \frac{\sqrt{a}}{\cosh(2ax)},$$

where $a > 0$ is a known parameter. Find the potential acting on the particle. Calculate the uncertainty in the momentum.

4.5 A particle of mass m is subject to a constant force $\mathbf{F}$. Calculate the expectation values $\langle \mathbf{r} \rangle_t$ and $\langle \mathbf{p} \rangle_t$ at any time $t > 0$ in terms of the initial expectation values $\langle \mathbf{r} \rangle_0$ and $\langle \mathbf{p} \rangle_0$ which can be computed from the initial wave function of the particle

$$\psi(\mathbf{r}, 0) = N\, x\, e^{-\alpha(x^2+y^2+z^2)+iqz},$$

where $\alpha > 0$ and q known real parameters.

4.6 Consider a hermitian operator $\hat{A}$ and a normalized wave function $\psi(\mathbf{r})$. If $\chi(\mathbf{r}) = \hat{A}\psi(\mathbf{r})$, show that

$$\langle A^2 \rangle = \int d^3r \, |\chi(\mathbf{r})|^2$$

and

$$(\Delta A)^2 = \int d^3r \, \left| \chi(\mathbf{r}) - \psi(\mathbf{r}) \int d^3r' \psi^*(\mathbf{r}')\chi(\mathbf{r}') \right|^2 .$$

4.7 Consider a set of three hermitian operators $\hat{A}$, $\hat{B}$, $\hat{C}$, satisfying the following commutation relations

$$\left[\hat{A}, \hat{B} \right] = i\hbar\hat{C}, \quad \left[\hat{B}, \hat{C} \right] = i\hbar\hat{A}, \quad \left[\hat{C}, \hat{A} \right] = i\hbar\hat{B}.$$

Show that

$$\left[\hat{A}, \hat{A}^2 + \hat{B}^2 + \hat{C}^2 \right] = \left[\hat{B}, \hat{A}^2 + \hat{B}^2 + \hat{C}^2 \right] = \left[\hat{C}, \hat{A}^2 + \hat{B}^2 + \hat{C}^2 \right] = 0 .$$

What is the maximal subset of mutually commuting operators among the $\hat{A}$, $\hat{B}$, $\hat{C}$ and the sum of their squares? Assume now that the physical system described by these operators has a Hamiltonian $\hat{H} = \omega \hat{A}$, where ω is a known parameter. Derive the time-evolved expectation values $\langle A \rangle_t$, $\langle B \rangle_t$ and $\langle C \rangle_t$ in terms of the corresponding initial ($t = 0$) expectation values $\langle A \rangle_0$, $\langle B \rangle_0$ and $\langle C \rangle_0$ for any state of the system.

4.8 Consider a particle of mass m moving in one dimension and subject to a harmonic force arising from the potential $\frac{1}{2}m\omega^2x^2$ (harmonic oscillator). The initial ($t = 0$) wave function of the system is

$$\psi(x,0) = \frac{1}{\sqrt{2}} \left(\frac{m\omega}{\hbar\pi} \right)^{1/4} \left(1 + x\sqrt{\frac{2m\omega}{\hbar}} \right) e^{-\frac{m\omega}{2\hbar}x^2} .$$

Find the expectation values $\langle x \rangle_t$ and $\langle p \rangle_t$ for $t > 0$.

4.9 The *exponential* of an operator is defined through the corresponding series expansion as

$$e^{\hat{A}} = \sum_{n=0}^{\infty} \frac{\left(\hat{A} \right)^n}{n!} = 1 + \hat{A} + \frac{1}{2}\hat{A}^2 + \cdots .$$

Evaluate the commutator

$$\left[e^{\frac{i}{\hbar}\mathbf{a}\cdot\mathbf{p}}, x_j \right] ,$$

where $\mathbf{p}$ is the momentum operator. The operator $\hat{T}(\mathbf{a}) = e^{\frac{i}{\hbar}\mathbf{a}\cdot\mathbf{p}}$ is not hermitian. Find its eigenfunctions and eigenvalues. For the latter show that they are complex numbers of measure one. Show that the hermitian conjugate of $\hat{T}$ coincides with its *inverse* satisfying $\hat{T}^{\dagger}\hat{T} = 1$.

4.10 A hermitian operator $\hat{A}$ has a *degenerate* spectrum. Thus, the pair of eigenfunctions $\psi_{\alpha}^{(1)}$ and $\psi_{\alpha}^{(2)}$ correspond to one and the same eigenvalue α, namely

$$\hat{A}\psi_{\alpha}^{(1)} = \alpha\psi_{\alpha}^{(1)}, \quad \hat{A}\psi_{\alpha}^{(2)} = \alpha\psi_{\alpha}^{(2)}.$$

These eigenfunctions are not orthogonal, i.e., $\int dx\, \psi_{\alpha}^{(1)*}\, \psi_{\alpha}^{(2)} \neq 0$. Out of this pair constructs an orthonormal pair.

References

1. J.D. Jackson, *Mathematics for Quantum Mechanics* (Dover, Mineola, 1990)
2. J. Mathews, R. Walker, *Mathematical Methods of Physics*, 2nd edn. (Pearson, 1970)
3. G. Arfken, H. Weber, R. Harris, *Mathematical Methods for Physicists*, 7th edn. (Academic Press, Cambridge, 2013)
4. P. Dennery, A. Krzywicki, *Mathematics for Physicists* (Dover, Mineola, 1996)
5. A. Messiah, *Quantum Mechanics*. (Dover publications, Mineola, 1958). Single-volume reprint of the Wiley, New York, two-volume 1958 edn
6. E. Merzbacher, *Quantum Mechanics*, 3rd edn. (Wiley, New York, 1998), pp. 18–22
7. F. Levin, *An Introduction to Quantum Theory* (Cambridge University Press, Cambridge, 2002)

Chapter 5
Basic Principles of Quantum Mechanics

5.1 Basic Postulates

As we have remarked in the very beginning of our discussion of Quantum Mechanics, the concept of the *state* of a quantum system is very different than that of a classical one. In the latter case, the state of the system is determined by a set of variables $x(t)$, $\dot{x}(t)$, ... which are directly measurable and can be known at any instant of time. In the case of a quantum system the state of the system is determined by a function of certain of its variables $\psi(x, \dots)$ related to the probability distribution of these variables. This is the *wave function* of the system. In contrast to the case of a classical system the connection of measurements and the state represented by ψ is very indirect.

One of the basic outcomes of the various experiments that established the wave function as the basic concept characterizing quantum systems is the fact that wave functions obey a *Superposition Principle*. This means that the *space of states* is endowed with an addition operation, thus, being a *vector space*. In contrast, the states of a classical system do not correspond to any such mathematical structure. Shortly we shall go into this mathematical concept of vector space of states with considerable detail. For the moment we will just state the general principles or axioms on which the analysis of any quantum system is based.[1]

1. **"Axiom"-1**. *Each dynamical state of a quantum system corresponds to an element of an abstract vector space.*

---

[1]In what follows we shall use by convention the term *"axioms"*, borrowed from rigorous mathematics, since each of these principles has a mathematical content. These often go by the name *"Dirac-Von Neumann Axioms"* [1]. Nevertheless, our presentation of these principles will be limited to a non-rigorous level.

© Springer Nature Switzerland AG 2019
K. Tamvakis, *Basic Quantum Mechanics*, Undergraduate Texts in Physics,
https://doi.org/10.1007/978-3-030-22777-7_5

The wave functions obeying the superposition principle correspond to these abstract vectors as the coordinates correspond to a position vector in standard position space.

What about the various physical quantities like position, momentum, energy,.. etc.? These quantities are the objects that we want to measure. They will be referred to as *physical observables*. In Quantum Mechanics these will correspond to *Hermitian operators* acting in the above introduced vector space of states. This is the second general principle stated as

2. **"Axiom"-2**. *Physical observables are represented by Hermitian operators acting in the vector space of states.*

The above two *"axioms"* refer to vectors and operators. These abstract objects will be related to the numbers that are the outcomes of experiments either through inner products of vectors or through expectation values of operators. We know already that the Hermitian operator corresponding to each physical observable has its unique set of eigenvalues and eigenstates. Furthermore its expectation values are expressible in terms of its eigenvalues. In this framework, we state our third *"axiom"* as

3. **"Axiom"-3**. *The only outcomes of the measurement of a physical observable $\mathcal{A}$ are its eigenvalues $\{\alpha_1, \alpha_2, \ldots\}$.*

According to *"axiom"-3* the outcome of an experiment will be a collection of different eigenvalues of the measured quantity $\mathcal{A}$, say $\{\alpha_1, \alpha_2, \ldots\}$, the relative number of times each eigenvalue appears being its probability. These probabilities determine the state of the system *before* the measurement. This is the fourth *"axiom"* stated as

4 **"Axiom"-4**. *The state of the system before the measurement of $\mathcal{A}$ is a superposition of $\mathcal{A}$-eigenstates $\psi = \sum_\alpha C_\alpha \psi_\alpha$, where $|C_\alpha|^2$ is the probability of each measured eigenvalue.*

What is the state of the system *after* a particular eigenvalue α has been measured? This is answered by the fifth *"axiom"* stated as

5 **"Axiom"-5**. *The state of the system after a measurement of $\mathcal{A}$ that has given the eigenvalue α is the corresponding eigenstate ψ_α. Any subsequent measurement of $\mathcal{A}$ will yield exactly the same eigenvalue α and the system will continue to occupy the same eigenstate ψ_α.*

The previous three *"axioms"* refer to the process of a quantum measurement and could have been combined into one *"axiom"*. What about the evolution of the system when *no measurement is performed on it*? We already know that its time evolution is determined by the Schroedinger equation. This can be stated again as a final *"axiom"*

6 **"Axiom"-6**. *When no measurement is performed on the system the wave function evolves with time following the Schroedinger equation*

$$\hat{\mathcal{H}}\psi(t) = i\hbar\frac{\partial\psi(t)}{\partial t}. \tag{5.1}$$

Note that the evolution of the system according to the last axiom is purely deterministic in contrast to what happens as a result of measurement when the only possible predictions are statistical in nature. The fact that quantum systems behave under measurements in this indeterministic fashion has been the cause of endless ongoing debates aiming at a correct interpretation.

Since the measurement process embodies the concept of probabilities and that of the acausal evolution of the state, being both the main departures from classical theory, it is important to clarify further its basic aspects. Consider a quantum system in an arbitrary state $\Psi(x)$ and an apparatus designed to measure a particular observable $\mathcal{Q}$ of the system. Whatever is the state of the system at the instant prior to the measurement the measured value of $\mathcal{Q}$ is bound to be one of its eigenvalues q. That means that the system will necessarily occupy one of the eigenstates $\psi_q(x)$ of $\mathcal{Q}$. This is consistent with the fact that the eigenstates are the only states with vanishing uncertainty $\Delta\mathcal{Q}$. Any subsequent measurements of $\mathcal{Q}$ will yield the same eigenvalue q, meaning that the system continues to occupy the same eigenstate $\psi_q(x)$. It is said that the system has been *prepared* in the state $\psi_q(x)$.

Consider now that instead of a single measurement of $\mathcal{Q}$ that yielded the eigenvalue q we performed a series of measurements *under the exact same conditions* or equivalently measurements on a statistical *ensemble* of identical copies of the system. The result would be a series of, different in general, eigenvalues of $\mathcal{Q}$

$$q', q'', \ldots \tag{5.2}$$

For a large number of measurements the frequency (probability) of each particular eigenvalue $\mathcal{P}(q)$ can be used to determine the state $|\psi\rangle$ of the system prior to measurement through the rule

$$\mathcal{P}_q = |C_q|^2 = \left|\int dx\psi_q^*(x)\Psi(x)\right|^2. \tag{5.3}$$

The fact that as a result of the measurement the system has made a transition from a superposition of eigenstates to a single eigenstate

$$\Psi(x) = \sum_{q'} C_{q'}\psi_{q'}(x) \implies \psi_q(x) \tag{5.4}$$

has been termed *collapse of the wave function*.

5.2   The Hilbert Space of States

The states of a quantum system correspond to the elements of an abstract *vector space* $\mathcal{E}$ [2–5]. We shall use an abstract notation for these *state vectors*, namely $\{|\psi\rangle, |\chi\rangle, \ldots\}$. Sometimes we shall use for these state vectors the name *"ket"* given to them by P.A.M. Dirac.[2] Shortly we will see their precise relation to our familiar wave functions. The definition of a vector space is that of a set endowed with two operations, namely

(1) **Addition** $|a\rangle, |b\rangle \in \mathcal{E} \implies |c\rangle = |a\rangle + |b\rangle \in \mathcal{E}$
with the following properties

$$|a\rangle + |b\rangle = |b\rangle + |a\rangle$$

$$|a\rangle + (|b\rangle + |c\rangle) = (|a\rangle + |b\rangle) + |c\rangle \tag{5.5}$$

$$|a\rangle + 0 = 0 + |a\rangle = |a\rangle.$$

Note that the addition of states is the mathematical expression of the superposition principle.

(2) **Multiplication by a complex number** $\lambda \in \mathcal{C}, |a\rangle \in \mathcal{E} \implies \lambda|a\rangle \in \mathcal{E}$
with the following properties

$$\lambda(|a\rangle + |b\rangle) = \lambda|a\rangle + \lambda|b\rangle$$

$$(\lambda + \mu)|a\rangle = \lambda|a\rangle + \mu|a\rangle$$

$$1|a\rangle = |a\rangle \tag{5.6}$$

$$0|a\rangle = 0.$$

Note that the position of the complex number is purely conventional and $\lambda|a\rangle = |a\rangle\lambda$. Note also that given a vector $|\psi\rangle$, all vectors $\lambda|\psi\rangle$, obtained by multiplying $|\psi\rangle$ by a complex number (called a *ray* of the vector space) are thought to represent the same state.

In addition to these two operations, the vector space of quantum states is also endowed with a third operation that promotes it into a *Euclidean Vector Space*. This operation is the *inner product* of two vectors and allows us to define a *norm* for each vector. The Euclidean Vector Space of quantum state vectors is called *the Hilbert space of states*.

---

[2]Dirac in his monumental textbook on Quantum Mechanics [6] gave the "ket" to the state vectors $|\psi\rangle$ and the name "bra" to their duals $\langle\chi|$, obviously inspired by the "bracket" symbol for the inner product of the two vectors $\langle\chi|\psi\rangle$.

(3) The **Inner Product**, defined as $|a\rangle, |b\rangle \in \mathcal{E} \implies \langle a|b\rangle \in \mathcal{C}$, has the following properties

$$\langle a|b\rangle = \langle b|a\rangle^*$$

$$|c\rangle = \lambda|a\rangle + \mu|b\rangle \implies \langle c|d\rangle = \lambda^*\langle a|d\rangle + \mu^*\langle b|d\rangle$$

$$|c\rangle = \lambda|a\rangle + \mu|b\rangle \implies \langle e|c\rangle = \lambda\langle e|a\rangle + \mu\langle e|b\rangle \tag{5.7}$$

$$\langle a|a\rangle \geq 0$$

$$|\langle a|b\rangle|^2 \leq \langle a|a\rangle \langle b|b\rangle.$$

The last property goes by the name *Schwartz Inequality*.

A set of kets are said to be *linearly independent* if no linear relation can be written in terms of them. For example, no linear relation can be written between the **i**, **j** and **k** orthonormal unit vectors in the standard three-dimensional Euclidean space. The maximum number of linearly independent vectors of a space is called the *dimensionality N* of the space. A set of linearly independent vectors $\{|e_1\rangle, |e_2\rangle, \ldots, |e_N\rangle\}$ equal in number to the dimensionality N of the space can be used to write linear relations of them with any other vector. This is because the set of the N linearly independent vectors plus any additional vector $|\psi\rangle$ will be a set of $N + 1$ linearly *dependent* vectors among which a linear relation of the form

$$|\psi\rangle = c_1|e_1\rangle + c_2|e_2\rangle + \ldots + c_n|e_N\rangle \tag{5.8}$$

is possible.

A set of linearly independent vectors equal in number to the dimensionality of the space is said to be a *basis*. The linear relation (5.8) can be thought as an *expansion* of $|\psi\rangle$ in terms of the above basis vectors $\{|e_j\rangle\}$. Note that the dimensionality of the Hilbert space can be infinite. Furthermore, a set of basis vectors could be labeled in terms of a *continuous index*. In that case, even for a finite interval of values of this index, the number of basis vectors is infinite and so is the space dimensionality.

Among the various possible bases, the so-called *orthonormal bases* are particularly useful. By definition the elements of an orthonormal basis satisfy orthonormality relations

$$\langle e_j|e_k\rangle = \delta_{jk}. \tag{5.9}$$

Writing

$$|\psi\rangle = \sum_j \psi_j|e_j\rangle \tag{5.10}$$

we may compute the coefficients ψ_j by taking the inner product with one of the basis vectors, say $|e_k\rangle$. Then, we obtain

$$\langle e_k|\psi\rangle = \sum_j \psi_j \langle e_k|e_j\rangle = \sum_j \psi_j \delta_{jk} = \psi_k \implies \psi_j = \langle e_j|\psi\rangle. \qquad (5.11)$$

For a continuous orthonormal basis the orthonormality is expressed with the help of the delta function and these relations take the form

$$|\psi\rangle = \int_{\alpha_1}^{\alpha_2} d\alpha \, \psi(\alpha)|\alpha\rangle \quad \left(\langle\alpha|\alpha'\rangle = \delta(\alpha - \alpha')\right)$$

$$\langle\alpha'|\psi\rangle = \int_{\alpha_1}^{\alpha_2} d\alpha \, \psi(\alpha) \delta(\alpha' - \alpha) \implies \psi(\alpha) = \langle\alpha|\psi\rangle. \qquad (5.12)$$

Note that a basis is also by definition *complete* because any vector belonging to the Hilbert space can be expressed in terms of it. Although this and analogous statements seem trivial, their rigorous realization in infinite dimensionality spaces often has to circumvent convergence questions. Starting from the expansion of an arbitrary ket in terms of the basis vectors $|\psi\rangle = \sum_i \langle e_i|\psi\rangle |e_i\rangle$, we may arrive at the following formal expression of completeness

$$\sum_i |e_i\rangle \langle e_i| = 1 \quad \textbf{or} \quad \int d\alpha \, |\alpha\rangle \langle\alpha| = 1. \qquad (5.13)$$

Note that the above expressions feature *outer products* of vectors $|a\rangle\langle b|$ which stand for operators in contrast to inner products $\langle a|b\rangle$ that are numbers.

Given an orthonormal basis $\{|e_j\rangle\}$ the arbitrary abstract ket $|\psi\rangle$ can be represented by the column $\psi_i = \langle e_i|\psi\rangle$

$$|\psi\rangle \implies \begin{bmatrix} \psi_1 \\ \psi_2 \\ \vdots \\ \psi_N \end{bmatrix}. \qquad (5.14)$$

This is exactly the same with what is done in ordinary Euclidean space where the position vector $\mathbf{r}$ in a given basis is represented by a triplet of coordinates (x, y, z), the coordinates being just the inner products $x = (\mathbf{r} \cdot \mathbf{i})$, $y = (\mathbf{r} \cdot \mathbf{j})$, $z = (\mathbf{r} \cdot \mathbf{k})$. All operations can be carried in terms of the coordinate columns without reference to the abstract vectors. For example, the inner product between two kets $\langle\chi|\psi\rangle$ can be written

$$\langle\chi|\psi\rangle = \sum_j \chi_j^* \langle e_j|\psi\rangle = \sum_j \sum_k \chi_j^* \psi_k \langle e_j|e_k\rangle = \sum_j \chi_j^* \psi_j$$

or

$$\langle \chi | \psi \rangle = \left(\chi_1^*, \chi_2^*, \ldots, \chi_N^* \right) \begin{pmatrix} \psi_1 \\ \psi_2 \\ \cdot \\ \cdot \\ \psi_N \end{pmatrix} = \chi^\dagger \psi \,. \tag{5.15}$$

It is interesting that the norm takes the following familiar pythagorean form

$$\langle \psi | \psi \rangle = \psi^\dagger \psi = |\psi_1|^2 + |\psi_2|^2 + \ldots + |\psi_N|^2 \,. \tag{5.16}$$

For a continuous basis $\{|\xi\rangle\}$ the ket $|\psi\rangle$ will be represented not by a column but by a *function*

$$|\psi\rangle \implies \langle \xi | \psi \rangle = \psi(\xi) \,. \tag{5.17}$$

Inner product expressions take the form of integrals

$$\langle \chi | \psi \rangle = \int d\xi \, \chi^*(\xi) \psi(\xi), \quad \langle \psi | \psi \rangle = \int d\xi |\psi(\xi)|^2 \,. \tag{5.18}$$

Notice that the orthogonality between two states takes the familiar orthogonality form of wave functions encountered earlier

$$\langle \psi_1 | \psi_2 \rangle = 0 \implies \int d\xi \, \psi_1^*(\xi) \psi_2(\xi) = 0 \,. \tag{5.19}$$

5.3   Operators in the Hibert Space

Operators are by definition objects that can act on vectors. The result of the action of an operator on a vector will be another vector. The *matrix element* of an operator $\hat{A}$ between the states $|\psi_1\rangle$ and $|\psi_2\rangle$ is the inner product of $|\psi_1\rangle$ with the ket that results from the action of $\hat{A}$ on $|\psi_2\rangle$. It is depicted as

$$\langle \psi_1 | \hat{A} | \psi_2 \rangle$$

and the operator is thought to act on the right.

We restrict ourselves to *linear operators*, namely operators with the property[3]

$$\hat{A} \left(\lambda | \chi \rangle + \mu | \psi \rangle \right) = \lambda \hat{A} | \chi \rangle + \mu \hat{A} | \psi \rangle \,. \tag{5.20}$$

---

[3] There will be only one case we shall encounter *antilinear* operators with the property $\hat{A}\lambda|\psi\rangle = \lambda^* \hat{A}|\psi\rangle$. That will be the case of of the *Time Reversal Operator*.

The *Hermitian Conjugate* $\hat{A}^\dagger$ of an operator $\hat{A}$ is defined as follows

$$\langle\psi_1|\hat{A}^\dagger|\psi_2\rangle = \left(\langle\psi_2|\hat{A}|\psi_1\rangle\right)^* . \tag{5.21}$$

Note that, if we denote with $\hat{A}|\psi_1\rangle = |\psi_3\rangle$ in the above relation, we get

$$\langle\psi_1|\hat{A}^\dagger|\psi_2\rangle = \langle\psi_2|\psi_3\rangle^* = \langle\psi_3|\psi_2\rangle$$

or

$$\hat{A}|\psi_1\rangle = |\psi_3\rangle \implies \langle\psi_1|\hat{A}^\dagger = \langle\psi_3| . \tag{5.22}$$

Thus, Hermitian conjugation transforms kets to bras and operators to the Hermitian conjugates (and, of course, numbers to their complex conjugates).

Consider an operator $\hat{A}$ and a given orthonormal basis $\{|e_j\rangle\}$ of the Hilbert space. The quantities

$$\mathcal{A}_{ij} = \langle e_i|\hat{A}|e_j\rangle \tag{5.23}$$

will be the *matrix elements* of the observable $\hat{A}$. They form a square matrix that represents the observable in the same way that the column vectors represent the kets. Any relation between operators and kets can be translated into an equivalent relation between the matrices (5.23) and the columns. For example, a relation $\hat{A}|\psi\rangle = |\chi\rangle$ gives

$$\sum_j \psi_j\hat{A}|e_j\rangle = \sum_j \chi_j|e_j\rangle \implies \langle e_k|\left(\sum_j \psi_j\hat{A}|e_j\rangle\right) = \sum_j \chi_j\langle e_k|e_j\rangle = \chi_k$$

or

$$\hat{A}|\psi\rangle = |\chi\rangle \iff \sum_j \mathcal{A}_{kj}\psi_j = \chi_k \iff \mathcal{A}\psi = \chi . \tag{5.24}$$

In the case of a continuous basis the matrix elements of an operator are defined in an analogous fashion

$$\mathcal{A}(\alpha, \alpha') = \langle\alpha|\hat{A}|\alpha'\rangle \tag{5.25}$$

and operator relations are translated to convoluted integrals

$$\hat{A}|\psi\rangle = |\chi\rangle \implies \int d\alpha' \, \mathcal{A}(\alpha, \alpha')\psi(\alpha') = \chi(\alpha) . \tag{5.26}$$

Physical observables are represented by Hermitian (i.e., self-conjugate) operators. The corresponding matrices (with respect to an orthonormal basis) $\mathcal{A}_{ij}$ are hermitian matrices satisfying

$$A_{ij} = \langle e_i|\hat{A}|e_i\rangle = \langle e_j|\hat{A}|e_i\rangle^* = A_{ji}^*$$

or

$$A = A^\dagger = (A^T)^*. \tag{5.27}$$

The *eigenvalue problem* for a Hermitian operator is set up as

$$\hat{A}|\psi_a\rangle = a|\psi_a\rangle. \tag{5.28}$$

We can prove the orthogonality of eigenstates and the reality of eigenvalues in the framework of the abstract Hilbert space following the steps of the corresponding proof for wave functions. Taking the inner product of (5.28) with one of the eigenvectors we obtain

$$\langle\psi_{a'}|\hat{A}|\psi_a\rangle = a\langle\psi_{a'}|\psi_a\rangle.$$

Using Hermiticity in the left-hand side, we get

$$\langle\psi_a|\hat{A}|\psi_{a'}\rangle^* = a\langle\psi_{a'}|\psi_a\rangle \implies a'^*\langle\psi_a|\psi_{a'}\rangle^* = a\langle\psi_{a'}|\psi_a\rangle$$

or

$$(a - a'^*)\langle\psi_{a'}|\psi_a\rangle. \tag{5.29}$$

For $a' = a$ the relation (5.29) gives

$$(a - a^*)\langle a|a\rangle = 0 \implies a = a^*, \tag{5.30}$$

meaning that the eigenvalues are real. Note that $\langle a|a\rangle$ cannot be zero, since that would mean it is the *null* state, i.e., the zero of the Hilbert space, which has no physical content as a state of a physical system.

Now, since the eigenvalues have been proven to be real the relation (5.29) reads

$$(a - a')\langle\psi_{a'}|\psi_a\rangle = 0. \tag{5.31}$$

If $a \neq a'$, the only way to satisfy it is by $\langle\psi_a|\psi_{a'}\rangle = 0$. Thus, the eigenvectors corresponding to different eigenvalues are orthogonal.[4] Since, they can always be normalized, we can write an orthonormality relation

$$\langle a|a'\rangle = \delta_{aa'}. \tag{5.32}$$

It is clear that the eigenvectors of Hermitian operators that represent the various physical observables of a quantum system provide orthonormal bases of the Hilbert space, discrete or continuous-or mixed- depending on the particular observ-

---

[4] Again, we have assumed for simplicity that there is no degeneracy.

able. These bases are not only orthonormal but also *complete*. There is an elegant expression of completeness as an operator relation. For any ket we may write

$$|\psi\rangle = \sum_a \langle a|\psi\rangle |a\rangle = \sum_a |a\rangle \langle a|\psi\rangle$$

or[5]

$$\sum_a |a\rangle \langle a| = \mathbf{I}. \tag{5.33}$$

This is the *Completeness Relation*, which, together with the *Orthonormality Relation* above, are satisfied by the eigenvectors of any Hermitian operator and define a complete and orthonormal basis of the Hilbert space of states.

The basis of the energy eigenfunctions. The energy is of particular importance among observables since the Hamiltonian appears in the Schroedinger equation that determines the time evolution of any system. In fact, the time-independent Schroedinger equation is nothing else but the energy eigenvalue equation

$$\hat{H}|E\rangle = E|E\rangle. \tag{5.34}$$

We have symbolized with $|E\rangle$ the energy eigenstates. As a result of hermiticity we automatically have

$$\langle E|E'\rangle = \delta_{EE'}, \quad \sum_E |E\rangle \langle E| = 1. \tag{5.35}$$

The Schroedinger equation for the state of the system at time t is

$$i\hbar \frac{d|\psi(t)\rangle}{dt} = \hat{H}|\psi(t)\rangle. \tag{5.36}$$

Expanding $|\psi(t)\rangle$ in the basis $\{|E\rangle\}$ as

$$|\psi(t)\rangle = \sum_E \psi_E(t)|E\rangle \tag{5.37}$$

and substituting in the Schroedinger equation

$$0 = \sum_E \left(i\hbar \dot{\psi}_E(t) - \psi_E(t)\hat{H} \right) |E\rangle = \sum_E \left(i\hbar \dot{\psi}_E(t) - \psi_E(t)E \right) |E\rangle,$$

we obtain

$$\psi_E(t) = \psi_E(0) e^{-\frac{i}{\hbar}Et}. \tag{5.38}$$

---

[5]The unit operator $\mathbf{I}$ or $\mathbf{1}$, defined as $\mathbf{I}|\psi\rangle = |\psi\rangle$ on any vector, coincides with the real number 1. So, we could just write $\sum_a |a\rangle \langle a| = 1$.

Thus, the general solution to the Schroedinger equation reads

$$|\psi(t)\rangle = \sum_E \psi_E(0) e^{-\frac{i}{\hbar} E t} |E\rangle. \tag{5.39}$$

The coefficients $\psi_E(0)$ are just the projections of the initial state $|\psi(0)\rangle$ on the basis vectors

$$\psi_E(0) = \langle E|\psi(0)\rangle. \tag{5.40}$$

An additional property of this basis, related to the conservation of energy, is that any observable depending on the energy will have time-independent expectation values. Consider[6]

$$\langle \psi(t)|F(\hat{H})|\psi(t)\rangle = \sum_{E,E'} \psi_E^*(0)\psi_{E'}(0) e^{-\frac{i}{\hbar}(E'-E)t} \langle E|F(\hat{H})|E'\rangle$$

$$= \sum_{E,E'} \psi_E^*(0)\psi_{E'}(0) e^{-\frac{i}{\hbar}(E'-E)t} F(E)\delta_{EE'} = \sum_E |\psi_E(0)|^2 F(E).$$

For instance, for any state,

$$(\Delta E)_t^2 = \langle \psi(t)|\hat{H}^2|\psi(t)\rangle - \left(\langle \psi(t)|\hat{H}|\psi(t)\rangle\right)^2 = (\Delta E)_0^2.$$

Projection Operators. The completeness relation (5.33) of the eigenstates of a Hermitian operator is an example of the use of *"outer products"* of Hilbert space vectors. The result of such a product is an operator. The easiest way to see that for a product $|\psi\rangle\langle\chi|$ is to consider an orthonormal basis $\{|a\rangle\}$ and substitute in the place of kets and bras the columns and rows that represent them. Then,

$$|\psi\rangle\langle\chi| \implies \psi_a \chi_{a'}^* \implies \begin{pmatrix} \psi_1\chi_1* & \psi_1\chi_2^* & \cdots & \cdots \\ \psi_2\chi_1^* & \psi_2\chi_2^* & \cdots & \cdots \\ \cdot & \cdot & \cdots & \cdots \end{pmatrix}. \tag{5.41}$$

Thus, clearly, $|\psi\rangle\langle\chi|$ is an operator corresponding to a matrix with respect to a given basis.

---

[6]Functions of operators can be defined through a power series definition

$$F(\hat{A}) = \sum_n \frac{F^{(n)}(0)}{n!} \hat{A}^n.$$

Furthermore, any object $|b\rangle\langle b|$, where $|b\rangle$ is any normalized ket, is a *Projection Operator*

$$\hat{\Pi}_b = |b\rangle\langle b|, \tag{5.42}$$

having the property

$$\hat{\Pi}_b|\psi\rangle = \langle b|\psi\rangle|b\rangle, \tag{5.43}$$

which amounts to projecting any state vector $|\psi\rangle$ toward the direction of $|b\rangle$. If $|\psi\rangle$ were orthogonal to $|b\rangle$, the result would be zero. More than that a projection operator has no powers, since

$$\hat{\Pi}_b^2|\psi\rangle = \langle b|\psi\rangle\hat{\Pi}_b|b\rangle = \langle b|\psi\rangle|b\rangle$$

or

$$\hat{\Pi}_b^2 = \hat{\Pi}_b \quad or \quad \hat{\Pi}_b\left(\hat{\Pi}_b - \mathbf{I}\right) = 0. \tag{5.44}$$

Note also that any projection operator $|b\rangle\langle b|$ is a Hermitian operator. As such it is expected to have real eigenvalues and orthogonal eigenstates. If $|\varpi\rangle$ are the eigenvectors and ϖ are the corresponding eigenvalues, we have from (5.44)

$$\hat{\Pi}_b\left(\hat{\Pi}_b - \mathbf{I}\right)|\varpi\rangle = 0 \implies \varpi(\varpi - 1) = 0, \tag{5.45}$$

meaning that the only eigenvalues are 0 and 1.

With the help of the projection operators we may rewrite any Hermitian operator in terms of its eigenvectors and eigenvalues, namely

$$\hat{A} = \hat{A}\,\mathbf{I} = \hat{A}\sum_a |a\rangle\langle a| = \sum_a |a\rangle a\,\langle a| = \sum_a a\,\hat{\Pi}_a. \tag{5.46}$$

The projection operators satisfy

$$\sum_a \hat{\Pi}_a = \mathbf{I} \quad and \quad \hat{\Pi}_a\hat{\Pi}_b = \delta_{ab}\hat{\Pi}_a. \tag{5.47}$$

Projection operators can be used to describe the measurement process [8, 9]. The basic axioms of Quantum Mechanics dictate that when the system is subject to a measurement of the observable Λ and the system is in a state $|\psi\rangle$,

(1) as a result of the measurement we shall obtain an eigenvalue λ_k with probability

$$p_k = |\langle\psi|k\rangle|^2 = \langle\psi|\hat{\Pi}_k|\psi\rangle \tag{5.48}$$

(2) the system will make a transition to the eigenstate $|k\rangle$

$$|\psi\rangle \;\rightarrow\; |k\rangle = \frac{1}{\sqrt{p_k}}\hat{\Pi}_k|\psi\rangle. \tag{5.49}$$

This is the so-called *collapse of the wavefunction*.

Unitary Operators. Not all operators acting on the Hilbert space vectors need to represent observables. A set of useful operators are those that describe *transformations* of states and, in particular, transformations that conserve inner products, i.e., probability amplitudes. These are the *Unitary Operators* defined so that their Hermitian adjoint equals their inverse

$$\hat{U}\hat{U}^\dagger = \hat{U}^\dagger\hat{U} = \mathbf{I}. \tag{5.50}$$

Their action on states is such that inner products are conserved, namely

$$\hat{U}|\psi\rangle = |\psi'\rangle \;\Longrightarrow\; \langle\psi_1'|\psi_2'\rangle = \langle\psi_1|\hat{U}^\dagger\hat{U}|\psi_2\rangle = \langle\psi_1|\psi_2\rangle. \tag{5.51}$$

Writing down the eigenvalue problem for a unitary operator, namely

$$\hat{U}|u\rangle = u\,|u\rangle, \quad \langle u|\hat{U}^\dagger = u^*\langle u|, \tag{5.52}$$

and taking the inner product, we obtain

$$\langle u|\hat{U}^\dagger\hat{U}|u\rangle = |u|^2\langle u|u\rangle \;\Longrightarrow\; |u|^2 = 1, \tag{5.53}$$

which means that the eigenvalues of a unitary operator are always of the form $e^{i\alpha}$.

A unitary operator can always be written as the exponential of a Hermitian operator times an "i", namely[7]

$$\hat{U} = e^{i\hat{A}} \quad \text{where} \quad \hat{A}^\dagger = \hat{A}. \tag{5.54}$$

5.4 General Proof of the Heisenberg Inequality

Consider two physical observables $\hat{A}$ and $\hat{B}$. For an arbitrary normalized state $|\psi\rangle$, the corresponding uncertainties are

$$(\Delta A)^2 = \langle\psi|\left(\hat{A} - \langle\hat{A}\rangle\right)^2|\psi\rangle, \quad (\Delta B)^2 = \langle\psi|\left(\hat{B} - \langle\hat{B}\rangle\right)^2|\psi\rangle. \tag{5.55}$$

Next, we may introduce the operators[8]

$$\bar{A} \equiv \hat{A} - \langle\hat{A}\rangle, \quad \bar{B} \equiv \hat{B} - \langle\hat{B}\rangle, \tag{5.56}$$

---

[7] Obviously $\hat{U}\hat{U}^\dagger = e^{i\hat{A}}e^{-i\hat{A}} = \mathbf{I}.$

[8] We have omitted the "hat" for reasons of notational economy.

which are obviously hermitian as well. Then, we have

$$(\Delta A)^2 = \langle \bar{A}^2 \rangle = \left| \left| \bar{A} |\psi\rangle \right| \right|^2, \quad (\Delta B)^2 = \langle \bar{B}^2 \rangle = \left| \left| \bar{B} |\psi\rangle \right| \right|^2. \qquad (5.57)$$

Applying the *Schwartz inequality* (5.7) we obtain

$$\left| \left| \bar{A} |\psi\rangle \right| \right| \, \left| \left| \bar{B} |\psi\rangle \right| \right| \geq \left| \langle \psi | \bar{A} \bar{B} | \psi \rangle \right| \qquad (5.58)$$

or

$$(\Delta A)^2 (\Delta B)^2 \geq \left| \langle \bar{A} \bar{B} \rangle \right|^2. \qquad (5.59)$$

The product of two operators can always be written in terms of their commutator and their *anticommutator*, the latter defined as $\left\{ \hat{A}, \, \hat{B} \right\} \equiv \hat{A}\hat{B} + \hat{B}\hat{A}$. Thus, we have

$$\bar{A}\bar{B} = \frac{1}{2} \left[\bar{A}, \, \bar{B} \right] + \frac{1}{2} \left\{ \bar{A}, \, \bar{B} \right\}. \qquad (5.60)$$

Note however that $\left[\bar{A}, \, \bar{B} \right] = \left[\hat{A}, \, \hat{B} \right]$. Note also that the expectation value of the commutator of two hermitian operators will always be a purely imaginary number. This goes as follows:

$$\left\langle \left[\hat{A}, \, \hat{B} \right] \right\rangle = \langle \hat{A}\hat{B} \rangle - \langle \left(\hat{A}\hat{B} \right)^{\dagger} \rangle = \langle \hat{A}\hat{B} \rangle - \langle \hat{A}\hat{B} \rangle^* = 2i \, Im \langle \hat{A}\hat{B} \rangle. \qquad (5.61)$$

In contrast, the expectation value of the anticommutator of these operators is always a real number, namely

$$\left\langle \left\{ \bar{A}, \, \bar{B} \right\} \right\rangle = \langle \bar{A}\bar{B} \rangle + \langle \left(\bar{A}\bar{B} \right)^{\dagger} \rangle = \langle \bar{A}\bar{B} \rangle + \langle \bar{A}\bar{B} \rangle^* = 2 Re \langle \bar{A}\bar{B} \rangle. \qquad (5.62)$$

Therefore, we may write

$$\left| \langle \bar{A}\bar{B} \rangle \right|^2 = \frac{1}{4} \left| \langle [\bar{A}, \, \bar{B}] \rangle \right|^2 + \frac{1}{4} \left| \langle \{\bar{A}, \, \bar{B}\} \rangle \right|^2. \qquad (5.63)$$

Obviously,

$$\left| \langle \bar{A}\bar{B} \rangle \right|^2 \geq \frac{1}{4} \left| \langle [\bar{A}, \, \bar{B}] \rangle \right|^2. \qquad (5.64)$$

Thus, we conclude that for any pair of hermitian operators and any normalizable state, the following inequality is always true

$$(\Delta A)^2 (\Delta B)^2 \geq \frac{1}{4} \left| \langle [\bar{A}, \, \bar{B}] \rangle \right|^2. \qquad (5.65)$$

This is *Heisenberg's Inequality*, incorporating the *Uncertainty Principle*.

5.5 The Position and Momentum Representations

We started our discussion of the basic quantum concepts considering the simple system of a particle moving in one space dimension. Our central object was the wave function of the particle $\psi(x)$ representing its probability amplitude to be found at a position x. In the framework of Hilbert space a definite position should correspond to an eigenstate of the position operator $\hat{x}$. The position eigenstates are defined by[9]

$$\hat{x}|x'\rangle = x'|x'\rangle \quad with \quad \begin{cases} \langle x'|x''\rangle = \delta(x' - x'') \\ \int_{-\infty}^{+\infty} dx' |x'\rangle \langle x'| = \mathbf{I}. \end{cases} \quad (5.66)$$

The eigenvalue spectrum is continuous.

An arbitrary state $|\psi\rangle$ can be expanded in this basis as

$$|\psi(x)\rangle = \int_{-\infty}^{+\infty} dx\, \psi(x)\, |x\rangle, \quad (5.67)$$

where

$$\psi(x) = \langle x|\psi\rangle. \quad (5.68)$$

This is the *wave function* of the system. In the language of the Hilbert space, it is just the projection of the state vector $|\psi\rangle$ on a position eigenstate $|x\rangle$. All inner products are readily expressible in terms of wave functions with the help of (5.67)

$$\langle\psi_1|\psi_2\rangle = \int dx \int dx'\, \psi_1^*(x)\psi_2(x')\langle x|x'\rangle = \int_{-\infty}^{+\infty} dx\, \psi_1^*(x)\psi_2(x). \quad (5.69)$$

Special case of this is the familiar normalization of the wave function

$$\langle\psi|\psi\rangle = 1 \implies \int_{-\infty}^{+\infty} dx\, |\psi(x)|^2 = 1. \quad (5.70)$$

The momentum operator $\hat{p}$ and its action in this basis can be defined either postulating the fundamental commutation relation

$$[\hat{x}, \hat{p}] = i\hbar \quad (5.71)$$

or by specifying the transformation elements $\langle x|p\rangle$ to the momentum eigenstates. Following the former course of quantization by postulating the commutator we act by it on a bra $\langle x|$ and a ket $|x'\rangle$. We get

$$\langle x|\left(\hat{x}\hat{p} - \hat{p}\hat{x}\right)|x'\rangle = i\hbar\delta(x - x') \implies (x - x')\langle x|\hat{p}|x'\rangle = i\hbar\,\delta(x - x').$$

---

[9]See also [7–9].

This equation gives the matrix elements $\langle x|\hat{p}|x'\rangle$ of the momentum operator in the position basis and it is solved by the substitution[10]

$$\langle x|\hat{p}|x'\rangle = -i\hbar\frac{\partial}{\partial x}\delta(x - x') = -i\hbar\frac{\partial}{\partial x}\langle x|x'\rangle. \tag{5.72}$$

Note that for any $|\psi\rangle$ we also have

$$\langle x|\hat{p}|\psi\rangle = \int dx'\, \psi(x')\langle x|\hat{p}|x'\rangle = -i\hbar\frac{\partial}{\partial x}\int dx'\, \delta(x - x')\psi(x') = -i\hbar\frac{\partial}{\partial x}\langle x|\psi\rangle$$

or just

$$\langle x|\hat{p} = -i\hbar\frac{\partial}{\partial x}\langle x|. \tag{5.73}$$

In a less formal notation we may conclude that in the $\{x\}$-representation

$$\hat{p} \implies -i\hbar\frac{\partial}{\partial x}. \tag{5.74}$$

Next, let's consider the momentum eigenstates, defined as

$$\hat{p}|p'\rangle = p'|p'\rangle \quad with \quad \begin{cases} \langle p'|p''\rangle = \delta(p' - p'') \\ \int dp'\, |p'\rangle \langle p'| = \mathbf{I}. \end{cases} \tag{5.75}$$

A state $|\psi\rangle$ can be expanded in momentum eigenstates as

$$|\psi\rangle = \int dp\, \tilde{\psi}(p)|p\rangle, \tag{5.76}$$

where $\tilde{\psi}(p)$ is the *momentum space wave function*, which we previously (Chaps. 1 and 2) denoted as $g(p)$

$$\tilde{\psi}(p) = \langle p|\psi\rangle. \tag{5.77}$$

In order to find the relation between $\psi(x)$ and $\tilde{\psi}(p)$ we need to know how the two bases transform to each other, namely the inner products $\langle x|p\rangle$. Actually these inner products are the spatial wave functions of the momentum eigenstates

$$\psi_p(x) = \langle x|p\rangle \quad \Longleftrightarrow \quad -i\hbar\frac{\partial}{\partial x}\psi_p(x) = p\psi_p(x), \tag{5.78}$$

---

[10]Substituting we get

$$(x - x')(-i\hbar)\frac{\partial}{\partial x}\delta(x - x') = -i\hbar\frac{\partial}{\partial x}\left((x - x')\delta(x - x')\right) + i\hbar\frac{\partial(x - x')}{\partial x}\delta(x - x') = i\hbar\delta(x - x')$$

We have used the property of the delta function $(x - x')\delta(x - x') = 0$.

from which, integrating, we deduce that

$$\langle x|p\rangle = \frac{e^{\frac{i}{\hbar}xp}}{(2\pi\hbar)^{1/2}}. \tag{5.79}$$

The normalization factor was chosen so that they satisfy the orthonormality condition

$$\int_{-\infty}^{+\infty} dx\, \psi_p^*(x)\psi_{p'}(x) = \delta(p - p'). \tag{5.80}$$

The relation between the wave functions $\psi(x)$ and $\tilde{\psi}(p)$ is now easily derived using the completeness formulas for either basis. We have

$$\psi(x) = \langle x|\psi\rangle = \int dp, \langle x|p\rangle\langle p|\psi\rangle = \int \frac{dp}{\sqrt{2\pi\hbar}}e^{\frac{i}{\hbar}xp}\tilde{\psi}(p)$$

and

$$\tilde{\psi}(p) = \langle p|\psi\rangle = \int dx\, \langle p|x\rangle\langle x|\psi\rangle = \int \frac{dx}{\sqrt{2\pi\hbar}}e^{-\frac{i}{\hbar}xp}\psi(x).$$

These expressions are the familiar Fourier transforms that we have previously encountered.

How does the position operator act on momentum eigenstates? Projecting $\hat{x}$ on a momentum eigenstate and using completeness, we obtain

$$\langle p|\hat{x} = \int dx'\, \langle p|x'\rangle\langle x'|\hat{x} = \int dx'\, \frac{e^{-\frac{i}{\hbar}x'p}}{(2\pi\hbar)^{1/2}}x'\langle x'| = +i\hbar\frac{\partial}{\partial p}\int dx'\, \frac{e^{-\frac{i}{\hbar}x'p}}{(2\pi\hbar)^{1/2}}|x'\rangle$$

$$= +i\hbar\frac{\partial}{\partial p}\int dx'\, \langle p|x'\rangle\langle x'| = +i\hbar\frac{\partial}{\partial p}\langle p|.$$

In a less formal language we conclude that in the $\{p\}$-representation the momentum acts as a number on the momentum space wave functions while the position operator acts according to

$$\hat{x} \implies +i\hbar\frac{\partial}{\partial p}. \tag{5.81}$$

5.6 Parity

Since the wave function is the central object of the spatial description of a quantum system it is expected that the symmetry properties of the wave functions will play a role in the dynamics. Among possible symmetry operations, the *space reflection* symmetry (*Parity*) is of particular importance. The parity transformation on the wave function is defined as

$$\hat{P}\psi(x) = \psi'(x) = \psi(-x). \tag{5.82}$$

In the ket notation this goes as

$$\hat{P}|\psi\rangle = |\psi'\rangle \quad such \ that \quad \langle x|\psi'\rangle = \langle -x|\psi\rangle.$$

Equivalently

$$\langle x|\hat{P} = \langle -x|. \tag{5.83}$$

Note that Parity is not just a transformation but it is also a physical observable, being a Hermitian operator. Indeed we have

$$\langle \psi_1|\hat{P}|\psi_2\rangle = \int_{-\infty}^{+\infty} dx \psi_1^*(x)\psi_2(-x) = \int_{-\infty}^{+\infty} dx' \psi_1^*(-x')\psi_2(x') = \langle \psi_2|\hat{P}|\psi_1\rangle^*,$$

which implies hermiticity

$$\hat{P}^\dagger = \hat{P}. \tag{5.84}$$

Parity is also a unitary operator, since

$$\hat{P}^2\psi(x) = \hat{P}\psi(-x) = \psi(x)$$

or

$$\hat{P}\hat{P}^\dagger = \hat{P}^2 = \mathbf{I}. \tag{5.85}$$

Parity *anticommutes* with the position operator since

$$\langle x|\hat{P}\hat{x} = \langle -x|\hat{x} = -x\langle -x| = -x\langle x|\hat{P} = -\langle x|\hat{x}\hat{P}$$

or

$$\hat{x}\hat{P} = -\hat{P}\hat{x} \quad \Longleftrightarrow \quad \hat{P}\hat{x}\hat{P} = -\hat{x}. \tag{5.86}$$

The action of Parity on momentum can be easily guessed from the effect of $x \to -x$ on the momentum operator in the $\{x\}$ representation $-i\hbar\frac{\partial}{\partial x} \to i\hbar\frac{\partial}{\partial x}$. The ket relation can also be shown in one line

$$\langle p|\hat{P} = \int_{-\infty}^{+\infty} dx \langle p|x\rangle\langle x|\hat{P} = \int_{-\infty}^{+\infty} dx \langle p|x\rangle\langle -x| = \int_{-\infty}^{+\infty} dx \langle -p| - x\rangle\langle -x|$$

or

$$\langle p|\hat{P} = \langle -p|. \tag{5.87}$$

From this we may easily also show

$$\hat{p}\hat{P} = -\hat{P}\hat{p} \quad \Longleftrightarrow \quad \hat{P}\hat{p}\hat{P} = -\hat{p}. \tag{5.88}$$

Next, let us consider the Parity eigenvalue problem. Denoting with ϖ the Parity eigenvalues, we have, in terms of wave functions

$$\hat{P}\psi(x) = \varpi\,\psi(x)$$
$$\hat{P}^2 = \mathbf{I} \qquad\qquad \Longrightarrow \quad \varpi^2 = 1. \tag{5.89}$$

As a result of hermiticity the possible eigenvalues are real and therefore $\varpi = \pm 1$, meaning that

$$\hat{P}\psi^{(+)}(x) = \psi^{(+)}(x), \quad \hat{P}\psi^{(-)}(x) = -\psi^{(-)}(x). \tag{5.90}$$

Thus, as expected, the eigenfunctions of Parity are the even and odd functions.

As it has been proven, if the Hamiltonian of a system commutes with Parity, the eigenfunctions of the energy will be necessarily eigenstates of Parity, i.e., even and odd functions. We have seen this occurring automatically in the solution of the time-independent Schroedinger equation for the square well.

Problems and Exercises

5.1 Consider the energy eigenstates $|\psi_n\rangle$ of the infinite square well. Show that the position matrix elements between states of the same parity vanish, namely

$$\langle\psi_{2k}|x|\psi_{2k'}\rangle = \langle\psi_{2k+1}|x|\psi_{2k'+1}\rangle = 0.$$

Calculate the matrix elements[11] $\langle\psi_{2k+1}|x|\psi_{2k'}\rangle$.

5.2 Set up the energy eigenvalue problem in a discrete orthonormal basis $\{|e_i\rangle\}$ for a system with Hamiltonian matrix

$$\mathcal{H}_{ij} = \langle e_i|\hat{H}|e_j\rangle.$$

Work out the special case of dimensionality $N = 3$ and

$$\mathcal{H} = \begin{pmatrix} 0 & \alpha & \alpha \\ \alpha & 0 & \alpha \\ \alpha & \alpha & 0 \end{pmatrix},$$

---

[11] You may use $\int dx\, x\, \sin(\alpha x) = \alpha^{-2}\sin(\alpha x) - x\alpha^{-1}\cos(\alpha x)$.

where α is a known real parameter, determining eigenvalues and eigenvectors.

5.3 Consider a system with a purely discrete energy spectrum and express the wave vector and the Hamiltonian of the system as a column and a square matrix in a given orthonormal basis. Derive the Schroedinger equation in matrix form. Specialize in the case of system with a Hilbert space of dimensionality $N = 2$ and assume the following properties of the basis $|1\rangle$, $|2\rangle$

$$\hat{H}|1\rangle = i\alpha\,|2\rangle, \quad \hat{H}|2\rangle = -i\alpha|1\rangle,$$

where α known real parameter and $\hat{H}$ the Hamiltonian operator. Solve the Schroedinger equation.

5.4 A system is described by two non-commuting observables which in a given orthonormal basis are represented by the matrices

$$A = \begin{pmatrix} 0 & 1 \\ 1 & 0 \end{pmatrix}, \quad B = \begin{pmatrix} 0 & -i \\ i & 0 \end{pmatrix}$$

A measurement of A is performed on a statistical ensemble of identical copies of the system. What are the possible outcomes? The value $A \to +1$ came up with probability $1/2$. What will be the resulting state of the system for this branch? What is the state prior to measurement? Assume now that on the branch of $A \to +1$ a measurement of B is performed. What are the possible outcomes and what will be their relative probabilities?

5.5 Consider again the system of the problem **5.4** and the state

$$|\psi\rangle = \frac{1}{\sqrt{1+\lambda^2}}\,(|1\rangle + i\lambda\,|2\rangle)\,,$$

where λ is a real parameter. Calculate the uncertainties (ΔA) and (ΔB) in this state and verify Heisenberg's inequality.

5.6 Consider two linearly independent state vectors $|\psi\rangle$, $|\chi\rangle$ which are normalizable but neither normalized nor orthogonal. Construct an equivalent pair of orthonormal kets. The following are given:

$$\langle\psi|\psi\rangle = 2, \quad \langle\chi|\chi\rangle = 3/2, \quad \langle\psi|\chi\rangle = i\,.$$

5.7 Prove that

$$\frac{d\langle\mathbf{p}\rangle_t}{dt} = -\frac{m}{\hbar^2}\langle\left[\hat{H},\left[\hat{H},\hat{\mathbf{r}}\right]\right]\rangle\,.$$

5.8 Show that for any system of a particle with Hamiltonian $\hat{H} = p^2/2m + V(\mathbf{r})$ the following is true

$$\left[\left[\hat{H}, \hat{x}_i\right], \hat{x}_j\right] = -\frac{\hbar^2}{m}\delta_{ij}.$$

Based on that, prove the identity

$$\sum_n |\langle\psi_n|\mathbf{r}|\psi_0\rangle|^2 (E_n - E_0) = \frac{\hbar^2}{2m},$$

where $|\psi_n\rangle$ are the eigenstates of energy E_n and $n = 0$ corresponds to the ground state. Verify this identity for the case of one-dimensional infinite square well.

5.9 Consider the Schroedinger equation for a particle moving in one dimension

$$\left(\frac{\hat{p}^2}{2m} + V(\hat{x})\right) |\psi(t)\rangle = i\hbar\frac{d|\psi(t)\rangle}{dt}$$

and write it in the $\{p\}$ representation applying the recipe[12] .

$$\hat{x} \rightarrow +i\frac{\partial}{\partial p}, \quad \hat{p} \rightarrow p.$$

5.10 The quantum mechanical treatment of any system starts by considering its classical analogue and replacing its canonical variables and momenta with the corresponding operators. This correspondence is not always straightforward and the ordering of operators has to be arranged so that hermiticity is valid. As an example consider a particle subject to velocity dependent forces with a classical Hamiltonian

$$H = \frac{1}{2m}(\mathbf{p} - m\omega \times \mathbf{r})^2,$$

where ω is a constant parametric vector. Consider the corresponding operator Hamiltonian $\hat{H}$ and show that it is automatically hermitian. Derive the time evolution equations for the expectation values $\langle\mathbf{r}\rangle_t$ and $\langle\mathbf{p}\rangle_t$.

References

1. J. Von Neumann, *The Mathematical Foundations of Quantum Mechanics*, ed. by N.A. Wheeler (Princeton University Press, Princeton, 2018)
2. J.D. Jackson, *Mathematics for Quantum Mechanics* (Dover, 1990)
3. J. Mathews, R. Walker, *Mathematical Methods of Physics*, 2nd edn. (Pearson, 1970)

---

[12]Note that $f(i\frac{d}{dk}) \int \frac{dx}{\sqrt{2\pi}}\psi(x)e^{-ikx} = \int \frac{dx}{\sqrt{2\pi}} f(x)\psi(x)e^{-ikx}$.

4. G. Arfken, H. Weber, R. Harris, *Mathematical Methods for Physicists*, 7th edn. (Academic, New York, 2013)
5. P. Dennery, A. Krzywicki, *Mathematics for Physicists* (Dover, 1996)
6. P.A.M. Dirac, *The Principles of Quantum Mechanics* (Oxford University Press, Oxford, 1999)
7. E. Merzbacher, *Quantum Mechanics*, 3rd edn. (Wiley, New York, 1998)
8. A. Messiah, *Quantum Mechanics* (Dover Publications, single-volume reprint of the Wiley, New York, two-volume 1958 edition)
9. K. Gottfried, T.-M. Yan, *Quantum Mechanics: Fundamentals* (Springer, Berlin, 2004)

Chapter 6
Time Evolution

6.1 The Time-Evolution Operator

As we saw in the previous chapter one of the postulates of Quantum Mechanics states that the time evolution of the state vector of a system obeys the Schroedinger equation.[1] An alternative way to think about the time evolution of state vectors is to consider time evolution as a *transformation* in terms of a *time-evolution operator* $\hat{U}(t)$

$$|\psi(t)\rangle = \hat{U}(t)|\psi(0)\rangle . \tag{6.1}$$

Imposing probability conservation

$$\langle\psi(t)|\psi(t)\rangle = \langle\psi(0)|\psi(0)\rangle ,$$

leads to

$$\langle\psi(0)|\hat{U}^{\dagger}(t)\hat{U}(t)|\psi(0)\rangle = \langle\psi(0)|\psi(0)\rangle ,$$

which forces $\hat{U}(t)$ to be unitary, i.e.,

$$\hat{U}^{\dagger}(t)\hat{U}(t) = \hat{U}(t)\hat{U}^{\dagger}(t) = \mathbf{I}. \tag{6.2}$$

Recall that in Classical Mechanics the conservation of the energy of a system is directly related to the *homogeneity of time*, i.e., to invariance in time translations $t \rightarrow t + \alpha$. If an analogue of this is to persist in the quantum world, it would mean that the time-evolution operator should somehow be related to the energy operator, i.e., the Hamiltonian $\hat{H}$. This is actually what is dictated by *"axiom"-6*, which states that the time evolution of the state vector (6.1) should obey the Schroedinger equation

---

[1] For the material in this chapter see also [1–3].

© Springer Nature Switzerland AG 2019

K. Tamvakis, *Basic Quantum Mechanics*, Undergraduate Texts in Physics,
https://doi.org/10.1007/978-3-030-22777-7_6

$$i\hbar \frac{d}{dt}|\psi(t)\rangle = \hat{H}|\psi(t)\rangle \implies i\hbar\frac{d\hat{U}}{dt} = \hat{H}\hat{U}. \tag{6.3}$$

Assuming that $\hat{U}$ depends only on $\hat{H}$ as we argued above, (6.3) can be integrated to give

$$\hat{U}(t) = e^{-\frac{i}{\hbar}\hat{H}t} \implies |\psi(t)\rangle = e^{-\frac{i}{\hbar}\hat{H}t}|\psi(0)\rangle. \tag{6.4}$$

This can be generalized to

$$|\psi(t)\rangle = e^{-\frac{i}{\hbar}\hat{H}(t-t_0)}|\psi(t_0)\rangle. \tag{6.5}$$

Equation (6.5) can be thought of as an operator solution of the Schroedinger equation.

All the above analysis was carried out in the case that the Hamiltonian is time independent. In the case of an open system with time-dependent Hamiltonian, we shall assume that the Schroedinger equation still holds. The evolution operator is still meaningful but the operator differential equation that it satisfies

$$i\hbar\frac{d\hat{U}(t)}{dt} = \hat{H}(t)\hat{U}(t). \tag{6.6}$$

does not have a simple solution.[2]

6.2 The Schroedinger Versus the Heisenberg Picture

In the previous section on the time-evolution operator, we implicitly assumed that all time dependence is carried by the state vectors while the physical observables do not depend on time. This is only one way of describing things. It is called the *Schroedinger "Picture"*. There is an alternative way according to which the state vectors do not depend on time while the physical observables carry all the time dependence. This description is named the *Heisenberg "Picture"*. The two "pictures" should predict *exactly the same matrix elements* of physical observables (numerically), since these are the quantities that are measured and not the abstract kets or operators.

Let's denote by

$$|\psi_S(t)\rangle, \ |\psi_H(t)\rangle, \ \hat{A}_S(t), \ \hat{A}_H(t)$$

---

[2]In many problems of time-dependent perturbations, it is useful to write down an operator equation like (6.6) for the perturbing part of the Hamiltonian and obtain a formal solution in terms of a so-called *time-ordered exponential* $\hat{U}(t) = \mathcal{T}\exp[-i\int^t dt' \hat{H}(t')/\hbar]$, to be explained in a subsequent section.

the states and observables in either picture. Taking an initial time t_0, we define

$$|\psi_S(t)\rangle = \hat{U}(t - t_0)|\psi_S(t_0)\rangle \quad \hat{A}_S(t) = \hat{A}_S(t_0)$$

$$|\psi_H(t)\rangle = |\psi_H(t_0)\rangle \qquad \hat{A}_H(t) \neq \hat{A}_H(t_0)$$

(6.7)

where $\hat{U}(t - t_0) = e^{-\frac{i}{\hbar}\hat{H}(t-t_0)}$. The time evolution of $\hat{A}_H(t)$ will be determined shortly. We can always assume that the two descriptions coincide at the initial time, namely,

$$|\psi_S(t_0)\rangle = |\psi_H(t_0)\rangle, \quad \hat{A}_S(t_0) = \hat{A}_H(t_0). \tag{6.8}$$

The two "pictures" have to agree at any time giving the same matrix elements, namely

$$\langle \chi_S(t)|\hat{A}_S(t)|\psi_S(t)\rangle = \langle \chi_H(t)|\hat{A}_H(t)|\psi_H(t)\rangle \quad (\forall t). \tag{6.9}$$

Substituting (6.7) into the left-hand side of (6.9), we obtain

$$\langle \chi_S(t_0)|\hat{U}^\dagger(t - t_0)\hat{A}_S\hat{U}(t - t_0)|\psi_S(t_0)\rangle = \langle \chi_H(t_0)|\hat{U}^\dagger(t - t_0)\hat{A}_H(t_0)\hat{U}(t - t_0)|\psi_H(t_0)\rangle$$

$$= \langle \chi_H(t)|\hat{U}^\dagger(t - t_0)\hat{A}_H(t_0)\hat{U}(t - t_0)|\psi_H(t)\rangle$$

and comparing it to the right-hand side we conclude that the invariance of matrix elements requires that the Heisenberg "picture" time-evolution law for observables be (see Table 6.1)

$$\hat{A}_H(t) = \hat{U}^\dagger(t - t_0)\hat{A}_H(t_0)\hat{U}(t - t_0). \tag{6.10}$$

This operator relation can be differentiated with respect to time and give as an *operator differential equation*

$$\frac{d\hat{A}_H}{dt} = \frac{i}{\hbar}\left[\hat{H}, \hat{A}_H\right] \quad \textit{Heisenberg Equation.} \tag{6.11}$$

In the case that an operator depends explicitly on time the above equation is modified to

$$\frac{d\hat{A}_H}{dt} = \frac{\partial \hat{A}_H}{\partial t} + \frac{i}{\hbar}\left[\hat{H}, \hat{A}_H\right]. \tag{6.12}$$

Table 6.1 Heisenberg versus Schroedinger picture

Schroedinger	$|\psi_S(t)\rangle = e^{-\frac{i}{\hbar}\hat{H}t}|\psi(0)\rangle$	$\hat{A}_S(t) = \hat{A}_S(0)$
Heisenberg	$|\psi_H(t)\rangle = |\psi_H(0)\rangle$	$\hat{A}_H(t) = \hat{U}^\dagger(t)\hat{A}_H(0)\hat{U}(t)$

Note that if we take the expectation value of Heisenberg's equation in any state $|\psi_H\rangle$, we obtain

$$\langle\psi_H|\frac{d}{dt}\hat{A}_H|\psi_H\rangle \;=\; \frac{i}{\hbar}\langle\psi_H|[\hat{H}, \hat{A}_H]|\psi_H\rangle$$

or, since the Heisenberg states do not evolve with time,

$$\frac{d}{dt}\langle\psi_H|\hat{A}_H|\psi_H\rangle \;=\; \frac{i}{\hbar}\langle\psi_H|[\hat{H}, \hat{A}_H]|\psi_H\rangle\,.$$

Then, since matrix elements are *picture independent*, we simply have

$$\frac{d\langle\hat{A}\rangle}{dt} \;=\; \frac{i}{\hbar}\langle[\hat{H}, \hat{A}]\rangle\,, \tag{6.13}$$

which holds in *any picture*. It should be no surprise then that (6.13) coincides with Eq. (4.67), previously derived in the Schroedinger description.

It is clear that if an observable commutes with the Hamiltonian, it will not change in the course of time, i.e.,

$$\left[\hat{H}, \hat{C}\right] = 0 \;\;\Longrightarrow\;\; \hat{C}_H(t) = \hat{C}(0)\,. \tag{6.14}$$

Such an observable is called a *constant of the motion*. For instance, the energy itself is a constant of the motion and $\hat{H}(t) = \hat{H}(0)$.

It should be noted that the fundamental *equal time commutation relation* of position and momentum is *picture invariant*, namely

$$\left[\hat{x}_H(t), \hat{p}_H(t)\right] =$$

$$\hat{U}^\dagger(t)\hat{x}_H(0)\hat{U}(t)\hat{U}^\dagger(t)\hat{p}_H(0)\hat{U}(t) - \hat{U}^\dagger(t)\hat{p}_H(0)\hat{U}(t)\hat{U}^\dagger(t)\hat{x}_H(0)\hat{U}(t) =$$

$$\hat{U}^\dagger\left[\hat{x}_H(0), \hat{p}_H(0)\right]\hat{U}(t) = \hat{U}^\dagger\left[\hat{x}_S, \hat{p}_S\right]\hat{U}(t) = i\hbar\hat{U}^\dagger(t)\hat{U}(t) = i\hbar$$

or

$$\left[\hat{x}(t), \hat{p}(t)\right] = i\hbar. \tag{6.15}$$

In contrast, other commutation relations change drastically. For example, $[\hat{x}(t), \hat{x}(0)] \neq 0$.

Example 6.1 (Solving the Heisenberg Equation For a Free Particle) For a free particle ($\hat{H} = \frac{\hat{p}^2}{2m}$) the Heisenberg equation of $\hat{x}_H(t)$ is

$$\frac{d\hat{x}(t)}{dt} = \frac{i}{\hbar}\left[\hat{H}(t), \hat{x}(t)\right] = \frac{i}{2\hbar m}\left(\hat{p}(t)\left[\hat{p}(t), \hat{x}(t)\right] + \left[\hat{p}(t), \hat{x}(t)\right]\hat{p}(t)\right) = \frac{i(-2i\hbar)}{2\hbar m}\hat{p}(t)$$

or

$$\frac{d\hat{x}(t)}{dt} = \frac{\hat{p}(t)}{m}. \tag{6.16}$$

Similarly, we have

$$\frac{d\hat{p}(t)}{dt} = \frac{i}{\hbar}\left[\hat{H}(t),\ \hat{p}(t)\right] = \frac{i}{2\hbar m}\left[\hat{p}^2(t),\ \hat{p}(t)\right] = 0$$

and

$$\hat{p}(t) = \hat{p}(0). \tag{6.17}$$

Integrating (6.16) we obtain

$$\hat{x}(t) = \hat{x}(0) + \frac{\hat{p}(0)}{m}t. \tag{6.18}$$

Note that

$$\left[\hat{x}(t),\ \hat{x}(0)\right] = \frac{t}{m}\left[\hat{p}(0),\ \hat{x}(0)\right] = -\frac{i\hbar t}{m}. \tag{6.19}$$

Example 6.2 Consider a particle moving on the two-dimensional plane under the influence of a two-dimensional delta function potential $V(\mathbf{r}) = \lambda\delta(\mathbf{r}) = \lambda\delta(x)\delta(y)$. Show that the operator

$$\hat{D} = \hat{H}\frac{t}{m} - \frac{1}{4m}(\mathbf{r}\cdot\mathbf{p} + \mathbf{p}\cdot\mathbf{r}) \tag{6.20}$$

is a constant of the motion.

We have

$$\left[\hat{D},\ \hat{H}\right] = \ldots = \frac{i\hbar}{2m^2}\hat{p}^2 - \frac{i\lambda\hbar}{2m}(\mathbf{r}\cdot\nabla)\delta(\mathbf{r}) =$$

$$\frac{i\hbar}{2m^2}\hat{p}^2 - \frac{i\lambda\hbar}{2m}(\nabla\cdot(\mathbf{r}\delta(\mathbf{r})) - (\nabla\cdot\mathbf{r})\delta(\mathbf{r})) = \frac{i\hbar}{m}\hat{H}.$$

We have used the facts that $\mathbf{r}\delta(\mathbf{r}) = 0$ and that in two dimensions $\nabla\cdot\mathbf{r} = 2$. Thus,

$$\frac{d\hat{D}}{dt} = \frac{\partial\hat{D}}{\partial t} + \frac{i}{\hbar}\left[\hat{H},\ \hat{D}\right] = \frac{\hat{H}}{m} - \frac{\hat{H}}{m} = 0.$$

6.3 The Time–Energy Uncertainty Relation

Consider a system characterized by its Hamiltonian operator $\hat{H}$ and a number of observables $\hat{A}_1$, $\hat{A}_2$, $\ldots$. Heisenberg's inequality for the Hamiltonian and any of these observables takes the form

$$(\Delta E)^2(\Delta\hat{A})^2 \geq \frac{1}{4}\left|\left\langle\left[\hat{A},\ \hat{H}\right]\right\rangle\right|^2, \tag{6.21}$$

ΔE standing for $(\Delta\hat{H})$. Using (6.13), we obtain

$$(\Delta E)(\Delta \hat{A}) \geq \frac{\hbar}{2} \left| \frac{d\langle \hat{A} \rangle}{dt} \right| \implies (\Delta E)\left(\frac{(\Delta \hat{A})}{\left| \frac{d\langle \hat{A} \rangle}{dt} \right|} \right) \geq \frac{\hbar}{2}. \tag{6.22}$$

Consider now the quantity

$$\tau_A = \frac{(\Delta \hat{A})}{\left| \frac{d\langle \hat{A} \rangle}{dt} \right|}, \tag{6.23}$$

appearing in (6.22). This quantity has the dimensions of time. It is large if the spread of the values $(\Delta A)_t$ is large in comparison to the rate of change of its mean value. τ_A is small when $(\Delta A)_t$ is small in comparison to the rate of change of $\langle A \rangle_t$. Thus, it represents the *characteristic time of change of the distribution of the observable* A. Among the observables that characterize the system there will be one that is the most sensitive and its characteristic time τ will be the smallest

$$\tau \leq \tau_A \quad (\forall A).$$

This will be the *characteristic time of change of the system as a whole.* Obviously,

$$(\Delta E)\tau_A \geq (\Delta E)\tau \geq \frac{\hbar}{2}.$$

Thus, we may write the *Time–Energy Uncertainty Relation*

$$(\Delta E)\tau \geq \frac{\hbar}{2}, \tag{6.24}$$

where

$$\tau = \min_{\{A\}} \left(\frac{(\Delta \hat{A})}{\left| \frac{d\langle \hat{A} \rangle}{dt} \right|} \right). \tag{6.25}$$

Example 6.3 A particle of mass m and electric charge q moves under the influence of a constant electric field $\vec{\mathcal{E}}$. The particle wave function at an initial time $t = 0$ is

$$\psi(\mathbf{r}, 0) = (\alpha/\pi)^{3/4} \, e^{-\alpha r^2/2}.$$

(a) Calculate the expectation values $\langle \mathbf{r} \rangle_t$, $\langle \mathbf{p} \rangle_t$, $\langle H \rangle$ and the uncertainties $(\Delta \mathbf{r})_t$, $(\Delta \mathbf{p})_t$, (ΔE) at a time $t > 0$.
(b) Determine the characteristic times of change for the observables of the system τ_r, τ_p.
(c) Write down and verify the "time–energy" uncertainty inequality.

Solving the Heisenberg equation, we obtain

$$\frac{d\mathbf{p}}{dt} = q\vec{\mathcal{E}} \implies \mathbf{p}(t) = \mathbf{p}(0) + q\vec{\mathcal{E}}t$$

$$\frac{d\mathbf{r}}{dt} = \frac{\mathbf{p}}{m} \implies \mathbf{r}(t) = \mathbf{r}(0) + \frac{\mathbf{p}(0)}{m}t + \frac{q\vec{\mathcal{E}}}{2m}t^2 \tag{6.26}$$

The initial expectation values vanish

$$\langle \mathbf{r} \rangle_0 = (\alpha/\pi)^{3/2} \int d^3r \, \vec{r} e^{-\alpha r^2} = 0$$

$$\langle \mathbf{p} \rangle_0 = -i\hbar(\alpha/\pi)^{3/2} \int d^3r \, e^{-\alpha r^2/2} \nabla e^{-\alpha r^2/2} = i\hbar\alpha(\alpha/\pi)^{3/2} \int d^3r \, \vec{r} e^{-\alpha r^2} = 0$$

because of oddness. We also have

$$\langle p^2 \rangle_0 = \tfrac{3}{2}\alpha\hbar^2$$

$$\langle r^2 \rangle_0 = \tfrac{3}{2\alpha}$$

$$\langle \mathbf{r} \cdot \mathbf{p} + \mathbf{p} \cdot \mathbf{r} \rangle_0 = \ldots = 0 \tag{6.27}$$

$$\langle \mathbf{p}^4 \rangle_0 = \tfrac{15}{4}\hbar^4\alpha^2$$

(For the calculation of these integrals see the section of the Appendix on Gaussian integrals). The expectation values of the energy do not depend on time and, therefore, we have

$$\langle H \rangle = \frac{\langle p^2 \rangle_0}{2m} - q\vec{\mathcal{E}} \cdot \langle \vec{r} \rangle_0 = \frac{3\alpha\hbar^2}{4m}$$

$$\langle H^2 \rangle = \frac{1}{4m^2}\langle p^4 \rangle_0 + q^2\mathcal{E}_i\mathcal{E}_j\langle x_i x_j \rangle_0 - \frac{q}{2m}\mathcal{E}_i\langle x_i p^2 + p^2 x_i \rangle_0 \tag{6.28}$$

The last term in the expression of $\langle H^2 \rangle$ vanishes because of parity, while in the second, due to the spherical symmetry of the initial wave function, we may have

$$\langle x_i x_j \rangle_0 = \frac{\delta_{ij}}{3}\langle r^2 \rangle_0 . \tag{6.29}$$

Thus, finally, we have

$$\langle H^2 \rangle = \frac{15}{16}\frac{\hbar^4\alpha^2}{m^2} + \frac{q^2\mathcal{E}^2}{2\alpha} \quad \text{and} \quad (\Delta E)^2 = \frac{3}{8}\frac{\hbar^4\alpha^2}{m^2} + \frac{q^2\mathcal{E}^2}{2\alpha} . \tag{6.30}$$

The time-evolved expectation values are

$$\langle \mathbf{p} \rangle_t = q\vec{\mathcal{E}}t$$

$$\langle \mathbf{r} \rangle_t = \frac{q\vec{\mathcal{E}}}{2m}t^2$$

$$\langle p^2 \rangle_t = \tfrac{3}{2}\alpha\hbar^2 + q^2\mathcal{E}^2 t^2 \tag{6.31}$$

$$\langle r^2 \rangle_t = \frac{3}{2\alpha} + \frac{q^2\mathcal{E}^2 t^2}{4m^2} + \frac{3\alpha\hbar^2 t^2}{2m^2}$$

The corresponding uncertainties are

$$(\Delta\mathbf{r})_t^2 = \frac{3}{2\alpha}\left(1 + \frac{\hbar^2\alpha^2 t^2}{m^2}\right) \tag{6.32}$$

$$(\Delta\mathbf{p})_t^2 = \tfrac{3}{2}\alpha\hbar^2$$

The characteristic response times of the position and momentum are

$$\tau_r = \frac{(\Delta \mathbf{r})_t}{\left| \frac{d\langle \mathbf{r} \rangle_t}{dt} \right|} = \frac{m}{q\mathcal{E}} \sqrt{\frac{3}{2\alpha} \left(\frac{1}{t^2} + \left(\frac{\hbar\alpha}{m} \right)^2 \right)}$$

$$\tau_p = \frac{(\Delta \mathbf{p})_t}{\left| \frac{d\langle \mathbf{p} \rangle_t}{dt} \right|} = \frac{\hbar}{q\mathcal{E}} \sqrt{\frac{3\alpha}{2}}$$

(6.33)

Note that always $\tau_p < \tau_r$. Thus, the characteristic response time of the system is

$$\tau = \frac{\hbar}{q\mathcal{E}} \sqrt{\frac{3\alpha}{2}} .$$

(6.34)

The "energy–time" uncertainty relation reads

$$(\Delta E)\tau = \frac{\hbar}{2} \sqrt{3} \sqrt{1 + \frac{3\alpha}{4} \left(\frac{\hbar^2\alpha}{q\mathcal{E}m} \right)^2} \geq \frac{\hbar}{2}$$

(6.35)

and is obviously verified.

6.4   The Interaction Picture

The Schroedinger and the Heisenberg pictures are two extreme cases among the many equivalent[3] descriptions of time evolution. It is often more convenient to use a description (i.e., "picture") in which part of the time evolution is carried out by the state vectors and part is carried out by the operators. This is the case of time-dependent Hamiltonians of the form $H = H_0 + H_{int}(t)$, composed of a *"known"* or *"free"* part H_0 and a time-dependent interaction part $H_{int}(t)$. The Schroedinger equation and its formal operator solution are

$$i\hbar \frac{d}{dt} |\psi(t)\rangle = \left(\hat{H}_0 + \hat{H}_{int}(t) \right) |\psi(t)\rangle \implies |\psi(t)\rangle = \hat{U}(t, t_0) |\psi(t_0)\rangle . \quad (6.36)$$

Since, in general $\left[\hat{H}_0, \hat{H}_{int}(t) \right] \neq 0$, the expression for $\hat{U}$ will not be a simple product of exponentials. Note however that, in view of

$$|\psi(t)\rangle = \hat{U}(t, t_0) |\psi(t_0)\rangle \implies \frac{d}{dt} |\psi(t)\rangle = \frac{d\hat{U}}{dt} |\psi(t_0)\rangle = \frac{d\hat{U}}{dt} \hat{U}^\dagger |\psi(t)\rangle$$

the operator $\hat{U}$ has to satisfy

---

[3]Different "pictures" are related through a unitary transformation and, therefore, give the same expectation values of physical observables.

$$\left(\frac{d}{dt}\hat{U}(t, t_0)\right)\hat{U}^\dagger(t, t_0) = -\frac{i}{\hbar}\left(\hat{H}_0 + \hat{H}_{int}(t)\right). \tag{6.37}$$

Nevertheless, the evolution operator can always be put in the form

$$\hat{U}(t, t_0) = \hat{U}_0(t, t_0)\,\hat{U}_I(t, t_0), \tag{6.38}$$

where $\hat{U}_0(t, t_0) = e^{-\frac{i}{\hbar}(t-t_0)\hat{H}_0}$ is the standard evolution operator corresponding to the known piece $\hat{H}_0$, while $\hat{U}_I(t, t_0)$ is an "unknown" piece that involves the interaction. Starting from the parametrization (6.38) the *"Interaction Picture"* is defined as follows. State vectors evolve in time through the action of the "interacting" part $\hat{U}_I(t, t_0)$, while the operators that correspond to physical observables evolve through the "known" or "free" part alone

$$|\psi_I(t)\rangle = \hat{U}_I(t, t_0)|\psi_I(t_0)\rangle$$
$$\hat{Q}_I(t) = \hat{U}_0^\dagger(t, t_0)\,\hat{Q}_I(t_0)\,\hat{U}_0(t, t_0). \tag{6.39}$$

We may assume that at the initial time t_0 the Interaction and Schroedinger pictures coincide

$$|\psi_I(t_0)\rangle = |\psi(t_0)\rangle, \quad \hat{Q}_I(t_0) = \hat{Q}(t_0). \tag{6.40}$$

The differential version of (6.39) can be obtained by differentiating with respect to time. We have

$$\frac{d}{dt}|\psi_I(t)\rangle = \left(\frac{d}{dt}\hat{U}_I(t, t_0)\right)|\psi_I(t_0)\rangle = \frac{d}{dt}\left(\hat{U}_0^\dagger(t, t_0)\hat{U}(t, t_0)\right)|\psi_I(t_0)\rangle$$

$$= \left(\frac{d\hat{U}_0^\dagger}{dt}\hat{U} + \hat{U}_0^\dagger\frac{d\hat{U}}{dt}\right)|\psi_I(t_0)\rangle = \left(\frac{d\hat{U}_0^\dagger}{dt}\hat{U} + \hat{U}_0^\dagger\frac{d\hat{U}}{dt}\hat{U}^\dagger\hat{U}\right)|\psi_I(t_0)\rangle$$

$$= \frac{i}{\hbar}\left(\hat{H}_0\hat{U}_0^\dagger - \hat{U}_0^\dagger\left(\hat{H}_0 + \hat{H}_{int}\right)\right)\hat{U}|\psi_I(t_0)\rangle = -\frac{i}{\hbar}\hat{U}_0^\dagger\hat{H}_{int}\hat{U}_0\hat{U}_I|\psi_I(t_0)\rangle$$

or, finally

$$i\hbar\frac{d}{dt}|\psi_I(t)\rangle = \hat{H}_I(t)\,|\psi_I(t)\rangle, \tag{6.41}$$

where

$$\hat{H}_I(t) = \hat{U}_0^\dagger(t, t_0)\,\hat{H}_{int}(t)\,\hat{U}_0(t, t_0). \tag{6.42}$$

Thus, the "interaction picture" states $|\psi_I(t)\rangle$ satisfy a Schroedinger equation with Hamiltonian the interacting part time evolved through the *"known"* part. On the other hand, differentiating the evolution equation for operators, we obtain

$$\frac{d\hat{Q}_I}{dt} = \frac{i}{\hbar}\left[\hat{H}_0, \hat{Q}_I\right],$$

(6.43)

which is Heisenberg's equation with the free part of the Hamiltonian. Therefore, in the Interaction Picture of time evolution, states evolve with the interacting part of the Hamiltonian, while observables with the free part. The relation of the interaction picture state vectors to the Schroedinger ones is quite simple. Indeed, we have

$$|\psi_I(t)\rangle = \hat{U}_I(t, t_0)|\psi_I(t_0)\rangle = \hat{U}_0^\dagger(t, t_0)\hat{U}(t, t_0)|\psi_S(t_0)\rangle = \hat{U}_0^\dagger(t, t_0)|\psi_S(t)\rangle.$$

(6.44)

6.5 Time-Ordered Exponentials

The Schroedinger equation for a time-dependent Hamiltonian has the form

$$i\hbar\frac{d}{dt}|\psi(t)\rangle = \hat{H}(t)|\psi(t)\rangle$$

(6.45)

in both the Schroedinger and the Interaction picture. It is easy to convert (6.45) into an integral equation. Integrating, we get

$$|\psi(t)\rangle = |\psi(t_0)\rangle - \frac{i}{\hbar}\int_{t_0}^t dt'\, \hat{H}(t')|\psi(t')\rangle.$$

(6.46)

A series of successive approximations to this equation is

$$|\psi(t)\rangle^{(0)} = |\psi(t_0)\rangle$$

$$|\psi(t)\rangle^{(1)} = |\psi(t_0)\rangle - \frac{i}{\hbar}\int_{t_0}^t dt'\hat{H}(t')\,|\psi(t_0)\rangle$$

$$|\psi(t)\rangle^{(2)} = |\psi(t_0)\rangle - \frac{i}{\hbar}\int_{t_0}^t dt'\hat{H}(t')\,|\psi(t_0)\rangle - \frac{1}{\hbar^2}\int_{t_0}^t dt'\int_{t_0}^{t'} dt''\,\hat{H}(t')\hat{H}(t'')|\psi(t_0)\rangle$$

$$\cdots$$

(6.47)

Considering $|\psi(t)\rangle^{(2)}$, we may rewrite the last term as

$$-\frac{1}{2\hbar^2}\int_{t_0}^t dt'\int_{t_0}^t dt''\,\Theta(t'-t'')\left(\hat{H}(t')\hat{H}(t'') + \hat{H}(t')\hat{H}(t'')\right)$$

$$= -\frac{1}{2\hbar^2}\int_{t_0}^t dt'\int_{t_0}^t dt''\left(\Theta(t'-t'')\hat{H}(t')\hat{H}(t'') + \Theta(t''-t')\hat{H}(t'')\hat{H}(t')\right)$$

The product of operators in parenthesis is called a *time-ordered product*, defined in general as

$$T\left(\mathcal{Q}_1(t)\mathcal{Q}_2(t')\right) = \begin{cases} \mathcal{Q}_1(t)\mathcal{Q}_2(t') & (t > t') \\ \\ \mathcal{Q}_2(t')\mathcal{Q}_1(t) & (t' > t) \end{cases} \tag{6.48}$$

Obviously, $T\left(\mathcal{Q}_1(t)\mathcal{Q}_2(t')\right) = T\left(\mathcal{Q}_2(t')\mathcal{Q}_1(t)\right)$.

Thus, the last term in $|\psi(t)\rangle^{(2)}$ is written as

$$-\frac{1}{\hbar^2}\int_{t_0}^{t} dt' \int_{t_0}^{t} dt'' \, T\left(\hat{H}(t')\hat{H}(t'')\right) .$$

An analogous time-ordered expression can be written for the nth order terms

$$\int_{t_0}^{t} dt_1 \int_{t_0}^{t_1} dt_2 \ldots \hat{H}(t_1)\hat{H}(t_2)\ldots\hat{H}(t_n) = \frac{1}{n!}\int_{t_0}^{t} dt_1 \int_{t_0}^{t} dt_2 \ldots T\left(\hat{H}(t_1)\ldots\hat{H}(t_n)\right)$$

All these terms, having an $1/n!$ coefficient can be summed to an exponential. Therefore, the series of successive terms is summed up to give the solution

$$|\psi(t)\rangle = T\left(e^{-\frac{i}{\hbar}\int_{t_0}^{t} dt' \hat{H}(t')}\right)|\psi(t_0)\rangle . \tag{6.49}$$

Although the expression (6.49) is just a formal solution of (6.46) it turns out that it is very useful when expanded in a series, e.g., when (6.45) is considered in the interaction picture and the interaction $\hat{H}_{int}$ is "*small*" in comparison to the "*free*" part. Then, the expansion of (6.49) is very likely to converge, since higher powers of the interaction get smaller and smaller.

Problems and Exercises

6.1 The Hamiltonian operator of a system is a sum of two terms $\hat{H} = \hat{H}_0 + \hat{H}_1$. Prove the identity

$$e^{-\frac{i}{\hbar}\hat{H}(t-t_0)} = e^{-\frac{i}{\hbar}\hat{H}_0(t-t_0)} - \frac{i}{\hbar}\int_{t_0}^{t} dt' \, e^{-\frac{i}{\hbar}\hat{H}_0(t-t')} \, \hat{H}_1 \, e^{-\frac{i}{\hbar}\hat{H}(t'-t_0)} .$$

6.2 A system is described by a Hilbert space of dimension $N = 3$. For a given orthonormal basis $|1\rangle$, $|2\rangle$, $|3\rangle$ the Hamiltonian of the system is represented by the square matrix

$$\mathcal{H} = \begin{pmatrix} 0 & \alpha & 0 \\ \alpha & 0 & \alpha \\ 0 & \alpha & 0 \end{pmatrix},$$

where α a known real parameter. If the initial ($t = 0$) state of the system is

$$|\psi(0)\rangle = \frac{1}{\sqrt{3}} (|1\rangle + |2\rangle + |3\rangle),$$

calculate the probability to find the system at a time $t > 0$ in this state. What is the minimum time that this probability becomes certainty?

6.3 Show that if the commutator $[\hat{A}, \hat{B}]$ of the two operators $\hat{A}$ and $\hat{B}$ is just a complex number, the following is true

$$e^{\hat{A}+\hat{B}} = e^{\hat{A}} e^{\hat{B}} e^{-\frac{1}{2}[\hat{A}, \hat{B}]}.$$

Then, consider a system with a Hamiltonian $\hat{H} = \hat{H}_0 + \hat{H}_1$ for which the commutator $[\hat{H}_0, \hat{H}_1]$ is a just a complex number and apply this property to calculate the *interaction picture* evolution operator $\hat{U}_I(t)$.

6.4 Consider a particle of mass m and electric charge q subject to a homogeneous electric field **E**. Solve Heisenberg's equations for the position and momentum operators and calculate the commutators

$$\left[\hat{x}_i(t), \hat{x}_j(t')\right], \quad \left[\hat{p}_i(t), \hat{p}_j(t')\right], \quad \left[\hat{x}_i(t), \hat{p}_j(t')\right].$$

6.5 Consider a particle of mass m moving freely in one dimension. Calculate the *"matrix"* representing the time-evolution operator in the position representation $\{|x\rangle\}$

$$\mathcal{K}(x, x'; t) = \langle x|e^{-\frac{i}{\hbar}\hat{H}t}|x'\rangle.$$

What is the physical meaning of the probability amplitude $\mathcal{K}(x, x'; t)$?

6.6 Consider the hermitian operator $\hat{A}(t)$ in the framework of the Heisenberg picture. The corresponding eigenvalue problem will be stated as

$$\hat{A}(t)|\alpha, t\rangle = \alpha |\alpha, t\rangle.$$

Show that the states $|\alpha, t\rangle$ satisfy the orthonormality and completeness relations at any time t. Show also that they satisfy the equation

$$\hat{H}|\alpha, t\rangle = -i\hbar \frac{d|\alpha, t\rangle}{dt}$$

which has the opposite sign than the Schroedinger equation.

6.7 Derive the time-evolution operator for a two-state system with a time-dependent Hamiltonian

$$\mathcal{H} = \begin{pmatrix} \epsilon & \eta(t) \\ \eta(t) & \epsilon \end{pmatrix},$$

where ϵ is a given parameter and $\eta(t)$ is a given function of time.

6.8 A two-state system, in a given orthonormal basis, has a Hamiltonian matrix

$$\mathcal{H} = \begin{pmatrix} 0 & E \\ E & 0 \end{pmatrix},$$

where E a known real parameter. An observable of the system, in the same basis, is represented by the matrix

$$A = \begin{pmatrix} 0 & -i \\ i & 0 \end{pmatrix}.$$

Find the corresponding Heisenberg picture matrix $A_H(t)$.

6.9 A physical system of dimensionality $N = 2$ is characterized by the Hamiltonian $\hat{H}$ and two other physical observables $\hat{A}$ and $\hat{B}$. Using as a basis the eigenvectors of $\hat{A}$ the three observables are represented by the three matrices

$$\hat{\mathcal{H}} = \begin{pmatrix} 0 & E \\ E & 0 \end{pmatrix}, \quad \hat{A} = \begin{pmatrix} \alpha & 0 \\ 0 & -\alpha \end{pmatrix}, \quad \hat{B} = \begin{pmatrix} 0 & -i\beta \\ i\beta & 0 \end{pmatrix},$$

where E, α, and β real parameters. A measurement of the observable $\hat{A}$ is performed at time $t = 0$ yielding the value $-\alpha$. A subsequent measurement of the observable $\hat{B}$ is then performed on the system at a later time $t > 0$. What is the probability to find the system in the eigenstate of $\hat{A}$ corresponding to the eigenvalue $+\alpha$?

6.10 Consider the projection operator to a normalized state vector $|\psi\rangle$

$$\hat{\rho} = |\psi\rangle\langle\psi|.$$

This is the so-called *density matrix*. In the framework of the Schroedinger picture, study the time evolution of $\hat{\rho}$, on the one hand expressing it through the time-evolution operator and, on the other hand, deriving a differential equation of motion for it. As an example, consider a two-state system with Hamiltonian

$$\mathcal{H} = \begin{pmatrix} 0 & -iE \\ iE & 0 \end{pmatrix}$$

and derive the time-evolved density matrix that results from the choice

$$\psi_1(0) = \psi_2(0) = 1/\sqrt{2} \implies \rho(0) = \frac{1}{2} \begin{pmatrix} 1 & 1 \\ 1 & 1 \end{pmatrix}.$$

References

1. E. Merzbacher, *Quantum Mechanics*, 3rd edn. (Wiley, New York, 1998)
2. A. Messiah, *Quantum Mechanics* (Dover Publications, single-volume reprint of the Wiley, New York, two-volume 1958 edition)
3. K. Gottfried, T.-M. Yan *Quantum Mechanics: Fundamentals* (Springer, Berlin, 2004)

Chapter 7
Some More Simple Systems

7.1 The Harmonic Oscillator

Electrons and nuclei in atoms and molecules are in general subject to complicated forces. Nevertheless, since, at least classically, their equilibria correspond to the minima of the potential energy, if we are interested in studying their behavior for small displacements from their equilibrium positions, we may adopt a linear approximation to these forces, or, equivalently, a quadratic approximation to the potential. In the simplest possible case of a particle moving in one dimension, its potential can be Taylor-expanded around its minimum x_0, defined by $V'(x_0) = 0$, $V''(x_0) > 0$, as

$$V(x) \approx V(x_0) + \frac{1}{2}V''(x_0)(x - x_0)^2 + \frac{1}{6}V'''(x_0)(x - x_0)^3 + \cdots . \qquad (7.1)$$

As long as we stay in the vicinity of x_0, the dominant term is the quadratic term.[1] Keeping only the quadratic term amounts to the so-called *harmonic approximation* shown in Fig. 7.1.

A classical particle subject to such a quadratic potential performs oscillations around the equilibrium point, justifying the name *harmonic oscillator*. We can simplify the expression of the potential without any loss of generality taking the potential to vanish at the minimum and taking the coordinate system such that $x_0 = 0$. Then, introducing a parameter ω with the dimensions of frequency, we may set $V''(0) = m\omega^2$. As a result, the potential becomes just $\frac{1}{2}m\omega^2 x^2$ and the Hamiltonian for the particle will be

$$H = \frac{p^2}{2m} + \frac{1}{2}m\omega^2 x^2 . \qquad (7.2)$$

The harmonic oscillator potential is shown in Fig. 7.2.

---

[1]This can be quantified to $|x - x_0| << 3V''(x_0)/|V'''(x_0)|$.

© Springer Nature Switzerland AG 2019
K. Tamvakis, *Basic Quantum Mechanics*, Undergraduate Texts in Physics,
https://doi.org/10.1007/978-3-030-22777-7_7

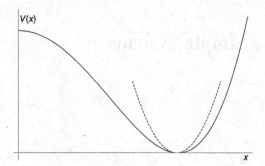

Fig. 7.1 Harmonic approximation to a general potential

7.1.1   Energy Eigenvalues of the Harmonic Oscillator

Consider the Hamilton operator[2]

$$\hat{H} = \frac{\hat{p}^2}{2m} + \frac{1}{2}m\omega^2\hat{x}^2 . \tag{7.3}$$

Notice that the Hamiltonian is the sum of two squares. It is not difficult to see that for any operator, written in terms of two Hermitian operators a and b as $z = a + ib$, the following is true:

$$a^2 + b^2 = z^{\dagger}z + \frac{1}{2}[z, z^{\dagger}] .$$

Applying this observation to our Hamiltonian, we see that, since z is a linear expression of the position and momentum, the commutator will be a constant and, thus, our Hamiltonian will be of the form $z^{\dagger}z$ up to an additive constant. This form has certain

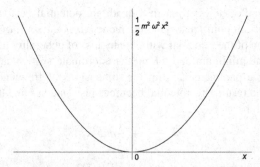

Fig. 7.2 The harmonic oscillator

---

[2]In solving the harmonic oscillator energy eigenvalue problem, we shall follow the algebraic method originally employed by [1]. The alternative differential equation method can be found, e.g., in [2].

advantages in comparison to the initial form, since now the solution of the eigenstate problem will be determined by the properties of a single operator, namely, z.

We therefore introduce the following operators:

$$\hat{a} = \sqrt{\frac{m\omega}{2\hbar}}\,\hat{x} + \frac{i}{\sqrt{2m\hbar\omega}}\,\hat{p}$$

$$\hat{a}^{\dagger} = \sqrt{\frac{m\omega}{2\hbar}}\,\hat{x} - \frac{i}{\sqrt{2m\hbar\omega}}\,\hat{p},$$

(7.4)

They obey the very simple commutation relation

$$\left[\hat{a},\,\hat{a}^{\dagger}\right] = 1.$$

(7.5)

The Hamiltonian is expressed in terms of them as

$$\hat{H} = \hbar\omega\left(\hat{a}^{\dagger}\hat{a} + \frac{1}{2}\right).$$

(7.6)

These operators will be given the name *creation* ($\hat{a}$) and *annihilation* operators ($\hat{a}^{\dagger}$) due to their properties of increasing or decreasing the energy eigenvalues. These properties will be analyzed shortly.

Thus, it is clear by (7.6) that the eigenvalue problem of the Hamiltonian corresponds to the eigenvalue problem of the Hermitian operator

$$\hat{N} = \hat{a}^{\dagger}\hat{a}.$$

(7.7)

Denoting by $|n\rangle$ the corresponding eigenstates, the eigenvalue problem reads

$$\hat{N}|n\rangle = n|n\rangle.$$

(7.8)

The eigenstates $|n\rangle$ are also eigenstates of the Hamiltonian $\hbar\omega\left(\hat{N} + 1/2\right)$, i.e., energy eigenstates. The energy eigenvalues are related to n as

$$E_n = \hbar\omega\left(n + \frac{1}{2}\right),$$

(7.9)

Assuming that the eigenstates $|n\rangle$ are normalized, we obtain from (7.8)

$$\langle n|\hat{N}|n\rangle = n \implies \langle n|\hat{a}^{\dagger}\hat{a}|n\rangle = n$$

or

$$n = \left|\left|\hat{a}|n\rangle\right|\right|^2 \geq 0.$$

(7.10)

Therefore, n *is nonnegative* and $E_n \geq \hbar\omega/2$. We shall be referring to the number n as *occupation number*. The eigenstate of lowest energy (*ground state*) will be the one corresponding to $n = 0$. It corresponds to the eigenvalue $E_0 = \hbar\omega/2$ and will be denoted as $|0\rangle$. Note that the lowest energy state (configuration) of a classical oscillator has vanishing energy, corresponding to a motionless particle sitting at the minimum of the potential. In contrast, the ground state energy of the quantum oscillator is nonzero. This fact is related to the uncertainty principle and will be discussed shortly.

Proceeding to restrict further the energy spectrum and determine the eigenstates, we may consider the commutators of $\hat{a}$ and $\hat{a}^\dagger$ with the Hamiltonian. We get

$$\left[\hat{H}, \hat{a}\right] = -\hbar\omega\,\hat{a}$$

$$\left[\hat{H}, \hat{a}^\dagger\right] = \hbar\omega\,\hat{a}^\dagger,$$
(7.11)

Acting with the first of these commutator relations on a state $|n\rangle$, we obtain

$$\hat{H}\hat{a}|n\rangle - \hat{a}E_n|n\rangle = -\hbar\omega|n\rangle \implies \hat{H}\left(\hat{a}|n\rangle\right) = \hbar\omega\left(n + \frac{1}{2} - 1\right)\left(\hat{a}|n\rangle\right)$$

or

$$\hat{H}\left(\hat{a}|n\rangle\right) = E_{n-1}\left(\hat{a}|n\rangle\right),$$
(7.12)

meaning that the state $\hat{a}|n\rangle$ is also an eigenstate of the energy but with occupation number reduced by one. Assuming that there is no degeneracy, this means that this state must be proportional to the state $|n-1\rangle$, namely,

$$\hat{a}|n\rangle = C_n|n-1\rangle.$$
(7.13)

Taking the norm of this state, we obtain $\langle n|\hat{a}^\dagger\hat{a}|n\rangle = \langle n|\hat{N}|n\rangle = |C_n|^2$ or $C_n = \sqrt{n}$. Thus, we have

$$\hat{a}|n\rangle = \sqrt{n}\,|n-1\rangle.$$
(7.14)

Further action with $\hat{a}$ on the ground state cannot take us to a state of lower energy as we obtain the null state of the Hilbert space

$$\hat{a}|0\rangle = 0.$$
(7.15)

Next, considering the commutation relation of the creation operator $\hat{a}^\dagger$ with the Hamiltonian and acting on a state $|n\rangle$,

$$\left[\hat{H}, \hat{a}^\dagger\right]|n\rangle = \hbar\omega\hat{a}^\dagger|n\rangle,$$

we obtain

$$\hat{H}\left(\hat{a}^\dagger|n\rangle\right) - E_n\left(\hat{a}^\dagger|n\rangle\right) = \hbar\omega\left(\hat{a}^\dagger|n\rangle\right)$$

or

$$\hat{H}\left(\hat{a}^{\dagger}|n\rangle\right) = E_{n+1}\left(\hat{a}^{\dagger}|n\rangle\right),\tag{7.16}$$

meaning that the state $\hat{a}^{\dagger}|n\rangle$ is an eigenstate corresponding to an eigenvalue $E_{n+1} = \hbar\omega(n+1+1/2)$. We may write

$$\hat{a}^{\dagger}|n\rangle = C_n'|n+1\rangle.$$

Taking the norm and recalling that $\hat{a}\hat{a}^{\dagger} = \hat{a}^{\dagger}\hat{a} + 1 = \hat{N} + 1$, we obtain $C_n' = \sqrt{n+1}$ and

$$\hat{a}^{\dagger}|n\rangle = \sqrt{n+1}|n+1\rangle.\tag{7.17}$$

Note that we can start from the ground state and obtain the states

$$\hat{a}^{\dagger}|0\rangle = |1\rangle, \quad \left(\hat{a}^{\dagger}\right)^2|0\rangle = \sqrt{2}|2\rangle, \quad \left(\hat{a}^{\dagger}\right)^3|0\rangle = \sqrt{3!}|3\rangle, \ldots$$

or

$$\frac{\left(\hat{a}^{\dagger}\right)^n}{\sqrt{n!}}|0\rangle = |n\rangle.\tag{7.18}$$

Assume now that we start from a state $|n\rangle$ and act with the annihilation operator m times, m being an integer equal to or just below n (i.e., $n - 1 < m \le n$). If n is not an integer, then $n - m$ would be a number between 0 and 1 and an additional action of the annihilation operator would take us to states labeled by a negative occupation number, which is impossible. Thus, the integer m has to be equal to n in order to end up at the ground state. Therefore, we conclude that the occupation number eigenvalues n are the nonnegative integers $n = 0, 1, 2, \ldots$ and the energy eigenvalues correspond to discrete *"quanta"*.

Summarizing our conclusions, we have obtained the energy eigenvalues

$$E_n = \hbar\omega\left(n + \frac{1}{2}\right) \quad (n = 0, 1, 2, \ldots)\tag{7.19}$$

and the energy eigenstates

$$|0\rangle, |1\rangle, |2\rangle, \ldots, |n\rangle, \ldots\tag{7.20}$$

which can be all obtained from the ground state through (7.18).

As we noted earlier the lowest energy *state* of a classical oscillator has vanishing energy, corresponding to a motionless particle sitting at the minimum of the potential. In contrast, the ground state energy of the quantum oscillator has the nonzero value $\hbar\omega/2$. This fact is related to the uncertainty principle and should have been anticipated according to the following argument. Assume that the equilibrium position of the oscillator has some uncertainty d. The corresponding uncertainty in the momentum will be at least $\hbar/d$. Then, the energy of the oscillator will be

$$E(d) = \frac{\hbar^2}{2md^2} + \frac{1}{2}m\omega^2(d/2)^2$$

Minimizing this expression with respect to d, we obtain

$$E'(d_0) = 0 \implies d_0 = \sqrt{\frac{2\hbar}{m\omega}}.$$

This spatial uncertainty corresponds to a compromise between the kinetic energy and the potential energy of the oscillator. The corresponding minimum value of the energy will be

$$E(d_0) = \frac{\hbar\omega}{2},$$

which is the *"zero-point energy"* found above.

7.1.2   The $\{x\}$ Versus the $\{N\}$ Representation

The energy eigenstates define a discrete basis in Hilbert space in terms of which the various operators and observables are represented by infinite matrices. In particular, the operators $\hat{a}$ and $\hat{a}^\dagger$ are represented by the matrices

$$\langle n|\hat{a}|n'\rangle = \sqrt{n'}\langle n|n'-1\rangle = \sqrt{n'}\,\delta_{n,n'-1}$$

$$\langle n|\hat{a}^\dagger|n'\rangle = \sqrt{n'+1}\langle n|n'+1\rangle = \sqrt{n'+1}\,\delta_{n,n'+1}$$

(7.21)

or

$$a = \begin{pmatrix} 0 & 1 & 0 & \dots & \dots \\ 0 & 0 & \sqrt{2} & \dots & \dots \\ 0 & 0 & 0 & \sqrt{3} & \dots \\ \dots & \dots & \dots & \dots \end{pmatrix}, \quad a^\dagger = \begin{pmatrix} 0 & 0 & 0 & \dots & \dots \\ 1 & 0 & 0 & \dots & \dots \\ 0 & \sqrt{2} & 0 & \dots & \dots \\ 0 & 0 & \sqrt{3} & \dots & \dots \\ \dots & \dots & \dots & \dots & \dots \end{pmatrix}. \quad (7.22)$$

The Hamiltonian matrix can be obtained multiplying these matrices and comes out correctly as a diagonal matrix

$$\mathcal{H}_{nn'} = \hbar\omega \left(\sum_{n''} \sqrt{n''+1}\,\delta_{n,n''+1}\sqrt{n'}\,\delta_{n'',n'-1} + \frac{1}{2}\delta_{nn'} \right)$$

$$= \hbar\omega\left(n + \frac{1}{2}\right)\delta_{nn'} = E_n\delta_{nn'}\,.$$

In order to build the $\{x\}$ representation, we need the inner products $\langle x|n\rangle$ that will take us from one basis to the other. These inner products are the energy *eigenfunctions*

$$\psi_n(x) = \langle x|n\rangle\,. \tag{7.23}$$

We may start from the ground state that has the property

$$\hat{a}|0\rangle = 0 \implies \langle x|\hat{a}|0\rangle = 0$$

or

$$\langle x|\left(\sqrt{\frac{m\omega}{2\hbar}}\,\hat{x} + \frac{i}{\sqrt{2m\hbar\omega}}\,\hat{p}\right)|0\rangle = 0 \implies \sqrt{\frac{m\omega}{2\hbar}}x\langle x|0\rangle + \frac{\hbar}{\sqrt{2m\hbar\omega}}\frac{d}{dx}\langle x|0\rangle = 0 \tag{7.24}$$

or

$$\frac{d\psi_0(x)}{dx} = -x\left(\frac{m\omega}{\hbar}\right)\psi_0(x)\,. \tag{7.25}$$

Integrating, we obtain

$$\psi_0(x) = \psi_0(0)\,e^{-\frac{m\omega}{2\hbar}x^2}\,.$$

The coefficient $\psi_0(0)$ is obtained by normalization to be

$$1 = |\psi_0(0)|^2 \int_{-\infty}^{+\infty} dx\, e^{-\frac{m\omega}{\hbar}x^2} = |\psi_0(0)|^2\sqrt{\frac{\hbar\pi}{m\omega}} \implies \psi_0(0) = \left(\frac{m\omega}{\hbar\pi}\right)^{1/4}\,.$$

Thus, the ground state wave function is

$$\psi_0(x) = \left(\frac{m\omega}{\hbar\pi}\right)^{1/4} e^{-\frac{m\omega}{2\hbar}x^2}\,. \tag{7.26}$$

The excited state eigenfunctions can be readily obtained through

$$\psi_n(x) = \langle x|\frac{(\hat{a}^\dagger)^n}{\sqrt{n!}}|0\rangle \tag{7.27}$$

or

$$\psi_n(x) = \frac{1}{\sqrt{n!}} \left(\sqrt{\frac{m\omega}{2\hbar}} x - \sqrt{\frac{\hbar}{2m\omega}} \frac{d}{dx} \right)^n \psi_0(x). \tag{7.28}$$

Thus, we obtain

$$\psi_1(x) = \left(\frac{m\omega}{\hbar\pi} \right)^{1/4} \sqrt{\frac{2m\omega}{\hbar}} \, x \, e^{-\frac{m\omega}{2\hbar}x^2}$$

$$\psi_2(x) = \frac{1}{\sqrt{2}} \left(\frac{m\omega}{\hbar\pi} \right)^{1/4} \left(\frac{2m\omega}{\hbar} x^2 - 1 \right) e^{-\frac{m\omega}{2\hbar}x^2} \tag{7.29}$$

$$\cdots$$

Note that, as expected, since the Hamiltonian is even and, therefore, commutes with parity, the energy eigenfunctions are automatically parity eigenstates, i.e., even and odd. We have

$$\hat{P}\psi_n(x) = (-1)^n \psi_n(x). \tag{7.30}$$

The wave functions (7.29) are in the form of a polynomial times the exponential $e^{-\frac{m\omega}{2\hbar}x^2}$. In the literature they are usually expressed in terms of the *Hermite Polynomials* $H_n(\xi)$ as

$$\psi_n(x) = \left(\frac{m\omega}{\hbar\pi} \right)^{1/4} \frac{H_n(\xi)}{\sqrt{2^n n!}} e^{-\xi^2/2} \quad \left(\xi \equiv \sqrt{\frac{m\omega}{\hbar}} x \right). \tag{7.31}$$

The Hermite polynomials can be generated from the *Rodrigues formula*

$$H_n(\xi) = (-1)^n e^{\xi^2} \frac{d^n}{d\xi^n} e^{-\xi^2}. \tag{7.32}$$

The first few are

$$H_0(\xi) = 1$$

$$H_1(\xi) = 2\xi$$

$$H_2(\xi) = 4\xi^2 - 2$$

$$H_3(\xi) = 8\xi^3 - 12\xi$$

$$H_4(\xi) = 16\xi^4 - 48\xi^2 + 12$$

$$\tag{7.33}$$

$$\cdots$$

Note also

$$H_{2n+1}(0) = 0, \quad H_{2n}(0) = (-1)^n \frac{(2n)!}{n!}, \tag{7.34}$$

and the very beautiful completeness relation

$$\sum_{n=0}^{\infty} \frac{z^n}{n!} H_n(\xi) H_n(\xi') = \frac{1}{\sqrt{1-4z^2}} e^{-\frac{4z^2}{1-4z^2}(\xi^2+\xi'^2)+\frac{4z}{1-4z^2}\xi\xi'}. \qquad (7.35)$$

A meaningful question that may be asked in the framework of the $\{x\}$ representation is about the spatial and momentum uncertainty of the energy eigenstates $(\Delta x)_n$ and $(\Delta p)_n$. In order to answer it, we must calculate the expectation values $\langle n|\hat{x}|n\rangle$, $\langle n|\hat{x}^2|n\rangle$, $\langle n|\hat{p}|n\rangle$, and $\langle n|\hat{p}^2|n\rangle$. Note, however, that the matrix elements $\langle n|\hat{x}|n\rangle$ and $\langle n|\hat{p}|n\rangle$ vanish because of parity. Indeed we have, thanks to (5.86),

$$\langle n|\hat{x}|n\rangle = -\langle n|\hat{P}\hat{x}\hat{P}|n\rangle = -(-1)^n(-1)^n\langle n|\hat{x}|n\rangle = -\langle n|\hat{x}|n\rangle$$

and, therefore, $\langle \hat{x}\rangle_n = 0$. Similarly for $\langle \hat{p}\rangle_n$.

From the definition of creation/annihilation operators in (7.4), we obtain that

$$\hat{x} = \sqrt{\frac{\hbar}{2m\omega}} \left(\hat{a} + \hat{a}^\dagger\right)$$

$$\hat{p} = -i\sqrt{\frac{m\hbar\omega}{2}} \left(\hat{a} - \hat{a}^\dagger\right). \qquad (7.36)$$

From these, we have

$$\langle n|\hat{x}^2|n\rangle = \frac{\hbar}{2m\omega} \left\| \left(\hat{a} + \hat{a}^\dagger\right)|n\rangle \right\|^2 = \frac{\hbar}{2m\omega} \left\| \sqrt{n}|n-1\rangle + \sqrt{n+1}|n+1\rangle \right\|^2$$

or

$$\langle n|\hat{x}^2|n\rangle = \frac{\hbar}{2m\omega}(2n+1). \qquad (7.37)$$

Similarly we get

$$\langle n|\hat{p}^2|n\rangle = \frac{m\hbar\omega}{2} \left\| \left(\hat{a} - \hat{a}^\dagger\right)|n\rangle \right\|^2 = \frac{m\hbar\omega}{2} \left\| \sqrt{n}|n-1\rangle - \sqrt{n+1}|n+1\rangle \right\|^2$$

or

$$\langle n|\hat{p}^2|n\rangle = \frac{m\hbar\omega}{2}(2n+1). \qquad (7.38)$$

Thus, the corresponding uncertainties are

$$(\Delta x)_n^2 = \frac{\hbar}{2m\omega}(2n+1)$$
$$\implies (\Delta x)_n^2(\Delta p)_n^2 = \frac{\hbar^2}{4}(2n+1)^2. \qquad (7.39)$$
$$(\Delta p)_n^2 = \frac{m\hbar\omega}{2}(2n+1)$$

Note that the ground states saturates the lower limit of Heisenberg's inequality having

$$(\Delta x)_0 (\Delta p)_0 = \hbar/2$$

and being a state of *minimum uncertainty*.

Example 7.1 A particle of mass m and electric charge q is moving in one dimension, being subject to a harmonic force $-m\omega^2 x$ and a constant electric force $q\mathcal{E}$, arising from a homogeneous electric field $\mathcal{E}$. Find the eigenvalues and eigenfunctions of the energy.

The Hamiltonian of the system is

$$\hat{H} = \frac{\hat{p}^2}{2m} + \frac{1}{2}m\omega^2 x^2 - q\mathcal{E}x. \tag{7.40}$$

An equivalent expression for $\hat{H}$ is obtained by combining the last two terms in a square as

$$\hat{H} = \frac{\hat{p}^2}{2m} + \frac{1}{2}m\omega^2 \left(x - \frac{q\mathcal{E}}{m\omega^2} \right)^2 - \frac{q^2\mathcal{E}^2}{2m\omega^2}. \tag{7.41}$$

The time-independent Schroedinger equation is

$$\left\{ -\frac{\hbar^2}{2m}\frac{d^2}{dx^2} + \frac{1}{2}m\omega^2 \left(x - \frac{q\mathcal{E}}{m\omega^2} \right)^2 - \frac{q^2\mathcal{E}^2}{2m\omega^2} \right\} \tilde{\psi}_E(x) = E\,\tilde{\psi}_E(x) \tag{7.42}$$

or

$$\left\{ -\frac{\hbar^2}{2m}\frac{d^2}{dx'^2} + \frac{1}{2}m\omega^2 x'^2 \right\} \tilde{\psi}_E(x) = \left(E + \frac{q^2\mathcal{E}^2}{2m\omega^2} \right) \tilde{\psi}_E(x), \tag{7.43}$$

where

$$x' = x - \frac{q\mathcal{E}}{m\omega^2}.$$

It is clear from the form of the Schroedinger equation for $\tilde{\psi}_E$ that

$$\tilde{\psi}_E(x) = \psi_n(x') = \psi_n(x - q\mathcal{E}/m\omega^2) \quad and \quad E = \hbar\omega\left(n + \frac{1}{2} \right) - \frac{q^2\mathcal{E}^2}{2m\omega^2}, \tag{7.44}$$

where $\psi_n(x)$ are the energy eigenfunctions of the standard harmonic oscillator centered at $x = 0$.

Example 7.2 A simple harmonic oscillator is initially ($t = 0$) in a state with wave function

$$\psi(x, 0) = \frac{1}{\sqrt{2}} \left(1 - \frac{x}{|x|} \right) f(x), \tag{7.45}$$

where $f(x)$ is a real, normalized, odd function

$$f(-x) = -f(x), \qquad \int_{-\infty}^{+\infty} dx\, f^2(x) = 1. \tag{7.46}$$

(a) Is $\psi(x, 0)$ normalized?

(b) What is the probability at $t = 0$ to find the particle anywhere in the region $x \leq 0$? What about the region $x \geq 0$?

(c) Is there some time t_1 at which the particle will be with certainty in the region $x \geq 0$?

(d) Is there some time t_2 at which the particle will be with certainty in the region $x \leq 0$?

(e) Is there some time t_3 at which the probability to find the particle in the $x \geq 0$ region will be equal to the probability to find the particle in the $x \leq 0$ region?

(a) Obviously, from $\psi(x, 0) = \begin{cases} (x < 0) & \sqrt{2}\, f(x) \\ \\ (x > 0) & 0 \end{cases}$ we get

$$\int_{-\infty}^{+\infty} dx\, |\psi(x, 0)|^2 = 2 \int_{-\infty}^{0} dx\, f^2(x) = 2 \frac{1}{2} \int_{-\infty}^{+\infty} dx\, f^2(x) = 1. \tag{7.47}$$

(b) The probability to find the particle initially ($t = 0$) in the region $x \leq 0$ is 1 (certainty), while the probability to find it anywhere in the $x \geq 0$ region vanishes.

(c) The wave function at a later time $t > 0$ is

$$\psi(x, t) = e^{-i\omega t/2} \sum_{n=0}^{\infty} C_n e^{-in\omega t}\, \psi_n(x), \tag{7.48}$$

where (Note that the harmonic oscillator energy eigenfunctions are real)

$$C_n = \int_{-\infty}^{+\infty} dx\, \psi_n(x)\, \psi(x, 0) = \sqrt{2} \int_{-\infty}^{0} dx\, \psi_n(x)\, f(x) = C_n^*. \tag{7.49}$$

The corresponding probability density is

$$|\psi(x, t)|^2 = \left| \sum_{n=0}^{\infty} C_n e^{-in\omega t}\, \psi_n(x) \right|^2. \tag{7.50}$$

Notice that for $t_1 = \pi/\omega$ we have $\exp[-in\omega t_1] = \exp[-i\pi n] = (-1)^n$. Thus, we have

$$|\psi(x, t_1)|^2 = \left| \sum_{n=0}^{\infty} C_n (-1)^n \psi_n(x) \right|^2 = \left| \sum_{n=0}^{\infty} C_n \psi_n(-x) \right|^2 = |\psi(-x, 0)|^2, \tag{7.51}$$

making use of the fact that $\psi_n(-x) = (-1)^n \psi_n(x)$. Since

$$\psi(-x, 0) = \begin{cases} (-x < 0) & \sqrt{2}\, f(-x) \\ \\ (-x > 0) & 0 \end{cases} = \begin{cases} (x > 0) & -\sqrt{2}\, f(x) \\ \\ (x < 0) & 0 \end{cases}$$

at time $t_1 = \pi/\omega$ we have a particle being exclusively in the $x \geq 0$ region.

(d) Notice that for $t_2 = 2\pi/\omega$ we have $\exp[-in\omega t_2] = \exp[-i2\pi n] = 1$. Thus, we have

$$|\psi(x, t_1)|^2 = \left| \sum_{n=0}^{\infty} C_n \psi_n(x) \right|^2 = |\psi(x, 0)|^2, \tag{7.52}$$

which describes a particle exclusively in the $x \leq 0$ region.

(e) At the time $t_3 = \pi/2\omega$ we have

$$|\psi(x, t_3)|^2 = \left| \sum_{n=0}^{\infty} C_n (-i)^n \psi_n(x) \right|^2 \tag{7.53}$$

and

$$|\psi(-x, t_3)|^2 = \left| \sum_{n=0}^{\infty} C_n (-i)^n (-1)^n \psi_n(x) \right|^2 = \left| \sum_{n=0}^{\infty} C_n i^n \psi_n(x) \right|^2 = |\psi^*(x, t_3)|^2 \tag{7.54}$$

or

$$|\psi(-x, t_3)|^2 = |\psi(x, t_3)|^2 . \tag{7.55}$$

Thus, at $t = t_3 = \pi/2\omega$ the probability density is even.

7.1.3 Coherent States

As we saw the ground state of the harmonic oscillator is a state of minimal spreading for which the position/momentum Heisenberg relation becomes an equality. It turns out that we can define suitable superpositions of the energy eigenstates that share this property. Actually, these states are characterized by quite a few remarkable properties. They are called *coherent states*.[3] They may be introduced as

$$|z\rangle = e^{-|z|^2/2} \sum_{n=0}^{\infty} \frac{z^n}{\sqrt{n!}} |n\rangle , \tag{7.56}$$

in terms of the complex parameter z. Note that the exponential factor in front has been chosen so that $\langle z|z \rangle = 1$. An equivalent closed expression can be obtained if we make use of the fact that $|n\rangle = \frac{(\hat{a}^\dagger)^n}{\sqrt{n!}} |0\rangle$. Then, (7.56) becomes

$$e^{-|z|^2/2} \sum_{n=0}^{\infty} \frac{(z\hat{a}^\dagger)^n}{n!} |0\rangle = e^{-|z|^2/2} e^{z\hat{a}^\dagger} |0\rangle$$

or

$$|z\rangle = e^{-|z|^2/2} e^{z\hat{a}^\dagger} |0\rangle . \tag{7.57}$$

The states (7.56) have been constructed so that they are eigenstates of the annihilation operator $\hat{a}$, namely,

$$\hat{a}|z\rangle = z|z\rangle . \tag{7.58}$$

This is shown as follows:

---

[3]More on the subject of coherent states in various quantum systems can be found in [3].

$$\hat{a}|z\rangle = e^{-|z|^2/2} \sum_{n=0}^{\infty} \frac{z^n}{\sqrt{n!}} \sqrt{n}|n-1\rangle = e^{-|z|^2/2} \sum_{n'=n-1=0}^{\infty} \frac{z^{n'+1}}{\sqrt{n'!}}|n'\rangle = z|z\rangle .$$

As we mentioned in the beginning of this subsection, the property that makes coherent states interesting is the fact that they exhibit minimal position and momentum uncertainties. Furthermore, what makes this property even more interesting is that this happens at all times. Therefore, in order to compute the time-evolved uncertainties in these states, we consider the relevant Heisenberg operators. From the Heisenberg equation for $\hat{a}(t)$, we obtain

$$\frac{d\hat{a}(t)}{dt} = \frac{i}{\hbar}\left[\hat{H}(t), \hat{a}(t)\right] = i\omega\left[\hat{a}^\dagger(t)\hat{a}(t), \hat{a}(t)\right] = -i\omega\,\hat{a}(t)$$

which, integrated, gives

$$\hat{a}(t) = \hat{a}(0)\,e^{-i\omega t}, \quad \hat{a}^\dagger(t) = \hat{a}^\dagger(0)\,e^{i\omega t} . \tag{7.59}$$

Then, from (7.36) we obtain

$$\hat{x}(t) = \sqrt{\frac{\hbar}{2m\omega}}\left(\hat{a}(0)e^{-i\omega t} + \hat{a}^\dagger(0)e^{i\omega t}\right)$$
$$\hat{p}(t) = -i\sqrt{\frac{m\hbar\omega}{2}}\left(\hat{a}(0)e^{-i\omega t} - \hat{a}^\dagger(0)e^{i\omega t}\right), \tag{7.60}$$

where $\hat{a}(0)$, $\hat{a}^\dagger(0)$ are equal to the Schroedinger operators with the previously studied lowering and raising properties. Then, thanks to (7.58), we get

$$\langle z|\hat{x}(t)|z\rangle = \sqrt{\frac{\hbar}{2m\omega}}\left(z\,e^{-i\omega t} + z^*\,e^{i\omega t}\right)$$
$$\langle z|\hat{p}(t)|z\rangle = -i\sqrt{\frac{m\hbar\omega}{2}}\left(z\,e^{-i\omega t} - z^*\,e^{i\omega t}\right) \tag{7.61}$$

and

$$\langle z|\hat{x}^2(t)|z\rangle = \frac{\hbar}{2m\omega}\left(1 + 2|z|^2 + z^2 e^{-2i\omega t} + z^{*2}e^{2i\omega t}\right)$$
$$\langle z|\hat{p}^2(t)|z\rangle = \frac{m\hbar\omega}{2}\left(1 - 2|z|^2 - z^2 e^{-2i\omega t} - z^{*2}e^{2i\omega t}\right) \tag{7.62}$$

giving finally

$$\begin{matrix}(\Delta x)_t^2 = \frac{\hbar}{2m\omega} \\ (\Delta p)_t^2 = \frac{\hbar m\omega}{2}\end{matrix} \quad \Longrightarrow \quad (\Delta x)_t^2(\Delta p)_t^2 = \frac{\hbar^2}{4} . \tag{7.63}$$

A valid question that could be asked is *what is the physical meaning of the parameter* z? This could be clarified if we calculate the average number of quanta in a coherent state. This can be seen in the relation

$$\langle N \rangle = \langle z|\hat{a}^\dagger \hat{a}|z \rangle = |z|^2 . \tag{7.64}$$

Next, writing down the uncertainty in the number of quanta

$$(\Delta N)^2 = \langle z|N^2|z \rangle - \langle z|N|z \rangle^2 = \langle z|\hat{a}^\dagger \hat{a}\hat{a}^\dagger z|z \rangle - |z|^4 =$$

$$z \langle z|\hat{a}^\dagger \left(1 + \hat{a}^\dagger \hat{a}\right)|z \rangle - |z|^4 = |z|^4 + |z|^2 - |z|^4$$

or

$$(\Delta N)^2 = |z|^2 . \tag{7.65}$$

Thus, z is related to the uncertainty in the number of quanta. We may further conclude that

$$(\Delta N) = \sqrt{\langle N \rangle} . \tag{7.66}$$

Resetting this formula in the form

$$\frac{(\Delta N)}{\langle N \rangle} = \frac{1}{\sqrt{\langle N \rangle}} , \tag{7.67}$$

we may conclude that *as the number of quanta increases, their relative uncertainty decreases*. In the limit $\langle N \rangle \to \infty$ we get $(\Delta N)/\langle N \rangle \to 0$.

Example 7.3 Prove that the coherent states $|z\rangle \equiv e^{-|z|^2/2} e^{z\hat{a}^\dagger}|0\rangle$ satisfy the "*completeness relation*"

$$\frac{1}{\pi} \int dz \int d\bar{z} \, |z\rangle\langle z| = 1 . \tag{7.68}$$

Using the polar description for the complex number $z = \rho e^{i\phi}$, we have

$$\frac{1}{\pi} \int_0^\infty d\rho \, \rho \int_0^{2\pi} d\phi \, e^{-\rho^2} \sum_{n,n'=0}^\infty \frac{\rho^{n+n'}}{\sqrt{n!n'!}} e^{i(n-n')\phi} |n\rangle\langle n'| = \frac{1}{\pi} \int_0^\infty d\rho \, \rho \, e^{-\rho^2} \sum_{n=0}^\infty \frac{\rho^{2n}}{n!} |n\rangle\langle n|$$

where we used the fact $\int_0^{2\pi} d\phi \, e^{i(n-n')\phi} = 2\pi \, \delta_{nn'}$. Doing the integral $\int_0^\infty d\rho \, \rho \, e^{-\rho^2} \rho^{2n} = \frac{1}{2} \int_\infty d\xi \, \xi^n \, e^{-\xi} = \frac{n!}{2}$, we obtain

$$\frac{1}{\pi} \sum_{n=0}^\infty (2\pi) \frac{n!}{2} \frac{1}{n!} |n\rangle\langle n| = \sum_{n=0}^\infty |n\rangle\langle n| = 1 .$$

7.2 The Ammonia Molecule

The *ammonia molecule* is a characteristic example of a system with an approximate two-dimensional—and therefore quite simple—Hilbert space of states. The ammonia molecule, consisting of one nitrogen and three hydrogen atoms, has many states but, in a given state of rotation, there are two possible configurations of the molecule, corresponding to different positions of the nitrogen atom with respect to the plane of the hydrogen atoms, that amount to two different states of the molecule (Fig. 7.3.).

The characteristic energy of states corresponding to electron excitations within the atoms of the molecule is in the ultraviolet or optical range, while the characteristic energies of vibrations of the molecule are in the infrared and, finally, those of rotational excitations in the far infrared. In contrast, the energy difference between the configurational states of the molecule mentioned above is much, much less than those, namely, in the microwave region. Thus, when we investigate phenomena in that energy region, the ammonia molecule behaves effectively as a *two-state system*.

Let's denote the two different states of the system by the two orthogonal kets $|1\rangle$, $|2\rangle$, taken also to be normalized as well. Thus, we have

$$\langle i|j\rangle = \delta_{ij} \quad (i, j = 1, 2). \tag{7.69}$$

The state of the system at any time t can be expanded in this basis as

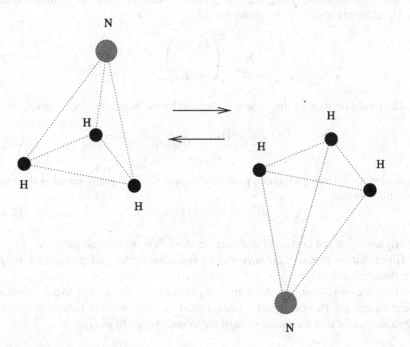

Fig. 7.3 The ammonia molecule

$$|\psi(t)\rangle = \psi_1(t)|1\rangle + \psi_2(t)|2\rangle \quad\Longrightarrow\quad \begin{pmatrix} \psi_1(t) \\ \psi_2(t) \end{pmatrix}. \qquad (7.70)$$

Normalization corresponds to

$$\langle\psi(t)|\psi(t)\rangle = 1 \quad\Longrightarrow\quad |\psi_1(t)|^2 + |\psi_2(t)|^2 = 1. \qquad (7.71)$$

Note that this holds at all times. The Hamiltonian of the system with respect to the above basis of states will correspond to a matrix of the form

$$\mathcal{H}_{ij} = \langle i|\hat{H}|j\rangle \quad\Longrightarrow\quad \mathcal{H} = \begin{pmatrix} E_{11} & E_{12} \\ E_{12}^* & E_{22} \end{pmatrix}, \qquad (7.72)$$

in terms of the real numbers E_{11}, E_{22} and the complex number E_{12}. This is the most general form dictated by Hermiticity. For the ammonia molecule under study, the measured values of these parameters are such that it suffices to take

$$E_{11} \approx E_{22} = E_0 \quad and \quad E_{12} = E_{12}^* = A \ll E_0,$$

although in order to keep this analysis more general we shall not need to make use of the last approximation. Thus, we start with the phenomenological assumption that the Hamiltonian matrix in the above basis is

$$\mathcal{H} = \begin{pmatrix} E_0 & A \\ A & E_0 \end{pmatrix}. \qquad (7.73)$$

The time evolution of the system is dictated by the Schroedinger equation

$$i\hbar\frac{d|\psi(t)\rangle}{dt} = \hat{H}|\psi(t)\rangle. \qquad (7.74)$$

Equivalent to the Schroedinger equation is its operator solution in terms of the time-evolution operator

$$|\psi(t)\rangle = \hat{U}(t)|\psi(0)\rangle. \qquad (7.75)$$

The dynamics of the system is embodied in the following central question:

Given that the system is initially ($t = 0$) in a state $|\psi(0)\rangle$, what is its state $|\psi(t)\rangle$ in a future time $t > 0$?

There are more than one alternative ways to arrive at the answer to this question. We shall analyze two of them for pedagogical reasons. Without loss of generality, we shall assume that the initial state of the system is $|\psi(0)\rangle = |1\rangle$.

(1) Solution through the Schroedinger equation. One rather straightforward approach to the problem is to transform the Schroedinger equation into a matrix equation in the given basis and solve the resulting system of differential equations. This goes as follows:

$$i\hbar \langle i| \frac{d}{dt} |\psi(t)\rangle = \langle i|\hat{H}|\psi(t)\rangle \implies$$

$$i\hbar \langle i| \frac{d}{dt} \left(\sum_j |j\rangle\langle j| \right) |\psi(t)\rangle = \langle i|\hat{H} \left(\sum_j |j\rangle\langle j| \right) |\psi(t)\rangle$$

or

$$i\hbar \frac{d\psi_i(t)}{dt} = \sum_j \mathcal{H}_{ij} \psi_j(t), \tag{7.76}$$

where $\psi_i(t) = \langle i|\psi(t)\rangle$. Substituting the explicit form of $\mathcal{H}$, this becomes

$$\begin{pmatrix} \frac{d\psi_1(t)}{dt} \\ \frac{d\psi_2(t)}{dt} \end{pmatrix} = -\frac{i}{\hbar} \begin{pmatrix} E_0 & A \\ A & E_0 \end{pmatrix} \begin{pmatrix} \psi_1(t) \\ \psi_2(t) \end{pmatrix}. \tag{7.77}$$

This is equivalent to the system of equations

$$\frac{d\psi_1}{dt} = -\frac{i}{\hbar} E_0 \psi_1 - \frac{i}{\hbar} A \psi_2$$
$$\frac{d\psi_2}{dt} = -\frac{i}{\hbar} E_0 \psi_2 - \frac{i}{\hbar} A \psi_1. \tag{7.78}$$

Adding and subtracting the above, we obtain the equivalent pair

$$\dot{\psi}_1 + \dot{\psi}_2 = -\frac{i}{\hbar}(E_0 + A)(\psi_1 + \psi_2)$$
$$\dot{\psi}_1 - \dot{\psi}_2 = -\frac{i}{\hbar}(E_0 - A)(\psi_1 - \psi_2)$$

or, introducing the alternative components[4]

$$\psi_\pm = \frac{1}{\sqrt{2}}(\psi_1 \pm \psi_2) \qquad \psi_{1,2} = \frac{1}{\sqrt{2}}(\psi_+ \pm \psi_-), \tag{7.79}$$

we obtain two decoupled equations

---

[4]The $1/\sqrt{2}$ factor guarantee that the $\psi_\pm$ obey the same normalization condition

$$|\psi_+(t)|^2 + |\psi_-(t)|^2 = 1$$

as the $\psi_{1,2}$.

$$\dot{\psi}_{\pm} = -\frac{i}{\hbar}(E_0 \pm A)\psi_{\pm} \tag{7.80}$$

that can be readily integrated to give

$$\psi_{\pm}(t) = \psi_{\pm}(0)\,e^{-\frac{i}{\hbar}(E_0 \pm A)t}. \tag{7.81}$$

Using the inverse relations giving our original components, we have

$$\psi_1(t) = \frac{1}{\sqrt{2}}\psi_+(0)e^{-\frac{i}{\hbar}(E_0+A)t} + \frac{1}{\sqrt{2}}\psi_-(0)e^{-\frac{i}{\hbar}(E_0-A)t}$$

$$\psi_2(t) = \frac{1}{\sqrt{2}}\psi_+(0)e^{-\frac{i}{\hbar}(E_0+A)t} - \frac{1}{\sqrt{2}}\psi_-(0)e^{-\frac{i}{\hbar}(E_0-A)t}. \tag{7.82}$$

Our initial state corresponds to

$$\psi_1(0) = \langle 1|\psi(0)\rangle = \langle 1|1\rangle = 1, \quad \psi_2(0) = \langle 2|\psi(0)\rangle = \langle 2|1\rangle = 0.$$

Therefore,

$$\psi_+(0) = \psi_-(0) = \frac{1}{\sqrt{2}}.$$

Thus, finally the evolved state is

$$\begin{pmatrix} \psi_1(t) \\ \psi_2(t) \end{pmatrix} = e^{-\frac{i}{\hbar}E_0 t} \begin{pmatrix} \cos(At/\hbar) \\ -i\,\sin(At/\hbar) \end{pmatrix} \tag{7.83}$$

or in ket language

$$|\psi(t)\rangle = e^{-\frac{i}{\hbar}E_0 t}[\,\cos(At/\hbar)|1\rangle - i\,\sin(At/\hbar)|2\rangle\,]. \tag{7.84}$$

This is the state vector of the system at any time t. We may answer a number of possible questions in terms of it. First of all, we see that the probability for the molecule to occupy any particular state $|\chi\rangle$ is *periodic in time*. Indeed, we have

$$P_\chi(t) = |\langle \chi|\psi(t)\rangle|^2 = |\chi_1^*\cos(At/\hbar) - i\chi_2^*\sin(At/\hbar)|^2.$$

Since the system is periodic, there is nonzero probability that at a certain time it will return to the initial state. The corresponding probability is

$$P_1(t) = |\langle 1|\psi(t)\rangle|^2 = \cos^2(At/\hbar).$$

This means that at the times

$$t_n = \frac{n\hbar\pi}{A} \quad \text{with} \quad n = 1, 2, \ldots$$

the system will return to its initial state.

We may also ask whether the system ever goes to the orthogonal state $|2\rangle$. Indeed, since

$$P_2(t) = |\langle 2|\psi(t)\rangle|^2 = \sin^2(At/\hbar),$$

this happens at the times

$$t_n^* = (2n+1)\frac{\pi\hbar}{2A}.$$

(2) Solution through the energy eigenvalue problem. An alternative way to determine the evolution of the system is to first solve the energy eigenvalue problem and then employ the general solution of the Schroedinger equation $|\psi(t)\rangle$ expanded in energy eigenfunctions (stationary states).

The energy eigenvalue problem consists of determining the eigenvectors $|E\rangle$ and eigenvalues E satisfying

$$\hat{H}|E\rangle = E|E\rangle. \tag{7.85}$$

Again we transform this operator relation into a matrix equation projecting on the given basis

$$\langle i|\hat{H}\left(\sum_j |j\rangle\langle j|\right)|E\rangle = E\langle i|E\rangle$$

or

$$\sum_j \mathcal{H}_{ij} C_{Ej} = E C_{Ei}, \tag{7.86}$$

where we denoted

$$C_{Ej} \equiv \langle j|E\rangle. \tag{7.87}$$

For our specific molecular Hamiltonian, this is

$$\begin{pmatrix} E_0 & A \\ A & E_0 \end{pmatrix} \begin{pmatrix} C_{E1} \\ C_{E2} \end{pmatrix} = E \begin{pmatrix} C_{E1} \\ C_{E2} \end{pmatrix}. \tag{7.88}$$

This matrix equation is really a homogeneous system of linear equations

$$(E_0 - E)C_{E1} + AC_{E2} = 0$$

$$AC_{E1} + (E_0 - E)C_{E2} = 0. \tag{7.89}$$

As is known from the theory of linear systems of equations, in order for this system to possess a nontrivial solution, its determinant of coefficients should be zero, namely,

$$\begin{vmatrix} E_0 - E & A \\ A & E_0 - E \end{vmatrix} = 0 . \tag{7.90}$$

This amounts to an equation determining the eigenvalues[5]

$$(E_0 - E)^2 - A^2 = 0 \implies E_\pm = E_0 \pm A . \tag{7.91}$$

For each of these eigenvalues, we have a corresponding eigenvector represented by the column $\begin{pmatrix} C_1^{(\pm)} \\ C_2^{(\pm)} \end{pmatrix}$. The eigenvectors are determined by inserting into (7.88) the values of the energy eigenvalues. We have for the "+" eigenvalue

$$\begin{pmatrix} E_0 & A \\ A & E_0 \end{pmatrix} \begin{pmatrix} C_1^{(+)} \\ C_2^{(+)} \end{pmatrix} = (E_0 + A) \begin{pmatrix} C_1^{(+)} \\ C_2^{(+)} \end{pmatrix} \implies C_2^{(+)} = C_1^{(+)} .$$

Similarly for the "−" eigenvalue

$$\begin{pmatrix} E_0 & A \\ A & E_0 \end{pmatrix} \begin{pmatrix} C_1^{(-)} \\ C_2^{(-)} \end{pmatrix} = (E_0 - A) \begin{pmatrix} C_1^{(-)} \\ C_2^{(-)} \end{pmatrix} \implies C_2^{(-)} = -C_1^{(-)} .$$

Note that, since the system of equations is homogeneous, one of the coefficients will be undetermined and will be fixed by normalizing the eigenvectors. Summarizing, we obtain

$$E_+ = E_0 + A \implies |E_+\rangle \implies \frac{1}{\sqrt{2}} \begin{pmatrix} 1 \\ 1 \end{pmatrix}$$

$$\tag{7.92}$$

$$E_- = E_0 - A \implies |E_-\rangle \implies \frac{1}{\sqrt{2}} \begin{pmatrix} 1 \\ -1 \end{pmatrix}$$

We have normalized the eigenvectors taking $C_1^{(\pm)} = 1/\sqrt{2}$. The eigenvectors can also be written in ket language as

$$|E_\pm\rangle = \frac{1}{\sqrt{2}} (|1\rangle \pm |2\rangle) , \tag{7.93}$$

---

[5]The equation determining the eigenvalues is $\det (\mathcal{H} - E \mathbf{I}) = 0 $.

the inverse relations being

$$|1\rangle = \frac{1}{\sqrt{2}}(|E_+\rangle + |E_-\rangle), \quad |2\rangle = \frac{1}{\sqrt{2}}(|E_+\rangle - |E_-\rangle). \qquad (7.94)$$

Note that, as eigenvectors of a Hermitian operator, they are automatically orthogonal

$$\langle E_+|E_-\rangle = \frac{1}{2}(1, 1)\begin{pmatrix} 1 \\ -1 \end{pmatrix} = 0.$$

Having determined the energy eigenvectors and eigenvalues, we are in a position to use them as an alternative basis. The evolved state of the system, written in terms of the evolution operator, is

$$|\psi(t)\rangle = \hat{U}(t)|\psi(0)\rangle = e^{-\frac{i}{\hbar}\hat{H}t}|\psi(0)\rangle$$

$$= e^{-\frac{i}{\hbar}\hat{H}t}\left(\sum_E |E\rangle\langle E|\right)|\psi(0)\rangle = \sum_E \langle E|\psi(0)\rangle e^{-\frac{i}{\hbar}Et}|E\rangle \qquad (7.95)$$

and using the fact that $|\psi(0)\rangle = |1\rangle$,

$$|\psi(t)\rangle = C_1^{(+)} e^{-\frac{i}{\hbar}(E_0+A)t}|E_+\rangle + C_1^{(-)} e^{-\frac{i}{\hbar}(E_0-A)t}|E_-\rangle =$$

$$\frac{1}{\sqrt{2}}e^{-\frac{i}{\hbar}E_0 t}\left(e^{-\frac{i}{\hbar}At}\frac{1}{\sqrt{2}}(|1\rangle + |2\rangle) + e^{\frac{i}{\hbar}At}\frac{1}{\sqrt{2}}(|1\rangle - |2\rangle)\right)$$

or

$$|\psi(t)\rangle = e^{-\frac{i}{\hbar}E_0 t}[\cos(At/\hbar)|1\rangle - i\sin(At/\hbar)|2\rangle]. \qquad (7.96)$$

Example 7.4 Consider a two-state system described by its Hamiltonian $\hat{H}$ and another physical observable $\hat{P}$ (parity). In a given orthonormal basis, these observables take the form

$$\mathcal{H} = \begin{pmatrix} E_0 & A \\ A & E_0 \end{pmatrix}, \quad \mathcal{P} = \begin{pmatrix} 1 & 0 \\ 0 & -1 \end{pmatrix}. \qquad (7.97)$$

A measurement of the parity of the system is performed at $t = 0$ and the parity is determined to be $+1$. Subsequently, at time $t > 0$, a measurement of the energy is performed. What are the possible outcomes and what is the probability of each?

Since parity comes out as a diagonal matrix in the given basis $|1\rangle$, $|2\rangle$, these vectors are the eigenvectors of parity, $|1\rangle$ being the eigenstate of parity $+1$ and $|2\rangle$ being the eigenstate of parity -1. The measurement of parity at $t = 0$ prepares the system in the initial state $|\psi(0)\rangle = |1\rangle$. Using the calculation above, we know that the evolved state at time $t > 0$ will be

$$|\psi(t)\rangle = e^{-\frac{i}{\hbar}E_0 t}[\cos(At/\hbar)|1\rangle - i\sin(At/\hbar)|2\rangle]. \qquad (7.98)$$

Introducing back the energy eigenstates, this is written as

$$|\psi(t)\rangle = \frac{1}{\sqrt{2}} \left[e^{-\frac{i}{\hbar}E_+t}|E_+\rangle + e^{-\frac{i}{\hbar}E_-t}|E_-\rangle \right] . \tag{7.99}$$

Thus, the probability of each energy eigenvalue is 50%.

Example 7.5 For the system of previous example (7.4.) construct the evolution operator $\hat{U}(t)$ and determine the Heisenberg operator for the parity $\hat{P}(t)$.

We have

$$\hat{U}(t) = e^{-\frac{i}{\hbar}t\hat{H}} = e^{-\frac{i}{\hbar}t\hat{H}} \sum_a |E_a\rangle\langle E_a| = e^{-\frac{i}{\hbar}E_+t}|E_+\rangle\langle E_+| + e^{-\frac{i}{\hbar}E_-t}|E_-\rangle\langle E_-| \tag{7.100}$$

or

$$\mathcal{U}(t) = \frac{1}{2}e^{-\frac{i}{\hbar}E_+t} \begin{pmatrix} 1 \\ 1 \end{pmatrix}(1,\ 1) + e^{-\frac{i}{\hbar}E_-t} \begin{pmatrix} 1 \\ -1 \end{pmatrix}(1,\ -1)$$

$$= \frac{1}{2}e^{-\frac{i}{\hbar}E_0t} \begin{pmatrix} \cos(At/\hbar) & -i\sin(At/\hbar) \\ -i\sin(At/\hbar) & \cos(At/\hbar) \end{pmatrix} . \tag{7.101}$$

The parity operator is

$$\hat{P}(t) = \hat{U}^\dagger(t)\,\hat{P}(0)\,\hat{U}(t) \tag{7.102}$$

or

$$\mathcal{P}(t) = \begin{pmatrix} \cos(2At/\hbar) & -i\sin(2At/\hbar) \\ i\sin(2At/\hbar) & -\cos(2At/\hbar) \end{pmatrix} . \tag{7.103}$$

Example 7.6 Consider a three-state system described by a Hamiltonian H. Let $|E_1\rangle$, $|E_2\rangle$, $|E_3\rangle$ be three eigenstates of the energy corresponding to the eigenvalues E_1, E_2, E_3. An observable A of the system has the following properties:

$$\hat{A}|E_1\rangle = \alpha\,(|E_1\rangle + |E_3\rangle), \quad \hat{A}|E_2\rangle = \alpha|E_2\rangle$$

$$\hat{A}|E_3\rangle = \alpha\,(|E_1\rangle + |E_3\rangle) . \tag{7.104}$$

Assume that initially ($t = 0$) the system occupies the state

$$|\psi(0)\rangle = \frac{1}{\sqrt{3}}\,(|E_1\rangle + |E_2\rangle + |E_3\rangle) . \tag{7.105}$$

(a) Calculate the expectation value $\langle E \rangle$ and the uncertainty (ΔE) of the energy.

(b) Calculate the expectation value $\langle A \rangle$ and the uncertainty (ΔA) at any time $t > 0$.

(c) Verify the inequality

$$(\Delta E)^2(\Delta A)_t^2 \geq \frac{1}{4}\left|\langle\left[\hat{H},\,\hat{A}\right]\rangle_t\right|^2 \tag{7.106}$$

in the case $E_1 = E_2$.

(a) Note that all expectation values of the Hamiltonian or its powers are time-independent. Thus, we have

$$\langle \mathbf{H} \rangle = \frac{1}{3} (1, 1, 1) \begin{pmatrix} E_1 & 0 & 0 \\ 0 & E_2 & 0 \\ 0 & 0 & E_3 \end{pmatrix} \begin{pmatrix} 1 \\ 1 \\ 1 \end{pmatrix} = \frac{1}{3}(E_1 + E_2 + E_3) \tag{7.107}$$

and

$$(\Delta E)^2 = \langle \mathbf{H}^2 \rangle - (\langle \mathbf{H} \rangle)^2 = \frac{1}{3}\left(E_1^2 + E_2^2 + E_3^2 \right) - \frac{1}{9}(E_1 + E_2 + E_3)^2$$

$$= \frac{1}{9}\left((E_1 - E_2)^2 + (E_1 - E_3)^2 + (E_2 - E_3)^2 \right). \tag{7.108}$$

(b) From the above relations satisfied by $\hat{A}$, we conclude that

$$A = \begin{pmatrix} \alpha & 0 & \alpha \\ 0 & \alpha & 0 \\ \alpha & 0 & \alpha \end{pmatrix}, \quad A^2 = \begin{pmatrix} 2\alpha^2 & 0 & 2\alpha^2 \\ 0 & \alpha^2 & 0 \\ 2\alpha^2 & 0 & 2\alpha^2 \end{pmatrix} \tag{7.109}$$

and

$$\langle \hat{A} \rangle_t = \frac{1}{3}\left(e^{\frac{i}{\hbar}E_1 t}, e^{\frac{i}{\hbar}E_2 t}, e^{\frac{i}{\hbar}E_3 t} \right) \begin{pmatrix} \alpha & 0 & \alpha \\ 0 & \alpha & 0 \\ \alpha & 0 & \alpha \end{pmatrix} \begin{pmatrix} e^{-\frac{i}{\hbar}E_1 t} \\ e^{-\frac{i}{\hbar}E_2 t} \\ e^{-\frac{i}{\hbar}E_3 t} \end{pmatrix}$$

$$= \frac{\alpha}{3}\left(3 + 2\cos((E_1 - E_3)t/\hbar) \right). \tag{7.110}$$

Similarly

$$\langle \hat{A}_t^2 \rangle = \frac{1}{3}\left(e^{\frac{i}{\hbar}E_1 t}, e^{\frac{i}{\hbar}E_2 t}, e^{\frac{i}{\hbar}E_3 t} \right) \begin{pmatrix} 2\alpha^2 & 0 & 2\alpha^2 \\ 0 & \alpha^2 & 0 \\ 2\alpha^2 & 0 & 2\alpha^2 \end{pmatrix} \begin{pmatrix} e^{-\frac{i}{\hbar}E_1 t} \\ e^{-\frac{i}{\hbar}E_2 t} \\ e^{-\frac{i}{\hbar}E_3 t} \end{pmatrix}$$

$$= \frac{\alpha^2}{3}\left(5 + 4\cos((E_1 - E_3)t/\hbar) \right). \tag{7.111}$$

Therefore,

$$(\Delta A)_t^2 = \frac{\alpha^2}{9}\left(6 - 4\cos^2((E_1 - E_3)t/\hbar) \right). \tag{7.112}$$

(c) We have

$$[\mathcal{H}, \mathcal{A}] = \alpha(E_1 - E_3) \begin{pmatrix} 0 & 0 & 1 \\ 0 & 0 & 0 \\ -1 & 0 & 0 \end{pmatrix}$$

and

$$\langle [\mathcal{H}, \mathcal{A}] \rangle = \frac{1}{3}\alpha(E_1 - E_3) \left(e^{\frac{i}{\hbar}E_1 t}, e^{\frac{i}{\hbar}E_2 t}, e^{\frac{i}{\hbar}E_3 t} \right) \begin{pmatrix} 0 & 0 & 1 \\ 0 & 0 & 0 \\ -1 & 0 & 0 \end{pmatrix} \begin{pmatrix} e^{-\frac{i}{\hbar}E_1 t} \\ e^{-\frac{i}{\hbar}E_2 t} \\ e^{-\frac{i}{\hbar}E_3 t} \end{pmatrix}$$

$$= \frac{2i\alpha}{3}(E_1 - E_3)\sin((E_1 - E_3)t/\hbar).$$

Therefore $(\Delta E)^2(\Delta A)_t^2 \geq \frac{1}{4}|\langle [A, H] \rangle_t|^2$ amounts to

$$\frac{1}{9}\left((E_1 - E_2)^2 + (E_1 - E_3)^2 + (E_2 - E_3)^2\right)\left(6 - 4\cos^2((E_1 - E_3)t/\hbar)\right)$$

$$\geq (E_1 - E_3)^2 \sin^2((E_1 - E_3)t/\hbar). \tag{7.113}$$

In the case $E_1 = E_2$ we have

$$\frac{2}{9}\left(6 - 4\cos^2(E_1 - E_3)\right) \geq \sin^2(E_1 - E_3) \rightarrow \frac{1}{9}\cos^2(E_1 - E_3) \geq -\frac{1}{3},$$

which is always true.

7.3   The One-Dimensional Lattice

A very simple model that mimics the behavior of electrons in solids is a one-dimensional periodic array of identical atoms (positively charged ions, really) and an electron that can sit at the location of the atoms but can also make transitions between neighboring atoms (ultimately due to tunneling phenomena through the potential barriers between atoms). Let the lattice of atoms consist of N atoms at the fixed locations

$$a, 2a, 3a, \ldots, Na$$

a being the constant distance between successive atoms. In place of position eigenvectors $|x\rangle$, we may consider the states $|n\rangle$ each of which corresponds to the electron having a fixed position at the nth atom. The Hilbert space of states is an N-dimensional vector space spanned by these states which, as mutually exclusive, are assumed to be orthogonal and, therefore, to constitute an orthonormal basis. If no transitions between atoms were possible, the Hamiltonian in this basis would be a diagonal matrix

$$\mathcal{H}_0 = \begin{pmatrix} E_0 & 0 & 0 & \dots & \dots \\ 0 & E_0 & 0 & \dots & \dots \\ 0 & 0 & E_0 & 0 & \dots, \\ \dots & \dots & \dots & \dots & \dots \\ \dots & \dots & \dots & \dots \end{pmatrix}. \tag{7.114}$$

Since all atomic sites are equivalent, the energy eigenvalues E_0 are equal. The corresponding Hamilton operator $\hat{H}_0$ can be expressed in terms of the projection operators $\Pi_n = |n\rangle\langle n|$ as

$$\hat{H}_0 = \sum_{n=1}^{N} E_0 |n\rangle\langle n|. \tag{7.115}$$

This is the so-called *tight-binding approximation* in which transitions between atomic sites do not occur. We may include now the possibility for transitions by adding to (7.114) non-diagonal terms between first neighbors as

$$\hat{H} = \sum_{n=1}^{N} E_0 |n\rangle\langle n| - g \sum_{n=1}^{N-1} (|n\rangle\langle n+1| + |n+1\rangle\langle n|), \tag{7.116}$$

where g is a parameter measuring the transition probability between neighboring sites. In matrix form (7.116) is

$$\mathcal{H} = \begin{pmatrix} E_0 & -g & 0 & \dots & \dots \\ -g & E_0 & -g & \dots & \dots \\ 0 & -g & E_0 & -g & \dots, \\ \dots & \dots & -g & \dots & \dots \\ \dots & \dots & \dots & \dots \end{pmatrix}. \tag{7.117}$$

The eigenvalue problem of (7.116) is

$$\hat{H}|\psi_E\rangle = E|\psi_E\rangle. \tag{7.118}$$

Expanding the eigenstate in the $\{n\}$ basis

$$|\psi_E\rangle = \sum_{n=1}^{N} \psi_n |n\rangle \tag{7.119}$$

'and substituting it into (7.118), we obtain

$$\sum_{n'=1}^{N} \left(E_0 |n'\rangle \langle n'| - g|n'+1\rangle\langle n'| - g|n'\rangle\langle n'+1| \right) \sum_{n=1}^{N} \psi_n |n\rangle = E \sum_{n=1}^{N} \psi_n |n\rangle$$

or

$$\sum_{n=1}^{N} (E_0\, \psi_n - E\, \psi_n - g\psi_{n+1}) \, |n\rangle - g \sum_{n'=1}^{N} \psi_{n'} |n'+1\rangle = 0$$

or

$$\sum_{n=1}^{N} (E_0\, \psi_n - E\, \psi_n - g\psi_{n+1}) \, |n\rangle - g \sum_{n=2}^{N} \psi_{n-1} |n\rangle = 0$$

or

$$(E - E_0)\psi_n + g\,(\psi_{n+1} + \psi_{n-1}) = 0 \quad with \ n = 1, 2, \ldots, N, \tag{7.120}$$

keeping in mind that, since the lattice starts up at a and ends up at Na, $\psi_0 = \psi_{N+1} = 0$.

Let us now look for a solution to (7.120) in the form of a plane wave

$$\psi(x) = \frac{e^{ikx}}{\sqrt{L}} \implies \psi_n = \frac{e^{ik(na)}}{\sqrt{N}} \tag{7.121}$$

with k a wave number parameter to be related to the energy eigenvalues E. Indeed, substituting (7.121) into (7.120), we get

$$E = E_0 - 2g\cos(ka). \tag{7.122}$$

Note, however, that if we change the wave number by an integer times $2\pi/a$, i.e.,

$$k \to k + \frac{2\pi}{a}, \tag{7.123}$$

the eigenfunction and the eigenvalue do not change. It suffices to restrict the values of k in the range

$$-\frac{\pi}{a} \leq k < \frac{\pi}{a}. \tag{7.124}$$

This region has the name *Brillouin Zone*.[6] Positive k corresponds to electron motion toward the right while negative k motion toward the left. Notice that spatial discreteness has led us to restricting the momentum in the Brillouin zone. This is the inverse of what we have encountered in the case of the infinite square well where spatial compactness has led to discreteness of momentum.

At low momenta $k << 1/a$ or, equivalently, in the $a \to 0$ limit (continuum), the energy eigenvalues are

$$E \approx E_0 - g + \frac{g}{2}(ka)^2 = \frac{\hbar^2 k^2}{2m^*} + constant, \qquad (7.125)$$

where $m^* \equiv \frac{\hbar^2}{ga^2}$ is an *effective mass* of the low-energy electrons.

Example 7.7 Calculate the uncertainty in position for any state $|\psi_E\rangle$ and compare with the localized states $|n\rangle$, which have $(\Delta x)_n = 0$.

We have

$$\langle \psi_E | \hat{x} | \psi_E \rangle = \frac{a}{N} \sum_{n=1}^{N} n = \frac{a}{2}(N+1)$$

and

$$\langle \psi_E | \hat{x}^2 | \psi_E \rangle = \frac{a^2}{N} \sum_{n=1}^{N} n^2 = \frac{a^2}{6}(N+1)(2N+1).$$

Thus, we obtain

$$(\Delta x)_E^2 = \frac{1}{12}(N^2 - 1). \qquad (7.126)$$

This means that the energy eigenstates are extremely delocalized extending over the entire lattice $((\Delta x)_E \sim O(N))$, in contrast to the position eigenstates $|n\rangle$.

Example 7.8 Consider a *periodic lattice* for which $\psi_{N+1} = \psi_1$ and show that there is a further restriction on k imposed by periodicity, namely, $k_j = j\frac{2\pi}{Na}$ ($j = 1, \ldots, N$) and there are N eigenstates corresponding to N energy eigenvalues. Demonstrate the orthonormality of these eigenstates.

The periodicity condition $\psi_1 = \psi_{N+1}$ implies

$$e^{ika} = e^{i(N+1)ka} \implies k = \frac{2\pi}{Na} j$$

with $j = 1, \ldots, N$. The orthonormality of the states $|\psi_{E_j}\rangle$ follows in a straightforward fashion. We have[7]

$$\langle \psi_{E_j} | \psi_{E_{j'}} \rangle = \begin{cases} (j \neq j') & \frac{1}{N} \sum_{n=1}^{N} e^{\frac{2\pi i n}{N}(j'-j)} = 0 \\ (j' = j) & \frac{1}{N} \sum_{n=1}^{N} = 1 \end{cases} \qquad (7.127)$$

---

[6]See any of the standard solid state textbooks, e.g., [4].

[7]$\sum_{n=1}^{N} b^n = b\frac{(1-b^N)}{(1-b)}$.

7.4 Periodic Potentials

In the case of a free particle, the energy eigenstates are simultaneously momentum eigenstates since $[\hat{H}, \hat{p}] = 0$. This is related to the *translational invariance* of the free-particle system. This is no longer true in the presence of forces since $[\hat{H}, \hat{p}] \neq 0$ in general. Nevertheless, in a system like the previously considered lattice, a discrete version of translational invariance exists, namely, invariance under $x \rightarrow x + a$, where a is the lattice spacing. This is in contrast to the continuous translational invariance $x \rightarrow x + \beta$ of the free-particle system where β is a continuous parameter.

The Translation Operator. Let us introduce a *translation operator* $\hat{T}(\beta)$ defined by its action on a wave function as

$$\hat{T}(\beta)\psi(x) = \psi(x + \beta). \tag{7.128}$$

It is not difficult to see that such an operator has the property

$$\hat{T}(\beta)\,\hat{T}(\gamma) = \hat{T}(\beta + \gamma). \tag{7.129}$$

Of course, $\hat{T}(0) = \mathbf{I}$ and, therefore,

$$\hat{T}(\beta)\,\hat{T}(-\beta) = \mathbf{I} \implies \hat{T}(-\beta) = \hat{T}^{-1}(\beta). \tag{7.130}$$

Next, we can show that $\hat{T}(\beta)$ is a *unitary operator*. Starting from the definition of Hermitian conjugation $\langle\psi_1|\hat{T}(\beta)|\psi_2\rangle = \langle\psi_2|\hat{T}^{\dagger}(\beta)|\psi_1\rangle^*$, we obtain

$$\int dx\,\psi_1^*(x)\psi_2(x + \beta) = \int dx\,\psi_2(x)\left(\hat{T}^{\dagger}(\beta)\psi_1(x)\right)^* \implies$$

$$\int dx\,\psi_2^*(x)\hat{T}^{\dagger}(\beta)\psi_1(x) = \int dx\,\psi_2^*(x + \beta)\psi_1(x) = \int dx\,\psi_2^*(x)\psi_1(x - \beta)$$

or

$$\hat{T}^{\dagger}(\beta)\psi_1(x) = \psi_1(x - \beta) \implies \hat{T}^{\dagger}(\beta) = \hat{T}(-\beta) \tag{7.131}$$

and

$$\hat{T}(\beta)\,\hat{T}^{\dagger}(\beta) = \hat{T}^{\dagger}(\beta)\,\hat{T}(\beta) = \mathbf{I}. \tag{7.132}$$

The definition (7.128) of the translation operator can also be written in terms of a Taylor expansion in β as

$$\hat{T}(\beta)\psi(x) = \psi(x + \beta) = \sum_{n=0}^{\infty} \frac{\beta^n}{n!}\psi^{(n)}(x) = \sum_{n=0}^{\infty} \frac{\beta^n}{n!}\frac{d^n}{dx^n}\psi(x). \tag{7.133}$$

Nevertheless, the derivative can be replaced by the momentum operator as $\frac{d}{dx} = \frac{i}{\hbar}\hat{p}$ and

$$\sum_{n=0}^{\infty} \frac{\beta^n}{n!} \frac{d^n}{dx^n} \psi(x) = \sum_{n=0}^{\infty} \frac{i^n \beta^n}{\hbar^n n!} \hat{p}^n \psi(x) = e^{\frac{i}{\hbar}\beta\hat{p}} \psi(x).$$

Thus, we may write

$$\hat{T}(\beta)\psi(x) = e^{\frac{i}{\hbar}\beta\hat{p}} \psi(x) \implies \hat{T}(\beta) = e^{\frac{i}{\hbar}\beta\hat{p}}. \tag{7.134}$$

All the above can be straightforwardly carried over to three dimensions

$$\hat{T}(\mathbf{a}) = e^{\frac{i}{\hbar}\mathbf{a}\cdot\hat{\mathbf{p}}}. \tag{7.135}$$

It is evident that a translationally invariant system must obey

$$\left[\hat{H}, \hat{T}(\beta) \right] = 0. \tag{7.136}$$

This is, however, equivalent to $[\hat{H}, \hat{p}] = 0$ or momentum conservation. Of course, this is not true in general in the case of interactions.

Bloch's Theorem. In the case of a system with a periodic potential,[8] the Hamiltonian obeys discrete translational invariance by the given length a

$$x \rightarrow x + a. \tag{7.137}$$

Therefore, the Hamiltonian must commute with $\hat{T}(a)$

$$\left[\hat{H}, \hat{T}(a) \right] = 0 \tag{7.138}$$

and the energy eigenstates will also be eigenstates of $\hat{T}(a)$. Since $\hat{T}(a)$ is a unitary operator, its eigenvalues will be of the form $e^{i\theta}$. We may label its eigenstates with some wave number parameter k and have

$$\hat{T}(a)\psi_k(x) = e^{ika} \psi_k(x). \tag{7.139}$$

Nevertheless, this implies also

$$\psi_k(x+a) = e^{ika} \psi_k(x). \tag{7.140}$$

Note that there is an arbitrariness in labeling the eigenstates with the wave number k, since $k + n(2\pi)/a$ gives the same eigenvalue. This degeneracy is avoided if we

---

[8] An introduction to periodic potentials and Bloch's theorem in considerable detail can be found in standard solid state textbooks like [4].

restrict the value range of k in the region

$$-\frac{\pi}{a} \le k < \frac{\pi}{a}. \tag{7.141}$$

This the *Brillouin Zone* that we also encountered in the one-dimensional lattice.

Let us return to (7.140) and write the wave functions as

$$\psi_k(x) = e^{ikx} u_k(x). \tag{7.142}$$

Substituting it in (7.140) we obtain that $u_k(x)$ has to be periodic, namely,

$$u_k(x + a) = u_k(x). \tag{7.143}$$

This is *Bloch's Theorem* which states:

In a periodic potential $V(x) = V(x + a)$ all energy eigenstates can be written as $\psi_k(x) = e^{ikx}u_k(x)$, where $u_k(x) = u_k(x + a)$ is periodic and k lies in the Brillouin Zone.

Energy Bands and Gaps. Among the most important properties of periodic potentials is that they predict the distinction between conductors and insulators. This follows directly from the existence of gaps in the energy spectrum. The proof is very general.

If $\psi(x)$ is a solution of the time-independent Schroedinger equation with a periodic potential $(V(x + a) = V(x))$, so is $\psi(x + a)$. Since we have a second-order differential equation, we shall have two independent solutions $\psi_1(x)$, $\psi_2(x)$. This pair will have to be linearly related to $\psi_1(x + a)$, $\psi_2(x + a)$ through a relation

$$\psi_i(x + a) = \mathcal{F}_{ij}(E)\,\psi_j(x), \tag{7.144}$$

where $\mathcal{F}(E)$ will be a 2×2 matrix dependent on the energy. This is the *"Flocket Matrix"*. There will also be an analogous relation to (7.144) for the derivatives

$$\psi_i'(x + a) = \mathcal{F}_{ij}(E)\,\psi_j'(x). \tag{7.145}$$

We can now introduce the matrix

$$\mathcal{W}(x) = \begin{pmatrix} \psi_1(x) & \psi_1'(x) \\ \psi_2(x) & \psi_2'(x) \end{pmatrix}. \tag{7.146}$$

Using the Schroedinger equation, it is straightforward to prove that the determinant of $\mathcal{W}$ is a constant

$$\det(\mathcal{W}(x)) = \psi_1(x)\psi_2'(x) - \psi_2(x)\psi_1'(x) \quad \Longrightarrow \quad (\det(\mathcal{W}))' = 0. \tag{7.147}$$

From (7.144) and (7.145), we can write the matrix equation

$$W(x + a) = \mathcal{F}(E) W(x) \tag{7.148}$$

and taking its derivative arrive at

$$\det(\mathcal{F}) = 1 . \tag{7.149}$$

The solution of (7.144) will depend on the eigenvalues $f_\pm$ of $\mathcal{F}(E)$. These are determined by the equation

$$\det(\mathcal{F} - f\mathbf{I}) = 0 \implies f^2 - Tr(\mathcal{F})f + \det(\mathcal{F}) = 0 \tag{7.150}$$

or just

$$f^2 - Tr(\mathcal{F})f + 1 = 0 \implies f_\pm = \frac{1}{2}\left(Tr(\mathcal{F}) \pm \sqrt{(Tr(\mathcal{F}))^2 - 4} \right). \tag{7.151}$$

Note that, because $\det(\mathcal{F}) = 1$, we must have $f_+ f_- = 1$.

For $|Tr(\mathcal{F})| < 2$, the solutions will be complex. Then, they can always be written in the form

$$f_\pm = e^{\pm ika} \tag{7.152}$$

in terms of a parameter k. If $\eta_\pm$ are the eigenvectors corresponding to $f_\pm$, we can obtain from (7.144)

$$\eta_\pm^{(T)} \psi(x + a) = \eta_\pm^{(T)} \mathcal{F}(E)\psi(x) = f_\pm \eta_\pm^{(T)} \psi(x)$$

or

$$\psi_\pm(x + a) = e^{\pm ika} \psi_\pm(x), \tag{7.153}$$

where $\psi_\pm = \eta_\pm^{(T)} \psi = \eta_{\pm,i} \psi_i$. These states correspond to the *bands* of the energy spectrum.

For $|Tr(\mathcal{F})| > 2$, the eigenvalues will be of the form

$$f_\pm = e^{\pm qa} \tag{7.154}$$

for some real q. Through analogous steps, we can arrive at

$$\psi_\pm(x + a) = e^{\pm qa} \psi_\pm(x). \tag{7.155}$$

However, these states are not admissible since they diverge either at $+\infty$ or $-\infty$. Therefore, they must be rejected. This is the proof that there will be gaps in the energy spectrum.

The Kronig–Penney Model. As an example of a solvable periodic potential, we may consider an infinite periodic series of delta functions

$$V(x) = \sum_{n=-\infty}^{+\infty} V_0\, \delta(x - na). \tag{7.156}$$

The solution of the Schroedinger equation in the region $(0, a)$, where the potential vanishes, will be

$$\psi(x) = Ae^{iqx} + Be^{-iqx} \quad (E = \frac{\hbar^2 q^2}{2m}). \tag{7.157}$$

Note that we have used the symbol q for the wave number in order to avoid confusion with the symbol k appearing in Bloch's theorem. Now, Bloch's theorem dictates that

$$\psi(0) = u(0) = u(a) = \psi(a)\, e^{-ika}, \tag{7.158}$$

which for our particular wave function gives

$$A + B = \left(Ae^{iqa} + Be^{-iqa}\right) e^{-ika}. \tag{7.159}$$

In addition to this, we have the discontinuity of the derivative at the location of the delta function. At the point $x = 0$, we have

$$-\frac{\hbar^2}{2m} \left(\psi'(+0) - \psi'(-0)\right) + V_0 \psi(0) = 0 \tag{7.160}$$

or

$$-\frac{\hbar^2}{2m} \left(u'(+0) - u'(-0)\right) + V_0\, u(0) = 0. \tag{7.161}$$

Using the fact that $u'(x + a) = u'(x)$ at the point $x = 0 - \epsilon$ gives $u'(-\epsilon) = u'(a - \epsilon)$, the above relation becomes

$$-\frac{\hbar^2}{2m} \left(u'(\epsilon) - u'(a - \epsilon)\right) + V_0\, u(0) = 0. \tag{7.162}$$

Since ϵ and $a - \epsilon$ lie in the region where the expression (7.157) holds, we obtain

$$-iq\frac{\hbar^2}{2m} \left(A - B - Ae^{i(q-k)a} + Be^{-i(q+k)a}\right) + V_0(A + B) = 0. \tag{7.163}$$

The Eqs. (7.159) and (7.163) are just a homogeneous two-equation system leading to a value for the ratio B/A and to the condition on the energy

$$\cos(ka) = \cos(qa) + \frac{mV_0}{\hbar^2 q} \sin(qa). \tag{7.164}$$

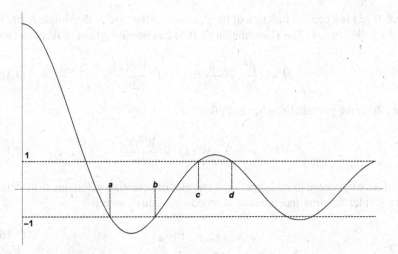

Fig. 7.4 Energy bands and gaps in the Kronig–Penney model

This equation gives the allowed values of k in terms of the energy E. Whenever the value of q is such that the right-hand side exceeds 1, these values of energy are not allowed (gaps). Due to the periodicity of the right-hand side in qa, for a given value of k in the Brillouin zone $-\pi/a < k < \pi/a$, there is an infinite number of corresponding values of qa satisfying this equation. In Fig. 7.4, we have plotted the right-hand side of (7.164) in terms of qa for $mV_0a/\hbar^2 = 5$. Note the gaps between the energy bands, corresponding to the intervals where the values of the plotted function exceed the allowed values of a cosine (e.g., the gap between $qa = c$ and $qa = d$).

7.5 Other Potentials

We saw that in the case of the harmonic oscillator the Hamilton operator had the positive definite form $\hat{H} = \hbar\omega\,\hat{a}^\dagger\hat{a}$ with $\hat{a} = \hat{x}\sqrt{m\omega/2\hbar} + i\,\hat{p}/\sqrt{2m\hbar\omega}$ (up to a constant $\hbar\omega/2$). This form led to a number of properties. We may explore this by allowing for a general position dependence of $\hat{a}$ and investigate Hamiltonians of the form[9]

$$\hat{H} = \hat{Q}^\dagger\hat{Q} = \frac{1}{2m}\left(-i\hat{p} + W(\hat{x})\right)\left(i\hat{p} + W(\hat{x})\right). \qquad (7.165)$$

The operator $\hat{Q}$ stands for

$$\hat{Q} = \frac{1}{\sqrt{2m}}\left(i\hat{p} + W(\hat{x})\right), \qquad (7.166)$$

---

[9]For more examples, see [5].

where $W(x)$ is a general function of the position. In the case of the simple harmonic oscillator $W = m\omega x$. The Hamiltonian (7.165) can be worked out to take the form

$$\hat{H} = \frac{\hat{p}^2}{2m} - \frac{\hbar}{2m} W'(x) + \frac{W^2(x)}{2m} \qquad (7.167)$$

from which the potential can be read off to be

$$V(x) = -\frac{\hbar}{2m} W'(x) + \frac{W^2(x)}{2m} . \qquad (7.168)$$

Before we proceed to study specific examples of potentials belonging to this class, let's consider the time-independent Schroedinger equation. It is

$$\hat{Q}^\dagger \hat{Q} |\psi_E\rangle = E |\psi_E\rangle . \qquad (7.169)$$

It is clear that the energy eigenvalues have to be nonnegative, i.e.,

$$E \geq 0 . \qquad (7.170)$$

Depending on the value of the potential at infinity, the energy spectrum can be composed of scattering states as well as discrete states. In the case that the potential vanishes at infinity, the only possible bound state is one of the zero energy. Actually, for a vanishing energy solution, we can immediately obtain that the corresponding eigenfunction satisfying the Schroedinger equation with $E = 0$ is

$$\psi_0(x) = \psi_0(0) \, e^{-\frac{1}{\hbar} \int_0^x dx' \, W(x')} . \qquad (7.171)$$

The Volcano potential. Consider a particle moving in a potential of the type (7.168) with

$$W(x) = N\hbar \left(\frac{x}{x^2 + a^2} \right) . \qquad (7.172)$$

We have

$$V(x) = \frac{\hbar^2 N}{2m} \frac{\left((N+1)x^2 - a^2 \right)}{\left(x^2 + a^2 \right)^2} . \qquad (7.173)$$

This potential is plotted in Fig. 7.5.

The potential vanishes at infinity. There is a zero energy bound state with a corresponding normalizable wave function

$$\psi_0(x) = \psi_0(0) \, e^{-\frac{1}{\hbar} \int_0^x dx' W(x')} = \frac{\psi_0(0)}{\left(x^2/a^2 + 1 \right)^N} . \qquad (7.174)$$

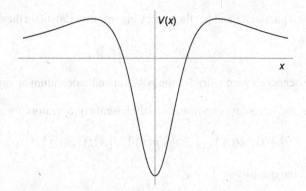

Fig. 7.5 Volcáno potential $(N = 1)$

Problems and Exercises

7.1 Consider the system of a quantum particle of mass m moving in one dimension and subject to a harmonic force $F = -m\omega^2 x$. The system is initially $(t = 0)$ in the state with wave function $\psi(x, 0) = \Theta(x)\,\psi_1(x)$, where $\psi_1(x)$ is the energy eigenfunction of the first excited state. Calculate the probability to find the system in the ground state at any time $t > 0$.

7.2 Find the energy eigenfunctions and eigenvalues for the system of a particle in the potential

$$V(x) = \begin{cases} +\infty & (x < 0) \\ \frac{1}{2}m\omega^2 x^2 & (x > 0) \end{cases}$$

7.3 A particle of mass m moves in one dimension subject to a potential

$$V(x) = \frac{1}{2}m\omega^2 x^2 - \lambda x,$$

where λ and ω are known parameters.
   (a) Find the energy eigenstates and eigenvalues of the system.
   (b) Assume that initially $(t = 0)$ the system occupies the state with wave function

$$\psi(x, 0) = \left(\frac{m\omega}{\hbar\pi}\right)^{1/4} e^{-\frac{m\omega}{2\hbar}x^2}.$$

Calculate the probability to find the particle in the state of lowest energy.

7.4 A harmonic oscillator $(H = p^2/2m + m\omega^2 x^2/2)$ is initially $(t = 0)$ in the state

$$|\psi(0)\rangle = \frac{1}{\sqrt{2}}\left(|0\rangle + e^{i\alpha}|1\rangle\right),$$

where α is a real parameter and $|n\rangle$ the energy eigenstates. Calculate the expectation values

$$\langle x \rangle_t, \quad \langle p \rangle_t, \quad \langle x^2 \rangle_t, \quad \langle p^2 \rangle_t$$

and verify Heisenberg's inequality for the position and momentum at any time.

7.5 Calculate the following commutators of Heisenberg operators for a harmonic oscillator:

$$\left[\hat{x}(t), \, \hat{x}(t') \right], \quad \left[\hat{p}(t), \, \hat{p}(t') \right], \quad \left[\hat{x}(t), \, \hat{p}(t') \right].$$

7.6 Calculate the quantity

$$\langle 1 | e^{\frac{i}{\hbar} \lambda \hat{p}} | 1 \rangle$$

for a harmonic oscillator. $|1\rangle$ is the first excited energy eigenstate and λ a real parameter. (*Hint: You may use the operator identity $e^{A+B} = e^A e^B e^{-\frac{1}{2}[A,B]}$, valid whenever the commutator $[A, B]$ is a number.*)

7.7 A two-state system, in the framework of an orthonormal basis $|1\rangle$, $|2\rangle$, has a Hamiltonian matrix

$$\mathcal{H} = \begin{pmatrix} 0 & E_0 \\ E_0 & 0 \end{pmatrix},$$

with E_0 a real parameter. The initial state of the system is

$$|\psi(t)\rangle = \frac{1}{\sqrt{2}} \left(|1\rangle + e^{i\alpha} |2\rangle \right)$$

with α a real parameter. Find the probability to return in this state as a function of the time $t > 0$.

7.8 Prove that a harmonic oscillator ($H = p^2/2m + m\omega^2 x^2/2$) in an arbitrary state will have an uncertainty $(\Delta x)_t^2$ which will be a periodic function of time with period $(2\omega)^{-1}$.

7.9 Consider a harmonic oscillator ($H = p^2/2m + m\omega^2 x^2/2$) in the state

$$|\psi\rangle = e^{\frac{i}{\hbar} \lambda \hat{p}} |n\rangle,$$

where $|n\rangle$ is an energy eigenstate and λ is a real parameter. Calculate the uncertainties in position and momentum and verify the Heisenberg inequality.

7.10 Consider a two-state system characterized by the three physical observables $\hat{A}, \hat{B}, \hat{C}$. For a given orthonormal basis $|1\rangle$, $|2\rangle$ the corresponding operators are represented by the matrices

$$A = \begin{pmatrix} 0 & 1 \\ 1 & 0 \end{pmatrix}, \quad B = \begin{pmatrix} 0 & -i \\ i & 0 \end{pmatrix}, \quad C = \begin{pmatrix} 1 & 0 \\ 0 & -1 \end{pmatrix}.$$

(a) Find the eigenstates and eigenvalues of $\hat{A}$ and $\hat{B}$.

(b) Assume that a measurement of $\hat{C}$ is performed and the outcome is the value +1. Immediately afterward, the system is subject to a measurement of $\hat{B}$. What is the probability of an eigenvalue +1? Assume that the outcome of the second measurement is the eigenvalue +1. Immediately afterward, the system is subject to a third measurement of $\hat{C}$ again. What is the probability of an eigenvalue -1?

7.11 A particle of mass m moves in one dimension subject to the potential

$$V(x) = -\frac{\hbar}{2m} W'(x) + \frac{W^2(x)}{2m}.$$

Verify that there exists a zero energy bound state and write down its wave function for $W(x) = m\omega x + \lambda x^3$ (with ω and λ real parameters).

7.12 For a particle moving in one dimension, the probability amplitude to make a transition from a point x to a point x' in time T is given by

$$K(x', x; T) = \langle x' | e^{-\frac{i}{\hbar} T \hat{H}} | x \rangle.$$

If the particle is subject to a periodic potential $V(x + a) = V(x)$, show that

$$K(x' + a, x + a; T) = K(x', x; T).$$

Show that the opposite is also true, namely, if K is periodic, the potential will have to be periodic as well.

7.13 For the periodic potential $V(x) = \sum_{-\infty}^{+\infty} V_0 \delta(x - na)$ (Kronig–Penney model) the allowed values of the Bloch wave number k are given in terms of the energy wave number $q = \sqrt{2mE/\hbar^2}$ by the equation

$$\cos(ka) = \cos(qa) + \frac{m V_0}{\hbar^2 q} \sin(qa).$$

Show that at the end points of the Brillouin zone $k_n = n\pi/a$ with $n\pi$ very large

$$q(k_n) \approx \frac{n\pi}{a} + \epsilon_n$$

with $\epsilon_n \ll 1$. Determine the possible values of ϵ_n.

7.14 Consider the *"double harmonic oscillator"*, i.e., a particle of mass m moving in the potential

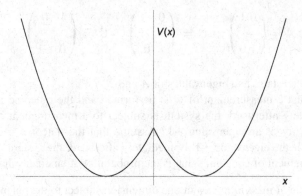

Fig. 7.6 Double oscillator

$$V(x) = \frac{1}{2}m\omega^2 (|x| - a)^2$$

shown in Fig. 7.6.

You may assume that the distance between the two minima $2a$ is much larger than the characteristic oscillator length $a >> \sqrt{\hbar/m\omega}$. It is expected that the ground state will have the property $\psi(x) \approx \psi_0(x + a)$ for $x << 0$ and $\psi(x) \approx \psi_0(x - a)$ for $x >> 0$, where $\psi_0(x)$ is the ground state wave function of the simple harmonic oscillator. Therefore, a reasonable estimate for the ground state of the system is $\psi(x) = N\,(\psi_0(x + a) + \psi_0(x - a))$. Calculate the expectation value of the energy in this state. Do the same for the state $\tilde{\psi}(x) = \tilde{N}\,(\psi_0(x + a) - \psi_0(x - a))$. The following approximation is assumed to be valid $\int_0^\infty d\xi\,\xi\,e^{-(\xi + a\sqrt{m\omega/\hbar})^2} \approx \frac{\hbar}{4m\omega a^2} e^{-\frac{m\omega}{\hbar}a^2}$.
Hint: You may use the identities $\hat{H}\psi_0(x \pm a) = \frac{\hbar\omega}{2}\psi_0(x \pm a)$ for $x < 0$ and $x > 0$ correspondingly.

7.15 Consider the simple harmonic oscillator and rewrite the corresponding creation and annihilation operators in terms of the number operator $\hat{N}$ and a *"phase"* operator $\hat{\phi}$ as

$$\hat{a} = \left(\hat{N} + 1\right)^{1/2} e^{i\hat{\phi}}, \quad \hat{a} = e^{-i\hat{\phi}} \left(\hat{N} + 1\right)^{1/2}.$$

Prove the commutation relations[10]

$$\left[e^{i\hat{\phi}}, \hat{N}\right] = e^{i\hat{\phi}}, \quad \left[e^{-i\hat{\phi}}, \hat{N}\right] = -e^{-i\hat{\phi}}, \quad [\cos\hat{\phi}, \hat{N}] = i\sin\hat{\phi}, \quad [\sin\hat{\phi}, \hat{N}] = -i\cos\hat{\phi}.$$

Finally, show that for any state the following is true:

---

[10]The trigonometric phase operators are defined as

$$\cos\hat{\phi} = \frac{1}{2}\left(e^{i\hat{\phi}} + e^{-i\hat{\phi}}\right), \quad \sin\hat{\phi} = \frac{1}{2i}\left(e^{i\hat{\phi}} - e^{-i\hat{\phi}}\right).$$

$$(\Delta N)^2 (\Delta \cos \phi)^2 \geq \frac{1}{4} \, |\langle \sin \phi \rangle|^2 \ .$$

References

1. P.A.M. Dirac, *The Principles of Quantum Mechanics* (Oxford University Press, Oxford, 1999)
2. E. Merzbacher, *Quantum Mechanics*, 3rd edn. (Wiley, New York, 1998)
3. J. Klauder, B.-S. Skagerstam, *Coherent States* (World Scientific, Singapore, 1985)
4. N. Ashcroft, N.D. Mermin, *Solid State Physics* (Holt-Saunders International, 1976)
5. F. Cooper, A. Khare, U. Sukhatme, *Supersymmetry in Quantum Mechanics* (World Scientific, Singapore, 2001)

Chapter 8
Angular Momentum

8.1 Angular Momentum as a Quantum Observable

A particle moving in three-dimensional space, apart from its position and momentum, is also characterized by its *angular momentum*.[1,2]

$$\hat{\mathbf{L}} = \hat{\mathbf{r}} \times \hat{\mathbf{p}}. \tag{8.1}$$

In terms of components, this is[3]

$$\hat{L}_i = \epsilon_{ijk}\hat{x}_j\hat{p}_k \implies \begin{cases} \hat{L}_x = \hat{y}\hat{p}_z - \hat{z}\hat{p}_y \\ \hat{L}_y = \hat{z}\hat{p}_x - \hat{x}\hat{p}_z \\ \hat{L}_z = \hat{x}\hat{p}_y - \hat{y}\hat{p}_x \end{cases} \tag{8.2}$$

Since $\hat{\mathbf{L}}$ is an observable, it has to be a Hermitian operator. Indeed, we have

$$\hat{L}_i^\dagger = \epsilon_{ijk}\left(\hat{x}_j\hat{p}_k\right)^\dagger = \epsilon_{ijk}\hat{p}_k\hat{x}_j = \epsilon_{ijk}\left(\hat{x}_j\hat{p}_k - i\hbar\delta_{jk}\right) = \hat{L}_i + \epsilon_{ijj} = \hat{L}_i$$

---

[1]Sometimes, it is called *orbital angular momentum* in order to be differentiated from the *spin* or *intrinsic angular momentum*.

[2]The *cross product* of vector operators is defined in the same way as the cross products of *c-number* vectors, taking into account the non-commutativity of operators. For example, $\left(\hat{\mathbf{A}} \times \hat{\mathbf{B}}\right)_i = \epsilon_{ijk}\hat{A}_j\hat{B}_k \neq \epsilon_{ijk}\hat{B}_k\hat{A}_j$.

[3]Note that here and elsewhere we have adopted the *"Einstein convention"* according to which a repeated index signifies summation, unless otherwise stated, e.g., $A_{ij}a_j$ stands for $\sum_j A_{ij}a_j$.

© Springer Nature Switzerland AG 2019
K. Tamvakis, *Basic Quantum Mechanics*, Undergraduate Texts in Physics,
https://doi.org/10.1007/978-3-030-22777-7_8

or
$$\hat{\mathbf{L}}^{\dagger} = \hat{\mathbf{L}}. \tag{8.3}$$

In contrast to the position or momentum operators of which different components are commuting, in the case of the angular momentum different components do not commute. After some manipulations[4] we obtain

$$\left[\hat{L}_x, \hat{L}_y\right] = i\hbar \hat{L}_z$$

$$\left[\hat{L}_y, \hat{L}_z\right] = i\hbar \hat{L}_x \tag{8.4}$$

$$\left[\hat{L}_z, \hat{L}_x\right] = i\hbar \hat{L}_y$$

or collectively
$$\left[\hat{L}_i, \hat{L}_j\right] = i\hbar \epsilon_{ijk} \hat{L}_k. \tag{8.5}$$

Note however that the square of the angular momentum operator

$$\hat{\mathbf{L}}^2 = \hat{L}_x^2 + \hat{L}_y^2 + \hat{L}_z^2 \tag{8.6}$$

commutes with all components. Without much effort we can show that

$$\left[\hat{\mathbf{L}}^2, \hat{L}_i\right] = 0. \tag{8.7}$$

The fact that $\left[\hat{L}_x, \hat{L}_y\right] \neq 0$ means that L_x and L_y cannot have common eigenstates. Therefore, they cannot be determined simultaneously.[5] In fact they satisfy a nontrivial uncertainty relation

$$(\Delta L_x)^2 (\Delta L_y)^2 \geq \frac{\hbar^2}{4} \left|\langle \hat{L}_z \rangle\right|^2. \tag{8.8}$$

---

[4]We have

$$[\hat{L}_x, \hat{L}_y] = [\hat{y}\hat{p}_z - \hat{z}\hat{p}_y, \hat{z}\hat{p}_x - \hat{x}\hat{p}_z] = [\hat{y}\hat{p}_z, \hat{z}\hat{p}_x] - [\hat{y}\hat{p}_z, \hat{x}\hat{p}_z] - [\hat{z}\hat{p}_y, \hat{z}\hat{p}_x] + [\hat{z}\hat{p}_y, \hat{x}\hat{p}_z]$$

$$= \hat{y}[\hat{p}_z, \hat{z}]\hat{p}_x + \hat{p}_y[\hat{z}, \hat{p}_z]\hat{x} = i\hbar \left(\hat{x}\hat{p}_y - \hat{y}\hat{p}_x\right) = i\hbar \hat{L}_z$$

and similarly for the rest.

[5]There is a trivial exception to this, namely, when we are in a state with vanishing angular momentum $\mathbf{L}|\psi\rangle = 0$ and all three components vanish simultaneously.

8.2 Central Forces

In Classical Mechanics, the topic of *central forces* between macroscopic objects (particles or planets) is a very well studied subject. By definition, central forces are directed toward an attraction or repulsion center and depend only on the distance from it. Such forces, like the universal gravitational attraction or the Coulomb force are of the type $\mathbf{F} = \hat{\mathbf{r}} F(r) = -\hat{\mathbf{r}} V'(r)$, arising from potentials $V(r)$ that depend only on the radius r. The fact that the trajectories of classical particles or even planets moving in a central potential are planar is intimately connected to the conservation of angular momentum, which is predicted to be a constant vector perpendicular to the plane of motion. These properties have their quantum analogue for microscopic particles moving in central potentials.

In order to investigate whether the angular momentum of a quantum particle moving in a central potential $V(r)$ is a constant of the motion it is most convenient to consider this question in the framework of the Heisenberg picture of time evolution. The Heisenberg equation for the angular momentum operator is[6]

$$\frac{d\mathbf{L}}{dt} = \frac{i}{\hbar}\left[\hat{H}, \hat{\mathbf{L}}\right] = \frac{i}{2m\hbar}\left[\hat{p}^2, \mathbf{r}\right] \times \mathbf{p} + \frac{i}{\hbar}\mathbf{r} \times \left[V(r), \mathbf{p}\right]. \qquad (8.9)$$

Proceeding to calculate these commutators we note that the Hamiltonian, as a constant of the motion, does not depend on time. Thus, the momentum and position operators within the Hamiltonian can be taken to be at time t. Next, we recall that the equal-time commutator of the position and momentum is the same as the Schroedinger picture commutator. This last property leads to the fact that $[\hat{\mathbf{p}}, V(r)] = -i\hbar\nabla V(r)$, where $\hat{\mathbf{p}}$ and r are Heisenberg operators. These manipulations give for the first commutator in (8.9) $[\hat{p}^2, x_j] = -2i\hbar\hat{p}_j$ that leads to no contribution since

$$[\hat{p}^2, \mathbf{r}] \times \mathbf{p} = -2i\hbar\mathbf{p} \times \mathbf{p} = 0.$$

The second commutator is $[V(r), \mathbf{p}] = i\hbar\nabla V(r) = -i\hbar\mathbf{F}$ and does not contribute either since, for a central force $\nabla V = \hat{\mathbf{r}} V'(r)$ or $\mathbf{F} = -\hat{\mathbf{r}} V'(r)$, and

$$\mathbf{r} \times [V(r), \mathbf{p}] = -i\hbar\mathbf{r} \times \mathbf{F} = 0.$$

Thus, we have

$$\frac{d\mathbf{L}}{dt} = 0 \implies \mathbf{L} = constant \qquad (8.10)$$

and *the angular momentum is a constant of the motion for a particle that moves in a central potential.*

---

[6]We have dropped the "hat" from the position operators in order to avoid confusion with the corresponding "hat" on the position unit vector. Note also that all appearing commutators are equal-time commutators.

8.3   The Angular Momentum in the $\{x\}$ Representation

In the position representation, the angular momentum operator is

$$\mathbf{L} = \mathbf{r} \times \mathbf{p} \implies -i\hbar\,(\mathbf{r} \times \nabla)\,. \tag{8.11}$$

It is immediately evident that *position rescalings*

$$x_j \rightarrow \alpha\,x_j$$

by a parameter α will leave the angular momentum unaffected because the position variables enter in the combinations $\epsilon_{ijk}x_j\frac{\partial}{\partial x_k}$. Thus, the angular momentum operator in (8.11) cannot depend on the radius r. If we employ *spherical coordinates* (see the appropriate section of the Mathematical Appendix), the angular momentum operator will depend only on the angles θ and ϕ. The appearing gradient differential operator has the following form:

$$\nabla = \hat{r}\frac{\partial}{\partial r} + \frac{\hat{\theta}}{r}\frac{\partial}{\partial \theta} + \frac{\hat{\phi}}{r\sin\theta}\frac{\partial}{\partial \phi}\,. \tag{8.12}$$

Inserting this in (8.11), we get

$$\mathbf{L} \rightarrow -i\hbar\left(\hat{\phi}\frac{\partial}{\partial \theta} - \frac{\hat{\theta}^{\cdot}}{\sin\theta}\frac{\partial}{\partial \phi}\right)\,. \tag{8.13}$$

The corresponding Cartesian coordinates are

$$\hat{L}_x \rightarrow i\hbar\left(\sin\phi\frac{\partial}{\partial \theta} + \cos\phi\cot\theta\frac{\partial}{\partial \phi}\right)$$

$$\hat{L}_y \rightarrow -i\hbar\left(\cos\phi\frac{\partial}{\partial \theta} - \sin\phi\cot\theta\frac{\partial}{\partial \phi}\right) \tag{8.14}$$

$$\hat{L}_z \rightarrow -i\hbar\frac{\partial}{\partial \phi}\,.$$

The square of the angular momentum is

$$\hat{L}^2 \rightarrow -\hbar^2\left(\frac{1}{\sin\theta}\frac{\partial}{\partial \theta}\sin\theta\frac{\partial}{\partial \theta} + \frac{1}{\sin^2\theta}\frac{\partial^2}{\partial \phi^2}\right)\,. \tag{8.15}$$

In the framework of spherical coordinates, the separation of the kinetic energy of a particle into translational and rotational parts becomes quite transparent. Starting from

$$\hat{L}^2 = -\hbar^2\,(\mathbf{r} \times \nabla)^2 = -\hbar^2\epsilon_{ijk}\epsilon_{i\ell m}x_j\nabla_k x_\ell\nabla_m$$

$$= -\hbar^2 \left(\delta_{j\ell}\delta_{km} - \delta_{jm}\delta_{k\ell} \right) x_j \nabla_k x_\ell \nabla_m = -\hbar^2 \left(x_j \nabla_k x_j \nabla_k - x_j \nabla_k x_k \nabla_j \right)$$

$$= -\hbar^2 \left(x_j \delta_{kj} \nabla_k + r^2 \nabla^2 - 3x_j \nabla_j - x_j x_k \nabla_k \nabla_j \right) = -\hbar^2 \left(r^2 \nabla^2 - 2x_j \nabla_j - x_j x_k \nabla_k \nabla_j \right)$$

It can be shown—by brute force—that

$$\frac{1}{r} \left(\frac{\partial}{\partial r} \right)^2 r = \frac{1}{r} \left(\frac{\mathbf{r}}{r} \cdot \nabla \right)^2 r = \frac{1}{r} \left(\frac{x_k}{r} \nabla_k \frac{x_j}{r} \nabla_j \right) r = 2\frac{x_j}{r^2} \nabla_j + \frac{x_j x_k}{r^2} \nabla_j \nabla_k .$$

Thus, we have[7]

$$\frac{\hat{L}^2}{r^2} = -\hbar^2 \nabla^2 + \hbar^2 \frac{1}{r} \left(\frac{\partial}{\partial r} \right)^2 r$$

or equivalently

$$- \hbar^2 \nabla^2 = \frac{\hat{L}^2}{r^2} - \hbar^2 \left(\frac{1}{r} \frac{\partial}{\partial r} r \right) \left(\frac{1}{r} \frac{\partial}{\partial r} r \right) . \tag{8.16}$$

This expression calls for a definition of a *radial momentum* operator

$$\hat{p}_r \equiv= -i\hbar \left(\frac{1}{r} \frac{\partial}{\partial r} r \right) = -i\hbar \left(\frac{\partial}{\partial r} + \frac{1}{r} \right) . \tag{8.17}$$

In terms of it

$$\hat{p}^2 = \frac{\hat{L}^2}{r^2} + \hat{p}_r^2 \tag{8.18}$$

and the kinetic energy separates in a *translational part* (or radial) part $\frac{\hat{p}_r^2}{2m}$ and a rotational part $\frac{\hat{L}^2}{2mr^2}$.

Note that p_r as defined above is a Hermitian operator.[8] The proof is straightforward

$$\int d^3r \, \psi_1^* \left(-i\hbar \frac{1}{r} \frac{\partial}{\partial r} r \right)^\dagger \psi_2 = \left(\int d^3r \, \psi_2^* \left(-i\hbar \frac{1}{r} \frac{\partial}{\partial r} r \right) \psi_1 \right)^*$$

$$= i\hbar \int d^3r \, \psi_2 \frac{1}{r} \frac{\partial}{\partial r} \left(r\psi_1^* \right) = i\hbar \int d\Omega \int_0^\infty dr \, r \, \psi_1^* \frac{\partial}{\partial r} \left(r\psi_2 \right)$$

$$= \underbrace{i\hbar \int d\Omega \, r\psi_1^* r\psi_2 \Big|_0^\infty}_{=0} - i\hbar \int d\Omega \int dr \, r\psi_1^* \frac{\partial}{\partial r} \left(r\psi_2 \right)$$

---

[7]Since the angular momentum depends only on angles and therefore commutes with the radius r^2, it makes no difference to write $\frac{1}{r^2} \hat{L}^2 = \hat{L}^2 \frac{1}{r^2} = \frac{\hat{L}^2}{r^2}$.

[8]In contrast, the operator $-i\hbar \frac{\partial}{\partial r}$ is *not*.

Fig. 8.1 Rotation around the $\hat{z}$-axis

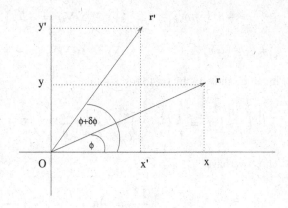

or

$$\int d^3r \psi_1^* \left(-i\hbar \frac{1}{r} \frac{\partial}{\partial r} r\right)^\dagger \psi_2 = \int d^3r \, \psi_1 * \left(-i\hbar \frac{1}{r} \frac{\partial}{\partial r} r\right) \psi_2 \implies \hat{p}_r^\dagger = \hat{p}_r \, .$$
(8.19)

8.4 Angular Momentum and Rotations

A *rotation* is a transformation of the position coordinates that preserves the lengths of vectors. It can be represented by a matrix $\mathcal{R}$ as[9]

$$x_j \rightarrow x_j' = \mathcal{R}_{jk} x_k \, .$$
(8.20)

Demanding that lengths are conserved, namely, $r^2 = x_j x_j = x_j' x_j' = r'^2$, corresponds to restricting the matrix $\mathcal{R}$ to be *orthogonal*[10] $\mathcal{R}^\perp \mathcal{R} = \mathcal{R}\mathcal{R}^\perp = \mathbf{I}$, which means that its *transpose* ($\mathcal{R}_{ij}^\perp = \mathcal{R}_{ji}$) is equal to its inverse.

Let's consider a rotation around the z-axis by an angle ϕ, as shown in Fig. 8.1. We have

$$\begin{cases} x' = x \cos\phi - y \sin\phi \\ y' = x \sin\phi + y \cos\phi \\ z' = z \end{cases} \implies \mathcal{R} = \begin{pmatrix} \cos\phi & -\sin\phi & 0 \\ \sin\phi & \cos\phi & 0 \\ 0 & 0 & 1 \end{pmatrix}$$
(8.21)

---

[9]See also [1–4].

[10]An *orthogonal* matrix is a *unitary* matrix with real elements.

For an infinitesimal rotation by a small angle $\delta\phi$, we have $\sin\delta\phi \approx \delta\phi$ and $\cos\delta\phi \approx 1 - O(\delta\phi^2)$. It is represented, up to first order in $\delta\phi$, in terms of vectors as follows:

$$\begin{cases} x' \approx x - y\,\delta\phi \\ y' \approx y + x\,\delta\phi \end{cases} \implies \mathbf{r}' = \mathbf{r} + \delta\phi\,(\hat{z} \times \mathbf{r}) \tag{8.22}$$

or in terms of a matrix as

$$x'_i = \mathcal{R}_{ij}x_j \quad \text{with} \quad \mathcal{R}_{ij} \approx \delta_{ij} - \epsilon_{3ij}\,(\delta\phi). \tag{8.23}$$

Consider now such an infinitesimal rotation acting on the quantum system of a particle and transforming a wave function ψ into a *rotated wave function* ψ'. Demanding that probability amplitudes are invariant under it corresponds to a statement that the rotated wave function has to satisfy

$$\psi'(\mathbf{r}') = \psi(\mathbf{r}) \tag{8.24}$$

or

$$\psi'(\mathbf{r} + \delta\mathbf{r}) = \psi(\mathbf{r}) \implies \psi'(\mathbf{r}) = \psi(\mathbf{r} - \delta\mathbf{r})$$

with $\delta\mathbf{r} = \delta\phi\,(\hat{z} \times \mathbf{r})$. Taylor-expanding we obtain

$$\psi'(\mathbf{r}) \approx \psi(\mathbf{r}) - \delta\mathbf{r} \cdot \nabla\psi(\mathbf{r}) = \psi(\mathbf{r}) - \delta\phi\,(\hat{z} \times \mathbf{r}) \cdot \nabla\psi(\mathbf{r})$$

$$= \psi(\mathbf{r}) - \delta\phi\,\hat{z} \cdot (\mathbf{r} \times \nabla)\,\psi(\mathbf{r}) = \psi(\mathbf{r}) - \frac{i}{\hbar}\delta\phi\,\hat{z} \cdot \hat{\mathbf{L}}\psi(\mathbf{r}).$$

The result

$$\psi'(\mathbf{r}) \approx \psi(\mathbf{r}) - \frac{i}{\hbar}\delta\phi\,\hat{L}_z\psi(\mathbf{r}) \tag{8.25}$$

can be extended to a finite rotation by an angle ϕ around the $\hat{z}$-axis as

$$\psi'(\mathbf{r}) = e^{-\frac{i}{\hbar}\phi\hat{L}_z}\psi(\mathbf{r}) \implies |\psi'\rangle = e^{-\frac{i}{\hbar}\phi\hat{L}_z}|\psi\rangle. \tag{8.26}$$

The *rotation operator*

$$\hat{U}(\phi) = e^{-\frac{i}{\hbar}\phi\hat{L}_z} \tag{8.27}$$

is a unitary operator

$$\hat{U}(\phi)\,\hat{U}^\dagger(\phi) = \hat{U}^\dagger(\phi)\hat{U}(\phi) = \mathbf{I}$$

conserving the inner products. The fact that the rotation operator depends directly on the orbital angular momentum is sometimes referred to as *"the angular momentum being the generator of rotations"*. What is easy to show is that invariance under rotations is directly related to angular momentum conservation. Indeed, we have

$$\hat{U}\hat{H}\hat{U}^\dagger = \hat{H} \implies [\hat{H}, \hat{U}] = 0 \implies [\hat{H}, \hat{\mathbf{L}}] = 0 \implies \hat{\mathbf{L}} = const. \qquad (8.28)$$

Scalar, Vector, and Axial Vector Operators. Let's consider the expectation value of the position operator in the rotated state $|\psi'\rangle = \hat{U}(\mathbf{a})|\psi$, where $\hat{U}(\mathbf{a}) = e^{-\frac{i}{\hbar}\mathbf{a}\cdot\hat{\mathbf{L}}}$ is a rotation operator corresponding to a rotation by an angle a around an axis defined by the direction of the constant vector $\mathbf{a}$,

$$\langle\psi'|\mathbf{r}|\psi'\rangle = \langle\psi|\hat{U}^\dagger(\mathbf{a})\,\mathbf{r}\,\hat{U}(\mathbf{a})|\psi\rangle. \qquad (8.29)$$

Demanding that this object transforms under rotations as the c-number position coordinates

$$\langle x_i \rangle' = \mathcal{R}_{ij}\langle x_j \rangle \qquad (8.30)$$

leads to the following transformation law under rotations for the position operator x_i

$$\hat{U}^\dagger(\mathbf{a})\,x_i\,\hat{U}(\mathbf{a}) = \mathcal{R}_{ij}\,x_j. \qquad (8.31)$$

For an infinitesimal rotation around the $\hat{z}$-axis this law becomes

$$\left(\mathbf{I} + \frac{i}{\hbar}\delta a\,\hat{L}_z\right) x_i \left(\mathbf{I} - \frac{i}{\hbar}\delta a\,\hat{L}_z\right) = x_i + \frac{i}{\hbar}\delta a\left[\hat{L}_z, x_i\right] = x_i - \delta a\,\epsilon_{3ik}x_k$$

or

$$\left[\hat{L}_z, x_i\right] = i\hbar\epsilon_{3ik}x_k. \qquad (8.32)$$

This is easily generalized to

$$\left[\hat{L}_i, x_j\right] = i\hbar\epsilon_{ijk}x_k. \qquad (8.33)$$

This can be checked to be true explicitly. *Any operator* $\hat{V}_i$ *that has the same type of commutation relation with* $\hat{\mathbf{L}}$ *as the position operator, namely*

$$\left[\hat{L}_i, \hat{V}_j\right] = i\hbar\epsilon_{ijk}\hat{V}_k \qquad (8.34)$$

is called a "vector" operator. Note that this commutation relation is equivalent to the relation

$$\hat{U}^\dagger(\mathbf{a})\,\hat{V}_i\,\hat{U}(\mathbf{a}) = \mathcal{R}_{ij}\,\hat{V}_j. \qquad (8.35)$$

Apart from the position, the momentum and the angular momentum itself are vector operators since

$$\left[\hat{L}_i, \hat{p}_j\right] = i\hbar\epsilon_{ijk}\hat{p}_k, \quad \left[\hat{L}_i, \hat{L}_j\right] = i\hbar\epsilon_{ijk}\hat{L}_k. \qquad (8.36)$$

Operators that commute with the angular momentum or, equivalently, are invariant under rotations are called *scalars*

$$\hat{U}^\dagger(\mathbf{a})\,\hat{S}\,\hat{U}(\mathbf{a}) = \hat{S} \quad \leftrightarrow \quad \left[\hat{\mathbf{L}}, \hat{S}\right] = 0. \tag{8.37}$$

The squares of vector operators are easily shown to be scalars

$$\left[r^2, \hat{\mathbf{L}}\right] = \left[\hat{p}^2, \hat{\mathbf{L}}\right] = \left[\hat{L}^2, \hat{\mathbf{L}}\right] = 0. \tag{8.38}$$

There is a further differentiation of vector operators according to their behavior under spatial reflection (parity). Vector operators that change sign under parity are called *polar vectors* or, simply, just *vectors*. Such are the position and the momentum

$$\hat{P}\,\mathbf{r}\,\hat{P} = -\mathbf{r}, \quad \hat{P}\,\hat{\mathbf{p}}\,\hat{P} = -\hat{\mathbf{p}}. \tag{8.39}$$

In contrast, vector operators that commute with parity are called *axial vectors* or *pseudovectors*. An example of axial vector is the angular momentum

$$\hat{P}\hat{\mathbf{L}}\hat{P} = \hat{\mathbf{L}} \quad \leftrightarrow \quad \left[\hat{P}, \hat{\mathbf{L}}\right] = 0. \tag{8.40}$$

Problems and Exercises

8.1 Show that the angular momentum commutation relations can be written in the form
$$\hat{\mathbf{L}} \times \hat{\mathbf{L}} = i\hbar\hat{\mathbf{L}}.$$

8.2 Show that for any representation $\{|n\rangle\}$ for which $\hat{L}_x$ and $\hat{L}_y$ are represented by real matrices, $\hat{L}_z$ will be represented by a purely imaginary matrix.

8.3 Show that if an operator commutes with two of the components of angular momentum, it will necessarily commute with the third component also.

8.4 A particle moves in a potential $V = V(x^2 + y^2)$ that does not depend on the azimuthal angle ϕ. The initial wave function of the particle is spherically symmetric, i.e. $\psi(\mathbf{r}, t = 0) = \psi(r)$. What is the angular momentum of the particle in the $\hat{z}$-direction at times $t > 0$?

8.5 Consider the operator

$$\hat{Q}_{ij} \equiv \hat{V}_i\hat{V}_j - \frac{1}{3}\delta_{ij}\hat{V}^2,$$

where $\hat{V}_i$ is a *vector* operator. This is an example of a *tensor operator*. Show that the following commutation relation is true

$$\left[\hat{Q}_{ij}, \hat{L}_k \right] = i\hbar \left(\hat{Q}_{i\ell}\epsilon_{jk\ell} + \hat{Q}_{j\ell}\epsilon_{ik\ell} \right).$$

8.6 The wave function of a particle at time $t = 0$ is a function of the radius, i.e., $\psi(\mathbf{r}, t = 0) = \psi(r)$. The Hamiltonian of the system is unspecified. It is known that at a time t_1 the wave function is

$$\psi(\mathbf{r}, t_1) = \left(1 + \alpha\frac{z}{r} \right) f(r),$$

where $f(r)$ a function of the radius and α a constant. Use this information in order to conclude whether or not the commutator $\left[\hat{H}, \hat{\mathbf{L}} \right]$ vanishes.

8.7 Let $\hat{\mathbf{V}}$ be a vector operator satisfying the definition $[\hat{L}_i, \hat{V}_j] = i\hbar\epsilon_{ijk}\hat{V}_k$ in terms of the angular momentum $\hat{\mathbf{L}}$. Prove the identity

$$\left[\hat{L}^2, \hat{\mathbf{L}} \times \hat{\mathbf{V}} \right] = 2i\hbar^2 \left(\hat{L}^2\hat{\mathbf{V}} - \left(\hat{\mathbf{L}} \cdot \hat{\mathbf{V}} \right) \hat{\mathbf{L}} \right).$$

8.8 Consider the operator $\hat{P}_x = \hat{P} e^{-\frac{i}{\hbar}\pi\hat{L}_x}$, consisting of a rotation by π around the $\hat{x}$-axis and a parity transformation. Show that this operator describes reflection on the (y, z)-plane. Show that the following identity is true

$$\hat{P}_x^\dagger \hat{L}_i \hat{P}_x = \hat{L}_i - 2\hat{L}_x\delta_{ix}.$$

Show also that

$$\hat{P}_x^\dagger \hat{V}_i \hat{P}_x = \hat{L}_i - 2\hat{V}_x\delta_{ix}$$

for any vector operator (e.g. $\hat{\mathbf{r}}, \hat{\mathbf{p}}, \dots$).

8.9 A particle has a Hamiltonian of the form

$$\hat{H} = \frac{\hat{p}^2}{2m} + V(r) - \alpha\hat{L}_z,$$

where $V(r)$ is spherically symmetric and α stands for a known parameter. Determine the expectation value of the angular momentum $\langle\hat{\mathbf{L}}\rangle_t$ at time t in terms of the expectation value $\langle\hat{\mathbf{L}}\rangle_0$ at $t = 0$. Show that, if the system occupies at $t = 0$ a state with real wave function, the expectation value of the angular momentum vanishes at all times.

8.10 For a state represented by the wave function

$$\psi(\mathbf{r}) = N x e^{-\alpha r^2},$$

where $\alpha > 0$ and N normalization constant to be computed, calculate the uncertainty in angular momentum $(\Delta \mathbf{L})^2 = \langle \hat{L}^2 \rangle - \left(\langle \hat{\mathbf{L}} \rangle \right)^2$.

References

1. A. Messiah, *Quantum Mechanics* (Dover Publications, single-volume reprint of the Wiley, New York, two-volume 1958 edition)
2. E. Merzbacher, *Quantum Mechanics*, 3rd edn. (Wiley, New York, 1998)
3. K. Gottfried, T.-M. Yan, *Quantum Mechanics: Fundamentals* (Springer, Berlin, 2004)
4. W. Greiner, B. Müller, *Quantum Mechanics Symmetries*, 2nd edn. (Springer, Berlin, 1992)

Chapter 9
Eigenstates of the Angular Momentum

9.1 The Angular Momentum Eigenvalue Problem

From the angular momentum commutation relations (8.5) and (8.7) it is clear that the *maximal subset of commuting observables* of angular momentum consists of the pair of L^2 and one of the components. Taking this component to be L_z, the operators that have common eigenvectors and are knowable with extreme accuracy are $\hat{L}^2$, $\hat{L}_z$. Before we write down the corresponding eigenvalue equations it should be noted that the eigenvalues of the operator $\hat{L}^2$ are nonnegative. Denoting with λ the eigenvalues of $\hat{L}^2$ and with $|\lambda\rangle$ the corresponding eigenstates, we have

$$\hat{L}^2|\lambda\rangle = \lambda|\lambda\rangle \implies \langle\lambda|\hat{L}^2|\lambda\rangle = \lambda\langle\lambda|\lambda\rangle$$

or

$$\lambda\langle\lambda|\lambda\rangle = \langle\lambda|\hat{L}_x^2|\lambda\rangle + \langle\lambda|\hat{L}_y^2|\lambda\rangle + \langle\lambda|\hat{L}_z^2|\lambda\rangle = \langle\lambda|\hat{L}_x^\dagger\hat{L}_x|\lambda\rangle + \langle\lambda|\hat{L}_y^\dagger\hat{L}_y|\lambda\rangle + \langle\lambda|\hat{L}_z^\dagger\hat{L}_z|\lambda\rangle$$

or

$$\left\|\hat{L}_x|\lambda\rangle\right\|^2 + \left\|\hat{L}_y|\lambda\rangle\right\|^2 + \left\|\hat{L}_z|\lambda\rangle\right\|^2 = \lambda\langle\lambda|\lambda\rangle \implies \lambda \geq 0.$$

Thus, the eigenvalues of $\hat{L}^2$ are nonnegative. In what follows we shall denote these eigenvalues in terms of a quantum number ℓ as

$$\lambda = \hbar^2\ell(\ell+1).$$

It is clear that ℓ has to be nonnegative. The justification for this seemingly lopsided parametrization will be clear when the quantum number ℓ will prove to obtain integer values.

Let us denote the eigenvalues of $\hat{L}_z$ and $\hat{L}^2$ as $\hbar m$ and $\hbar^2\ell(\ell+1)$ in terms of the two parameters (quantum numbers) m and ℓ, of which we already know that $\ell \geq 0$. We also denote the common eignvectors as $|\ell, m\rangle$. We have

© Springer Nature Switzerland AG 2019

K. Tamvakis, *Basic Quantum Mechanics*, Undergraduate Texts in Physics,

https://doi.org/10.1007/978-3-030-22777-7_9

$$\hat{L}^2|\ell, m\rangle = \hbar^2 \ell(\ell+1)|\ell, m\rangle, \quad \hat{L}_z|\ell, m\rangle = \hbar m|\ell, m\rangle. \tag{9.1}$$

We proceed to determine the allowed values for the quantum numbers ℓ, m and the corresponding eigenstates. We shall follow an algebraic method similar to the one employed in the harmonic oscillator which will make no explicit reference to the position-momentum definition of the angular momentum but will be based exclusively to the commutation relations and the hermiticity of the angular momentum operators.[1] To emphasize that we shall use the symbol $\mathbf{J}$ for the angular momentum operators and the symbols j, m for the corresponding eigenvalues. We have already demonstrated that j has to be nonnegative. Our only assumptions will be

$$\left[\hat{J}_i, \hat{J}_j\right] = i\hbar \epsilon_{ijk} \hat{J}_k$$

$$\hat{J}^2|j, m\rangle = \hbar^2 j(j+1) \quad and \quad \mathbf{J}^\dagger = \mathbf{J} \tag{9.2}$$

$$\hat{J}_z|j, m\rangle = \hbar m|j, m\rangle.$$

As a result, our conclusions will be applicable in the case of *spin*, a quantum observable that will be introduced shortly. Spin satisfies the same commutation algebra but is entirely unrelated to position and momentum, having no classical analogue.

Let us define the operators

$$\hat{J}_\pm = \hat{J}_x \pm i\hat{J}_y \quad (\hat{J}_\pm^\dagger = \hat{J}_\mp). \tag{9.3}$$

Note that

$$\hat{J}_+\hat{J}_- = \hat{J}_x^2 + \hat{J}_y^2 - i\left[\hat{J}_x, \hat{J}_y\right] = \hat{J}^2 - \hat{J}_z^2 + \hbar\hat{J}_z$$

$$\hat{J}_-\hat{J}_+ = \hat{J}_x^2 + \hat{J}_y^2 + i\left[\hat{J}_x, \hat{J}_y\right] = \hat{J}^2 - \hat{J}_z^2 - \hbar\hat{J}_z. \tag{9.4}$$

Subtracting, we also obtain

$$\left[\hat{J}_+, \hat{J}_-\right] = 2\hbar\hat{J}_z. \tag{9.5}$$

Similarly, we obtain

$$\left[\hat{J}_\pm, \hat{J}_z\right] = \mp\hbar\hat{J}_\pm. \tag{9.6}$$

Acting with $\hat{J}_\pm$ on the $\hat{J}_z$-eigenstate equation we get

$$\hat{J}_\pm\left(\hat{J}_z|j, m\rangle\right) = \hbar m\hat{J}_\pm|j, m\rangle$$

---

[1] See also [1–4].

or, using (9.6), we obtain

$$\left(\hat{J}_z\hat{J}_\pm \mp \hbar\hat{J}_z\right)|j, m\rangle = \hbar m \hat{J}_\pm|j, m\rangle)$$

or

$$\hat{J}_z\left(\hat{J}_\pm|j, m\rangle\right) = \hbar(m \pm 1)\left(\hat{J}_\pm|j, m\rangle\right). \tag{9.7}$$

This means that $\hat{J}_\pm|jm\rangle$ is an eigenstate of $\hat{J}_z$ corresponding to an eigenvalue $\hbar(m \pm 1)$. We may write

$$\hat{J}_\pm|j, m\rangle = C_\pm|j, m \pm 1\rangle. \tag{9.8}$$

Thus, $\hat{J}_\pm$ act on the $\hat{J}_z$-eigenstates as *raising and lowering operators*, increasing or decreasing eigenvalues by one unit of $\hbar$. The coefficient $C_\pm$ can be fixed by normalization as follows:

$$\langle j, m|\hat{J}_\mp\hat{J}_\pm|j, m\rangle = |C_\pm|^2\langle j, m \pm 1|j, m \pm 1\rangle = |C_\pm|^2$$

$$\langle j, m|\hat{J}_\mp\hat{J}_\pm|j, m\rangle = \langle j, m|\left(\hat{J}^2 - \hat{J}_z^2 \mp \hbar\hat{J}_z\right)|j, m\rangle = |C_\pm|^2$$

or

$$C_\pm = \hbar\sqrt{j(j + 1) - m(m \pm 1)}. \tag{9.9}$$

Positivity of the expression under the square root symbol constrains the quantum numbers j, m by the conditions

$$m^2 \pm m - j(j + 1) \le 0$$

or

$$\begin{cases} (j + m)(j + 1 + m) \ge 0 \\ (j - m)(j + 1 - m) \ge 0 \end{cases} \implies -j \le m \le +j. \tag{9.10}$$

Therefore, we can write

$$\hat{J}_\pm|j, m\rangle = \hbar\sqrt{j(j + 1) - m(m \pm 1)}\,|j, m \pm 1\rangle. \tag{9.11}$$

Acting with $\hat{J}_+$ on the state of maximum $m = j$, namely $|j, j\rangle$, we obtain

$$\hat{J}_+|j, j\rangle = 0 \tag{9.12}$$

due to vanishing of the coefficient. Similarly, acting with $\hat{J}_-$ on the state of minimum $m = -j$, namely $|j, -j\rangle$, we get

$$\hat{J}_-|j, -j\rangle = 0. \tag{9.13}$$

Thus, the possible $\hat{J}_z$ eigenstates are

$$|j, -j\rangle, \ |j, -j+1\rangle, \ \ldots \ \ldots, \ |j, \ j-1\rangle, \ |j, \ j\rangle .$$

Starting from the eigenstate $|j, j\rangle$, corresponding to the maximum $\hat{J}_z$-eigenvalue $m = j$ and acting with the lowering operator $\hat{J}_-$ successively we can end up in the eigenstate $|j, -j\rangle$ corresponding to the minimum $\hat{J}_z$-eigenvalue $m = -j$ only if our number of successive steps is $2j + 1$, i.e., if $2j + 1$ is an integer. This means that in principle j can take up the integer or half-integer values

$$j = 0, \ \frac{1}{2}, \ 1, \ \frac{3}{2}, \ 2, \ \ldots \quad and \quad -j \le m \le j . \tag{9.14}$$

Let us now restrict ourselves to the case of orbital angular momentum $\mathbf{L} = \mathbf{r} \times \mathbf{p}$. In this case, we know that there is an intimate relation with spatial rotations. We may consider an eigenstate $|\ell, m\rangle$ and apply on it a rotation by 2π. Of course, we expect to end up with the same state, i.e.,

$$e^{\frac{i}{\hbar}2\pi \hat{L}_z}|\ell, m\rangle = |\ell, m\rangle \implies e^{im(2\pi)}|\ell, m\rangle = |\ell, m\rangle .$$

Thus, m can only be an integer and, therefore, for the orbital angular momentum the half-integer eigenvalues are excluded, namely

$$\ell = 0, \ 1, \ 2, \ \ldots \quad and \quad -\ell \le m \le \ell . \tag{9.15}$$

Example 9.1 Rigid macroscopic bodies, subject to rotational motion around an axis, are described by a Hamiltonian $H = \frac{L^2}{2I}$, where $\mathbf{L}$ is their angular momentum and I their moment of inertia. This picture (Rigid Rotator) is also useful for the description of the rotational states of molecules. Consider such a system starting at $t = 0$ in the state

$$|\psi(0)\rangle = \frac{1}{\sqrt{6}} (|1, -1\rangle + |2, 1\rangle - 2|2, 0\rangle) . \tag{9.16}$$

(a) Find the state of the system at times $t > 0$ and calculate the probability of return to its original state.

(b) Calculate the uncertainties (ΔL_x), (ΔL_y) and (ΔL_z) and verify the uncertainty relation $(\Delta L_x)(\Delta L_z) \ge \hbar|\langle L_y\rangle|/2$.

(a) We have

$$|\psi(t)\rangle = \frac{1}{\sqrt{6}} \left(e^{-\frac{i\hbar}{I}t}|1, -1\rangle + e^{-\frac{3i\hbar}{2I}t}|2, 1\rangle - 2e^{-\frac{3i\hbar}{2I}t}|2, 0\rangle \right) . \tag{9.17}$$

The return probability is

$$|\langle\psi(0)|\psi(t)\rangle|^2 = \frac{1}{36} \left| e^{-\frac{i\hbar}{I}t} + e^{-\frac{3i\hbar}{2I}t} + 4e^{-\frac{3i\hbar}{2I}t} \right|^2$$

$$= \frac{1}{36}\left|1 + 5e^{-\frac{i\hbar}{7}t}\right|^2 = -\frac{5}{9}\sin^2(\hbar t/2I).$$ (9.18)

(b) The expectation values of $\hat{\mathbf{L}}$ and $\hat{L}^2$ will be independent of time, since the angular momentum commutes with the Hamiltonian and it is a constant of the motion. We have

$$\hat{L}_+|\psi(0)\rangle = \frac{\hbar}{\sqrt{6}}\left(\sqrt{2}|1,0\rangle + 2|2,2\rangle - 2\sqrt{6}|2,1\rangle\right), \quad \hat{L}_-|\psi(0)\rangle = \hbar\left(|2,0\rangle - 2|2,-1\rangle\right)$$

and

$$\hat{L}_x|\psi(0)\rangle = \frac{\hbar}{2}\left(\frac{1}{\sqrt{3}}|1,0\rangle + \sqrt{\frac{2}{3}}|2,2\rangle - 2|2,1\rangle + |2,0\rangle - 2|2,-1\rangle\right)$$

$$\hat{L}_y|\psi(0)\rangle = -\frac{i}{2}\left(\frac{1}{\sqrt{3}}|1,0\rangle + \sqrt{\frac{2}{3}}|2,2\rangle - 2|2,1\rangle - |2,0\rangle + 2|2,-1\rangle\right)$$ (9.19)

$$\hat{L}_z|\psi(0)\rangle = \frac{\hbar}{\sqrt{6}}\left(-|1,-1\rangle + |2,1\rangle\right)$$

Note also that $\langle\psi|\hat{L}_j^2|\psi\rangle = \langle\psi|\hat{L}_j^\dagger\hat{L}_j|\psi\rangle = \left\|\hat{L}_j|\psi\rangle\right\|^2$. From these we obtain

$$\langle\hat{L}_x\rangle = \langle\hat{L}_z\rangle = 0, \quad \langle\hat{L}_y\rangle = -\frac{2\hbar}{\sqrt{6}} \quad \text{and} \quad \langle\hat{L}_x^2\rangle = \langle\hat{L}_y^2\rangle = \frac{5\hbar^2}{2}, \quad \langle\hat{L}_z^2\rangle = \frac{\hbar^2}{3}$$

and, finally

$$(\Delta L_x)^2 = \frac{5\hbar^2}{2}, \quad (\Delta L_y)^2 = \frac{11\hbar^2}{6}, \quad (\Delta L_z)^2 = \frac{\hbar^2}{3}.$$ (9.20)

Thus, the inequality $(\Delta L_x)(\Delta L_z) \geq \hbar|\langle L_y\rangle|/2$ reduces to $5 > 1$, which is obviously true.

9.2 Spherical Harmonics

In this section, we shall consider the angular momentum eigenstates in the $\{x\}$ representation. The corresponding eigenfunctions go by the name *Spherical Harmonics*

$$Y_{\ell m}(\theta, \phi) = \langle\mathbf{r}|\ell, m\rangle.$$ (9.21)

where θ and ϕ are the polar and azimuthal angles of spherical coordinates. Sometimes we shall collectively refer to the pair of these angles in terms of the corresponding solid angle Ω and denote the spherical harmonics as $Y_{\ell m}(\Omega)$. Using the expressions of the angular momentum operators in this basis, given by (8.14), (8.15), we may cast the relevant eigenvalue equations

$$\hat{L}^2|\ell, m\rangle = \hbar^2\ell(\ell + 1)|\ell, m\rangle \quad \hat{L}_z|\ell, m\rangle = \hbar m|\ell, m\rangle$$ (9.22)

in the form

$$\frac{1}{\sin\theta}\frac{\partial}{\partial\theta}\left(\sin\theta\frac{\partial Y_{\ell m}}{\partial\theta}\right) + \frac{1}{\sin^2\theta}\frac{\partial^2 Y_{\ell m}}{\partial\phi^2} = -\ell(\ell + 1)Y_{\ell m}$$

$$-i\frac{\partial Y_{\ell m}}{\partial\phi} = m Y_{\ell m}$$ (9.23)

The second of these equations can be readily integrated to give

$$Y_{\ell m}(\theta, \phi) = Y_{\ell m}(\theta, 0) \, e^{im\phi} .\tag{9.24}$$

Next, let's consider the maximal $\hat{L}_z$-eigenfunction $Y_{\ell\ell}$ which satisfies the equation

$$\hat{L}_+|\ell, \ell\rangle = 0 \implies -i\hbar e^{i\phi} \left(i\frac{\partial}{\partial\theta} - \cot\theta \frac{\partial}{\partial\phi} \right) Y_{\ell\ell}(\theta, \phi) = 0$$

or

$$\left(\frac{d}{d\theta} - \ell\cot\theta \right) Y_{\ell\ell}(\theta, 0) = 0\tag{9.25}$$

with solution

$$Y_{\ell\ell}(\theta, \phi) = C_\ell \, e^{i\ell\phi} \, (\sin\theta)^\ell .\tag{9.26}$$

The coefficient C_ℓ will be fixed at the end by normalization. The rest of the eigenfunctions are determined by the repeated action of the lowering operator $\hat{L}_-$. For example

$$Y_{\ell,\ell-1}(\theta, \phi) = \hat{L}_- Y_{\ell\ell}(\theta, \phi) = i\hbar e^{i\phi} \left(-i\frac{\partial}{\partial\theta} - \cot\theta \frac{\partial}{\partial\phi} \right) Y_{\ell\ell}(\theta, \phi)$$

$$= C_\ell \sqrt{2\ell} \, (\sin\theta)^{\ell-1} \cos\theta \, e^{i(\ell-1)\phi} .\tag{9.27}$$

A general formula can be derived by induction as

$$Y_{\ell m}(\theta, \phi) = \mathcal{N}_{\ell m} \frac{e^{im\phi}}{(\sin\theta)^m} \left(\frac{d}{d\cos\theta} \right)^{\ell-m} (\sin\theta)^{2\ell} \quad for\ m \geq 0,\tag{9.28}$$

where we have absorbed the constant C_ℓ in the overall factor $\mathcal{N}_{\ell m}$ which is fixed by the normalization condition[2]

$$\int_0^{2\pi} d\phi \int_{-1}^1 d\cos\theta \, Y_{\ell m}^*(\theta, \phi) \, Y_{\ell'm'}(\theta, \phi) = \delta_{\ell\ell'}\delta_{mm'}\tag{9.29}$$

to be

$$\mathcal{N}_{\ell m} = \frac{(-1)^\ell}{2^\ell \ell!} \left\{ \frac{(2\ell+1)}{4\pi} \frac{(\ell+m)!}{(\ell-m)!} \right\}^{1/2} .\tag{9.30}$$

---

[2]Note that the *Completeness* relation is

$$\sum_{\ell=0}^\infty \sum_{m=-\ell}^\ell Y_{\ell m}(\Omega) Y_{\ell m}^*(\Omega') = \delta(\cos\theta - \cos\theta')\delta(\phi - \phi') .$$

Table 9.1 Spherical harmonics

$Y_{00} = \frac{1}{\sqrt{4\pi}}$

$Y_{1,-1} = \sqrt{\frac{3}{8\pi}}\, e^{-i\phi} \sin\theta, \quad Y_{10} = \sqrt{\frac{3}{4\pi}} \cos\theta, \quad Y_{1,1} = -\sqrt{\frac{3}{8\pi}}\, e^{i\phi} \sin\theta$

$Y_{2,-1} = \sqrt{\frac{15}{8\pi}}\, e^{-i\phi} \cos\theta \sin\theta, \quad Y_{2,0} = -\sqrt{\frac{5}{16\pi}}\left(1 - 3\cos^2\theta\right), \quad Y_{2,1} = -\sqrt{\frac{15}{8\pi}}\, e^{i\phi} \cos\theta \sin\theta$

$Y_{2,-2} = \sqrt{\frac{15}{32\pi}}\, e^{-2i\phi} \sin^2\theta, \quad Y_{2,2} = \sqrt{\frac{15}{32\pi}}\, e^{2i\phi} \sin^2\theta$

The negative m spherical harmonics can be obtained by observing that

$$-i\hbar \frac{\partial}{\partial\phi} Y_{\ell m}(\theta, \phi) = \hbar m\, Y_{\ell m}(\theta, \phi) \implies -i\hbar \frac{\partial}{\partial\phi} Y_{\ell m}^*(\theta, \phi) = -\hbar m\, Y_{\ell m}^*(\theta, \phi),$$

which implies that up to a phase factor N, we have $Y_{\ell,-m} = N\, Y_{\ell m}^*$. This phase factor is taken by convention to be $(-1)^m$ and we have

$$Y_{\ell,-m}(\theta, \phi) = (-1)^m\, Y_{\ell m}^*(\theta, \phi). \tag{9.31}$$

The first few spherical harmonics are listed in Table 9.1

Spherical harmonics with $m = 0$ can be seen to be independent of the azimuthal angle ϕ. From the general expression (9.28) we get

$$Y_{\ell,0}(\theta) = \mathcal{N}_{\ell 0} \left(\frac{d}{d\cos\theta}\right)^\ell (\sin\theta)^{2\ell} = \frac{(-1)^\ell}{2^\ell \ell!} \left(\frac{2\ell+1}{4\pi}\right)^{1/2} \left(\frac{d}{d\cos\theta}\right)^\ell \left(1 - \cos^2\theta\right)^\ell \tag{9.32}$$

This is however proportional to the so-called *Legendre Polynomials*[3] $P_\ell(x)$, defined as

$$P_\ell(x) \equiv \frac{(-1)^\ell}{2^\ell \ell!} \left(\frac{d}{dx}\right)^\ell \left(1 - x^2\right)^\ell. \tag{9.33}$$

Therefore, we have the relation

$$Y_{\ell,0}(\theta) = \sqrt{\frac{4\pi}{(2\ell+1)}}\, P_\ell(\cos\theta). \tag{9.34}$$

The first few Legendre Polynomials are $P_0(x) = 1$, $P_1(x) = x$, $P_2 = \frac{1}{2}(3x^2 - 1)$.

---

[3]The Legendre Polynomials are a complete orthonormal set of functions in the interval $[-1, 1]$. Note the orthonormality relation $\int_{-1}^{+1} dx\, P_\ell(x) P_{\ell'}(x) = \frac{2\delta_{\ell\ell'}}{(2\ell+1)}$.

Closing this section it is important to note that, since the angular momentum is invariant under spatial reflection (i.e., parity)

$$\mathbf{r} \to -\mathbf{r}, \quad \mathbf{p} \to -\mathbf{p} \quad \Longrightarrow \quad \mathbf{L} \to \mathbf{L} \tag{9.35}$$

the parity operator will commute with it, namely, $\left[\hat{P}, \hat{\mathbf{L}}\right] = 0$, and the two observables will have common eigenvectors. In other words, the angular momentum eigenfunctions (spherical harmonics) will be eigenfunctions of parity, i.e., will be even and odd under spatial reflection. In spherical coordinates, the parity operation reads

$$\mathbf{r} \to -\mathbf{r} \quad \Longrightarrow \quad \begin{cases} r \to r \\ \theta \to \pi - \theta \\ \phi \to \pi + \phi. \end{cases} \tag{9.36}$$

Under this reflection, the general formula for spherical harmonics (9.28) gives

$$Y_{\ell m}(\pi - \theta, \pi + \phi) = (-1)^\ell \, Y_{\ell m}(\theta, \phi) \tag{9.37}$$

Thus, the parity of each $|\ell, m\rangle$ eigenstate is $(-1)^\ell$.

Example 9.2 A rigid rotator is in a state with wave function

$$\psi(\theta) = \sqrt{\frac{3}{4\pi}} \cos^2 \theta . \tag{9.38}$$

Calculate the uncertainty $(\Delta \mathbf{L})^2$.

Note that

$$\psi(\theta) = \frac{1}{\sqrt{3}} \left(Y_{00} + \frac{2}{\sqrt{5}} Y_{20} \right) \tag{9.39}$$

We have

$$\hat{L}^2 \psi(\theta) = 4\hbar^2 \sqrt{\frac{3}{5}} Y_{20}(\theta) \quad and \quad \langle \psi | \hat{L}^2 | \psi \rangle = \frac{8\hbar^2}{5} . \tag{9.40}$$

We also have

$$\hat{\mathbf{L}}\psi = \left(\hat{x}\frac{1}{2} \left(\hat{L}_+ + \hat{L}_- \right) + \hat{y}\frac{1}{2i} \left(\hat{L}_+ - \hat{L}_- \right) + \hat{z}\hat{L}_z \right) \psi(\theta)$$

$$= \hat{x}\hbar\sqrt{\frac{2}{5}} \left(Y_{21}(\Omega) + Y_{2,-1}(\Omega) \right) - i\hat{y}\hbar\sqrt{\frac{2}{5}} \left(Y_{21}(\Omega) - Y_{2,-1}(\Omega) \right) . \tag{9.41}$$

From this we obtain

$$\langle \psi | \hat{\mathbf{L}} | \psi \rangle = 0 \quad and \quad (\Delta \mathbf{L})^2 = \langle \psi | \hat{L}^2 | \psi \rangle - \langle \psi | \hat{\mathbf{L}} | \psi \rangle^2 = \frac{8\hbar^2}{5} . \tag{9.42}$$

Problems and Exercises

9.1 Show that the uncertainty $(\Delta L_x)^2$ in an eigenstate $|\ell, m\rangle$ is

$$(\Delta L_x)^2 = \frac{\hbar^2}{2} \left(\ell(\ell + 1) - m^2 \right) .$$

9.2 A system is in a state

$$|\psi\rangle = \frac{1}{\sqrt{2}}|1, 1\rangle + \frac{1}{2}|1, 0\rangle + \frac{1}{2}|1, -1\rangle ,$$

where the states $|\ell, m\rangle$ are the eigenstates of angular momentum.

(a) Calculate the expectation values $\langle \hat{L}_x \rangle$ and $\langle \hat{L}_x^2 \rangle$.
(b) What are the possible values of L^2 and L_z that can be the outcome of measurements and what are the corresponding probabilities of each combination.

9.3 Consider a system with angular momentum $\ell = 1$. Construct the matrices $\mathcal{L}_x$, $\mathcal{L}_y$, $\mathcal{L}_z$ that represent $\hat{L}_x$, $\hat{L}_y$, $\hat{L}_z$ in the basis $|1, 1\rangle$, $|1, 0\rangle$, $|1, -1\rangle$ and find the corresponding eigenvalues and eigenvectors. Verify the following matrix equality

$$e^{-\frac{i}{\hbar}\beta \mathcal{L}_y} = 1 - \frac{i}{\hbar}\mathcal{L}_y \sin \beta - \frac{1}{\hbar^2}\mathcal{L}_y^2(1 - \cos \beta) .$$

9.4 The Hamiltonian of a particle can be written in the form

$$\hat{H} = \alpha \hat{L}^2 + \beta \hat{L}_- \hat{L}_+^2 \hat{L}_- ,$$

where α, β positive numbers. Verify that this Hamiltonian is hermitian and find its eigenvalues and eigenstates.

9.5 Consider a system characterized only by its angular momentum with Hamiltonian $\hat{H} = \hat{L}^2/2I$, where I is a moment of inertia parameter (rigid rotor). The system is in a state
$$|\psi\rangle = N \left(|1, 1\rangle + a|1, 0\rangle + |1, -1\rangle \right) ,$$

where N is a normalization constant and a a parameter. What are the possible values of L^2, L_z and what is their corresponding probability in the case of a measurement of these quantities? Calculate $\langle L_z \rangle$ and (ΔL_z). Find the value of the parameter a for which $|\psi\rangle$ becomes an eigenstate of $\hat{L}_x$.

9.6 A system of angular momentum $j = 1$ is described by a Hamiltonian

$$\hat{H} = a + b \hat{J}_x ,$$

with a, b real parameters. At a time $t = 0$ the system is subject to a measurement of J_z with the value $+\hbar$ as outcome. Later on at a time $t > 0$ the system is subject to a second measurement, this time of J_y. What is the probability of a possible outcome $-\hbar$?

9.7 A system of angular momentum $j = 1$ is described by a Hamiltonian

$$\hat{H} = \frac{a}{\hbar} \hat{J}_z^2 + \frac{b}{\hbar} \left(\hat{J}_x^2 - \hat{J}_y^2 \right) ,$$

where a, b real parameters. If the system occupies initially a state $|\psi(0)\rangle = |1, 1\rangle$, find the state of the system $|\psi(t)\rangle$ at a later time $t > 0$. What is the probability for the system to return to the initial state?

9.8 A *rigid rotor* with moment of inertia I, described by the Hamiltonian $\hat{H} = \frac{\hat{L}^2}{2I}$, is in a state with wave function

$$\psi(\theta) = N (1 + \cos\theta)^2 ,$$

where N normalization constant. Express this wave function in terms of spherical harmonics. What are the possible outcomes of energy measurements? What is the probability of each outcome.

9.9 A *rigid rotor* with moment of inertia I, described by the Hamiltonian $\hat{H} = \frac{\hat{L}^2}{2I}$, is initially ($t = 0$) in a state with wave function

$$\psi(\theta) = \frac{1}{\sqrt{8\pi}} \left(1 + \sqrt{3} \sin\theta \cos\phi \right) .$$

Find the wave function at a later time $t > 0$ and calculate the probability to find the system again in the initial state.

9.10 A quantum system has total angular momentum $j = 3/2$. Construct the angular momentum matrices in the basis of eigenvalues $|jm\rangle$ of $\hat{J}^2$ and $\hat{J}_z$.

References

1. E. Merzbacher, *Quantum Mechanics*, 3rd edn. (Wiley, New York, 1998)
2. A. Messiah, *Quantum Mechanics* (Dover publications, Mineola, 1958). Single-volume reprint of the John Wiley & Sons, New York, two-volume 1958 edition
3. W. Greiner, B. Müller, *Quantum Mechanics Symmetries*, 2nd edn. (Springer, Berlin, 1992)
4. D. Griffiths, *Introduction to Quantum Mechanics*, 2nd edn. (Cambridge University Press, Cambridge, 2017)

Chapter 10
Spin

10.1 The Stern–Gerlach Experiment

Not all physical observables of microscopic systems have a classical analogue. The most important such property is the *spin* of elementary particles. The experiment carried out in 1922 by Otto Stern and Walther Gerlach played a decisive role in the discovery of the spin of the electron. The essential part of the Stern–Gerlach apparatus is a beam of neutral atoms (Silver) deflected by the inhomogeneous magnetic field produced by an electromagnet. Silver atoms consist[1] of 47 electrons, 46 of which form a sphere of zero angular momentum. All the angular momentum of the atom comes exclusively from the forty-seventh electron. For the present discussion of the Stern–Gerlach phenomena the whole atom can be essentially replaced by this single electron. A classical particle having electric charge and angular momentum would also have a *magnetic dipole moment*. Thus, the electron would be expected to have a magnetic dipole moment $\mathbf{m} = \frac{e\mathbf{L}}{2m_e c}$, where m_e and e are its mass and charge. The presence of the speed of light c is due to the particular system of units used. In the presence of a magnetic field $\mathbf{B}$, there will be a contribution to the potential energy $\Delta H = -\mathbf{m} \cdot \mathbf{B}$ and, for an inhomogeneous magnetic field, a resulting force $\mathbf{F} = \nabla (\mathbf{m} \cdot \mathbf{B})$. The atoms of the Stern–Gerlach beam, to a very good approximation, can be considered as a beam of single electrons. Assuming that these electrons are in an angular momentum eigenstate $|\ell, m\rangle$, if the gradient of the magnetic field is mostly along the $+\hat{z}$-direction, the force exerted on them simplifies to

$$\mathbf{F} \approx \hat{z} \frac{e\hbar m}{2m_e c} \left(\frac{\partial B_z}{\partial z} \right) . \tag{10.1}$$

Atoms in an $\ell = 1$ state would be expected to be separated into three branches, namely, those with $m = 1$ that will feel an upward force, those with $m = 0$ that will feel no force at all, and those with $m = -1$ that will feel a downward force. Never-

---

[1] Apart from the nucleus that plays no role in the phenomena to be discussed.

© Springer Nature Switzerland AG 2019
K. Tamvakis, *Basic Quantum Mechanics*, Undergraduate Texts in Physics,
https://doi.org/10.1007/978-3-030-22777-7_10

theless, the results of the Stern–Gerlach experiment are quite different. The atomic beam exhibits an additional splitting in *two* branches, an upper branch attributable to a term $e\hbar B'_z/2m_e c$ and a downward branch attributable to a term $-e\hbar B'_z/2m_e c$.

The explanation of the results of the Stern–Gerlach experiment became ultimately possible only with the hypothesis of the existence of a purely quantum attribute of the electron without a classical analogue, namely, the electron *spin*. The electron spin consists of three Hermitian operators $\hat{S}_x$, $\hat{S}_y$, $\hat{S}_z$ that satisfy the commutation relation

$$\left[\hat{S}_x, \, \hat{S}_y \right] = i\hbar \, \hat{S}_z \tag{10.2}$$

which is the same as the corresponding commutation relation for the orbital angular momentum. The general eigenvalue problem of these operators is

$$\hat{S}^2|s, m\rangle = \hbar^2 s(s + 1)|s, m\rangle, \quad \hat{S}_z|s, m\rangle = \hbar m|s, m\rangle \tag{10.3}$$

and in the case of the electron is

$$\hat{S}^2|1/2, m\rangle = 3\hbar^2/4|1/2, m\rangle, \quad \hat{S}_z|1/2, m\rangle = \hbar m|1/2, m\rangle, \tag{10.4}$$

corresponding to the principle quantum number being $s = 1/2$ and the azimuthal quantum number being $m = \pm 1/2$. The explanation of the Stern–Gerlach phenomena requires the existence of an *electron magnetic dipole moment due to spin*

$$\mathbf{m}_s = \frac{g e \mathbf{S}}{2 m_e c}. \tag{10.5}$$

The constant g is called *Lande factor* and for the electron it is to a very good approximation $g = 2$. The corresponding interaction energy will be

$$\Delta H = -\mathbf{m}_s \cdot \mathbf{B} \sim -\frac{e}{m_e c} S_z B_z. \tag{10.6}$$

If the original beam consists of silver atoms with one active electron (the forty-seventh) in a spherically symmetric state of zero angular momentum, then it will only split in two branches, one resulting from atoms in the $S_z = +\hbar/2$ state feeling an upward force $+\frac{e\hbar m}{2m_e c}\left(\frac{\partial B_z}{\partial z}\right)$ and a second resulting from atoms in the $S_z = -\hbar/2$ state feeling a downward force $-\frac{e\hbar m}{2m_e c}\left(\frac{\partial B_z}{\partial z}\right)$.

It turns out that not only the electron but all elementary particles have spin. The value of the s quantum number ($s = 1/2$ for the electron) is part of the identity of each particle, together with other fundamental attributes like their mass. Although many types of particles have spin $s = 1/2$ (electron, proton, neutrino, ...), there exist particles with other values, integers like $s = 0, 1, 2, \ldots$ (pion, photon, graviton, ...), or half-integers like $3/2, 5/2, \ldots$ (exotic hadrons).

10.2 Spin 1/2

The case of particles with $s = 1/2$ is not only the simplest realization of spin but also the most common, since most of ordinary matter (electrons, protons, etc.) is made up from spin-1/2 particles. It is important to keep in mind that since the spin is not related to the degrees of freedom that have to do with the spatial motion of the system, the spin operators $\hat{\mathbf{S}}$ commute with the position, momentum, orbital angular momentum, and all related operators[2]

$$\left[\hat{\mathbf{S}}, \mathbf{r}\right] = \left[\hat{\mathbf{S}}, \hat{\mathbf{p}}\right] = \left[\hat{\mathbf{S}}, \hat{\mathbf{L}}\right] = 0. \tag{10.7}$$

The spin operators $\hat{\mathbf{S}}$ satisfy the angular momentum commutation algebra

$$\left[\hat{S}_i, \hat{S}_j\right] = i\hbar \epsilon_{ijk} \hat{S}_k, \tag{10.8}$$

with the indices i, j, k taking up any of the 1, 2, 3 values. The eigenvalue equations are

$$\hat{S}^2|1/2, m\rangle = \frac{3}{4}\hbar^2 |1/2, m\rangle, \quad \hat{S}_z|1/2, m\rangle = \hbar m|1/2, m\rangle \tag{10.9}$$

and m can have only two values, since

$$-1/2 \le m \le 1/2 \implies m = -1/2, +1/2. \tag{10.10}$$

We shall symbolize the spin eigenstates as[3]

$$|\pm\rangle = |1/2, \pm 1/2\rangle \quad \begin{cases} \hat{S}^2|\pm\rangle = \frac{3}{4}\hbar^2|\pm\rangle \\ \hat{S}_z|\pm\rangle = \pm\frac{\hbar}{2}|\pm\rangle. \end{cases} \tag{10.11}$$

All the operator machinery used in solving the general angular momentum eigenvalue problem in Chap. 9 is in our disposal here. For example, raising and lowering operators are defined as $\hat{S}_\pm = \hat{S}_x \pm i\hat{S}_y$ with the properties

$$\hat{S}_\pm|+\rangle = 0, \quad \hat{S}_-|-\rangle = 0, \quad \hat{S}_+|-\rangle = \hbar|+\rangle, \quad \hat{S}_-|+\rangle = \hbar|-\rangle. \tag{10.12}$$

In the framework of the spin-eigenstate basis $|+\rangle$, $|-\rangle$, which is automatically orthonormal, the spin operators are represented by 2×2 matrices

$$\mathbf{S}_{ab} = \langle a|\hat{\mathbf{S}}|b\rangle \quad \text{with} \quad a, b = \pm. \tag{10.13}$$

---

[2]For the general theory of spin 1/2, see also [1–3].

[3]An alternative notation that can be more convenient at times is $|\uparrow\rangle = |+\rangle$ and $|\downarrow\rangle = |-\rangle$.

Starting from the simplest case of $\hat{S}^2$ and $\hat{S}_z$, we have

$$\langle\pm|\hat{S}^2|\pm\rangle = \tfrac{3}{4}\hbar^2\langle\pm|\pm\rangle = \tfrac{3}{4}\hbar^2$$
$$\Longrightarrow S^2 = \tfrac{3}{4}\hbar^2 \begin{pmatrix} 1 & 0 \\ 0 & 1 \end{pmatrix}$$
$$\langle\pm|\hat{S}^2|\mp\rangle = \tfrac{3}{4}\hbar^2\langle\pm|\mp\rangle = 0$$

$$\langle\pm|\hat{S}_z|\pm\rangle = \pm\tfrac{\hbar}{2}\langle\pm|\pm\rangle = \pm\tfrac{\hbar}{2}$$
$$\Longrightarrow S_z = \tfrac{\hbar}{2} \begin{pmatrix} 1 & 0 \\ 0 & -1 \end{pmatrix}$$
$$\langle\pm|\hat{S}_z|\mp\rangle = \mp\tfrac{\hbar}{2}\langle\pm|\mp\rangle = 0$$

$$(10.14)$$

as expected, since the matrix representing any observable in the basis of its eigenstates is diagonal, having as diagonal elements the corresponding eigenvalues. For S_x, we have

$$\langle+|\hat{S}_x|+\rangle = \tfrac{1}{2}\langle+|\left(\hat{S}_+ + \hat{S}_-\right)|+\rangle = \tfrac{\hbar}{2}\langle+|-\rangle = 0$$

$$\langle-|\hat{S}_x|-\rangle = \tfrac{1}{2}\langle-|\left(\hat{S}_+ + \hat{S}_-\right)|-\rangle = \tfrac{\hbar}{2}\langle-|+\rangle = 0$$
$$\Longrightarrow S_x = \frac{\hbar}{2}\begin{pmatrix} 0 & 1 \\ 1 & 0 \end{pmatrix}.$$
$$\langle+|\hat{S}_x|-\rangle = \tfrac{1}{2}\langle+|\left(\hat{S}_+ + \hat{S}_-\right)|-\rangle = \tfrac{\hbar}{2}\langle+|+\rangle = \tfrac{\hbar}{2}$$

$$\langle-|\hat{S}_x|+\rangle = \tfrac{1}{2}\langle-|\left(\hat{S}_+ + \hat{S}_-\right)|+\rangle = \tfrac{\hbar}{2}\langle-|-\rangle = \tfrac{\hbar}{2}$$

$$(10.15)$$

Similarly, for S_y, we get

$$\langle+|\hat{S}_y|+\rangle = \tfrac{1}{2i}\langle+|\left(\hat{S}_+ - \hat{S}_-\right)|+\rangle = i\tfrac{\hbar}{2}\langle+|-\rangle = 0$$

$$\langle-|\hat{S}_y|-\rangle = \tfrac{1}{2}\langle-|\left(\hat{S}_+ - \hat{S}_-\right)|-\rangle = i\tfrac{\hbar}{2}\langle-|+\rangle = 0$$
$$\Longrightarrow S_y = \frac{\hbar}{2}\begin{pmatrix} 0 & -i \\ i & 0 \end{pmatrix}.$$
$$\langle+|\hat{S}_y|-\rangle = \tfrac{1}{2i}\langle+|\left(\hat{S}_+ - \hat{S}_-\right)|-\rangle = -i\tfrac{\hbar}{2}\langle+|+\rangle = -i\tfrac{\hbar}{2}$$

$$\langle-|\hat{S}_y|+\rangle = \tfrac{1}{2i}\langle-|\left(\hat{S}_+ - \hat{S}_-\right)|+\rangle = i\tfrac{\hbar}{2}\langle-|-\rangle = i\tfrac{\hbar}{2}$$

$$(10.16)$$

On summarizing, we have

$$S_x = \frac{\hbar}{2}\begin{pmatrix} 0 & 1 \\ 1 & 0 \end{pmatrix}, \quad S_y = \frac{\hbar}{2}\begin{pmatrix} 0 & -i \\ i & 0 \end{pmatrix}, \quad S_z = \frac{\hbar}{2}\begin{pmatrix} 1 & 0 \\ 0 & -1 \end{pmatrix}. \qquad (10.17)$$

The matrices corresponding to $\hat{S}_{\pm}$ are

$$S_+ = \begin{pmatrix} 0 & 2 \\ 0 & 0 \end{pmatrix}, \quad S_- = \begin{pmatrix} 0 & 0 \\ 2 & 0 \end{pmatrix}. \tag{10.18}$$

The Pauli Matrices. The spin matrices obtained above are most conveniently expressed in terms of the three so-called *Pauli Matrices* denoted by σ_1, σ_2, σ_3 or σ_x, σ_y, σ_z

$$\sigma_x = \begin{pmatrix} 0 & 1 \\ 1 & 0 \end{pmatrix}, \quad \sigma_y = \begin{pmatrix} 0 & -i \\ i & 0 \end{pmatrix}, \quad \sigma_z = \begin{pmatrix} 1 & 0 \\ 0 & -1 \end{pmatrix}. \tag{10.19}$$

Any 2×2 matrix with complex elements can be written as a superposition of σ_1, σ_2, σ_3 and the 2×2-unit matrix I

$$A = \begin{pmatrix} a & b \\ c & d \end{pmatrix} = \frac{a}{2}(\sigma_z + I) + \frac{b}{2}(\sigma_x + i\sigma_y) + \frac{c}{2}(\sigma_x - i\sigma_y) + \frac{d}{2}(I - \sigma_z)$$

or

$$A = \frac{1}{2}(a + d)I + \frac{1}{2}(b + c)\sigma_x + \frac{i}{2}(b - c)\sigma_y + \frac{1}{2}(a - d)\sigma_z. \tag{10.20}$$

Thus, the four matrices I, σ_x, σ_y, σ_z are a basis in an abstract vector space defined by all 2×2 complex matrices.

The Pauli matrices have a number of very useful properties summarized in a few identities[4]

$$\begin{cases} \{\sigma_i, \sigma_j\} = 0 \text{ for } i \neq j \\ [\sigma_i, \sigma_j] = 2i\epsilon_{ijk}\sigma_k \quad\quad \Longrightarrow \quad \sigma_i\sigma_j = \delta_{ij}I + i\epsilon_{ijk}\sigma_k. \\ \sigma_x^2 = \sigma_y^2 = \sigma_z^2 = I \end{cases} \tag{10.21}$$

In addition, we also have the important property

$$(\mathbf{a} \cdot \sigma)^2 = a^2 \tag{10.22}$$

for any c-number vector $\mathbf{a}$. This is proven as

$$(\mathbf{a} \cdot \sigma)^2 = a_i a_j \sigma_i \sigma_j = a_i a_j \left(\delta_{ij}I + i\epsilon_{ijk}\sigma_k\right) = a^2 + (\mathbf{a} \times \mathbf{a}) \cdot \sigma = a^2.$$

---

[4]With $\{A, B\} = AB + BA$ we symbolize the *anticommutator* of two matrices or operators.

With the help of the last property, we may compute

$$e^{i\mathbf{a}\cdot\sigma} = \sum_{n=0}^{\infty} \frac{i^n}{n!}(\mathbf{a}\cdot\sigma)^n = \sum_{k=0}^{\infty} \frac{i^{2k}}{(2k)!}(\mathbf{a}\cdot\sigma)^{2k} + \sum_{k=0}^{\infty} \frac{i^{2k+1}}{(2k+1)!}(\mathbf{a}\cdot\sigma)^{2k+1}$$

$$= \sum_{k=0}^{\infty} \frac{(-1)^k}{(2k)!}a^{2k} + i\,(\hat{\mathbf{a}}\cdot\sigma)\sum_{k=0}^{\infty} \frac{(-1)^k}{(2k+1)!}a^{2k+1}$$

or

$$e^{i\mathbf{a}\cdot\sigma} = \cos a + i\,(\hat{\mathbf{a}}\cdot\sigma)\,\sin a, \tag{10.23}$$

where $\hat{\mathbf{a}}$ is the unit vector $\mathbf{a}/a$.

10.3 Spinors

Consider the system of a particle of spin 1/2. Any state $|\Psi\rangle$ of the system can be expanded in spin eigenstates as

$$|\Psi\rangle = \Psi_+|+\rangle + \Psi_-|-\rangle, \tag{10.24}$$

and represented in terms of a column *"spinor"*

$$|\Psi\rangle \implies \Psi = \begin{pmatrix} \Psi_+ \\ \Psi_- \end{pmatrix}. \tag{10.25}$$

The spin eigenstates themselves are represented by column spinors, namely,

$$\chi_+ = \begin{pmatrix} 1 \\ 0 \end{pmatrix}, \quad \chi_- = \begin{pmatrix} 0 \\ 1 \end{pmatrix}. \tag{10.26}$$

An arbitrary spinor is written in terms of them as

$$\chi = \begin{pmatrix} \alpha \\ \beta \end{pmatrix} = \alpha\,\chi_+ + \beta\,\chi_-. \tag{10.27}$$

Normalization of such a spinor reads

$$\chi^\dagger\chi = 1 \quad \leftrightarrow \quad (\alpha^*, \beta^*)\begin{pmatrix} \alpha \\ \beta \end{pmatrix} = |\alpha|^2 + |\beta|^2 = 1. \tag{10.28}$$

If we are describing the spatial degrees of freedom of the system in the $\{x\}$ representation, then we have a spinor wave function

$$|\Psi\rangle \implies \Psi(\mathbf{r}) = \begin{pmatrix} \Psi_+(\mathbf{r}) \\ \Psi_-(\mathbf{r}) \end{pmatrix}. \tag{10.29}$$

Normalization of the ket that describes a system with both spatial and spin degrees of freedom reads

$$\langle\Psi|\Psi\rangle = 1 \implies \int d^3r\, \Psi^\dagger(\mathbf{r})\Psi(\mathbf{r}) = \int d^3r\, \left(|\Psi_+(\mathbf{r})|^2 + |\Psi_-(\mathbf{r})|^2\right) = 1 \tag{10.30}$$

or

$$\int d^3r\, |\Psi_+(\mathbf{r})|^2 + \int d^3r\, |\Psi_-(\mathbf{r})|^2 = 1. \tag{10.31}$$

10.4 Spin Measurements

Although we have not gone into a detailed discussion of measurement processes in quantum systems, the basic rules for quantum measurements were included among the *"Axioms"*[5] of Quantum Mechanics stated in Chap. 5. If we put aside the spatial degrees of freedom, the behavior of a particle with spin under spin measurements is particularly simple and yet highly nontrivial and very instructive.

Let us consider such a system of a particle with spin (an electron) and assume that we have a Stern–Gerlach type of apparatus that measures the spin in any desired direction. Suppose that the apparatus has been arranged so that it measures the spin in the $\hat{x}$-direction and we feed it with a beam of electrons in the state $|+\rangle$, i.e., having spin $S_z \to +\hbar/2$. After the measurement, the only possible states will be eigenstates of S_x and the only possible values will be the eigenvalues of S_x. Let's determine the eigenstates and eigenvalues of S_x by solving the corresponding eigenvalue problem. Working in our standard S_z-eigenvector basis we may denote the eigenvalues by λ and the eigenvectors by a collumn vector η. We have

$$\eta = \begin{bmatrix} a \\ b \end{bmatrix} \implies S_x \eta = \lambda \eta \implies \frac{\hbar}{2} \begin{pmatrix} 0 & 1 \\ 1 & 0 \end{pmatrix} \begin{pmatrix} a \\ b \end{pmatrix} = \lambda \begin{pmatrix} a \\ b \end{pmatrix} \tag{10.32}$$

---

[5] **1.** The only outcomes of the measurement of an observable are its eigenvalues. **2.** The state of the system before the measurement of an observable A is a superposition of A-eigenstates $\psi = \sum_\alpha C_\alpha \psi_\alpha$, where $|C_\alpha|^2$ is the probability of each measured eigenvalue. **3.** The state of the system after a measurement of A that has given the eigenvalue α is the corresponding eigenstate ψ_α. Any subsequent measurement of A will yield exactly the same eigenvalue α and the system will continue to occupy the state ψ_α.

and

$$\begin{vmatrix} -\lambda & \hbar/2 \\ \hbar/2 & -\lambda \end{vmatrix} = 0 \implies \lambda = \pm\hbar/2 \text{ and } \eta_\pm = \frac{1}{\sqrt{2}} \begin{pmatrix} 1 \\ \pm 1 \end{pmatrix}. \tag{10.33}$$

Thus, the S_x-eigenstates are

$$|S_x, \pm\hbar/2\rangle = \frac{1}{\sqrt{2}} (|+\rangle \pm |-\rangle). \tag{10.34}$$

The probability to measure $S_x \to +\hbar/2$ is

$$\mathcal{P}(S_x = \hbar/2) = |\langle+|S_x, +\hbar/2\rangle|^2 = \frac{1}{2}. \tag{10.35}$$

Assume now that we perform instantaneously a measurement of S_z on the branch corresponding to spin $S_x = +\hbar/2$. What is the probability to measure $S_z = -\hbar/2$? According to our rules, it must be

$$\mathcal{P}(S_z = -\hbar/2) = |\langle-|S_x, +\hbar/2\rangle|^2 = \frac{1}{2}. \tag{10.36}$$

10.5 Time Evolution in a Homogeneous Magnetic Field

Consider the system of a particle of spin 1/2 under the influence of a homogeneous magnetic field **B** and let's ignore the spatial degrees of freedom. The system could be just an electron or an atomic nucleus. If the particle is in the presence of a homogeneous magnetic field, the magnetic dipole moment associated with spin[6] $\mathbf{m}_s = \frac{eg\mathbf{S}}{2Mc}$ will contribute to the energy with a term $-\mathbf{m}_s \cdot \mathbf{B}$. Having put aside the spatial degrees of freedom, assuming that they will not play any role in the phenomena to be discussed, the Hamiltonian of the system will be

$$\hat{H} = -\frac{eg}{2Mc}\hat{\mathbf{S}} \cdot \mathbf{B} = -\frac{egB}{2Mc}\mathbf{S} \cdot \hat{n}, \tag{10.37}$$

where the unit vector $\hat{n}$ corresponds to the direction of the homogeneous magnetic field. We may write $\hat{n}$ in terms of two angles as

---

[6]In the case of a nucleus e, g, and M stand for the corresponding charge, Lande factor ($g \neq 2$), and mass. Note, however, that even electrically neutral particles (e.g., the neutron) can have nonzero magnetic dipole moment due to their spin.

$$\hat{n} = n_x\,\hat{x} + n_y\,\hat{y} + n_z\,\hat{z} \;\; with \;\; \begin{cases} n_x = \sin\alpha\,\cos\beta \\[4pt] n_y = \sin\alpha\,\sin\beta \\[4pt] n_z = \cos\alpha \end{cases} \tag{10.38}$$

and, introducing the parameter[7]

$$\omega = \frac{|e|gB}{2Mc}, \tag{10.39}$$

we obtain the Hamiltonian matrix, expressed in the basis of $\hat{S}^2$, $\hat{S}_z$ eigenstates as

$$\mathcal{H} = \frac{\hbar\omega}{2}\,(\vec{\sigma}\cdot\hat{n}) = \frac{\hbar\omega}{2}\begin{pmatrix} n_z & n_x - in_y \\[4pt] n_x + in_y & -n_z \end{pmatrix}. \tag{10.40}$$

Let's assume now that the system is initially ($t = 0$) in a state $|\psi(0)\rangle$. The evolved state will be

$$|\psi(t)\rangle = e^{-\frac{i}{\hbar}\hat{H}t}|\psi(0)\rangle. \tag{10.41}$$

Expressing the state of the system in the above basis, we get

$$\begin{bmatrix} \psi_1(t) \\ \psi_2(t) \end{bmatrix} = e^{-\frac{i}{2}\omega t(\vec{\sigma}\cdot\hat{n})}\begin{bmatrix} \psi_1(0) \\ \psi_2(0) \end{bmatrix} = \left[\cos(\omega t/2) - i\,(\vec{\sigma}\cdot\hat{n})\sin(\omega t/2)\right]\begin{bmatrix} \psi_1(0) \\ \psi_2(0) \end{bmatrix}$$

$$= \begin{bmatrix} \cos(\omega t/2) - i\cos\alpha\,\sin(\omega t/2) & -i\sin\alpha\,e^{-i\beta}\,\sin(\omega t/2) \\[6pt] -i\sin\alpha\,e^{i\beta}\,\sin(\omega t/2) & \cos(\omega t/2) + i\cos\alpha\,\sin(\omega t/2) \end{bmatrix}\begin{bmatrix} \psi_1(0) \\ \psi_2(0) \end{bmatrix}$$

$$= \begin{bmatrix} (\cos(\omega t/2) - i\cos\alpha\,\sin(\omega t/2))\,\psi_1(0) - i\sin\alpha\,e^{-i\beta}\,\sin(\omega t/2)\,\psi_2(0) \\[6pt] -i\sin\alpha\,e^{i\beta}\,\sin(\omega t)\,\psi_1(0) + (\cos(\omega t) + i\cos\alpha\,\sin(\omega t/2))\,\psi_2(0) \end{bmatrix}. \tag{10.42}$$

Note that $|\psi(t + 2\pi/\omega)\rangle = -|\psi(t)\rangle$, the spinor $|\psi(t)\rangle$ being periodic in time with a period $4\pi/\omega$.

There is a number of interesting questions that we may answer based on the expression (10.42) of the time-evolved spinor wave vector. For example, we may ask what is the probability to return at time $t > 0$ to the initial state $|\psi(0)\rangle$. This will be

---

[7] ω is the so-called *Larmor frequency* for the electron ($e < 0$, $g \approx 2$).

$$P(t) = |\langle \psi(0)|\psi(t)\rangle|^2 \,. \tag{10.43}$$

We can always take one of the initial spinor components real, namely, $\psi_1(0) = C_1$, and set $\psi_2(0) = C_2 e^{i\gamma}$ with C_1, $C_2 = \sqrt{1 - C_1^2}$ real. Then, after some manipulations, we obtain

$$P(t) = \cos^2(\omega t/2) + \sin^2(\omega t/2) \left(\cos \alpha \left(C_1^2 - C_2^2\right) + 2C_1 C_2 \sin \alpha \cos(\beta - \gamma)\right)^2 \,. \tag{10.44}$$

For a spinor initially prepared in a $|+\rangle$ or a $|-\rangle$ state, the return probability is

$$P(t) = 1 - \sin^2(\omega t/2) \sin^2 \alpha \,. \tag{10.45}$$

What about the expectation value $\langle \mathbf{S}\rangle_t$ of the spin itself? The initial value of the spin operator expectation value depends on the initial state $|\psi(0)\rangle$. It is

$$\langle \mathbf{S}\rangle_0 = \langle \psi(0)|\mathbf{S}|\psi(0)\rangle = \left(\psi_1^*(0), \, \psi_2^*(0)\right) \begin{pmatrix} \frac{\hbar}{2}\hat{z} & \frac{\hbar}{2}(\hat{x} - i\hat{y}) \\ \frac{\hbar}{2}(\hat{x} + i\hat{y}) & -\frac{\hbar}{2}\hat{z} \end{pmatrix} \begin{pmatrix} \psi_1(0) \\ \psi_2(0) \end{pmatrix}$$

or, in terms of the $\psi_1(0) = C_1$, $\psi_2(0) = C_2 e^{i\gamma}$ parametrization,

$$\langle \mathbf{S}\rangle_0 = \frac{\hbar}{2} \left(\hat{z}(C_1^2 - C_2^2) + 2C_1 C_2 \left(\hat{x} \cos \gamma + \hat{y} \sin \gamma\right)\right) \,. \tag{10.46}$$

In order to obtain $\langle \mathbf{S}\rangle_t$, it is equally convenient but more instructive to work in the Heisenberg picture. The Heisenberg equation is

$$\frac{dS_i}{dt} = \frac{i}{\hbar} \left[\omega \left(\mathbf{S} \cdot \hat{n}\right), S_i\right] = -\omega \, \epsilon_{jik} S_k \hat{n}_j = \omega \left(\hat{n} \times \mathbf{S}\right)_i \tag{10.47}$$

or just

$$\frac{d\mathbf{S}}{dt} = \omega \hat{n} \times \mathbf{S} \,. \tag{10.48}$$

The spin vector operator is the sum of two orthogonal pieces

$$\mathbf{S} = \mathbf{S}_{\parallel} + \mathbf{S}_{\perp} \,, \tag{10.49}$$

where

$$\mathbf{S}_{\parallel} \equiv \hat{n} \left(\hat{n} \cdot \mathbf{S}\right), \quad \mathbf{S}_{\perp} \equiv \hat{n} \times \mathbf{S} \,. \tag{10.50}$$

On multiplying (10.48) by $\hat{n}$, we obtain $\hat{n} \cdot \dot{\mathbf{S}} = 0$ and conclude that $\mathbf{S}_{\parallel}$ is a constant of the motion, i.e.,

$$\hat{n} \cdot \mathbf{S}(t) = \hat{n} \cdot \mathbf{S}(0) \,. \tag{10.51}$$

The equation for $\mathbf{S}_\perp$ is

$$\frac{d\mathbf{S}_\perp}{dt} = \omega\,\hat{n} \times \mathbf{S}_\perp \implies \frac{d^2\mathbf{S}_\perp}{dt^2} = \omega^2\,\hat{n} \times \left(\hat{n} \times \mathbf{S}_\perp\right) = -\omega^2 \mathbf{S}_\perp \qquad (10.52)$$

with solution

$$\mathbf{S}_\perp(t) = \mathbf{S}_\perp(0)\cos(\omega t) + \left(\hat{n} \times \mathbf{S}(0)\right)\sin(\omega t). \qquad (10.53)$$

The total spin is

$$\mathbf{S}(t) = \hat{n}\left(\hat{n} \cdot \mathbf{S}(0)\right)(1 - \cos(\omega t)) + \mathbf{S}(0)\cos(\omega t) + \left(\hat{n} \times \mathbf{S}(0)\right)\sin(\omega t) \qquad (10.54)$$

and its expectation value is

$$\langle\mathbf{S}\rangle_t = \hat{n}\left(\hat{n} \cdot \langle\mathbf{S}\rangle_0\right)(1 - \cos(\omega t)) + \langle\mathbf{S}\rangle_0\cos(\omega t) + \left(\hat{n} \times \langle\mathbf{S}\rangle_0\right)\sin(\omega t). \qquad (10.55)$$

A thorough examination of (10.55) reveals that this solution describes a vector $\langle\mathbf{S}\rangle_0$ that has a constant projection on the $\hat{n}$-axis, while its vertical part ($\langle\mathbf{S}_\perp\rangle_0$) describes a circle with angular velocity ω. Thus, $\langle\mathbf{S}\rangle_0$ sweeps a cone having $\hat{n}$ as its principal symmetry axis. This has been sketched in Fig. 10.1. The expectation value $\langle\mathbf{S}\rangle_t$ is periodic with period $T = 2\pi/\omega$. This is half the period of the Schroedinger picture spinor.

Example 10.1 Consider a spin 1/2 system subject to a measurement of the quantity $\Sigma = S_x + \alpha S_y$, with α a known parameter. At a later time, a second measurement, of S_x this time, is performed. The Hamiltonian of the system is $\hat{H} = -\omega\,\hat{S}_z$. Calculate the probability of each outcome.

The possible outcomes of the first measurement will be the eigenvalues of Σ. In the standard $|\pm\rangle$ basis of S_z eigenvectors, the eigenvalue problem of Σ is

$$\begin{pmatrix} 0 & 1 - i\alpha \\ 1 + i\alpha & 0 \end{pmatrix}\begin{pmatrix} \eta_1 \\ \eta_2 \end{pmatrix} = \sigma\begin{pmatrix} \eta_1 \\ \eta_2 \end{pmatrix} \qquad (10.56)$$

with eigenvalues

$$\begin{vmatrix} -\sigma & 1 - i\alpha \\ 1 + i\alpha & -\sigma \end{vmatrix} = 0 \implies \sigma = \pm\sqrt{1 + \alpha^2} \qquad (10.57)$$

and eigenvectors

$$\eta = \frac{1}{\sqrt{2}}\begin{pmatrix} 1 \\ \pm\frac{\sqrt{1+\alpha^2}}{1-i\alpha} \end{pmatrix} \implies \begin{aligned} |\Sigma_1\rangle &= \frac{1}{\sqrt{2}}\left(|+\rangle + \frac{\sqrt{1+\alpha^2}}{1-i\alpha}|-\rangle\right) \\ |\Sigma_2\rangle &= \frac{1}{\sqrt{2}}\left(|+\rangle - \frac{\sqrt{1+\alpha^2}}{1-i\alpha}|-\rangle\right). \end{aligned} \qquad (10.58)$$

Assume that the outcome is the eigenvalue $\sqrt{1 + \alpha^2}$. The state of the system at that instant (taken to be $t = 0$) will be

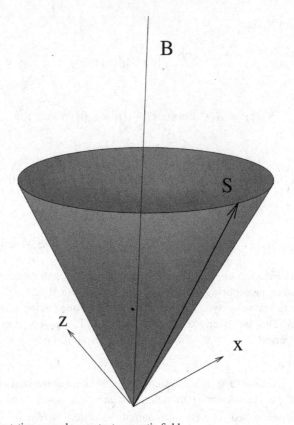

Fig. 10.1 Spin rotation around a constant magnetic field

$$|\psi(0)\rangle = \frac{1}{\sqrt{2}}\left(|+\rangle + \frac{\sqrt{1+\alpha^2}}{1-i\alpha}|-\rangle\right). \tag{10.59}$$

This state will involve in time and at the later time $t > 0$ will be

$$|\psi(t)\rangle = \frac{1}{\sqrt{2}}\left(e^{i\omega t/2}|+\rangle + \frac{\sqrt{1+\alpha^2}}{1-i\alpha}e^{-i\omega t/2}|-\rangle\right). \tag{10.60}$$

Right after the second measurement of S_x at a time t_0 the system will occupy one of the eigenstates $|S_x = \pm\hbar/2\rangle$. These states are easily found from the S_x-eigenvalue problem

$$\begin{pmatrix} 0 & 1 \\ 1 & 0 \end{pmatrix}\begin{pmatrix} \eta_1 \\ \eta_2 \end{pmatrix} = \sigma\begin{pmatrix} \eta_1 \\ \eta_2 \end{pmatrix} \implies \begin{vmatrix} -\sigma & 1 \\ 1 & -\sigma \end{vmatrix} = 0 \implies \sigma = \pm 1 \tag{10.61}$$

and

$$\begin{pmatrix} 0 & 1 \\ 1 & 0 \end{pmatrix}\begin{pmatrix} \eta_1 \\ \eta_2 \end{pmatrix} = \pm\begin{pmatrix} \eta_1 \\ \eta_2 \end{pmatrix} \implies \begin{aligned} |S_x = +\hbar/2\rangle &= \tfrac{1}{\sqrt{2}}\left(|+\rangle + |-\rangle\right) \\ |S_x = -\hbar/2\rangle &= \tfrac{1}{\sqrt{2}}\left(|+\rangle - |-\rangle\right). \end{aligned} \tag{10.62}$$

Therefore, the probability for each outcome will be

$$\mathcal{P}_{\pm}(t_0) = |\langle S_x = \pm\hbar/2|\psi(t_0)\rangle|^2 = \frac{1}{4}\left|e^{i\omega t_0/2} \pm \frac{\sqrt{1+\alpha^2}}{1-i\alpha}e^{-i\omega t_0/2}\right|^2$$

$$= \frac{1}{2}\left(1 \pm \frac{\cos(\omega t_0) - \alpha\sin(\omega t_0)}{\sqrt{1+\alpha^2}}\right) = \frac{1}{2}\left(1 \pm \cos(\omega t_0 + \beta)\right) = \begin{cases} \cos^2(\,(\omega t_0 + \beta)/2) \\ \sin^2(\,(\omega t_0 + \beta)/2), \end{cases} \quad (10.63)$$

where we have set $\alpha = \tan\beta$.

10.6 Spin in a Time-Dependent Magnetic Field

Consider the system of a particle with spin 1/2 (electron or nucleus) under the influence of a two-component magnetic field consisting of a homogeneous component (constant in time and space) and a time-dependent component perpendicular to it with a periodic time dependence, namely,[8]

$$\mathbf{B} = \hat{z}B_0 + B_1\left(\hat{x}\cos(\omega t) - \hat{y}\sin(\omega t)\right). \quad (10.64)$$

Assuming that the spatial degrees of freedom do not play any role in the phenomena to be discussed and can be ignored, the only relevant observable is the spin. The Hamiltonian operator will be

$$\hat{H} = \frac{-ge}{2Mc}\hat{\mathbf{S}}\cdot\mathbf{B}(t) = \frac{-ge}{2Mc}\left(B_0\hat{S}_z + B_1\hat{S}_x\cos(\omega t) - B_1\hat{S}_y\sin(\omega t)\right). \quad (10.65)$$

In terms of the Pauli matrices, it takes the form of the 2×2 matrix

$$\mathcal{H} = \begin{pmatrix} -\dfrac{ge\hbar B_0}{4Mc} & -\dfrac{ge\hbar B_1}{4Mc}e^{i\omega t} \\ -\dfrac{ge\hbar B_1}{4Mc}e^{-i\omega t} & \dfrac{ge\hbar B_0}{4Mc} \end{pmatrix}. \quad (10.66)$$

As in the previous section, where we have examined the spin motion in the presence of B_0 alone, we introduce the same frequency, using now the symbol[9] ω_0, and a new parameter η representing the relative ratio of the two magnetic fields, namely,

$$\omega_0 = \frac{geB_0}{2Mc} \quad \textbf{and} \quad \eta = \frac{B_1}{B_0}, \quad (10.67)$$

and have

---

[8]See also [1].

[9]We take $e > 0$ this time, having in mind applications in the case of nuclei.

$$\mathcal{H} = -\frac{1}{2}\hbar\omega_0 \begin{pmatrix} 1 & \eta\, e^{i\omega t} \\ \eta\, e^{-i\omega t} & -1 \end{pmatrix}. \tag{10.68}$$

Let us now assume that the initial state of the system is some state $|\psi(0)\rangle$. The time-evolved state will come out as a solution of the Schroedinger equation

$$i \begin{pmatrix} \dot{\psi}_1(t) \\ \dot{\psi}_2(t) \end{pmatrix} = -\frac{1}{2}\omega_0 \begin{pmatrix} 1 & \eta\, e^{i\omega t} \\ \eta\, e^{-i\omega t} & -1 \end{pmatrix} \begin{pmatrix} \psi_1(t) \\ \psi_2(t) \end{pmatrix}. \tag{10.69}$$

Note that since the Hamiltonian is time dependent, we do not have in our disposal the simple expression in terms of the exponential evolution operator which is only valid when the Hamiltonian is time independent. Therefore, we proceed to solve the Schroedinger equation as it stands. It is equivalent to the system of equations

$$\dot{\psi}_1 = i\frac{\omega_0}{2}\psi_1 + i\frac{\omega_0}{2}\eta e^{i\omega t}\psi_2$$
$$\dot{\psi}_2 = -i\frac{\omega_0}{2}\psi_2 + i\frac{\omega_0}{2}\eta e^{-i\omega t}\psi_1. \tag{10.70}$$

To simplify matters, we make the change of variables

$$\psi_1(t) = \phi_1(t)\,e^{i\omega_0 t/2}, \quad \psi_2(t) = \phi_2(t)\,e^{-i\omega_0 t/2} \tag{10.71}$$

and obtain

$$\begin{cases} \dot{\phi}_1 = \frac{i}{2}\eta\omega_0\, e^{i(\omega-\omega_0)t}\,\phi_2 \\ \dot{\phi}_2 = \frac{i}{2}\eta\omega_0\, e^{-i(\omega-\omega_0)t}\,\phi_1. \end{cases} \tag{10.72}$$

Differentiating once more and substituting back (10.72), we obtain the decoupled pair

$$\begin{cases} \ddot{\phi}_1 - i(\omega-\omega_0)\dot{\phi}_1 + \frac{1}{4}\eta^2\omega_0^2\phi_1 = 0 \\ \ddot{\phi}_2 + i(\omega-\omega_0)\dot{\phi}_2 + \frac{1}{4}\eta^2\omega_0^2\phi_2 = 0. \end{cases} \tag{10.73}$$

Inserting the trial solution $e^{i\lambda t}$, we obtain the following possibilities:

$$\phi_1 \propto e^{i\lambda_1 t} \implies \lambda_1^{(\pm)} = \frac{1}{2}(\omega - \omega_0 \pm \beta)$$
$$\phi_2 \propto e^{i\lambda_2 t} \implies \lambda_2^{(\pm)} = \frac{1}{2}(-\omega + \omega_0 \pm \beta) \tag{10.74}$$

with

$$\beta = \sqrt{(\omega - \omega_0)^2 + \eta^2\omega_0^2}. \tag{10.75}$$

Thus, the general solution is

$$\begin{cases} \phi_1(t) = e^{-\frac{i}{2}\omega_0 t}\left(A\,e^{\frac{i}{2}(\omega+\beta)t} + B\,e^{\frac{i}{2}(\omega-\beta)t} \right) \\[2mm] \phi_2(t) = e^{\frac{i}{2}\omega_0 t}\left(C\,e^{\frac{i}{2}(-\omega+\beta)t} + D\,e^{\frac{i}{2}(-\omega-\beta)t} \right) \end{cases} \tag{10.76}$$

or

$$\begin{cases} \psi_1(t) = A\,e^{\frac{i}{2}(\omega+\beta)t} + B\,e^{\frac{i}{2}(\omega-\beta)t} \\[2mm] \psi_2(t) = C\,e^{\frac{i}{2}(-\omega+\beta)t} + D\,e^{\frac{i}{2}(-\omega-\beta)t} . \end{cases} \tag{10.77}$$

Instead of solving for the most general initial conditions, let's simplify matters by assuming that initially the particle occupies one of the eigenstates of the *"undisturbed"* static system ($B_1 = 0$)

$$\psi_1(0) = 1, \ \psi_2(0) = 0. \tag{10.78}$$

Note that this implies also

$$\dot{\psi}_1(0) = i\omega_0/2, \quad \dot{\psi}_2(0) = i\omega_0\eta/2. \tag{10.79}$$

These initial conditions lead to

$$A + B = 1, \ C + D = 0, \ A - B = (\omega_0 - \omega)/\beta, \ C - D = \omega_0\eta/\beta$$

or

$$A = \frac{(\beta + \omega_0 - \omega)}{2\beta}, \ B = \frac{(\beta - \omega_0 + \omega)}{2\beta}, \ C = \frac{\omega_0\eta}{2\beta}, \ D = -\frac{\omega_0\eta}{2\beta} \tag{10.80}$$

and the solution is

$$\begin{pmatrix} \psi_1(t) \\ \psi_2(t) \end{pmatrix} = \begin{pmatrix} e^{i\omega t/2}\left(\cos(\beta t/2) + i\left(\frac{\omega_0-\omega}{\beta}\right)\sin(\beta t/2) \right) \\[3mm] i\frac{\omega_0\eta}{\beta}\,e^{-i\omega t/2}\sin(\beta t/2) \end{pmatrix}. \tag{10.81}$$

One of the questions of physical interest concerning the system is *what is the probability that the time-dependent magnetic field removes particles from the "up" state and populates the "down" state*. This probability is

$$P_-(t) = |\langle -|\psi(t)\rangle|^2 = \frac{\omega_0^2\eta^2}{\beta^2}\sin^2(\beta t/2) = \frac{(\omega_0\beta)^2\sin^2(\beta t/2)}{(\omega-\omega_0)^2 + (\beta\omega_0)^2}. \tag{10.82}$$

At the particular time $t_n = (2n + 1)\pi/\beta$, this probability attains its maximum value

$$\mathcal{P}_-^{(max)}(\omega) = \frac{(\omega_0 \beta)^2}{(\omega - \omega_0)^2 + (\beta \omega_0)^2}. \tag{10.83}$$

This is a function of the external frequency ω and clearly displays *resonance phenomena*, having a maximum at the Larmor frequency of the static system $\omega = \omega_0$. The resonance is sharpest in the limit $\eta \omega_0 \to 0$. Thus, the resonance phenomena are more intense for weak time-dependent magnetic fields $B_1 << B_0$.

The laboratory realization of these phenomena involves placing a given substance under study (e.g., water) in a strong magnetic field $\mathbf{B_0}$. At the same time, the alternating current of an electric coil can generate a weak rotating magnetic field $\mathbf{B_1}(t)$ of adjustable frequency ω. When we are close to $\omega \sim \omega_0$ the resonance phenomena become sharper and the absorption of energy increases. This is the *Nuclear Magnetic Resonance (NMR)* setup and can serve as a very accurate method of measuring the frequency ω_0, characteristic of each particular substance. The active part of the molecules is to a very good approximation their nuclear spin. The measured NMR frequencies are not only characteristic of the given molecule under study but also of the chemical environment it participates.

Example 10.2 Consider a particle of spin 1/2 with a Hamiltonian $\hat{H} = \hat{H}_0 + \hat{H}_s$ that consists of a spin-independent part $\hat{H}_0(\mathbf{r})$ and a spin-dependent part that can be parametrized as

$$\hat{H}_s = -\omega \hat{n} \cdot \mathbf{S} = -\frac{1}{2} \hbar \omega \begin{pmatrix} \cos \alpha & \sin \alpha \\ \sin \alpha & -\cos \alpha \end{pmatrix}. \tag{10.84}$$

Find the energy eigenfunctions and their corresponding eigenvalues in terms of the eigenfunctions and eigenvalues of the spin-independent part of the Hamiltonian.

By assumption $[\hat{H}_0, \hat{H}_s] = 0$. Substituting into the time-independent Schroedinger equation a trial product spinor wave function $\Psi = \psi_E(\mathbf{r}) \chi$ with $\chi = \begin{bmatrix} \chi_1 \\ \chi_2 \end{bmatrix}$, we obtain the solutions

$$\Psi_+(\mathbf{r}) = \psi_E(\mathbf{r}) \begin{pmatrix} \cos(\alpha/2) \\ \sin(\alpha/2) \end{pmatrix} \implies E_+ = E - \hbar\omega/2$$

$$\tag{10.85}$$

$$\Psi_-(\mathbf{r}) = \psi_E(\mathbf{r}) \begin{pmatrix} \sin(\alpha/2) \\ -\cos(\alpha/2), \end{pmatrix} \implies E_- = E + \hbar\omega/2$$

where $\hat{H}_0 \psi_E(\mathbf{r}) = E \psi_E(\mathbf{r})$.

Problems and Exercises

10.1 Consider an atomic nucleus of spin $s = 1/2$, participating in the lattice of a solid. If we ignore the spatial degrees of freedom, the nucleus is characterized only by its spin. The Hamiltonian of the system in the presence of a homogeneous magnetic field $\mathbf{B} = B\hat{y}$ is

$$\hat{H} = -\frac{ge_N}{2m_N c}\mathbf{B} \cdot \mathbf{S},$$

where $e_N > 0$ and m_N are the electric charge and the mass of the nucleus and g is the corresponding Lande factor. Verify the Heisenberg inequality

$$(\Delta S_x)^2 (\Delta S_y)^2 \geq \frac{\hbar^2}{4}\left|\langle \hat{S}_z \rangle\right|^2 \quad \text{given that} \quad |\psi(0)\rangle = \frac{1}{\sqrt{2}}\left(|+\rangle + e^{i\pi/3}|-\rangle\right)$$

for all times $t > 0$.

10.2 An electron is moving in the presence of a homogeneous magnetic field $\mathbf{B}$. The Hamiltonian of the system is

$$\hat{H} = \frac{1}{2m_e}\left(\mathbf{p} - \frac{e}{c}\mathbf{A}(\mathbf{r})\right)^2 - \frac{e}{m_e c}\mathbf{B} \cdot \mathbf{S},$$

where the electromagnetic vector potential is $\mathbf{A}(\mathbf{r}) = \frac{1}{2}\mathbf{B} \times \mathbf{r}$. Show that the Hamiltonian can be written in the form

$$\hat{H} = \frac{2}{\hbar^2 m_e}\left(\mathbf{S} \cdot \left(\mathbf{p} - \frac{e}{c}\mathbf{A}(\mathbf{r})\right)\right)^2.$$

10.3 A particle of spin $s = 1/2$ is in a normalized spinor state

$$|\psi\rangle = a|+\rangle + b|-\rangle,$$

where $|\pm\rangle$ are the eigenstates of $\hat{S}_z$. Show that you can determine the coefficients a, b from the expectation values $\langle \hat{S}_x \rangle$, $\langle \hat{S}_y \rangle$, $\langle \hat{S}_z \rangle$.

10.4 A system of spin 1/2 has a Hamiltonian

$$\hat{H} = a + b\hat{S}_y,$$

with a, b real constants. The system is subject at time $t = 0$ to a measurement of $\hat{S}_x$ with the outcome $+\hbar/2$. At a later time $t > 0$, a second measurement is performed, this time of $\hat{S}_z$. What is the probability of a value $-\hbar/2$?

10.5 For a system of spin 1/2 consider the operators

$$\hat{b} = \frac{1}{\hbar}\left(\hat{S}_x - i\hat{S}_y\right), \quad \hat{b}^\dagger = \frac{1}{\hbar}\left(\hat{S}_x + i\hat{S}_y\right).$$

Show that they are analogous to the *raising* and *lowering* operators of the harmonic oscillator—albeit with the commutators replaced by anticommutators—having the properties

$$\left\{\hat{b}, \hat{b}^\dagger\right\} = 1, \quad \left\{\hat{b}, \hat{b}\right\} = \left\{\hat{b}^\dagger, \hat{b}^\dagger\right\} = 0$$

and

$$\hat{b}^\dagger|-\rangle = |+\rangle, \quad \hat{b}|+\rangle = |-\rangle, \quad \hat{b}|-\rangle = \hat{b}^\dagger|+\rangle = 0.$$

10.6 Consider a system with spin **S** and orbital angular momentum **L**. Show that, if the definition of a vector operator is generalized to

$$\left[\hat{V}_i, \, \hat{J}_j\right] = i\hbar\epsilon_{ijk}\hat{V}_k,$$

with $\mathbf{J} \equiv \mathbf{L} + \mathbf{S}$, then the operators $\mathbf{r}, \hat{\mathbf{p}}, \hat{\mathbf{L}}, \hat{\mathbf{S}}, \hat{\mathbf{J}}$ are vectors. Show also that $\hat{\mathbf{J}}$ and $\hat{\mathbf{S}}$ are axial vectors.

10.7 Consider a system with spin $s = 1/2$ orbital angular momentum **L**. Show that

$$e^{\frac{i}{\hbar}\mathbf{a}\cdot\hat{\mathbf{J}}} \hat{\mathbf{S}} e^{-\frac{i}{\hbar}\mathbf{a}\cdot\hat{\mathbf{J}}} = \cos(a)\left(\hat{\mathbf{S}} - \hat{a}\left(\hat{a}\cdot\hat{\mathbf{S}}\right)\right) + \hat{a}\left(\hat{a}\cdot\hat{\mathbf{S}}\right) + \sin(a)\left(\hat{a}\times\hat{\mathbf{S}}\right),$$

where **a** a parametric vector, $\hat{a} = \mathbf{a}/a$ the corresponding unit vector, $\hat{\mathbf{S}}$ the spin operator, and $\mathbf{J} \equiv \mathbf{L} + \mathbf{S}$ the total angular momentum operator.

10.8 Consider a nucleus of mass m_N, positive charge e_N, and spin $s = 1$ in the presence of a homogeneous magnetic field $\mathbf{B} = \hat{z}B$. If we ignore the spatial degrees of freedom, the initial state of the system is the spinor

$$|\psi(0)\rangle = \cos(\alpha/2)|1, 1\rangle + \sin(\alpha/2)|1, -1\rangle,$$

where α is a known parameter. Calculate the time-evolved state $|\psi(t)\rangle$ and the uncertainty in spin $(\Delta\mathbf{S})$.

10.9 System of general spin s is described by a Hamiltonian $\hat{H} = -\omega\hat{S}_z$. Derive the expectation value $\langle\mathbf{S}\rangle_t$ for $t > 0$ and for an arbitrary state in terms of the expectation value $\langle\mathbf{S}\rangle_0$ at $t = 0$.

10.10 Consider a particle of spin $1/2$ with a Hamiltonian $\hat{H} = \hat{H}_0 + \hat{H}_s$ that consists of a spin-independent part $\hat{H}_0(\mathbf{r})$ and a spin-dependent part arising from an inhomogeneous magnetic field that can be parametrized as

$$\hat{H}_s = V_s(\mathbf{r})\begin{pmatrix} 1 & 0 \\ 0 & -1 \end{pmatrix}.$$

Consider the special case in which the particle is constrained to move only in the x-dimension, subject to harmonic forces $V_0(x) = \frac{1}{2}m\omega^2 x^2$ and $V_s = \lambda x$. Solve the energy eigenvalue problem.

References

1. D. Griffiths, *Introduction to Quantum Mechanics*, 2nd edn. (Cambridge University Press, Cambridge, 2017)
2. E. Merzbacher, *Quantum Mechanics*, 3rd edn. (Wiley, New York, 1998)
3. A. Messiah, *Quantum Mechanics* (Dover publications, Mineola, 1958). Single-volume reprint of the John Wiley & Sons, New York, two-volume 1958 edition

Chapter 11
Addition of Angular Momenta

11.1 Addition of Two Spins

In cases of many-particle systems or in the cases of composite particles like the nuclei, it is meaningful and useful to consider the *total spin* of the system. In the case of two independent spins $\mathbf{S}_1$ and $\mathbf{S}_2$, the total spin is

$$\hat{\mathbf{S}} = \hat{\mathbf{S}}_1 + \hat{\mathbf{S}}_2 \quad \left([\hat{\mathbf{S}}_1, \hat{\mathbf{S}}_2] = 0\right). \tag{11.1}$$

A maximal set of mutually commuting operators of the combined system is

$$\hat{S}_1^2, \hat{S}_{1z}, \hat{S}_2^2, \hat{S}_{2z}. \tag{11.2}$$

However, an equivalent set is the following:

$$\hat{S}_1^2, \hat{S}_2^2, \hat{S}^{2}, \hat{S}_z. \tag{11.3}$$

For $s_1 = s_2 = 1/2$, we have $\hat{S}_1^2 = \hat{S}_2^2 = 3\hbar^2/4$ and these two operators commute with everything. The fact that the remaining two of the second set commute can be shown easily as follows:

$$\left[\hat{S}^2, \hat{S}_z\right] = \left[\hat{S}_1^2 + \hat{S}_2^2 + 2\hat{\mathbf{S}}_1 \cdot \hat{\mathbf{S}}_2, \hat{S}_{1z} + \hat{S}_{2z}\right] = 2\left[\hat{\mathbf{S}}_1 \cdot \hat{\mathbf{S}}_2, \hat{S}_{1z} + \hat{S}_{2z}\right]$$

$$= \left[\hat{S}_{1j}, S_{1z}\right]\hat{S}_{2j} + \hat{S}_{1j}\left[\hat{S}_{2j}, \hat{S}_{2z}\right] = i\hbar\epsilon_{j3k}\hat{S}_{1k}\hat{S}_{2j} + i\hbar\epsilon_{j3k}\hat{S}_{1j}\hat{S}_{2k}$$

$$= i\hbar\epsilon_{j3k}\hat{S}_{1k}\hat{S}_{2j} + i\hbar\epsilon_{k3j}\hat{S}_{1k}\hat{S}_{2j} = i\hbar\left(\epsilon_{j3k} + \epsilon_{k3j}\right)\hat{S}_{1k}\hat{S}_{2j} = 0.$$

© Springer Nature Switzerland AG 2019
K. Tamvakis, *Basic Quantum Mechanics*, Undergraduate Texts in Physics,
https://doi.org/10.1007/978-3-030-22777-7_11

We shall denote the eigenstates of the first set of commuting operators (11.2) as $|s_1, m_1; s_2, m_2\rangle$. They are defined by

$$\hat{S}_1^2 |s_1, m_1; s_2, m_2\rangle = \tfrac{3}{4}\hbar^2 |s_1, m_1; s_2, m_2\rangle$$

$$\hat{S}_2^2 |s_1, m_1; s_2, m_2\rangle = \tfrac{3}{4}\hbar^2 |s_1, m_1; s_2, m_2\rangle$$

$$\hat{S}_{1z} |s_1, m_1; s_2, m_2\rangle = \hbar m_1 |s_1, m_1; s_2, m_2\rangle \qquad (11.4)$$

$$\hat{S}_{2z} |s_1, m_1; s_2, m_2\rangle = \hbar m_2 |s_1, m_1; s_2, m_2\rangle .$$

We may denote the eigenstates of the second set (11.3) as $|S, M\rangle$ in terms of the total spin quantum numbers S and M. The corresponding eigenvalue problem reads

$$\hat{S}_1^2 |S, M\rangle = \tfrac{3}{4}\hbar^2 |S, M\rangle$$

$$\hat{S}_2^2 |S, M\rangle = \tfrac{3}{4}\hbar^2 |S, M\rangle$$

$$\hat{S}^2 |S, M\rangle = \hbar^2 S(S+1) |S, M\rangle \qquad (11.5)$$

$$\hat{S}_z |S, M\rangle = \hbar M |S, M\rangle$$

Note that, since the total spin satisfies the angular momentum commutation algebra, its eigenvalues and eigenstates have the standard properties. Therefore,

$$-S \leq M \leq S . \qquad (11.6)$$

Let's consider the $\hat{z}$-component of the total spin and act on an eigenstate of the first set. We have

$$\hat{S}_z |s_1, m_1; s_2, m_2\rangle = \left(\hat{S}_{1z} + \hat{S}_{2z} \right) |s_1, m_1; s_2, m_2\rangle = \hbar(m_1 + m_2)|s_1, m_1; s_2, m_2\rangle .$$

From this we conclude that the S_z eigenvalues $\hbar M$ must be related to the $\hat{S}_{1z}$, $\hat{S}_{2z}$ eigenvalues $\hbar m_1$, $\hbar m_2$ by

$$M = m_1 + m_2 . \qquad (11.7)$$

Since the possible values of m_1, m_2 are $\pm 1/2$, we conclude that the possible values of M are

$$M = -1, 0, 1 . \qquad (11.8)$$

Furthermore, since $-S \leq M \leq S$, the appropriate values of S must be

$$S = 0, 1 . \qquad (11.9)$$

Therefore, we have four possible total spin eigenstates, namely,

$$|0,0\rangle, \ |1,-1\rangle, \ |1,0\rangle, \ |1,1\rangle. \tag{11.10}$$

Note that the multiplicity of these eigenstates is the same as the multiplicity of the other set of eigenstates $|s_1 = 1/2, m_1; s_2 = 1/2, m_2\rangle$, namely,[1]

$$|1/2; \ 1/2\rangle, \ |-1/2; \ 1/2\rangle, \ |1/2; \ -1/2\rangle, \ |-1/2; \ -1/2\rangle. \tag{11.11}$$

The total spin eigenstates can be easily deduced as linear combinations of the set (11.11). For example, the $M = 0$ eigenstates can only be combinations of $|\pm 1/2; \mp 1/2\rangle$, i.e.,

$$|1,0\rangle = a|1/2; \ -1/2\rangle + b|-1/2; \ 1/2\rangle \tag{11.12}$$

and

$$|0,0\rangle = c|1/2; \ -1/2\rangle + d|-1/2; \ 1/2\rangle. \tag{11.13}$$

The coefficients can be computed substituting these expressions in the $\hat{S}^2$ eigenvalue equation (see Example 11.1.). On the other hand, the $M = -1, 1$ can only be

$$|1,\pm 1\rangle = |\pm 1/2; \ \pm 1/2\rangle. \tag{11.14}$$

Thus, we finally have the *triplet* of symmetric combinations

$$|1,-1\rangle = |-1/2; \ -1/2\rangle$$

$$|1,0\rangle = \tfrac{1}{\sqrt{2}} (|1/2; \ -1/2\rangle + |-1/2; 1/2\rangle) \tag{11.15}$$

$$|1,1\rangle = |1/2; \ 1/2\rangle$$

and the antisymmetric *singlet*

$$|0,0\rangle = \frac{1}{\sqrt{2}} (|1/2; \ -1/2\rangle - |-1/2; 1/2\rangle). \tag{11.16}$$

The $1/\sqrt{2}$ factors arise from the normalization of the states.

The triplet and the singlet are also expressible in spinorial notation as

$$\begin{cases} \chi_{1,-1} = \chi_\downarrow^{(1)} \chi_\downarrow^{(2)} \\[2mm] \chi_{1,0} = \tfrac{1}{\sqrt{2}} \left(\chi_\uparrow^{(1)} \chi_\downarrow^{(2)} + \chi_\downarrow^{(1)} \chi_\uparrow^{(2)} \right) \\[2mm] \chi_{1,1} = \chi_\uparrow^{(1)} \chi_\uparrow^{(2)} \end{cases} \tag{11.17}$$

---

[1] Symbolized for notational economy as $|m_1; m_2\rangle$.

and

$$\chi_{0,0} = \frac{1}{\sqrt{2}} \left(\chi_\uparrow^{(1)} \chi_\downarrow^{(2)} - \chi_\downarrow^{(1)} \chi_\uparrow^{(2)} \right) . \tag{11.18}$$

Example 11.1 Determine the $|S, M = 0\rangle$ states in terms of the states $| \pm 1/2, \mp 1/2\rangle$.

Consider the expressions

$$|1, 0\rangle = a|1/2; -1/2\rangle + b| - 1/2; 1/2\rangle \tag{11.19}$$

and

$$|0, 0\rangle = c|1/2; -1/2\rangle + d| - 1/2; 1/2\rangle \tag{11.20}$$

with the coefficients a, b, c, d to be determined from the eigenvalue equations

$$\hat{S}^2|0, 0\rangle = 0, \quad \hat{S}^2|1, 0\rangle = 2\hbar^2|1, 0\rangle . \tag{11.21}$$

The operator $\hat{S}^2$ can be analyzed as follows:

$$\hat{S}^2 = \left(\hat{\mathbf{S}}_1 + \hat{\mathbf{S}}_2 \right)^2 = \hat{S}_1^2 + \hat{S}_2^2 + 2\hat{\mathbf{S}}_1 \cdot \hat{\mathbf{S}}_2 = \frac{3\hbar^2}{2} + 2\hat{S}_{1x}\hat{S}_{2x} + 2\hat{S}_{1y}\hat{S}_{2y} + 2\hat{S}_{1z}\hat{S}_{2z}$$

$$= \frac{3\hbar^2}{2} + \frac{1}{2} \left(\hat{S}_{1+} + \hat{S}_{1-} \right) \left(\hat{S}_{2+} + \hat{S}_{2-} \right) - \frac{1}{2} \left(\hat{S}_{1+} - \hat{S}_{1-} \right) \left(\hat{S}_{2+} - \hat{S}_{2-} \right) + 2\hat{S}_{1z}\hat{S}_{2z}$$

$$= \frac{3\hbar^2}{2} + \hat{S}_{1+}\hat{S}_{2-} + \hat{S}_{1-}\hat{S}_{2+} + 2\hat{S}_{1z}\hat{S}_{2z} . \tag{11.22}$$

Acting with it on $|0, 0\rangle$, we obtain for $\hat{S}^2|0, 0\rangle$

$$\hbar^2 \left(c|1/2; -1/2\rangle + d| - 1/2; 1/2\rangle \right) + \left(\hat{S}_{1+}\hat{S}_{2-} + \hat{S}_{1-}\hat{S}_{2+} \right) \left(c|1/2; -1/2\rangle + d| - 1/2; 1/2\rangle \right)$$

$$= \hbar^2 \left(c|1/2; -1/2\rangle + d| - 1/2; 1/2\rangle \right) + \hbar^2 \left(c| - 1/2; 1/2\rangle + d|1/2; -1/2\rangle \right)$$

or

$$\hat{S}^2|0, 0\rangle = (c + d)\hbar^2 \left(|1/2; -1/2\rangle + | - 1/2; 1/2\rangle \right) , \tag{11.23}$$

which implies $d = -c$. Similarly, we have

$$\hat{S}^2|1, 0\rangle = (a + b)\hbar^2 \left(|1/2; -1/2\rangle + | - 1/2; 1/2\rangle \right) . \tag{11.24}$$

Orthogonality is automatic. Normalizing these states, we obtain

$$\begin{cases} |1, 0\rangle = \frac{1}{\sqrt{2}} \left(|1/2; -1/2\rangle - | - 1/2; 1/2\rangle \right) \\ |0, 0\rangle = \frac{1}{\sqrt{2}} \left(|1/2; -1/2\rangle + | - 1/2; 1/2\rangle \right) . \end{cases} \tag{11.25}$$

11.2 Addition of Two Angular Momenta

We are ready now to attack the general problem of adding two arbitrary angular momenta $\mathbf{J}_1$ and $\mathbf{J}_2$, which could be spins or orbital angular momenta. The only requirement is that they satisfy the general angular momentum commutation algebra which is common to both spins and orbital angular momenta.[2] Assuming that they are independent ($[\mathbf{J}_1, \mathbf{J}_2] = 0$), their sum

$$\mathbf{J} = \mathbf{J}_1 + \mathbf{J}_2 \tag{11.26}$$

will also satisfy this algebra

$$\begin{aligned}
\left[\hat{J}_{1i}, \hat{J}_{1j}\right] &= i\hbar\epsilon_{ijk}\hat{J}_{1k} \\
\left[\hat{J}_{2i}, \hat{J}_{2j}\right] &= i\hbar\epsilon_{ijk}\hat{J}_{2k}
\end{aligned} \quad\Longrightarrow\quad \left[\hat{J}_i, \hat{J}_j\right] = i\hbar\epsilon_{ijk}\hat{J}_k. \tag{11.27}$$

As in the case of the addition of two spins, we have the following two complete sets of commuting operators:

$$\hat{J}_1^2, \, \hat{J}_2^2, \, \hat{J}_{1z}, \, \hat{J}_{2z} \tag{11.28}$$

and[3]

$$\hat{J}_1^2, \, \hat{J}_2^2, \, \hat{J}^2, \, \hat{J}_z \, . \tag{11.29}$$

The eigenvalue problem for each of these sets reads as follows[4]:

$$\begin{cases}
\hat{J}_1^2|j_1, j_2, m_1, m_2\rangle = \hbar^2 j_1(j_1+1)|j_1, j_2, m_1, m_2\rangle \\[2mm]
\hat{J}_2^2|j_1, j_2, m_1, m_2\rangle = \hbar^2 j_2(j_2+1)|j_1, j_2, m_1, m_2\rangle \\[2mm]
\hat{J}_{1z}|j_1, j_2, m_1, m_2\rangle = \hbar m_1|j_1, j_2, m_1, m_2\rangle \\[2mm]
\hat{J}_{2z}|j_1, j_2, m_1, m_2\rangle = \hbar m_2|j_1, j_2, m_1, m_2\rangle
\end{cases} \tag{11.30}$$

and

---

[2]For a detailed account of angular momenta addition see also [1–3].
[3]The proof that $[\hat{J}^2, \hat{J}_z] = 0$ proceeds as in the case of the addition of two spins.
[4]An alternative equivalent notation to $|j_1, j_2, m_1, m_2\rangle$ is $|j_1, m_1; j_2, m_2\rangle$.

$$\begin{cases} \hat{J}_1^2|j_1, j_2; j, m\rangle = \hbar^2 j_1(j_1+1)|j_1, j_2; j, m\rangle \\ \\ \hat{J}_2^2|j_1, j_2; j, m\rangle = \hbar^2 j_2(j_2+1)|j_1, j_2; j, m\rangle \\ \\ \hat{J}^2|j_1, j_2; j, m\rangle = \hbar^2 j(j+1)|j_1, j_2; j, m\rangle \\ \\ \hat{J}_z|j_1, j_2; j, m\rangle = \hbar m|j_1, j_2; j, m\rangle. \end{cases} \quad (11.31)$$

The multiplicity of the first set is $(2j_1+1)(2j_2+1)$. It will have to be equal to the multiplicity of the second set, namely, $\sum_j(2j+1)$.

Considering the $\hat{z}$-component of the total angular momentum and acting on an eigenstate of the first set, we obtain

$$\hat{J}_z|j_1, j_2, m_1, m_2\rangle = \left(\hat{J}_{1z}+\hat{J}_{2z}\right)|j_1, j_2, m_1, m_2\rangle, = \hbar(m_1+m_2)|j_1, j_2, m_1, m_2\rangle$$

from which we conclude that
$$m = m_1 + m_2. \quad (11.32)$$

Note now that the maximal values of m_1, m_2 are j_1 and j_2. Therefore, the maximal value of m must be $j_1 + j_2$ or

$$j_{max} = j_1 + j_2. \quad (11.33)$$

What is the minimal value j_{min}? This can be deduced from the requirement that the multiplicity of states has to be the same, namely,

$$(2j_1+1)(2j_2+1) = \sum_{j_{min}}^{j_{max}}(2j+1).$$

We may use the mathematical fact that $\sum_{n=N_1}^{n=N_2}(2n+1) = (N_2+1)^2 - N_1^2$. This implies

$$(2j_1+1)(2j_2+1) = (j_1+j_2+1)^2 - j_{min}^2 \implies j_{min}^2 = (j_1-j_2)^2$$

or, finally, that
$$|j_1 - j_2| \le j \le j_1 + j_2 \quad (11.34)$$

and, of course, $-j \le m \le j$.

The two sets of eigenstates, being both legitimate bases in the Hilbert subspace of angular momentum eigenstates, can be related by expanding either set in terms of the other. The total angular momentum eigenstates are expressed in terms of the individual angular momenta eigenstates as

$$|j_1, j_2; j, m\rangle = \sum_{j_1' j_2'} \sum_{m_1, m_2} \langle j_1' j_2' m_1 m_2 | j_1 j_2 j m \rangle \, |j_1', j_2', m_1, m_2\rangle . \qquad (11.35)$$

The expansion coefficients $\langle j_1' j_2' m_1 m_2 | j_1 j_2 j m \rangle$ are called *Clebsch–Gordan coefficients*. These coefficients are not all independent, satisfying a number of relations. Note also that some of them vanish identically, namely, those that correspond to $m \neq m_1 + m_2$. Indeed, acting on (11.35) by $\hat{J}_z = \hat{J}_{1z} + \hat{J}_{2z}$ we see that the two m_1, m_2 summations must satisfy $m = m_1 + m_2$, giving vanishing Clebsch–Gordan coefficients otherwise.

Apart from that, we have a number of further constraints. Consider the matrix element $\langle j_1'', j_2'', m_1'', m_2'' | \hat{J}_1^2 | j_1, j_2, j, m \rangle$ and substitute in it (11.35)

$$\langle j_1'', j_2'', m_1'', m_2'' | \hat{J}_1^2 | j_1, j_2, j, m \rangle =$$

$$\langle j_1'', j_2'', m_1'', m_2'' | \hat{J}_1^2 \sum_{j_1' j_2'} \sum_{m_1, m_2} \langle j_1' j_2' m_1 m_2 | j_1 j_2 j m \rangle \, |j_1'', j_2', m_1, m_2\rangle$$

$$= \sum_{j_1' j_2'} \sum_{m_1, m_2} \langle j_1' j_2' m_1 m_2 | j_1 j_2 j m \rangle \, \langle j_1'', j_2'', m_1'', m_2'' | \hat{J}_1^2 | j_1', j_2', m_1, m_2\rangle$$

$$= \begin{cases} \hbar^2 j_1(j_1 + 1)\langle j_1'', j_2'', m_1'', m_2'' | j_1, j_2, j, m\rangle \\[6pt] \hbar^2 j_1''(2j_1'' + 1)\langle j_1'', j_2'', m_1'', m_2'' | j_1, j_2, j, m\rangle, \end{cases}$$

which means that the coefficients with $j_1 \neq j_1''$ are zero. Similarly, considering the matrix element of $\hat{J}_2^2$, we conclude the same for the entry j_2. Thus, the summation on j_1', j_2' in (11.35) collapses and we have[5]

$$|j_1, j_2; j, m\rangle = \sum_{m_1 + m_2 = m} \langle j_1 j_2 m_1 m_2 | j_1 j_2 j m \rangle \, |j_1, j_2, m_1, m_2\rangle . \qquad (11.36)$$

We also have the inverse expansion

$$|j_1, j_2, m_1, m_2\rangle = \sum_{j=|j_1-j_2|}^{j_1+j_2} \sum_{m=-j}^{j} \langle j_1 j_2 j m | j_1 j_2 m_1 m_2 \rangle \, |j_1, j_2, j, m\rangle . \qquad (11.37)$$

Multiplying by $\langle j_1 j_2 m_1' m_2' |$, we obtain the *orthonormality condition*

$$\sum_{j=|j_1-j_2|}^{j_1+j_2} \sum_{m=-j}^{j} \langle j_1 j_2 m_1' m_2' | j_1 j_2 j m \rangle \, \langle j_1 j_2 j m | j_1 j_2 m_1 m_2 \rangle = \delta_{m_1 m_1'} \delta_{m_2 m_2'} . \qquad (11.38)$$

---

[5]For notational compactness, we may drop the commas as $\langle j_1, j_2, m_1, m_2 | j_1, j_2, j, m \rangle = \langle j_1 j_2 m_1 m_2 | j_1 j_2 j m \rangle$.

Another relation can also be obtained from the initial expansion (11.36) multiplying by $\langle j_1 j_2 j' m' |$. We get

$$\langle j_1 j_2 j' m' | j_1 j_2 j m \rangle = \sum_{m_1 + m_2 = m} \langle j_1 j_2 m_1 m_2 | j_1 j_2 j m \rangle \, \langle j_1 j_2 j' m' | j_1 j_2 m_1 m_2 \rangle$$

(11.39)

or

$$\sum_{m_1 + m_2 = m} \langle j_1 j_2 m_1 m_2 | j_1 j_2 j m \rangle \, \langle j_1 j_2 j' m' | j_1 j_2 m_1 m_2 \rangle = \delta_{jj'} \delta_{mm'} .$$

(11.40)

In order to calculate the Clebsch–Gordan coefficients, we make use of the raising and lowering operators $\hat{J}_\pm = \hat{J}_x \pm i \hat{J}_y$. As an example, consider

$$\langle j_1 j_2 m_1 m_2 | \hat{J}_\pm | j_1 j_2 j m \rangle = \hbar \sqrt{j(j+1) - m(m \pm 1)} \, \langle j_1 j_2 m_1 m_2 | j_1 j_2 j, m \pm 1 \rangle .$$

On the other hand, since $\hat{J}_\pm = \hat{J}_{1\pm} + \hat{J}_{2\pm}$, we have

$$\langle j_1 j_2 m_1 m_2 | \hat{J}_\pm | j_1 j_2 j m \rangle = \hbar \sqrt{j_1(j_1+1) - m_1(m_1 \mp 1)} \, \langle j_1 j_2, m_1 \mp 1, m_2 | j_1 j_2 j m \rangle$$

$$+ \hbar \sqrt{j_2(j_2+1) - m_2(m_2 \mp 1)} \, \langle j_1 j_2 m_1, m_2 \mp 1 | j_1 j_2 j m \rangle ,$$

(11.41)

which gives us the very useful relation

$$\langle j_1 j_2 m_1 m_2 | j_1 j_2 j, m \pm 1 \rangle = \begin{cases} \sqrt{\dfrac{j_1(j_1+1) - m_1(m_1 \mp 1)}{j(j+1) - m(m \pm 1)}} \, \langle j_1 j_2, m_1 \mp 1, m_2 | j_1 j_2 j m \rangle \\[2mm] + \\[2mm] \sqrt{\dfrac{j_2(j_2+1) - m_2(m_2 \mp 1)}{j(j+1) - m(m \pm 1)}} \, \langle j_1 j_2 m_1, m_2 \mp 1 | j_1 j_2 j m \rangle \end{cases}$$

(11.42)

Note that now the nonvanishing condition for the Clebsch–Gordan coefficient of the left-hand side is $m_1 + m_2 = m \pm 1$. The relation (11.42) will prove to be quite useful in the complete determination of Clebsch–Gordan coefficients. We shall not go any further along that line which is technically quite tedious. Nevertheless, we shall consider a special case of (11.42) for future use. For $m_2 = j_2$ and the lower sign, the second term of the right-hand side vanishes, and we obtain the relation

$$\langle j_1 j_2 m_1 j_2 | j_1 j_2 j, m - 1 \rangle = \sqrt{\frac{j_1(j_1+1) - m_1(m_1+1)}{j(j+1) - m(m-1)}} \, \langle j_1 j_2, m_1 + 1, j_2 | j_1 j_2 j m \rangle .$$

(11.43)

By the successive use of (11.42) and the various normalization relations we obtained
we can calculate all Clebsch–Gordan coefficients[6] [1]. A general formula for $m > 0$
and $j_1 > j_2$ is the following:

$$\langle j_1 j_2 jm | j_1 j_2 m_1 m_2 \rangle =$$

$$\delta_{m,m_1+m_2} \left(\frac{(2j+1)(j+j_1-j_2)!(j-j_1+j_2)!(j_1+j_2-j)!}{(j_1+j_2+j+1)!} \right)^{1/2}$$

$$\times \left((j+m)!(j-m)!(j_1-m_1)!(j_1+m_1)!(j_2-m_2)!(j_2+m_2)! \right)^{1/2}$$

$$\times \sum_k \frac{(-1)^k}{k!(j_1+j_2-j-k)!(j_1-m_1-k)!(j_2+m_2-k)!(j-j_2+m_1+k)!(j-j_1-m_2+k)!}.$$

(11.44)

The sum terminates for the k that makes the argument of any of the factorials negative.
For the coefficients with $m < 0$ and $j_1 < j_2$, we may use the following relations:

$$\langle j_1 j_2 jm | j_1 j_2 m_1 m_2 \rangle = (-1)^{j-j_1-j_2} \langle j_1 j_2, -m_1, -m_2 | j_1 j_2 j, -m \rangle$$

$$= (-1)^{j-j_1-j_2} \langle j_2 j_1 m_2 m_1 | j_2 j_1 jm \rangle. \quad (11.45)$$

The case of addition of spin 1/2 and orbital angular momentum is of particular
interest. For a physical system that possesses nonzero spin, we may define a *total
angular momentum* operator

$$\hat{\mathbf{J}} = \hat{\mathbf{L}} + \hat{\mathbf{S}}. \quad (11.46)$$

Since the spin commutes with the orbital angular momentum $\left(\left[\hat{\mathbf{L}}, \hat{\mathbf{S}} \right] = 0 \right)$, the
total angular momentum satisfies the standard angular momentum algebra as well.

According to our previous analysis, the total angular momentum quantum numbers will take the values

$$\ell - 1/2 \leq j \leq \ell + 1/2 \quad (11.47)$$

and, of course, $-j \leq m \leq j$.

The Clebsch–Gordan coefficients $\langle m_\ell, m_s | j = \ell \pm 1/2, m \rangle$ can be obtained from
the formula (11.44). For example,

$$\langle m_\ell = m - 1/2, m_s = 1/2 | j = \ell + 1/2, m \rangle = \sqrt{\frac{2\ell + 2m + 1}{2(2\ell + 1)}}. \quad (11.48)$$

---

[6]Note that they can be taken to be real.

11.3 Total Angular Momentum and Rotations

In the presence of spin, the rotation operator introduced in (8.27) in terms of the angular momentum can be modified by replacing the orbital angular momentum with the total angular momentum operator

$$\hat{\mathbf{J}} = \hat{\mathbf{L}} + \hat{\mathbf{S}}. \tag{11.49}$$

The rotation operator becomes

$$\hat{U}(\mathbf{a}) = e^{-\frac{i}{\hbar}\mathbf{a}\cdot\hat{\mathbf{J}}}. \tag{11.50}$$

A *vector operator* is now defined as an operator satisfying the transformation relation

$$\hat{U}^{\dagger}\hat{V}_i\hat{U} = \mathcal{R}_{ij}\hat{V}_j, \tag{11.51}$$

which, for an infinitesimal rotation, reduces to the commutation relation

$$\left[\hat{J}_i, \hat{V}_j\right] = i\hbar\epsilon_{ijk}\hat{V}_k. \tag{11.52}$$

This relation, applied for the position, momentum, angular momentum, and spin, is

$$\left[\hat{J}_i, x_j\right] = i\hbar\epsilon_{ijk}x_k, \quad \left[\hat{J}_i, \hat{p}_j\right] = i\hbar\epsilon_{ijk}\hat{p}_k$$
$$\left[\hat{J}_i, \hat{L}_j\right] = i\hbar\epsilon_{ijk}\hat{L}_k, \quad \left[\hat{J}_i, \hat{S}_j\right] = i\hbar\epsilon_{ijk}\hat{S}_k. \tag{11.53}$$

Thus, spin is also a vector operator. Nevertheless, although $\mathbf{r}$ and $\hat{\mathbf{p}}$ are true vectors, since they anticommute with parity, the angular momentum and spin commute with parity and are axial vectors

$$\left[\hat{\mathbf{J}}, \hat{P}\right] = \left[\hat{\mathbf{L}}, \hat{P}\right] = \left[\hat{\mathbf{S}}, \hat{P}\right] = 0. \tag{11.54}$$

Example 11.2 Protons and neutrons are elementary particle of spin $s_p = 1/2$ that make up the various nuclei of atoms. They interact strongly with other particles among which is the ρ-meson, an elementary particle of spin $s_\rho = 1$. Ignore spatial degrees of freedom, consider a proton-ρ-meson state

$$|p, \rho\rangle = |s_p = 1/2, m_p = -1/2; s_\rho = 1, m_\rho = 1\rangle \tag{11.55}$$

and express this state in terms of eigenstates of the total spin $|s_p = 1/2, s_\rho = 1, S, M\rangle$.

Applying the general formula (11.44) we obtain

$\langle s_p = 1/2, s_\rho = 1; S = 3/2, M = 1/2 | s_p = 1/2, m_p = -1/2; s_\rho = 1, m_\rho = 1 \rangle = \frac{1}{\sqrt{3}}$

$\langle s_p = 1/2, s_\rho = 1; S = 1/2, M = 1/2 | s_p = 1/2, m_p = -1/2; s_\rho = 1, m_\rho = 1 \rangle = \sqrt{\frac{2}{3}}.$

Therefore, we have

$$|p, \rho\rangle = |1/2, -1/2; 1, 1\rangle =$$
$$\langle 1/2, 1; S = 3/2, M = 1/2 | 1/2, -1/2; 1, 1 \rangle \, |1/2, 1; S = 3/2, M = 1/2\rangle$$
$$+ \langle 1/2, 1; S = 1/2, M = 1/2 | 1/2, -1/2; 1, 1 \rangle \, |1/2, 1; S = 1/2, M = 1/2\rangle$$
$$= \frac{1}{\sqrt{3}} |1/2, 1; S = 3/2, M = 1/2\rangle + \sqrt{\frac{2}{3}} |1/2, 1; S = 1/2, M = 1/2\rangle. \qquad (11.56)$$

Problems and Exercises

11.1 The system of two spin $1/2$ particles (not identical) has a Hamiltonian

$$\hat{H} = A \left(\hat{S}_{1z} + \hat{S}_{2z} \right) + B \left(\hat{\mathbf{S}}_1 \cdot \hat{\mathbf{S}}_2 \right),$$

with A and B are known parameters. Find and classify the energy eigenvalues.

11.2 The hydrogen atom, being the system of an electron and a proton, has a total spin $\hat{\mathbf{S}} = \hat{\mathbf{S}}_1 + \hat{\mathbf{S}}_2$, where $s_1 = 1/2$ is the spin of the electron and $s_2 = 1/2$ is the spin of the proton. Neglecting spatial degrees of freedom, assume that the Hamiltonian could be approximated by

$$\hat{H} = C_0 + C_1 \left(\hat{\mathbf{S}}_1 \cdot \hat{\mathbf{S}}_2 \right),$$

where C_0 and C_1 are known constants. If the system starts at $t = 0$ in a state $|1/2, -1/2\rangle$, find the time-evolved state $|\psi(t)\rangle$ and calculate the probability to be found in the state $|-1/2, 1/2\rangle$ at time $t > 0$.

11.3 The *positronium* system is a short-lived bound state of an electron and a positron[7] Assume that a positronium in an external homogeneous magnetic field, if we neglect the spatial degrees of freedom, could be described approximately with the Hamiltonian

$$\hat{H} = a \left(\hat{\mathbf{S}}_1 \cdot \hat{\mathbf{S}}_2 \right) + b \left(\hat{S}_{1z} - \hat{S}_{2z} \right),$$

where a is a parameter expressing the interaction between electron and positron and b is a parameter dependent on the external magnetic field.

(a) Show that the total spin eigenstates $|1, 1\rangle$ and $|1, -1\rangle$ are eigenstates of the Hamiltonian.

---

[7] A particle of the same mass and spin as the electron but opposite electric charge.

(b) The states $|1, 0\rangle$ and $|0, 0\rangle$ are not eigenstates of the Hamiltonian. Find two linear combinations of them that are eigenstates. Find also the corresponding eigenvalues.

11.4 Let $|j_1 j_2 j m\rangle$ be a common eigenstate of $\hat{J}_1^2$, $\hat{J}_2^2$, $\hat{J}^2$, $\hat{J}_z$ with eigenvalues $\hbar^2 j_1(j_1 + 1)$, $\hbar^2 j_2(j_2 + 1)$, $\hbar^2 j(j + 1)$, $\hbar m$. Verify that this state is also an eigenstate of the operator $\hat{\mathbf{J}}_1 \cdot \hat{\mathbf{J}}_2$ and express its eigenvalues in terms of j_1, j_2, j, and m. Do the same for the operators $\hat{\mathbf{J}} \cdot \hat{\mathbf{J}}_1$ and $\hat{\mathbf{J}} \cdot \hat{\mathbf{J}}_2$. Investigate the possible m-dependence of these eigenvalues.

11.5 Consider two spinless particles with orbital angular momentum $\ell_1 = \ell_2 = 1$. What are the possible values of the total orbital angular momentum $\hat{\mathbf{L}} = \hat{\mathbf{L}}_1 + \hat{\mathbf{L}}_2$? Calculate the relevant Clebsch–Gordan coefficients to prove that the state $|1, 1; \ell = 0, m = 0\rangle$, expressed in terms of $|\ell_1, \ell_2, m_1, m_2\rangle$ states, is

$$|1, 1; 0, 0\rangle = \frac{1}{\sqrt{3}} \left(|1, 1, 1, -1\rangle - |1, 1, 0, 0\rangle + |1, 1, -1, 1\rangle \right).$$

11.6 Show that the total angular momentum of a particle with spin $\hat{\mathbf{S}}$ and orbital angular momentum $\hat{\mathbf{L}}$ satisfies the angular momentum algebra. Calculate the expectation value

$$\langle \ell, 1/2, m_\ell, m_s | \hat{J}^2 | \ell, 1/2, m_\ell, m_s \rangle.$$

11.7 Consider the case of addition of the orbital angular momentum $\ell = 1$ and spin $s = 1/2$. The state with $j = 3/2$, $m_j = 1/2$ can only arise as a linear combination of $|1, 1/2, m_\ell = 1, m_s = -1/2\rangle$ and $|1, 1/2, m_\ell = 0, m_s = 1/2\rangle$. Compute these Clebsch–Gordan coefficients directly acting with $\hat{J}^2$ on $|1, 1/2, j = 3/2, m_j = 1/2\rangle$.

11.8 Consider a system of spin $\hat{\mathbf{S}}$ and orbital angular momentum $\hat{\mathbf{L}}$. Show that the total angular momentum $\hat{\mathbf{J}} = \hat{\mathbf{S}} + \hat{\mathbf{L}}$ satisfies the commutation relation

$$\left[\hat{J}^2, \hat{\mathbf{J}} \times \hat{\mathbf{S}} \right] = 2i\hbar \left(\hat{J}^2 \hat{\mathbf{S}} - \left(\hat{\mathbf{J}} \cdot \hat{\mathbf{S}} \right) \hat{\mathbf{J}} \right).$$

Show also

$$\langle jm' | \hat{\mathbf{S}} | jm \rangle = \frac{\langle jm' | \left(\hat{\mathbf{J}} \cdot \hat{\mathbf{S}} \right) \hat{\mathbf{J}} | jm \rangle}{\hbar^2 j(j + 1)}.$$

11.9 Consider an electron with orbital angular momentum ℓ and total angular momentum $j = \ell \pm 1/2$. Use the last formula of Problem 11.8. to calculate the expectation value

$$\langle \ell, 1/2, j = \ell \pm 1/2, m | \hat{S}_z | \ell, 1/2, j = \ell \pm 1/2, m \rangle.$$

Use this result to calculate the expectation value of the magnetic dipole moment $\mu = \frac{-e}{2m_e c}\left(\hat{L}_z + 2\hat{S}_z\right)$ in these states.

11.10 Consider the electron system of Problem 11.9. Show that the following is also true:

$$\langle jm'|\hat{\mathbf{S}}|jm\rangle = \frac{\langle jm'|\left(\hat{\mathbf{J}}\cdot\hat{\mathbf{S}}\right)|jm\rangle}{\hbar^2 j(j+1)}\langle jm'|\hat{\mathbf{J}}|jm\rangle .$$

References

1. W. Greiner, B. Müller, *Quantum Mechanics: Symmetries*, 2nd edn. (Springer, Berlin, 1992)
2. A. Messiah, *Quantum Mechanics*. Dover publications, single-volume reprint of the Wiley, New York, two-volume 1958 edn
3. S. Weinberg, *Lectures on Quantum Mechanics* (Cambridge University Press, Cambridge, 2015)

Chapter 12
Motion in Three Dimensions

12.1 Separation of Variables

We are already familiar from Chap. 2 with the Schroedinger equation for a particle moving freely in three-dimensional space

$$-\frac{\hbar^2}{2m}\nabla^2\psi(\mathbf{r}, t) = i\hbar\frac{\partial\psi(\mathbf{r}, t)}{\partial t} \implies \psi(\mathbf{r}, t) = \int \frac{d^3p}{(2\pi\hbar)^{3/2}} g(\mathbf{p}) e^{\frac{i}{\hbar}(\mathbf{p}\cdot\mathbf{r}-Et)}.$$

$$(12.1)$$

Its general solution above can be viewed as a superposition of *products* of one-dimensional plane wave solutions

$$\psi(\mathbf{r}, t) = \int d^3p\, g(\mathbf{p}) \left(\frac{e^{\frac{i}{\hbar}\left(p_x x - \frac{p_x^2}{2m}t\right)}}{(2\pi\hbar)^{1/2}}\right) \left(\frac{e^{\frac{i}{\hbar}\left(p_y y - \frac{p_y^2}{2m}t\right)}}{(2\pi\hbar)^{1/2}}\right) \left(\frac{e^{\frac{i}{\hbar}\left(p_z z - \frac{p_z^2}{2m}t\right)}}{(2\pi\hbar)^{1/2}}\right).$$

Note also that the Hamiltonian is a *sum* of three mutually commuting one-dimensional Hamiltonians

$$\hat{H} = -\frac{\hbar^2}{2m}\nabla^2 = \left(-\frac{\hbar^2}{2m}\frac{\partial^2}{\partial x^2}\right) + \left(-\frac{\hbar^2}{2m}\frac{\partial^2}{\partial y^2}\right) + \left(-\frac{\hbar^2}{2m}\frac{\partial^2}{\partial z^2}\right).$$

This applies also to the time-independent Schroedinger equation

$$\left\{\left(-\frac{\hbar^2}{2m}\frac{\partial^2}{\partial x^2}\right) + \left(-\frac{\hbar^2}{2m}\frac{\partial^2}{\partial y^2}\right) + \left(-\frac{\hbar^2}{2m}\frac{\partial^2}{\partial z^2}\right)\right\}\Psi_E(x, y, z) = E\Psi_E(x, y, z),$$

the corresponding eigenfunctions being

$$\Psi_E(x, y, z) = \left(\frac{e^{\frac{i}{\hbar}p_x x}}{(2\pi\hbar)^{1/2}}\right) \left(\frac{e^{\frac{i}{\hbar}p_y y}}{(2\pi\hbar)^{1/2}}\right) \left(\frac{e^{\frac{i}{\hbar}p_z z}}{(2\pi\hbar)^{1/2}}\right) \quad \text{with } E = \frac{p_x^2}{2m} + \frac{p_y^2}{2m} + \frac{p_z^2}{2m}.$$

© Springer Nature Switzerland AG 2019
K. Tamvakis, *Basic Quantum Mechanics*, Undergraduate Texts in Physics,
https://doi.org/10.1007/978-3-030-22777-7_12

This is the simplest case of a general circumstance: whenever the Hamiltonian is a sum of independent terms, there are partial Schroedinger equation solutions in the form of products of independent wave functions. This is the basis of the *Separation of Variables Method*.[1] This method is applicable for the solution of the Schroedinger equation whenever the Hamiltonian is a sum of commuting (i.e., independent) terms

$$\hat{H}(q_1, q_2, \ldots) = \sum_{a=1,2,\ldots} \hat{H}_a(q_a) \ \text{ with } \ \left[\hat{H}_a, \hat{H}_b\right] = 0. \tag{12.2}$$

Then the Schroedinger equation has the form

$$\left(\hat{H}_1(q_1) + \hat{H}_2(q_2) + \cdots\right)\Psi(q_1, q_2, \ldots) = E\,\Psi(q_1, q_2, \ldots). \tag{12.3}$$

Consider now a wave function in the form of a product

$$\Psi(q_1, q_2, \ldots) = \psi_1(q_1)\,\psi_2(q_2)\cdots \tag{12.4}$$

and substitute it in the time-independent Schroedinger equation as a trial solution for the energy eigenvalue problem. We get

$$\left(\hat{H}_1(q_1) + \hat{H}_2(q_2) + \hat{H}_3(q_3) + \cdots\right)\psi_1(q_1)\,\psi_2(q_2)\cdots =$$

$$\left(\hat{H}_1(q_1)\psi_1(q_1)\right)\psi_2(q_2)\psi_3(q_3)\cdots + \psi_1(q_1)\left(\hat{H}_2(q_2)\psi_2(q_2)\right)\psi_3(q_3)\cdots + \cdots$$

$$= E\,\psi_1(q_1)\,\psi_2(q_2)\,\psi_3(q_3)\cdots \tag{12.5}$$

If each of the factors satisfies an equation

$$\hat{H}_a\psi_a(q_a) = E_a\,\psi_a(q_a) \tag{12.6}$$

and the E_a's are such that

$$E = \sum_{a=1,2,\ldots} E_a, \tag{12.7}$$

then, the above product is a solution of the Schroedinger equation. Obviously, the most suitable type of coordinates to be used for the application of this method depends on the geometrical symmetry of the Hamiltonian.

A variant of the above method arises in the case that the Hamiltonian depends on a set of operators $\hat{Q}_1, \hat{Q}_2, \ldots, \hat{Q}_N$, among which one, say $\hat{Q}_N$, commutes with all the rest. Then, if $\psi_N(q_N)$ are the eigenfunctions of $\hat{Q}_N$ (i.e., $\hat{Q}_N\psi_N(q_N) = q_N\psi_N(q_N)$), then, the eigenvalue problem of the Hamiltonian admits solutions in the form of a product, namely,

---

[1]See any of Mathematical Methods textbooks, e.g., [1].

$$\hat{H}(\hat{Q}_1, \ldots, \hat{Q}_N)\Psi(q_1, \ldots, q_N) = E\Psi(q_1, \ldots, q_N)$$

$$\implies \hat{H}(\hat{Q}_1, \ldots, \hat{Q}_{N-1}, q_N)\tilde{\psi}(q_1, \ldots, q_{N-1}) = E\tilde{\psi}(q_1, \ldots, q_{N-1}) \quad (12.8)$$

with

$$\Psi(q_1, \ldots, q_N) = \tilde{\psi}(q_1, \ldots, q_{N-1})\,\psi_N(q_N). \quad (12.9)$$

For example, the eigenvalue problem $(\frac{\hat{p}^2}{2m} + V(\mathbf{r}) - \frac{e}{mc}B_z(\mathbf{r})S_z)\Psi(\mathbf{r}, \hat{S}_z) = E\Psi(\mathbf{r}, S_z)$ is reduced to the eigenvalue problem $(\frac{\hat{p}^2}{2m} + V(\mathbf{r}) \mp \frac{e\hbar}{2mc}B_z(\mathbf{r}))\tilde{\psi}_\pm(\mathbf{r}) = E_\pm\tilde{\psi}_\pm(\mathbf{r})$ in terms of a product eigenfunction $\tilde{\psi}_\pm(\mathbf{r})\chi_\pm$, where $\hat{S}_z\chi_\pm = \pm\frac{\hbar}{2}\chi_\pm$.

12.2 The Three-Dimensional Infinite Square Well

Consider a particle trapped inside a cube. The particle is free to move inside the cube but it experiences an infinite repulsion at the walls. The quantum analogue of such a situation is that of a particle moving in the potential

$$V(x, y, z) = \begin{cases} 0 & |x| < L, \ |y| < L, \ |z| < L \\ +\infty & |x| \geq L, \ |y| \geq L, \ |z| \geq L. \end{cases} \quad (12.10)$$

This is the three-dimensional version of the infinite square well we studied in one dimension. The time-independent Schroedinger equation is

$$\hat{H}(x, y, z)\,\psi_E(x, y, z) = E\,\psi_E(x, y, z) \quad (12.11)$$

or

$$-\frac{\hbar^2}{2m}\left(\frac{\partial^2}{\partial x^2} + \frac{\partial^2}{\partial y^2} + \frac{\partial^2}{\partial z^2}\right)\psi_E(x, y, z) = E\,\psi_E(x, y, z), \quad (12.12)$$

with the conditions at the boundary and beyond

$$\psi(x, y, z) = 0 \quad \forall \ |x| \geq L, \ |y| \geq L, \ |z| \geq L. \quad (12.13)$$

Note that the Hamiltonian is the sum of the three mutually commuting operators

$$\hat{H}_1(x) = -\frac{\hbar^2}{2m}\frac{\partial^2}{\partial x^2}, \quad \hat{H}_2(y) = -\frac{\hbar^2}{2m}\frac{\partial^2}{\partial y^2}, \quad \hat{H}_3(z) = -\frac{\hbar^2}{2m}\frac{\partial^2}{\partial z^2} \quad (12.14)$$

$$\left[\hat{H}_i, \hat{H}_j\right] = 0. \quad (12.15)$$

Therefore, we can apply the *separation of variables method* considering a trial solution of (12.12) in the form of a product

$$\psi(x, y, z) = X(x)\, Y(y)\, Z(z)\,. \tag{12.16}$$

Substituting it into the Schroedinger equation, we obtain

$$-\frac{\hbar^2}{2m}\frac{X''(x)}{X(x)} - \frac{\hbar^2}{2m}\frac{Y''(y)}{Y(y)} - \frac{\hbar^2}{2m}\frac{Z''(z)}{Z(z)} = E\,, \tag{12.17}$$

which features the sum of three terms being equal to a constant parameter E while each of them depends on a different variable. The only way this can be true is if each of these terms is a constant and the sum of these three constants equals E. Thus, we conclude

$$X''(x) = -k_x^2\, X(x) \implies X(x) = A_x\, e^{ik_x x} + B_x\, e^{-ik_x x}$$

$$Y''(y) = -k_y^2\, Y(y) \implies Y(y) = A_y\, e^{ik_y y} + B_y\, e^{-ik_y y} \tag{12.18}$$

$$Z''(z) = -k_z^2\, Z(z) \implies Z(z) = A_z\, e^{ik_z z} + B_z\, e^{-ik_z z}$$

and

$$\frac{\hbar^2 k_x^2}{2m} + \frac{\hbar^2 k_y^2}{2m} + \frac{\hbar^2 k_z^2}{2m} = E\,. \tag{12.19}$$

Applying the boundary condition of vanishing wave function on the walls, we obtain

$$B_x = -A_x\, e^{2ik_x L} = -A_x\, e^{-2ik_x L} \implies k_x = n_x \frac{\pi}{2L}$$

$$B_y = -A_y\, e^{2ik_y L} = -A_y\, e^{-2ik_y L} \implies k_y = n_y \frac{\pi}{2L} \tag{12.20}$$

$$B_z = -A_z\, e^{2ik_z L} = -A_z\, e^{-2ik_z L} \implies k_z \doteq n_z \frac{\pi}{2L}\,,$$

where

$$n_x,\, n_y,\, n_z = 1,\, 2,\, \ldots,\quad and \quad E_{n_x n_y n_z} = \frac{\hbar^2 \pi^2}{8mL^2}\left(n_x^2 + n_y^2 + n_z^2\right)\,. \tag{12.21}$$

The resulting solutions are

$$X_{n_x}(x) = A_x\left(e^{i\frac{n_x \pi}{2L}x} + (-1)^{n_x+1}e^{-i\frac{n_x \pi}{2L}x}\right)$$

$$Y_{n_y}(y) = A_y\left(e^{i\frac{n_y \pi}{2L}y} + (-1)^{n_y+1}e^{-i\frac{n_y \pi}{2L}y}\right) \tag{12.22}$$

$$Z_{n_z}(z) = A_z\left(e^{i\frac{n_z \pi}{2L}z} + (-1)^{n_z+1}e^{-i\frac{n_z \pi}{2L}z}\right)\,.$$

The coefficients A_x, A_y, A_z are determined by normalization of the solution

$$\int d^3r \, |X(x)Y(y)Z(z)|^2 = 1 \implies \left(\int_{-L}^{L} dx \, |X(x)|^2\right)\left(\int_{-L}^{L} dy|Y(y)|^2\right)\left(\int_{-L}^{L} dz|Z(z)|^2\right) = 1$$

satisfied by

$$\left(\int_{-L}^{L} dx \, |X(x)|^2\right) = \left(\int_{-L}^{L} dy|Y(y)|^2\right) = \left(\int_{-L}^{L} dz|Z(z)|^2\right) = 1 \quad (12.23)$$

and giving

$$A_x = A_y = A_z = \frac{1}{2\sqrt{L}} \quad (12.24)$$

up to an arbitrary phase. Thus, finally, we have the energy eigenfunctions

$$\psi_{n_x n_y n_z}(x, y, z) = X_n(x) \, Y_{n_y}(y) \, Z_{n_z}(z), \quad (12.25)$$

where

$$X_{n_x}(x) = \frac{1}{2\sqrt{L}} \left(e^{i\frac{n_x\pi}{2L}x} + (-1)^{n_x+1}e^{-i\frac{n_x\pi}{2L}x}\right)$$

$$Y_{n_y}(y) = \frac{1}{2\sqrt{L}} \left(e^{i\frac{n_y\pi}{2L}y} + (-1)^{n_y+1}e^{-i\frac{n_y\pi}{2L}y}\right) \quad (12.26)$$

$$Z_{n_z}(z) = \frac{1}{2\sqrt{L}} \left(e^{i\frac{n_z\pi}{2L}z} + (-1)^{n_z+1}e^{-i\frac{n_z\pi}{2L}z}\right).$$

The general solution will be a superposition of the above partial solution for different values of the positive integers n_x, n_y, n_z

$$\Psi(x, y, z) = \sum_{n_x=1}^{\infty}\sum_{n_y=1}^{\infty}\sum_{n_z=1}^{\infty} C_{n_x n_y n_z} \, X_{n_x}(x) \, Y_{n_y}(y) \, Z_{n_z}(z). \quad (12.27)$$

These energy eigenfunctions are also parity eigenfunctions

$$\hat{\mathcal{P}}\psi_{n_x n_y n_z}(x, y, z) = (-1)^{n_x+n_y+n_z+1} \, \psi_{n_x n_y n_z}(x, y, z). \quad (12.28)$$

Note that we can define *spatial reflection operators* that correspond to a reflection with respect to a plane. For example

$$\hat{\mathcal{P}}_x \, \Psi(x, y, z) = \Psi(-x, y, z). \quad (12.29)$$

Obviously, these operators share the same properties as the standard parity, i.e.,

$$\mathcal{P}_x^2 = \mathbf{I}, \quad \mathcal{P}_x^\dagger = \mathcal{P}_x. \quad (12.30)$$

The product of these reflection operators on the three Cartesian planes equals the standard parity

$$\hat{P}_x \hat{P}_y \hat{P}_z = \hat{P}. \tag{12.31}$$

Degeneracy. In contrast to the one-dimensional infinite square well, in the three-dimensional problem there are more than one states that correspond to the same energy. For example, although the lowest energy eigenvalue $E_{111} = \frac{3\hbar^2\pi^2}{8mL^2}$ corresponds to just one state

$$\psi_{111}(x, y, x) = L^{-3/2} \cos(\pi x/2L) \cos(\pi y/2L) \cos(\pi z/2L), \tag{12.32}$$

and, therefore, the ground state is unique, the first excited energy level is triply degenerate with three eigenfunctions corresponding to the same energy

$$\psi_{211}(x, y, z) = \frac{1}{L^{3/2}} \sin(\pi x/L) \cos(\pi y/2L) \cos(\pi z/2L)$$

$$\psi_{121}(x, y, z) = \frac{1}{L^{3/2}} \cos(\pi x/2L) \sin(\pi y/L) \cos(\pi z/2L) \implies E_{112} = \frac{3\hbar^2\pi^2}{4mL^2}.$$

$$\psi_{112}(x, y, z) = \frac{1}{L^{3/2}} \cos(\pi x/2L) \cos(\pi y/2L) \sin(\pi z/L)$$

$$\tag{12.33}$$

The same is true for the next level $E_{221} = 9\hbar^2\pi^2/8mL^2$ which has three states corresponding to it, namely, ψ_{122}, ψ_{212}, and ψ_{221}. The degree of degeneracy increases rapidly with energy. For example, the energy level $E_{123} = 7\hbar^2\pi^2/4mL^2$ has sixfold degeneracy, having the six states ψ_{123}, ψ_{213}, ψ_{132}, ψ_{312}, ψ_{321}, and ψ_{231}. The occurrence of degeneracy is always an indication of the existence of some symmetry. Here the degeneracy is a result of the symmetries of the cube. If the edges of the cube had different lengths $L_x \neq L_y \neq L_z$ the degeneracy would be entirely lifted and $E_{112} \neq E_{121} \neq E_{112}$.

Density of States. The integers n_x, n_y, n_z define an orthogonal lattice. Each lattice point defines a possible state of the system. Let's introduce now the quantity $\Gamma(E)$ defined as *"the number of states with energy smaller than a certain value E"*

$$\Gamma(E) = \sum_{n_x} \sum_{n_y} \sum_{n_z} \Bigg|_{\frac{\hbar^2\pi^2}{8mL^2}(n_x^2+n_y^2+n_z^2) \leq E}. \tag{12.34}$$

The relative distance between neighboring points of the lattice is

$$\frac{\Delta n_x}{n_x} = \frac{1}{n_x} \propto \sqrt{\frac{E_{111}}{E}}$$

which is a small number for energies that are large in comparison to the ground state energy. Thus, we may consider the *continuum limit* and replace the sums with integrals as

$$\Gamma(E) = \int dn_x \int dn_y \int dn_z \Bigg|_{\frac{\hbar^2\pi^2}{8mL^2}(n_x^2+n_y^2+n_z^2) \leq E} \tag{12.35}$$

or

$$\Gamma(E) = \int d\Omega \int_0^{\sqrt{8mL^2E/\hbar^2\pi^2}} dn\, n^2 . \tag{12.36}$$

Since n_x, n_y, n_z are positive, the angular integration is only over an octant. Thus,

$$\Gamma(E) = \frac{\pi}{2} \int_0^{\sqrt{8mL^2E/\hbar^2\pi^2}} dn\, n^2 = \frac{(2L)^3}{6\pi^2} k^3 , \tag{12.37}$$

where $k^2 = 2mE^2/\hbar$ the standard continuum wave number, or

$$\Gamma(E) = \frac{V}{6\pi^2} \left(\frac{2mE}{\hbar^2} \right)^{3/2} \tag{12.38}$$

with $V = (2L)^3$ the volume of the box. In terms of the quantity $\Gamma(E)$, we may define the *density of states* as

$$g(E) = \frac{d\Gamma}{dE} = \frac{V}{4\pi^2} \left(\frac{2m}{\hbar^2} \right)^{3/2} \sqrt{E} = \frac{mVk}{2\pi^2\hbar^2} . \tag{12.39}$$

Example 12.1 (Reflection and transmission in a two-dimensional potential step)
Solve the time-independent Schroedinger equation for a particle moving in the (x, y) plane under the influence of a step function potential

$$V(x, y) = \begin{cases} 0 & (x < 0, \forall y) \\ \\ V_0 & (x > 0, \forall y) \end{cases} \tag{12.40}$$

for $E > V_0$.

Introducing the product solution we have

$$\Psi(x, y) = X(x)Y(y) \implies \begin{cases} -\frac{\hbar^2}{2m} X''(x) = \left(\frac{\hbar^2 k_x^2}{2m} - V_0 \Theta(x) \right) X(x) \\ \\ -\frac{\hbar^2}{2m} Y''(y) = \frac{\hbar^2 k_y^2}{2m} Y(y) \end{cases} \tag{12.41}$$

with $E = \frac{\hbar^2 k^2}{2m} = \frac{\hbar^2 h_x^2}{2m} + \frac{\hbar^2 k_y^2}{2m} .$

The solution is

$$\Psi(x, y) = \begin{cases} X(x) = \begin{cases} Ae^{ik_xx} + Be^{-ik_xx} & (x < 0) \\ \\ C\,e^{iq_xx} & (x > 0) \ , \end{cases} \\ \\ Y(y) = e^{ik_yy} \end{cases} \tag{12.42}$$

where

$$\frac{\hbar^2 q_x^2}{2m} = \frac{\hbar^2 k_x^2}{2m} - V_0 \ . \tag{12.43}$$

The corresponding plane waves are shown in Fig. 12.1. Continuity at $x = 0$ gives

$$A + B = C, \quad k_x\,(A - B) = q_x\,C \implies \begin{cases} \frac{B}{A} = \frac{q_x - k_x}{q_x + k_x} \\ \\ \frac{C}{A} = \frac{2q_x}{q_x + k_x} \end{cases} \ . \tag{12.44}$$

Thus, the solution is

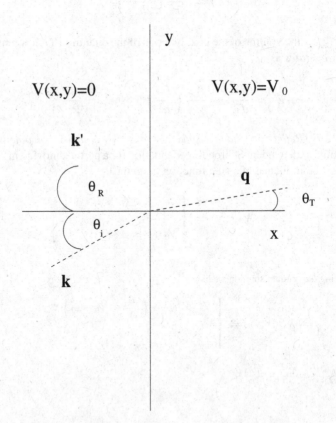

Fig. 12.1 Two-dimensional potential step

$$\Psi(x, y) = \begin{cases} X(x) = A \begin{cases} e^{ik_x x} + \left(\frac{q_x - k_x}{q_x + k_x}\right) e^{-ik_x x} & (x < 0) \\ \left(\frac{2q_x}{q_x + k_x}\right) e^{iq_x x} & (x > 0) \end{cases} \\ Y(y) = e^{ik_y y} \end{cases} \quad (12.45)$$

The solution parametrized above in terms of k_x, k_y, and q_x can also be parametrized by E and the *incidence angle*

$$\theta_i = \arctan(k_y/k_x), \quad (12.46)$$

in terms of which we have

$$k_x = \sqrt{2mE/\hbar^2} \cos\theta_i, \quad k_y = \sqrt{2mE/\hbar^2} \sin\theta_i \quad (12.47)$$

and

$$q_x = \sqrt{\frac{2m}{\hbar^2} \left(E \cos^2\theta_i - V_0\right)}. \quad (12.48)$$

Snell's law. We may introduce an *incident wave vector*

$$\mathbf{k}_i = (k \cos\theta_i, \, k \sin\theta_i), \quad (12.49)$$

a *reflected wave vector*

$$\mathbf{k}_r = \left(-k' \cos\theta_r, \, k' \sin\theta_r\right) \quad (12.50)$$

and a *transmitted wave vector*

$$\mathbf{k}_t = \left(k'' \cos\theta_t, \, k'' \sin\theta_t\right), \quad (12.51)$$

where θ_r and θ_t are the reflection and transmission angles. The solution can be written as

$$\Psi(\mathbf{r}) = \Theta(-x) \left(A \, e^{i\mathbf{k}_i \cdot \mathbf{r}} + B \, e^{i\mathbf{k}_r \cdot \mathbf{r}} \right) + \Theta(x) \, C \, e^{i\mathbf{k}_t \cdot \mathbf{r}}, \quad (12.52)$$

provided that

$$\theta_r = \theta_i, \quad k' = k = \sqrt{\frac{2mE}{\hbar^2}} \quad \text{(reflection laws)} \quad (12.53)$$

and

$$k'' \cos\theta_t = q_x, \quad k'' \sin\theta_t = k_y, \quad (12.54)$$

from which we obtain

$$(k'')^2 = q_x^2 + k_y^2 = \frac{2m}{\hbar^2} \left(E \cos^2\theta_i - V_0 + E \sin^2\theta_i\right) \implies k'' = \sqrt{\frac{2m}{\hbar^2}(E - V_0)} \quad (12.55)$$

and

$$k \sin\theta_i = k'' \sin\theta_t \quad \text{(refraction law)}. \quad (12.56)$$

The last relation is *Snell's law*

$$\frac{\sin\theta_t}{\sin\theta_i} = \frac{k}{k''} = \frac{1}{\sqrt{1 - \frac{V_0}{E}}}. \quad (12.57)$$

Note that for $V_0 > 0$, we have $\theta_t > \theta_i$, while for $V_0 < 0$, we have $\theta_t < \theta_i$.

12.3   The Three-Dimensional Harmonic Oscillator

General forces acting on a particle in three dimensions, described by a potential $V(\mathbf{r})$, can be approximated near equilibrium by a harmonic oscillator potential

$$V(\mathbf{r}) \approx V(\mathbf{r}_0) + \frac{1}{2} \sum_{i,j} V_{ij}(x_i - x_{0i})(x_j - x_{0j}) + \cdots , \qquad (12.58)$$

where the stable equilibrium point is defined by a minimum condition on the potential at the point $\mathbf{r}_0$, i.e., $\frac{\partial V}{\partial x_j}\big|_{\mathbf{r}_0} = 0$ and the requirement that all eigenvalues of the matrix $V_{ij} = \frac{\partial^2 V}{\partial x_i \partial x_j}\big|_{\mathbf{r}_0}$ are positive. Since a real symmetric matrix can always be diagonalized with a rotation that will leave the kinetic term unaffected, we may always cast the Hamiltonian in the form

$$\hat{H} = \frac{\hat{p}_x^2}{2m} + \frac{\hat{p}_y^2}{2m} + \frac{\hat{p}_z^2}{2m} + \frac{1}{2}m\omega_x^2 x^2 + \frac{1}{2}m\omega_y^2 y^2 + \frac{1}{2}m\omega_z^2 z^2 \qquad (12.59)$$

or

$$\hat{H} = \sum_{a=x,y,z} \hat{H}_a \quad \text{where} \quad \hat{H}_a = \frac{\hat{p}_a^2}{2m} + \frac{1}{2}m\omega_a^2 x_a^2 . \qquad (12.60)$$

Since

$$[\hat{H}_x, \hat{H}_y] = [\hat{H}_y, \hat{H}_z] = [\hat{H}_z, \hat{H}_x] = 0$$

the separation of variables method is applicable and the eigenvalue equation is satisfied by product solutions

$$\hat{H}\Psi_E(\mathbf{r}) = E\Psi_E(\mathbf{r}) \quad \text{with} \quad \Psi(\mathbf{r}) = \psi^{(x)}(x)\psi^{(y)}(y)\psi^{(z)}(z) ,$$

where

$$\begin{cases} \left(\frac{\hat{p}_x^2}{2m} + \frac{1}{2}m\omega_x^2 x^2\right)\psi^{(x)}(x) = E_x\,\psi^{(x)}(x) \\[2mm] \left(\frac{\hat{p}_y^2}{2m} + \frac{1}{2}m\omega_y^2 y^2\right)\psi^{(y)}(y) = E_y\,\psi^{(y)}(y) \\[2mm] \left(\frac{\hat{p}_z^2}{2m} + \frac{1}{2}m\omega_z^2 z^2\right)\psi^{(z)}(z) = E_z\,\psi^{(z)}(z) \end{cases} \qquad (12.61)$$

for

$$E_x + E_y + E_z = E . \qquad (12.62)$$

The eigenfunctions $\psi^{(a)}$ are the standard one-dimensional harmonic oscillator energy eigenfunctions, each corresponding to a different frequency ω_a and labeled by an integer $n_a = 0, 1, 2, \ldots$. The eigenvalues E_a are $\hbar\omega_a(n_a + 1/2)$. Thus,

$$E_{n_x n_y n_z} = \hbar\omega_x \left(n_x + \frac{1}{2}\right) + \hbar\omega_y \left(n_y + \frac{1}{2}\right) + \hbar\omega_z \left(n_z + \frac{1}{2}\right) \qquad (12.63)$$

with n_x, n_y, $n_z = 0, 1, 2, \ldots$. For example, the ground state is

$$\Psi_{000}(\mathbf{r}) = \left(\frac{m\omega_x}{\hbar\pi}\right)^{1/4} \left(\frac{m\omega_y}{\hbar\pi}\right)^{1/4} \left(\frac{m\omega_z}{\hbar\pi}\right)^{1/4} e^{-\frac{m\omega_x}{2\hbar}x^2} e^{-\frac{m\omega_y}{2\hbar}y^2} e^{-\frac{m\omega_z}{2\hbar}z^2} \qquad (12.64)$$

with the ground state energy

$$E_{000} = \frac{1}{2}\hbar\left(\omega_x + \omega_y + \omega_z\right). \qquad (12.65)$$

The isotropic harmonic oscillator. If the harmonic force is independent of the direction, i.e.,

$$\omega_x = \omega_y = \omega_z \qquad (12.66)$$

we have the so-called *isotropic harmonic oscillator* [2, 3]. Denoting by ω the common frequency, we have the eigenvalues

$$E_{n_x n_y n_z} = \hbar\omega \left(n_x + n_y + n_z + \frac{3}{2}\right) \quad with \quad \begin{array}{l} n_x = 0, 1, 2, \ldots \\ n_y = 0, 1, 2, \ldots \\ n_z = 0, 1, 2, \ldots \end{array} \qquad (12.67)$$

and the eigenfunctions

$$\Psi_{n_x n_y n_z}(\mathbf{r}) = \psi_{n_x}(x)\psi_{n_y}(y)\psi_{n_z}(z), \qquad (12.68)$$

where $\psi_{n_a}(x_a)$ are the one-dimensional harmonic oscillator energy eigenfunctions of the same frequency. For example, the ground state has energy $3\hbar\omega/2$ and eigenfunction

$$\Psi_{000}(r) = \left(\frac{m\omega}{\hbar\pi}\right)^{3/4} e^{-\frac{m\omega}{2\hbar}r^2}. \qquad (12.69)$$

The first excited level corresponds to the energy $5\hbar\omega/2$ but it has a *threefold degeneracy* corresponding to three different eigenfunctions, namely,

$$\Psi_{100}(\mathbf{r}) = x\sqrt{\frac{2m\omega}{\hbar}} \left(\frac{m\omega}{\hbar\pi}\right)^{3/4} e^{-\frac{m\omega}{2\hbar}r^2}$$

$$\Psi_{010}(\mathbf{r}) = y\sqrt{\frac{2m\omega}{\hbar}} \left(\frac{m\omega}{\hbar\pi}\right)^{3/4} e^{-\frac{m\omega}{2\hbar}r^2}$$

$$\Psi_{001}(\mathbf{r}) = z\sqrt{\frac{2m\omega}{\hbar}} \left(\frac{m\omega}{\hbar\pi}\right)^{3/4} e^{-\frac{m\omega}{2\hbar}r^2}.$$

The degeneracy is a common feature of all higher excited energy levels, which increases as we go to higher energies. There is however an alternative set of energy eigenstates of the isotropic oscillator which are common eigenstates of the Hamiltonian and the angular momentum. We shall analyze this set in a subsection of Chap. 13.

12.4 The Two-Dimensional Isotropic Oscillator

It is possible that harmonic forces are isotropic only within a particular plane and the corresponding Hamiltonian is of the form [3]

$$\hat{H} = \hat{H}_{||} + \hat{H}_{\perp} \tag{12.70}$$

with

$$\hat{H}_{||} = \frac{\hat{p}_z^2}{2m} + \frac{1}{2}m\omega_{||}^2 z^2$$
$$\hat{H}_{\perp} = \frac{\hat{p}_x^2}{2m} + \frac{\hat{p}_y^2}{2m} + \frac{1}{2}m\omega_{\perp}^2 (x^2 + y^2). \tag{12.71}$$

The energy eigenfunctions will be products of the *transverse* eigenfunctions $\psi^{(\perp)}(x, y)$ and the *longitudinal* ones $\psi^{(||)}(z)$, the latter being just the one-dimensional oscillator energy eigenfunctions. The energy eigenvalues will be

$$E = E_{\perp} + \hbar\omega_{||} \left(n_{||} + \frac{1}{2} \right) \quad (n_{||} = 0, 1, \ldots). \tag{12.72}$$

The *transverse* energy eigenvalues $E_{\perp}$ are determined by the *transverse* Schroedinger equation

$$\left(\frac{\hat{p}_x^2}{2m} + \frac{\hat{p}_y^2}{2m} + \frac{1}{2}m\omega_{\perp}^2 (x^2 + y^2) \right) \psi^{(\perp)}(x, y) = E_{\perp}\psi^{(\perp)}(x, y). \tag{12.73}$$

In what follows, we proceed to solve (12.73). Of course, we already know a set of solutions, namely, products of eigenfunctions $\psi_{n_x}(x)\psi_{n_y}(y)$ with same frequency $\omega_{\perp}$ and the eigenvalues parametrized as $E_{\perp} = \hbar\omega_{\perp}(n_x + n_y + 1)$. Nevertheless, here we are going to pursue a different solution consisting of eigenfunctions of the energy that are common eigenfunctions of the angular momentum $\hat{L}_z$.

Let us introduce the two operators

$$\hat{a}_{\pm} = \sqrt{\frac{m\omega_{\perp}}{4\hbar}} (x \pm iy) + \frac{i}{\sqrt{4m\hbar\omega_{\perp}}} (\hat{p}_x \pm i\hat{p}_y) \tag{12.74}$$

as well as their conjugates $\hat{a}_{\pm}^{\dagger}$. They satisfy the commutation relations

$$\left[\hat{a}_{\pm}, \hat{a}_{\pm}^{\dagger} \right] = 1 \quad and \quad \left[\hat{a}_{\pm}, \hat{a}_{\mp} \right] = 0. \tag{12.75}$$

In terms of them, the Hamiltonian is written as

$$\hat{H}_\perp = \hbar\omega_\perp \left(\hat{a}_+^\dagger \hat{a}_+ + \hat{a}_-^\dagger \hat{a}_- + 1 \right) \tag{12.76}$$

or, introducing the *"number operators"* $\hat{N}_\pm = \hat{a}_\pm^\dagger \hat{a}_\pm$, as

$$\hat{H}_\perp = \hbar\omega_\perp \left(\hat{N}_+ + \hat{N}_- + 1 \right). \tag{12.77}$$

The energy eigenstates are common eigenstates of the number operators, the eigenvalues of which, as can be shown in exactly the same fashion as in the one-dimensional oscillator, are the nonnegative integers n_+ and n_-. Then, the energy eigenvalues will be

$$E_\perp = \hbar\omega_\perp (n_+ + n_- + 1) \quad (n_+, n_- = 0, 1, \ldots). \tag{12.78}$$

The operators $\hat{a}_\pm$ satisfy the following commutation relations:

$$\left[\hat{H}_\perp, \hat{a}_\pm \right] = -\hbar\omega_\perp \hat{a}_\pm, \quad \left[\hat{H}_\perp, \hat{a}_\pm^\dagger \right] = \hbar\omega_\perp \hat{a}_\pm^\dagger. \tag{12.79}$$

We proceed denoting the energy eigenstates as $|n_+, n_-\rangle$. From the first commutator above, it is clear that the state of lowest energy ($n_+ = n_- = 0$) has to satisfy

$$\hat{a}_\pm |0, 0\rangle = 0. \tag{12.80}$$

Furthermore, the eigenstates satisfy

$$\hat{a}_+ |n_+, n_-\rangle = \sqrt{n_+} \, |n_+ - 1, n_-\rangle$$

$$\hat{a}_- |n_+, n_-\rangle = \sqrt{n_-} \, |n_+, n_- - 1\rangle$$

$$\hat{a}_+^\dagger |n_+, n_-\rangle = \sqrt{n_+ + 1} \, |n_+ + 1, n_-\rangle \qquad \Longrightarrow \quad |n_+, n_-\rangle = \frac{\left(\hat{a}_+^\dagger \right)^{n_+} \left(\hat{a}_-^\dagger \right)^{n_-}}{\sqrt{n_+! n_-!}} |0, 0\rangle.$$

$$\hat{a}_-^\dagger |n_+, n_-\rangle = \sqrt{n_- + 1} \, |n_+, n_- + 1\rangle$$

$$\tag{12.81}$$

Nevertheless, the Hamiltonian $\hat{H}_\perp$ is invariant under rotations in the (x, y)-plane. This symmetry should be reflected in the conservation of the angular momentum $\hat{L}_z$. Indeed, the angular momentum operator

$$\hat{L}_z = x\hat{p}_y - y\hat{p}_x = \hbar \left(\hat{a}_+^\dagger \hat{a}_+ - \hat{a}_-^\dagger \hat{a}_- \right) = \hbar \left(\hat{N}_+ - \hat{N}_- \right) \tag{12.82}$$

commutes with the Hamiltonian. The $\hat{L}_z$ eigenvalues are

$$\hat{L}_z |n_+, n_-\rangle = \hbar(n_+ - n_-)|n_+, n_-\rangle. \tag{12.83}$$

Problems and Exercises

12.1 Consider a particle moving in the x-y plane, subject to isotropic harmonic forces (*two-dimensional isotropic harmonic oscillator*). Construct the ground state wave function and the degenerate wave functions of the first excited state in the $\{x, y\}$-representation. Verify explicitly that they are eigenstates of the angular momentum $\hat{L}_z$.

12.2 Particles with energy $E > V_0$ move in a rectangular tube of square cross section with infinitely repulsive walls. This is described by the potential

$$V(x, y, z) = \begin{cases} 0 & (|x| < L, \ |y| < L, \ z < 0) \\ V_0 & (|x| < L, \ |y| < L, \ z > 0) \ . \\ +\infty & (|x| \geq L, \ |y| \geq L, \ \forall z) \end{cases}$$

Find the reflection coefficient of the potential step at $z = 0$.

12.3 A free particle of mass m is moving on the two-dimensional plane. Calculate the probability amplitude to make a transition from a point (x, y) to a point (x', y') in time t, namely,

$$\mathcal{K}(x', y'; x, y; t) = \langle x', y' | e^{-\frac{i}{\hbar} \hat{H} t} | x, y \rangle \ .$$

12.4 An electron is bound on the surface of a solid with a potential that is to a good approximation harmonic but anisotropic

$$V(x, y) = \frac{1}{2} m \omega_x^2 x^2 + \frac{1}{2} m \omega_y^2 y^2 \ .$$

When the system is subject to an external homogeneous electric field the potential is modified by the extra term

$$\Delta V = -eE \left(x \cos \theta + y \sin \theta \right) \ .$$

Solve the eigenvalue problem of the energy finding eigenvalues and eigenfunctions.

12.5 Consider the three-dimensional Schroedinger equation in a region without potential. Show that, apart from the known plane wave solution $e^{i \mathbf{k} \cdot \mathbf{r}}$, the spherical wave $\frac{e^{ikr}}{r}$ is also a solution (for $r \neq 0$). In contrast, show that e^{ikr} is not.

12.6 Consider a two-dimensional infinite square well. Write down the energy eigenfunctions and eigenvalues. Assume that at time $t = 0$, the system is in a state

$$\frac{1}{\sqrt{2}} \left(|n_x = 0\rangle |n_y = 1\rangle + e^{i\alpha} |n_x = 1\rangle |n_y = 0\rangle \right) \ ,$$

where α is a known real parameter. Calculate the uncertainties $(\Delta x)_t^2$ and $(\Delta y)_t^2$ at a time $t > 0$.

12.7 A particle of mass m is constrained to move on an infinite stripe of width $2L$ on the plane x-y. At the same time, the particle is subject to a harmonic force parallel to the direction of the stripe. This is equivalent to a potential

$$V(x, y) = \begin{cases} \frac{1}{2}m\omega^2 y^2 & (|x| < L, \; \forall y) \\ +\infty & (|x| \geq L, \; \forall y) \end{cases}.$$

Find the energy eigenfunctions and eigenvalues.

References

1. G. Arfken, H. Weber, R. Harris, *Mathematical Methods for Physicists*, 7th edn. (Academic, New York, 2013)
2. E. Merzbacher, *Quantum Mechanics*, 3rd edn. (Wiley, New York, 1998)
3. A. Messiah, *Quantum Mechanics* (Dover publications, Mineola, 1958). Single-volume reprint of the John Wiley & Sons, New York, two-volume 1958 edition

Chapter 13
Central Potentials

13.1 Particle Motion in a Central Field

Consider a particle of mass[1] μ moving in three-dimensional space under the influence of a central potential $V(r)$, depending solely on the distance from a given center, taken to be the origin of the coordinate system.[2] The Hamilton operator will be

$$\hat{H} = \frac{\hat{\mathbf{p}}^2}{2\mu} + V(r). \tag{13.1}$$

We may recall at this point that the operators $\hat{\mathbf{p}}^2$ and r commute with angular momentum

$$\left[\hat{\mathbf{p}}^2, \hat{\mathbf{L}}\right] = \left[r, \hat{\mathbf{L}}\right] = 0. \tag{13.2}$$

Thus, the Hamiltonian commutes with angular momentum and the energy eigenstates $|E, \ell, m\rangle$ are common eigenstates of $\hat{H}$, $\hat{L}^2$, $\hat{L}_z$

$$\left[\hat{H}, \hat{\mathbf{L}}\right] = 0 \implies \begin{cases} \hat{H}|E, \ell, m\rangle = E|E, \ell, m\rangle \\[2mm] \hat{L}^2|E, \ell, m\rangle = \hbar^2\ell(\ell+1)|E, \ell, m\rangle. \\[2mm] \hat{L}_z|E, \ell, m\rangle = \hbar m|E, \ell, m\rangle \end{cases} \tag{13.3}$$

In the $\{x\}$ representation, the energy eigenvalue equation (time-independent Schroedinger equation) has the form

---

[1] In this and subsequent chapters we shall denote the particle mass as μ in order to avoid confusion with the angular momentum quantum number m.

[2] For the material on central potentials treated in this chapter see also [1–4].

© Springer Nature Switzerland AG 2019
K. Tamvakis, *Basic Quantum Mechanics*, Undergraduate Texts in Physics,
https://doi.org/10.1007/978-3-030-22777-7_13

$$\left\{-\frac{\hbar^2}{2\mu}\nabla^2 + V(r)\right\}\psi_{E\ell m}(r,\theta,\phi) = E\,\psi_{E,\ell,m}(r,\theta,\phi). \tag{13.4}$$

However, we know from (8.18) that in spherical coordinates, we may separate the kinetic operator as

$$-\hbar^2\nabla^2 = -\frac{\hbar^2}{r}\frac{\partial^2}{\partial r^2}r + \frac{\hat{L}^2}{r^2}. \tag{13.5}$$

Thus, the Schroedinger equation takes the form

$$\left\{-\frac{\hbar^2}{2\mu}\frac{1}{r}\frac{\partial^2}{\partial r^2}r + \frac{\hat{L}^2}{2\mu r^2} + V(r)\right\}\psi_{E\ell m}(r,\theta,\phi) = E\psi_{E\ell m}(r,\theta,\phi). \tag{13.6}$$

Since $r^2\hat{H}$ consists of two commuting parts, namely, $\hat{L}^2$ that depends on angles and $-\hbar^2 r\frac{\partial^2}{\partial r^2}r$ that depends on the radius, we have grounds to apply the method of separation of variables and introduce a trial solution in the form of a product. Thus, we consider

$$\psi_{E\ell m}(r,\theta,\phi) = R_{E,\ell}(r)\,Y_{\ell m}(\theta,\phi)\,, \tag{13.7}$$

where $Y_{\ell m}(\theta,\phi)$ are the eigenfunctions of angular momentum (spherical harmonics). Substituting it into the Schroedinger equation, we get

$$Y_{\ell m}(\theta,\phi)\left(-\frac{\hbar^2}{2\mu}\frac{1}{r}\frac{d^2}{dr^2}\left(rR_{E,\ell}(r)\right) + (V(r) - E)\,R_{E,\ell}(r)\right)$$

$$+ R_{E,\ell}(r)\left(\frac{\hbar^2\ell(\ell+1)}{2\mu r^2}Y_{\ell,m}(\theta,\phi)\right) = 0 \tag{13.8}$$

or

$$-\frac{\hbar^2}{2\mu}\frac{1}{r}\frac{d^2(rR_{E\ell})}{dr^2} + \frac{\hbar^2\ell(\ell+1)}{2\mu r^2}R_{E\ell} + V(r)R_{E\ell} = ER_{E\ell}. \tag{13.9}$$

This is the *radial Schroedinger equation* and $R_{E,\ell}(r)$ is the *radial wave function*. The normalization of the radial wave functions follows directly from the normalization of the full wave function

$$\int d^3r\,|\psi_{E\ell m}(\mathbf{r})|^2 = 1 \implies \int d\Omega\,|Y_{\ell,m}(\Omega)|^2\int_0^\infty dr\,r^2\,|R_{E,\ell}(r)|^2 = 1$$

or

$$\int_0^\infty dr\,r^2\,|R_{E,\ell}(r)|^2 = 1. \tag{13.10}$$

The "one-dimensional" wave function. It is possible to introduce an alternative radial wave function as

$$R_{E,\ell}(r) = \frac{u_{E,\ell}(r)}{r} \quad or \quad u_{E,\ell}(r) = r\, R_{E,\ell}(r). \qquad (13.11)$$

Substituting (13.11) into the radial Schroedinger equation, it simplifies to

$$-\frac{\hbar^2}{2\mu}\frac{d^2 u_{E\ell}(r)}{dr^2} + \frac{\hbar^2 \ell(\ell+1)}{2\mu r^2} u_{E\ell}(r) + V(r)u_{E\ell}(r) = E\, u_{E\ell}(r) \qquad (13.12)$$

which has exactly the *one-dimensional form*. This is why $u_{E\ell}(r)$ is called the "*one-dimensional radial wave function*". Note however the modification in comparison to one-dimensional problems due to the presence of a repulsive *centrifugal potential* term arising from the nonzero angular momentum. Note also that $u_{E\ell}(r)$ has to satisfy a special boundary condition at the origin

$$R_{E\ell}(0) < \infty \implies u_{E\ell}(0) = 0. \qquad (13.13)$$

The normalization condition on $u_{E\ell}(r)$ simplifies to

$$\int_0^\infty dr\, |u_{E\ell}(r)|^2 = 1. \qquad (13.14)$$

Let's go back to (13.9) and see if we can draw any general conclusions about the behavior of its solutions. In all cases of physical interest, we have to deal with potentials that are bounded everywhere with a possible exception of the origin $r = 0$ where they may have at most a Coulomb-like singularity. Thus, we can assume that

$$\lim_{r\to 0} \{r^2 V(r)\} = 0. \qquad (13.15)$$

Then, the radial equation near the origin is dominated by the kinetic and the centrifugal terms and can be approximated by

$$-\frac{1}{r}\frac{d^2(rR)}{dr^2} + \frac{\ell(\ell+1)}{r^2}R \approx 0. \qquad (13.16)$$

Substituting a trial power ansatz $\sim r^\alpha$ we obtain two power solutions for $\alpha = \ell + 1$ and $\alpha = -\ell$, namely,

$$\lim_{r\to 0} \{R_{E\ell}(r)\} = A r^\ell + B r^{-(\ell+1)}. \qquad (13.17)$$

The second of these solutions is singular and, therefore, unacceptable as a wave function. Even a superposition of such solutions cannot lead to a square integrable wave function. As a result, only the regular solution is acceptable, behaving as r^ℓ near the origin and, therefore, being finite.

In many cases of physical interest, we will have a *short range* potential that vanishes in the asymptotic region like

$$\lim_{r \to \infty} \{r V(r)\} = 0. \tag{13.18}$$

For such a potential, the energy spectrum will have a continuous part with $E > 0$ corresponding to scattering states. Depending on the details of the potential in the near region, there might be also a discrete part in the energy spectrum corresponding to bound states. In general, the energy spectrum will be mixed, although there are cases where it could be fully continuous (e.g., a repulsive short range potential). A fully discrete spectrum is also possible for potentials that are increasing in the asymptotic region.

Each energy eigenvalue of the continuous spectrum is *infinitely degenerate*, since there are $\sum_{\ell=0}^{\infty} (2\ell + 1) = \infty$ eigenfunctions $\psi_{E\ell m}$ corresponding to it. The energy eigenvalues of the discrete spectrum, since they result from an equation that contains ℓ, will depend on ℓ but also on some additional quantum number that labels them, like what happens in one-dimensional bound state problems. Each of these eigenvalues, symbolized as $E_{n\ell}$, will have $(2\ell + 1)$-fold degeneracy corresponding to the $2\ell + 1$ different values of m (i.e., directions of L_z) that do not change the energy. For special potentials, depending on their symmetries, the degree of degeneracy may be much larger. For example, in the case of the discrete spectrum of the attractive Coulomb potential the energy eigenvalues depend only on the quantum number n and the degeneracy is $\sum_{\ell=0}^{n} (2\ell + 1) = n^2$.

13.2   The Free Particle

The problem of the three-dimensional motion of a free particle is a problem that has been solved at the very early stages of our discussion on Quantum Mechanics. There, we saw that the plane waves

$$\psi_{\mathbf{p}}(\mathbf{r}) = \frac{e^{\frac{i}{\hbar}\mathbf{p} \cdot \mathbf{r}}}{(2\pi\hbar)^{3/2}} \tag{13.19}$$

are eigenfunctions of the energy corresponding to the eigenvalues $E = \frac{p^2}{2\mu}$. These eigenfunctions are simultaneous eigenfunctions of the energy and the momentum $\hat{\mathbf{p}}$. Nevertheless, since a free-particle motion is a trivial case of a central force problem with $V(r) = 0$, we may apply on it the alternative description in terms of simultaneous eigenfunctions of the energy $\hat{H}_0 = \hat{p}^2/2\mu$ and the angular momentum $\hat{L}^2, \hat{L}_z$

$$\begin{cases} \hat{H}_0, \ \hat{\mathbf{p}} \implies \psi_{\mathbf{p}}(\mathbf{r}) = \frac{e^{\frac{i}{\hbar}\mathbf{p} \cdot \mathbf{r}}}{(2\pi\hbar)^{3/2}} & (plane \ waves) \\[4mm] \hat{H}_0, \ \hat{L}^2, \ \hat{L}_z \implies \psi_{E,\ell,m}(\mathbf{r}) = R_{E,\ell}(r) Y_{\ell m}(\Omega) & (spherical \ waves) \end{cases} \tag{13.20}$$

Therefore, setting $V = 0$ and applying the analysis of the previous section we obtain the set of common eigenfunctions of $\hat{L}^2$, $\hat{L}_z$ and $\hat{H}_0 = \hat{p}^2/2\mu$, denoting them as

$$\psi_{k\ell m}(r, \theta, \phi) = R_{k,\ell}(r)\, Y_{\ell,m}(\theta, \phi). \tag{13.21}$$

We have labeled them in terms of the wave number k, related to the energy through $E = \hbar^2 k^2/2\mu$, and the angular momentum quantum numbers. The radial wave function satisfies the radial Schroedinger equation

$$-\frac{\hbar^2}{2\mu r}\frac{d^2(rR_{k\ell})}{dr^2} + \frac{\hbar^2 \ell(\ell+1)}{2\mu r^2}R_{k\ell} = \frac{\hbar^2 k^2}{2\mu}R_{k\ell} \tag{13.22}$$

or

$$\frac{d^2 R_{k\ell}(r)}{dr^2} + \frac{2}{r}\frac{dR_{k\ell}(r)}{dr} - \frac{\ell(\ell+1)}{r^2}R_{k\ell}(r) + k^2 R_{k\ell}(r) = 0 \tag{13.23}$$

or

$$\frac{d^2 R_{k\ell}}{d(kr)^2} + \frac{2}{r}\frac{dR_{k\ell}}{d(kr)} - \frac{\ell(\ell+1)}{(kr)^2}R_{k\ell} + R_{k\ell} = 0. \tag{13.24}$$

Thus, the independent variable of this equation is not r but kr. Therefore, we may denote $R_{k,\ell}(r) = R_\ell(kr)$ and have

$$R_\ell''(kr) + \frac{2}{kr}R_\ell'(kr) - \frac{\ell(\ell+1)}{(kr)^2}R_\ell(kr) + R_\ell(kr) = 0. \tag{13.25}$$

This is a well-known differential equation

$$y_\ell''(x) + \frac{1}{x}y_\ell'(x) - \frac{\ell(\ell+1)}{x^2}y_\ell(x) + y_\ell(x) = 0 \tag{13.26}$$

with well-known solutions the so-called *spherical Bessel functions* $j_\ell(x)$ and the *spherical Neumann functions* $n_\ell(x)$, the latter being singular at the origin $x = 0$. Thus, for the free particle, only $j_\ell(kr)$ is acceptable as a radial wave function. Thus, the radial wave function of the free particle will be just $R_{k\ell}(r) = j_\ell(kr)$.

Digression on the spherical Bessel functions. The behavior of the spherical Bessel functions $j_\ell(x)$ and spherical Neumann functions $n_\ell(x)$ near the origin $x \sim 0$ is

$$j_\ell(x) \sim \frac{x^\ell}{1\cdot 3\cdot 5\cdots(2\ell+1)}, \quad n_\ell(x) \sim \frac{1\cdot 3\cdots(2\ell-1)}{x^{\ell+1}}. \tag{13.27}$$

Their asymptotic behavior ($x \sim \infty$) is

$$j_\ell(x) \sim \frac{\sin(x - \ell\pi/2)}{x}, \quad n_\ell(x) \sim -\frac{\cos(x - \ell\pi/2)}{x}. \tag{13.28}$$

Table 13.1 Spherical Bessel and Neumann functions

$j_0(x) = \frac{\sin x}{x}$	$n_0(x) = -\frac{\cos x}{x}$
$j_1(x) = \frac{\sin x}{x^2} - \frac{\cos x}{x}$	$n_1(x) = -\frac{\cos x}{x^2} - \frac{\sin x}{x}$
$j_2(x) = \left(\frac{3}{x^3} - \frac{1}{x}\right)\sin x - \frac{3}{x^2}\cos x$	$n_2(x) = -\left(\frac{3}{x^3} - \frac{1}{x}\right)\cos x - \frac{3}{x^2}\sin x$

In many cases, it is useful to introduce the *spherical Hankel functions*

$$h_\ell^{(\pm)}(x) = n_\ell(x) \pm i j_\ell(x) \tag{13.29}$$

with asymptotic behavior

$$h_\ell^{(\pm)}(x) \sim -(\pm i)^\ell \frac{e^{\mp ix}}{x}. \tag{13.30}$$

The first few of these spherical functions are shown in Table 13.1.

The spherical Bessel functions satisfy orthonormality relations

$$\int_0^\infty dr\, r^2\, j_\ell(kr) j_\ell(k'r) = \frac{\pi}{2k^2}\delta(k - k')$$

$$\int_0^\infty dk\, k^2\, j_\ell(kr) j_\ell(kr') = \frac{\pi}{2kr^2}\delta(r - r'). \tag{13.31}$$

Spherical wave analysis of a plane wave. We can always write down an expansion of a plane wave in terms of the complete set of states $|E, \ell, m\rangle$, i.e., in terms of spherical waves

$$\frac{e^{i\mathbf{k}\cdot\mathbf{r}}}{(2\pi)^{3/2}} = \sum_{\ell=0}^\infty \sum_{m=-\ell}^\ell C_{\ell,m}(\hat{k})\, Y_{\ell m}(\theta, \phi)\, j_\ell(kr). \tag{13.32}$$

In order to determine the coefficients $C_{\ell,m}(\hat{k})$ of this expansion, we shall use without proof the following identity[3]:

$$\int d^3r\, e^{i\mathbf{k}\cdot\mathbf{r}} Y_{\ell m}^*(\theta, \phi)\, j_\ell(k'r) = \frac{2\pi^2}{k^2}(-i)^\ell Y_{\ell m}(\theta_k, \phi_k)\delta(k - k'), \tag{13.33}$$

where θ_k, ϕ_k are the angles corresponding to the direction $\hat{k}$. Multiplying both sides of the expansion (13.32) with $Y_{\ell'm'}^*(\theta, \phi) j_{\ell'}(k'r)$, integrating with respect to $\mathbf{r}$ and using the orthonormality of spherical harmonics and spherical Bessel functions, we obtain

$$C_{\ell,m}(\hat{k}) = i^\ell \sqrt{\frac{2}{\pi}} Y_{\ell m}^*(\theta_k, \phi_k) \tag{13.34}$$

---

[3] See [1].

and, finally, the expansion takes the form

$$\frac{e^{i\mathbf{k}\cdot\mathbf{r}}}{(2\pi)^{3/2}} = \sqrt{\frac{2}{\pi}} \sum_{\ell=0}^{\infty} \sum_{m=-\ell}^{\ell} i^{\ell} Y_{\ell m}^{*}(\theta_k, \phi_k) Y_{\ell m}(\theta, \phi) j_{\ell}(kr). \tag{13.35}$$

In the case that the direction of $\mathbf{k}$ coincides with the $\hat{z}$-axis, this expression cannot depend on the azimuthal angle or, equivalently,

$$\hat{L}_z e^{ikz} = 0 \implies \frac{e^{ikz}}{(2\pi)^{3/2}} = \sqrt{\frac{2}{\pi}} \sum_{\ell=0}^{\infty} i^{\ell} Y_{\ell,0}(0) Y_{\ell,0}(\theta) j_{\ell}(kr)$$

or, using the relation to the Legendre polynomials, we obtain

$$Y_{\ell,0}(\theta) = \sqrt{(2\ell+1)4\pi} P_{\ell}(\cos\theta) \implies e^{ikz} = \sum_{\ell=0}^{\infty} i^{\ell}(2\ell+1) P_{\ell}(\cos\theta) j_{\ell}(kr).$$
$$\tag{13.36}$$

13.3 Examples of Central Potentials

13.3.1 A Spherical Cavity

Consider a particle trapped in an impenetrable spherical cavity. This can be described by the following central potential:

$$V(r) = \begin{cases} 0 & (0 \le r < a) \\ +\infty & (r \ge a) \end{cases}. \tag{13.37}$$

The radial equation inside the cavity is just the free-particle one

$$-\frac{\hbar^2}{2\mu}\frac{1}{r}\frac{d^2(rR_{k\ell})}{dr^2} + \frac{\hbar^2\ell(\ell+1)}{2\mu r^2}R_{k\ell} = \frac{\hbar^2 k^2}{2\mu}R_{k\ell} \tag{13.38}$$

with boundary condition

$$R_{k\ell}(a) = 0, \tag{13.39}$$

since the wave function has to vanish beyond the cavity. The solution of (13.38) is just

$$R_{\ell}(kr) = A j_{\ell}(kr) \tag{13.40}$$

with A fixed by normalization. Applying the boundary condition (13.39), we get a condition on the energy

$$j_{\ell}(ka) = 0. \tag{13.41}$$

The s-wave case ($\ell = 0$). Let's consider first the case of spherical waves ($\ell = 0$). In this case, the energy condition reads

$$\sin(ka) = 0 \implies k_n = n\frac{\pi}{a} \quad with \quad n = 1, 2, \ldots \tag{13.42}$$

and

$$E_n = \frac{\hbar^2 n^2 \pi^2}{2\mu a^2}. \tag{13.43}$$

The normalized eigenfunctions are

$$R_{n,0}(r) = \sqrt{\frac{2}{a}} \frac{\sin(n\pi r/a)}{r}. \tag{13.44}$$

The p-wave case ($\ell = 1$). In this case, we have

$$R_1(kr) = A\,j_1(kr) = A\left(\frac{\sin(kr)}{(kr)^2} - \frac{\cos(kr)}{(kr)}\right). \tag{13.45}$$

The boundary condition corresponds to the condition on the allowed energies

$$j_1(ka) = \frac{\sin(ka)}{(ka)^2} - \frac{\cos(ka)}{(ka)} = 0 \tag{13.46}$$

or

$$\tan(ka) = ka. \tag{13.47}$$

This is an equation that can be solved graphically as we did in the case of one-dimensional square well. The graphical solution is depicted in Fig. 13.1. The dotted

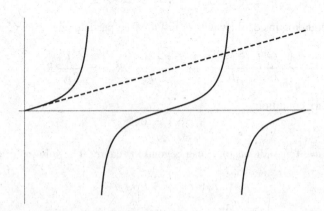

Fig. 13.1 Graphical solution of for p-waves

line corresponds to the right-hand side (ka) and we see that it intersects with all infinite branches of the left-hand side $(\tan(ka))$. Thus, we have an infinity of solutions.

Example 13.1 A particle of mass μ is trapped in a spherical region of inner radius a and outer radius $2a$. This situation can be described with a potential

$$
V(r) = \begin{cases} +\infty & (0 \le r \le a) \\ 0 & (a < r < 2a) \\ +\infty & (r \ge 2a) \end{cases}.
$$

Find the radial wave function and the energy spectrum for $\ell = 0$.

The particle satisfies the free Schroedinger equation in the region $(a, 2a)$ and its wave function is zero for $r \ge 2a$ and $r \le a$. The radial wave function in the inner region is

$$
R_\ell(kr) = A \, j_\ell(kr) + B \, n_\ell(kr). \tag{13.48}
$$

By continuity we must have

$$
A \, j_\ell(ka) + B \, n_\ell(ka) = 0
$$

$$
A \, j_\ell(2ka) + B \, n_\ell(2ka) = 0 \tag{13.49}
$$

or

$$
\begin{vmatrix} j_\ell(ka) & n_\ell(ka) \\ j_\ell(2ka) & n_\ell(2ka) \end{vmatrix} = 0 \implies j_\ell(ka) n_\ell(2ka) - j_\ell(2ka) n_\ell(ka) = 0. \tag{13.50}
$$

For $\ell = 0$, this amounts to

$$
\sin(ka) \cos(2ka) - \sin(2ka) \cos(ka) = 0
$$

or

$$
\sin(ka) \left(2 \cos^2(ka) - 1 \right) - 2 \sin(ka) \cos^2(ka) = 0
$$

or

$$
\sin(ka) = 0 \implies k_n = \frac{n\pi}{a} \quad (n = 1, 2, \ldots). \tag{13.51}
$$

Going back to the system of equations, we conclude that $B = 0$ and the solution is

$$
R_0^{(n)}(kr) = A \, \frac{\sin(n\pi r/a)}{n\pi r/a}, \quad E_n = \frac{\hbar^2 n^2 \pi^2}{2\mu a^2}. \tag{13.52}
$$

The coefficient A is determined by normalization

$$
\int_a^{2a} dr \, r^2 \, R_0^2(r) = 1 \implies A_n = n\pi \sqrt{2} a^{-3/2}. \tag{13.53}
$$

Finally, we get

$$
R_0^{(n)}(kr) = \sqrt{\frac{2}{a}} \, \frac{\sin(n\pi r/a)}{r}, \quad E_n = \frac{\hbar^2 n^2 \pi^2}{2\mu a^2}. \tag{13.54}
$$

13.3.2   Hard Sphere

Consider a spherical region of radius a that is impenetrable, exerting an infinite repulsive force. This situation can be described by a potential

$$V(r) = \begin{cases} +\infty & (0 \le r \le a) \\ 0 & (r > a) \end{cases}. \tag{13.55}$$

The wave function vanishes in the region $[0, a]$, while the particle moves freely in the outside region $r > a$, having a radial wave function ($E = \hbar^2 k^2 / 2\mu$)

$$R_\ell(kr) = A\, j_\ell(kr) + B\, n_\ell(kr), \tag{13.56}$$

with the boundary condition

$$R_\ell(ka) = 0 \implies B = -A\, \frac{j_\ell(ka)}{n_\ell(ka)}. \tag{13.57}$$

The wave functions

$$R_\ell(kr) \sim n_\ell(ka)\, j_\ell(kr) - j_\ell(ka)\, n_\ell(kr) \tag{13.58}$$

correspond to scattering states with a continuous energy spectrum. In the asymptotic region $r \sim \infty$, the radial wave function is

$$R_\ell(kr) \sim \frac{1}{kr}\left(n_\ell(ka)\, \sin(kr - \ell\pi/2) - j_\ell(ka)\, \cos(kr - \ell\pi/2) \right) \tag{13.59}$$

or

$$R_\ell(kr) = C\, \frac{\sin(kr - \ell\pi/2 + \delta_\ell)}{kr}, \tag{13.60}$$

where

$$\tan \delta_\ell = \frac{j_\ell(ka)}{n_\ell(ka)}. \tag{13.61}$$

The parameter δ_ℓ is called *the phase shift* and measures the difference in the phase of the outside spherical wave from what it would if the potential were absent.

In the case $\ell = 0$, things simplify a lot and we have

$$R_0(kr) = C\, \frac{\sin(k(r - a))}{kr} \quad \text{and} \quad \delta_0 = -ka. \tag{13.62}$$

For general ℓ at low energies, i.e., for $ka << 1$, we may use the behavior of the spherical Bessel functions near the origin and get

$$\delta_\ell \approx -\frac{(ka)^{2\ell+1}}{(2\ell+1)!!(2\ell-1)!!},$$ (13.63)

where $(2\ell \pm 1)!! = 1 \cdot 3 \cdot 5 \ldots (2\ell \pm 1)$.

13.3.3 The Spherical Well

Consider the potential

$$V(r) = \begin{cases} -V_0 & (0 \leq r \leq a) \\ 0 & (r > a) \end{cases}$$ (13.64)

with $V_0 > 0$. For a particle of energy E moving in this potential, there are in general solutions of the Schroedinger equation both for $E > 0$ (scattering states) and for $-V_0 < E < 0$ (bound states). First, we shall consider the case of scattering ($E > 0$).

Scattering states (E > 0). In this case, the radial Schroedinger equation is

$$-\frac{\hbar^2}{2\mu r}\frac{d^2(rR)}{dr^2} + \frac{\hbar^2\ell(\ell+1)}{2\mu r^2}R = \left(\frac{\hbar^2 k^2}{2\mu} + V_0\right)R = \frac{\hbar^2 q^2}{2\mu}R \quad (0 \leq r \leq a)$$ (13.65)

$$-\frac{\hbar^2}{2\mu r}\frac{d^2(rR)}{dr^2} + \frac{\hbar^2\ell(\ell+1)}{2\mu r^2}R = \frac{\hbar^2 k^2}{2\mu}R \qquad (r > a)$$

where

$$E = \frac{\hbar^2 k^2}{2\mu} \quad and \quad \frac{\hbar^2 q^2}{2\mu} = E + V_0.$$ (13.66)

The corresponding solutions are

$$R_{E\ell}(r) = \begin{cases} A\, j_\ell(qr) & (0 \leq r \leq a) \\ B\, j_\ell(kr) + C\, n_\ell(kr) & (r \geq a) \end{cases}$$ (13.67)

Note however that it is more convenient to express the outgoing part of the solution in terms of the alternative pair of Hankel functions

$$h_\ell^{(\pm)}(kr) = n_\ell(kr) \pm i\, j_\ell(kr).$$ (13.68)

The reason is that $h_\ell^{(\pm)}(kr)$ have a direct interpretation as outgoing ($h_\ell^{(-)}$) and incoming ($h_\ell^{(+)}$) spherical waves.[4] In contrast, j_ℓ and n_ℓ do not have any sense of propagation direction associated with them. Thus, we adopt the following expression for the solutions:

---

[4]The asymptotic behavior of the Hankel functions is $h_\ell^{(\pm)}(x) \sim \frac{e^{\mp ix}}{x}$.

$$R_{E\ell}(r) = \begin{cases} A\, j_\ell(qr) & (0 \le r \le a) \\ F\, h_\ell^{(-)}(kr) + G\, h_\ell^{(+)}(kr) & (r \ge a) \end{cases} \tag{13.69}$$

The ratios F/A and G/A are determined from the continuity of the wave function and its derivative at $r = a$ as

$$\frac{F}{A} h_\ell^{(-)}(ka) + \frac{G}{A} h_\ell^{(+)}(ka) = j_\ell(qa)$$

$$\frac{F}{A} \left.\frac{d}{dr} h_\ell^{(-)}\right|_{ka} + \frac{C}{A} \left.\frac{d}{dr} h_\ell^{(+)}\right|_{ka} = \left.\frac{d}{dr} j_\ell\right|_{qa}. \tag{13.70}$$

Let's be specific and concentrate on the simplest $\ell = 0$ case. Then, we have

$$j_0(kr) = \frac{\sin(kr)}{kr}, \quad h_0^{(+)}(kr) = -\frac{e^{-ikr}}{kr}, \quad h_0^{(-)}(kr) = -\frac{e^{ikr}}{kr}$$

and the wave function becomes

$$R_0(r) = \begin{cases} A\, \frac{\sin(qr)}{qr} & (0 \le r \le a) \\ -F\, \frac{e^{ikr}}{kr} - G\, \frac{e^{-ikr}}{kr} & (r \ge a) \end{cases}. \tag{13.71}$$

The continuity Equations (13.70) are

$$\frac{\sin(qa)}{qa} = -\frac{F/A}{ka} e^{ika} - \frac{G/A}{ka} e^{-ika}$$

$$\frac{\cos(qa)}{a} - \frac{\sin(qa)}{qa^2} = (F/A)\left(\frac{1}{ka^2} - \frac{i}{a}\right) e^{ika} + (G/A)\left(\frac{1}{ka^2} + \frac{i}{a}\right) e^{-ika}.$$

From these, we can obtain the coefficients F/A and G/A. These are solved to give

$$\begin{cases} F/A = -\frac{1}{2i} e^{-ika} \left(\cos(qa) + i\frac{k}{q} \sin(qa)\right) \\ G/A = \frac{1}{2i} e^{ika} \left(\cos(qa) - i\frac{k}{q} \sin(qa)\right) \end{cases}. \tag{13.72}$$

Note that the *radial probability current density*

$$\mathbf{J}_r = \hat{r} \frac{\hbar}{2\mu i} |Y_{\ell m}(\Omega)|^2 \left(R_{e\ell}^*(r) R'_{E\ell}(r) - R_{E\ell}(r) R'^{*}_{E\ell}(r)\right) \tag{13.73}$$

vanishes. The incoming part is equal to the outgoing one as expected from probability current conservation

$$J_r^{(out)} = J_r^{(in)} \implies \frac{\hbar k}{mr^2} |Y_{\ell m}(\Omega)|^2 |F|^2 = \frac{\hbar k}{mr^2} |Y_{\ell m}(\Omega)|^2 |G|^2, \tag{13.74}$$

since $G/A = (F/A)^*$ as it was derived above.

Note that we can always write the coefficients F and G using the polar expression for complex numbers as

$$F/A = -\frac{1}{2i}e^{-ika}\left[\cos^2(qa) + \frac{k^2}{q^2}\sin^2(qa)\right]^{1/2} e^{i\arctan((k/q)\tan(qa))}$$

$$G/A = \frac{1}{2i}e^{ika}\left[\cos^2(qa) + \frac{k^2}{q^2}\sin^2(qa)\right]^{1/2} e^{-i\arctan((k/q)\tan(qa))}$$

(13.75)

or, introducing

$$\delta_0 = \arctan((k/q)\tan(qa)) - ka \qquad (13.76)$$

$$F/A = -\frac{1}{2i}\left[\cos^2(qa) + \frac{k^2}{q^2}\sin^2(qa)\right]^{1/2} e^{i\delta_0}$$

$$G/A = \frac{1}{2i}\left[\cos^2(qa) + \frac{k^2}{q^2}\sin^2(qa)\right]^{1/2} e^{-i\delta_0}.$$

(13.77)

Then, the outside wave function takes the form

$$R_0(r) = A\left[\cos^2(qa) + \frac{k^2}{q^2}\sin^2(qa)\right]^{1/2} \frac{\sin(kr + \delta_0)}{kr}. \qquad (13.78)$$

Bound states (E < 0). In this case, we may define

$$E = -\frac{\hbar^2\kappa^2}{2\mu}, \quad E + V_0 = \frac{\hbar^2 q^2}{2\mu}. \qquad (13.79)$$

The radial Schroedinger equation takes the form

$$-\frac{1}{r}\frac{d^2(Rr)}{dr^2} + \frac{\ell(\ell+1)}{r^2}R = q^2 R \quad (0 \leq r \leq a)$$

$$-\frac{1}{r}\frac{d^2(Rr)}{dr^2} + \frac{\ell(\ell+1)}{r^2}R = -\kappa^2 R \quad (r \geq a).$$

(13.80)

The solution in the region of the potential is the same as in the previous case of scattering states. The solution in the outside region can be deduced also from the previous analysis with the replacement $k \to i\kappa$. Using the spherical Hankel functions again, we may write the outside solution as a linear combination of $h_\ell^{(+)}(i\kappa r)$ and $h_\ell^{(-)}(i\kappa r)$. Thus, we have

$$R_{E\ell}(r) = \begin{cases} A\,j_\ell(qr) & (0 \leq r \leq a) \\ B\,h_\ell^{(+)}(i\kappa r) + C\,h_\ell^{(-)}(i\kappa r) & (r \geq a) \end{cases}. \qquad (13.81)$$

Note however that from the asymptotic behavior of Hankel functions $h_\ell(x) \sim -(\pm i)^\ell e^{\mp ix}/x$, we have for $\kappa r \gg 1$

$$h_\ell^{(+)}(i\kappa r) \sim -i^{\ell-1}\frac{e^{\kappa r}}{\kappa r} \quad \text{and} \quad h_\ell^{(-)}(i\kappa r) \sim (-i)^{\ell-1}\frac{e^{-\kappa r}}{\kappa r}$$

and only $h_\ell^{(-)}(i\kappa r)$ is acceptable as a normalizable wave function. Thus, the radial wave function is

$$R_{E\ell}(r) = \begin{cases} A\, j_\ell(qr) & (0 \le r \le a) \\ \\ C\, h_\ell^{(-)}(i\kappa r) & (r \ge a) \end{cases} \tag{13.82}$$

From continuity, we have

$$A\, j_\ell(qa) = C\, h_\ell^{(-)}(i\kappa a)$$

$$A\frac{dj_\ell(qr)}{dr}\bigg|_{r=a} = C\frac{dh_\ell^{(-)}(i\kappa r)}{dr}\bigg|_{r=a}. \tag{13.83}$$

From these, we get one relation between the coefficients and a condition on the energy eigenvalues for which a negative energy solution is possible

$$\frac{A}{C} = \frac{h_\ell^{(-)}(i\kappa a)}{j_\ell(qa)}$$

$$\frac{h_\ell^{(-)}(i\kappa a)}{j_\ell(qa)}\frac{dj_\ell(qr)}{dr}\bigg|_{r=a} = \frac{dh_\ell^{(-)}(i\kappa r)}{dr}\bigg|_{r=a}. \tag{13.84}$$

We proceed by considering the simplest case of $\ell = 0$. In this case, the eigenvalue condition reduces to

$$\tan(qa) = -\frac{q}{\kappa}. \tag{13.85}$$

This is analogous to conditions we had in the case of the one-dimensional square well. In fact, it is the same condition we obtained there for eigenvalues corresponding to *odd eigenstates*. Recall that this condition, depending on the parameters of the potential, may or may not have a solution. Thus, in contrast to the one-dimensional well, the three-dimensional spherical well does not always have bound state solutions. In order to quantify all these, we may proceed to describe the graphical solution to (13.85). Defining

$$\xi = qa, \quad \beta = \sqrt{\frac{2mV_0a^2}{\hbar^2}} \implies \xi^2 = (qa)^2 = -(\kappa a)^2 + \beta^2, \tag{13.86}$$

the Eq. (13.85) becomes

$$\tan \xi = -\frac{\xi}{\sqrt{\beta^2 - \xi^2}}. \tag{13.87}$$

Fig. 13.2 Graphical solution for the bound states of the spherical well

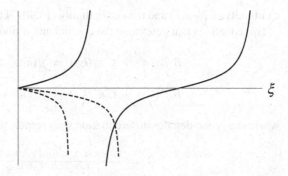

In Fig. 13.2, We have depicted its graphical solution. It is clear from the plot of the right-hand side and the left-hand side (dotted lines, plotted for two different values of β) of this equation that in order for them to intersect and, thus, supply us with a solution, the parameter β has to be larger than $\pi/2$.

13.3.4 The Delta-Shell Potential

Consider a particle moving freely in three dimensions with the exception of a very thin spherical shell where it experiences an infinite force. Such a situation can be modeled in terms of a *delta-shell potential*

$$V(r) = \frac{\hbar^2 \lambda}{2m} \delta(r - a) \tag{13.88}$$

parametrized by a strength parameter λ. The force experienced at $r = a$ is attractive for $\lambda < 0$ or repulsive for $\lambda > 0$.

Scattering (E > 0). The radial Schroedinger equation is

$$-\frac{1}{r^2} \frac{d^2}{dr^2} r^2 R_{E,\ell}(r) + \frac{\ell(\ell+1)}{r^2} R_{E,\ell}(r) + \lambda \delta(r-a) R_{e,\ell}(a) = k^2 R_{e,\ell}(r), \tag{13.89}$$

where $E = \hbar^2 k^2/2\mu$. Integrating from $a - \epsilon$ to $a + \epsilon$ with $\epsilon \to 0$, we obtain the discontinuity of the radial wave function derivative at $r = a$

$$R'_{E,\ell}(a + \epsilon) - R'_{E,\ell}(a - \epsilon) = \lambda R_{E,\ell}(a). \tag{13.90}$$

The radial wave function is

$$R_{E,\ell}(r) = \begin{cases} j_\ell(kr) & (0 \leq r \leq a) \\ B\,j_\ell(kr) + C\,n_\ell(kr) & (r \geq a) \end{cases} \tag{13.91}$$

Continuity of $R_{E,\ell}(r)$ and the discontinuity (13.90) of the derivative give us a system of two equations that determine the coefficients B and C

$$B\,j_\ell(ka) + C\,n_\ell(ka) = j_\ell(ka)$$

$$B\,j_\ell'(ka) + C\,n_\ell'(ka) = j_\ell'(ka) + \lambda\,j_\ell(ka), \tag{13.92}$$

where the prime denotes differentiation with respect to the radius r. The solution is

$$B = \frac{j_\ell(ka)n_\ell'(ka) - n_\ell(ka)\left(\lambda j_\ell(ka) + j_\ell'(ka)\right)}{j_\ell(ka)n_\ell'(ka) - j_\ell'(ka)n_\ell(ka)}$$

$$C = \frac{\lambda\,j_\ell^2(ka)}{j_\ell(ka)n_\ell'(ka) - j_\ell'(ka)n_\ell(ka)}. \tag{13.93}$$

Reparametrizing the wave function as

$$B = A\cos\delta_\ell, \quad C = -A\sin\delta_\ell, \tag{13.94}$$

we can write the outside ($r > a$) wave function as

$$R_{E,\ell}(r) = A\,(\cos\delta_\ell\,j_\ell(kr) - \sin\delta_\ell\,n_\ell(kr)) \tag{13.95}$$

which, in the asymptotic region $r \sim \infty$, is

$$R_{E,\ell}(r) \sim \frac{A}{kr}\sin(kr - \ell\pi/2 + \delta_\ell). \tag{13.96}$$

The parameter δ_ℓ is the *phase shift*. Using the derived expressions for the coefficients, we have

$$\tan\delta_\ell = -\frac{C}{B} = -\frac{\lambda\,j_\ell^2(ka)}{\left[j_\ell(ka)n_\ell'(ka) - n_\ell(ka)j_\ell'(ka) - \lambda n_\ell(ka)j_\ell(ka)\right]}. \tag{13.97}$$

In the technically simplest case of s-waves ($\ell = 0$), we have[5]

$$\tan\delta_0 = \frac{-\lambda\sin^2(ka)}{[k + \lambda\cos(ka)\sin(ka)]}. \tag{13.98}$$

Bound States (E < 0). Solutions of negative energy exist only in the case of attractive potential $\lambda < 0$. Setting $E = -\frac{\hbar^2\kappa^2}{2\mu}$, we have the solution

---

[5] $j_0'(ka) = -\sin(ka)/ka^2 + \cos(ka)/a$ and $n_0'(ka) = \cos(ka)/ka^2 + \sin(ka)/a$.

$$R_{R,\ell}(r) = \begin{cases} A\, j_\ell(i\kappa r) & (0 \le r \le a) \\ B\, h_\ell^{(-)}(i\kappa r) & (r \ge a) \end{cases}. \qquad (13.99)$$

No term with $h_\ell^{(+)}(i\kappa r)$ is allowed, since the asymptotic behavior of Hankel functions is $h_\ell^{(\pm)}(x) \propto e^{\mp ix}$ and, therefore, we would have to deal with $h_\ell^{(+)}(i\kappa r) \propto e^{+\kappa r}$. The wave function continuity condition reads

$$A\, j_\ell(i\kappa a) = B\, h_\ell^{(-)}(i\kappa a). \qquad (13.100)$$

In order to get the discontinuity relation at a, we integrate over an interval $[a - \epsilon, a + \epsilon]$ with $\epsilon \to 0$ and obtain

$$B\, h_\ell^{(-)'}(i\kappa a) - A\, j_\ell'(i\kappa a) + |\lambda| A\, j_\ell(i\kappa a) = 0. \qquad (13.101)$$

From Eqs. (13.100), (13.101), we obtain the condition

$$\frac{h_\ell^{(-)'}(i\kappa a)}{h_\ell^{(-)}(i\kappa a)} = -|\lambda| + \frac{j_\ell'(i\kappa a)}{j_\ell(i\kappa a)}. \qquad (13.102)$$

In the technically simplest case of $\ell = 0$, this condition reduces to

$$\frac{2\kappa a}{1 - e^{-2\kappa a}} = |\lambda| a. \qquad (13.103)$$

A graphical analysis of the above equation shows that there is one bound state solution for $|\lambda| a > 1$.

Example 13.2 Consider the case of a repulsive delta function shell of infinite strength ($\lambda \to \infty$) and show that the interior (i.e., $0 \le r \le a$) radial wave function coincides with that of a particle trapped in a spherical cavity, while the energy spectrum becomes discrete.

In the limit $\lambda \to \infty$, the exterior wave function becomes

$$R_{E,\ell}(r) = -\frac{\lambda\, j_\ell(ka)}{[j_\ell(ka)n_\ell'(ka) - j_\ell'(ka)n_\ell(ka)]}\, (n_\ell(ka)\, j_\ell(kr) - j_\ell(ka)n_\ell(kr))$$

and vanishes at $r = a$. The interior wave function is just $j_\ell(kr)$ as in the case of a particle in a spherical cavity. Continuity forces the interior wave function to satisfy the condition

$$j_\ell(ka) = 0$$

which is satisfied by a set of discrete values of the wave number. Note however that this condition forces the exterior wave function to vanish everywhere. Thus, in the limit $\lambda \to \infty$ the delta-shell potential can be mapped to the spherical cavity potential.[6]

13.3.5   The Isotropic Harmonic Oscillator

Up to now, we have considered examples of central forces that correspond to localized potentials that fall at infinity faster than the Coulomb potential. As an example of central forces that increase at large distances, we shall consider the case of *isotropic harmonic forces*. The three-dimensional harmonic oscillator has been discussed in a previous chapter. There, the Hamiltonian, expressed in Cartesian coordinates, being the sum of three commuting terms, each being a one-dimensional harmonic oscillator, led to a set of energy eigenfunctions which were common eigenfunctions of these terms. In the case that these oscillators have a common mass and frequency, we have the *isotropic harmonic oscillator*, for which, as a central potential problem, we may pursue the alternative approach of considering common energy and angular momentum eigenfunctions

$$\psi_{E\ell m}(\mathbf{r}) = R_{E\ell}(r)\, Y_{\ell m}(\theta, \phi)\,. \tag{13.104}$$

The radial Schroedinger equation is

$$\left\{ -\frac{\hbar^2}{2\mu}\left(\frac{d^2}{dr^2} + \frac{2}{r}\frac{d}{dr} \right) + \frac{\hbar^2 \ell(\ell+1)}{2\mu r^2} + \frac{1}{2}\mu\omega^2 r^2 \right\} R_{E\ell}(r) = E\, R_{E\ell}(r)\,. \tag{13.105}$$

The energy eigenvalues were found in the previous chapter to be $E_{n_x n_y n_z} = \hbar\omega(n_x + n_y + n_z + 3/2)$. We may write them in terms of a nonnegative integer quantum number $n = n_x + n_y + n_z$ as

$$E_n = \hbar\omega\left(n + \frac{3}{2} \right)\,. \tag{13.106}$$

There are $(n+2)(n+1)/2$ ways to obtain the same sum from three different nonnegative integers. Thus, each energy level is $(n+2)(n+1)/2$-fold degenerate.

Let us now look for possible solutions of the radial Schroedinger equation. In the neighborhood of the origin $r \sim 0$, the potential is negligible and only the first three terms are appreciable. Then, the equation is satisfied with a power law type of solution

$$R_{E\ell}|_{r\to 0} \sim r^\ell\,. \tag{13.107}$$

---

[6]There is an alternative conclusion that can be obtained in this limit: By multiplying the radial wave function by a constant before we take the above limit we could have the interior wave function to vanish, while the exterior wave function would coincide with that of the infinitely hard sphere.

In contrast, in the asymptotic region $r \sim \infty$, apart from the kinetic term only the potential r^2-term matters and we have an approximate solution

$$R_{E\ell}|_{r \to \infty} \sim e^{-\frac{\mu\omega}{2\hbar}r^2}. \tag{13.108}$$

Therefore, we may factor out the above behaviors at the two ends of the spatial range and write the radial wave function in the form

$$R_{E\ell}(r) = r^\ell e^{-\frac{\mu\omega}{2\hbar}r^2} \Lambda_{E\ell}(r^2). \tag{13.109}$$

Introducing (13.109) into the radial Schroedinger equation, after quite a bit of algebra, we arrive at

$$\left\{ r^2 \frac{d^2}{d(r^2)^2} + \left(\ell + \frac{3}{2} - \frac{\mu\omega}{\hbar}r^2 \right) \frac{d}{dr^2} + \frac{\mu\omega}{2\hbar}(n - \ell) \right\} \Lambda(r^2) = 0, \tag{13.110}$$

which is of the general form of the *Laguerre differential equation*

$$\frac{d^2 L_p^q}{dx^2} + \left(\frac{q+1}{x} - 1 \right) \frac{dL_p^q}{dx} + \frac{p}{x}L_p^q = 0 \tag{13.111}$$

with solutions the *generalized Laguerre polynomials*[7] $L_p^q(\mu\omega r^2/\hbar)$ with $q = \ell + \frac{1}{2}$ and $p = \frac{1}{2}(n - \ell)$. Thus, $\Lambda(r^2)$ stands for $L_{n/2-\ell/2}^{\ell+1/2}(\mu\omega r^2/\hbar)$. The values of the energy quantum number n are determined from the condition

$$n = 2p + \ell. \tag{13.112}$$

Since p is a nonnegative integer, for every value of n, we have

$$\begin{cases} n = even \implies \ell = 0, 2, \ldots, n - 2, n \\ n = odd \implies \ell = 1, 3, \ldots, n - 2, n \end{cases}. \tag{13.113}$$

For every given ℓ we have $(2\ell + 1)$ possible values of $m = -\ell, \ldots, +\ell$. As a result, the degeneracy of each n-level will be

---

[7]The first two are

$$L_0^p(x) = 1, \quad L_1^p(x) = 1 + p - x.$$

The rest can be obtained through the recursion formula

$$L_{k+1}^p(x) = \frac{1}{(k+1)} \left[(2k + 1 + p - x)L_k^p(x) - (k + p)L_{k-1}^p(x) \right].$$

$$\sum_{\ell = \ldots, n-2, n} (2\ell + 1) = \frac{1}{2}(n + 2)(n + 1).$$ (13.114)

Thus, finally, the radial energy eigenfunctions are

$$R_{n\ell}(r) = N_{n\ell}\, r^\ell\, e^{-\frac{\mu\omega}{2\hbar}r^2}\, L_{n/2-\ell/2}^{\ell+1/2}(\mu\omega r^2/\hbar),$$ (13.115)

where $N_{n\ell}$ is a normalization factor to meet $\int_0^\infty dr\, r^2 R^2 = 1$. The first few wave functions are

$$R_{00} = N_{00}\, e^{-\mu\omega r^2/2\hbar}$$

$$R_{11} = N_{11}\, r\, e^{-\mu\omega r^2/2\hbar}$$ (13.116)

$$R_{20} = N_{20}\left(3/2 - \mu\omega r^2/\hbar\right) e^{-\mu\omega r^2/2\hbar}$$

$$R_{22} = N_{22}\, r^2\, e^{-\mu\omega r^2/2\hbar}.$$

Problems and Exercises

13.1 Consider the spherical Bessel and spherical Neumann functions $j_\ell(x)$, $n_\ell(x)$.

(a) Prove that the expression $x^2\left(j_\ell(x)n'_\ell(x) - n_\ell(x)j'_\ell(x)\right)$ is a constant independent of x. Evaluate this constant.

(b) Show that the integral expression $j_\ell(x) = \frac{x^\ell}{2^{\ell+1}\ell!}\int_{-1}^1 ds(1 - s^2)^\ell e^{ixs}$ implies the relation $xj'_\ell(x) + (\ell + 1)j_\ell(x) = xj_{\ell-1}(x)$.

13.2 Consider a particle of mass μ and energy $E > 0$, subject to a central potential, in an energy/angular momentum eigenstate $\psi_{E\ell m}(\mathbf{r}) = R_{E\ell}(r)Y_{\ell m}(\Omega)$. Derive the expression for the *probability current density* $\mathbf{J}$ and show that its divergence vanishes. Consider the case of scattering in a *spherical potential well* and verify explicitly that the total radial current vanishes.

13.3 Consider the case of a particle of mass μ in the potential

$$V(r) = \begin{cases} +\infty & (0 \le r \le a) \\ -\frac{\hbar^2 g^2}{2\mu}\delta(r - b) & (r > a) \end{cases}.$$

Investigate the existence of bound states ($E < 0$) in the case $\ell = 0$.

13.4 Consider a particle of mass μ moving in the central potential

$$V(r) = V_0\Theta(a - r) + \frac{\hbar^2 g^2}{2\mu}\delta(r - a),$$

with $V_0 > 0$. Calculate the phase shift for $\ell = 0$.

13.5 Consider a particle of mass μ in the attractive *delta-shell potential* $V(r) = -\frac{\hbar^2 g^2}{2\mu}\delta(r-a)$ with $E < 0$. Use the identity shown in Problem 13.1. to rewrite the discontinuity condition of the derivative of the radial wave function in the form $1 = i\kappa a^2 g^2 h_\ell^{(-)}(i\kappa a) j_\ell(i\kappa a)$. Are there any solutions of angular momentum ℓ and zero binding energy? Show that if there is an integer ℓ_0 so that $g^2 a < 2\ell_0 + 1$, then, there will be bound states of angular momentum $\ell \le \ell_0$.

13.6 A particle of mass μ is subject to harmonic forces described by an isotropic harmonic oscillator potential $\mu\omega^2 r^2/2$. The particle occupies a state $\psi(\mathbf{r}) = \frac{1}{\sqrt{2}}$ $(\psi_{000} + \psi_{110})$, where $\psi_{n\ell m}$ are the eigenfunctions of the energy. Calculate the expectation values $\langle \mathbf{r} \rangle$, $\langle r \rangle$, $\langle r^2 \rangle$ and the uncertainties $(\Delta \mathbf{r})^2$ and $(\Delta r)^2$.

13.7 A particle of mass μ and zero angular momentum is bound in an attractive delta-shell potential. Calculate the probability to find the particle inside the shell.

13.8 Show that an attractive spherical well (of depth V_0 and width a) that can have a zero energy bound state with $\ell \ne 0$ will have to have suitable parameters that satisfy $j_{\ell-1}(q_0 a) = 0$, where $q_0^2 \hbar^2/2\mu = V_0$.

13.9 Investigate the existence of very small energy bound states ($\kappa a << 1$) with $\ell = 0$ in a spherical potential well. Show that the width a and the depth V_0 of the well have to satisfy

$$\frac{2\mu V_0 a^2}{\hbar^2} \approx \frac{\pi^2}{4} + 2\kappa a + \left(1 - \frac{4}{\pi^2}\right)(\kappa a)^2.$$

13.10 A particle of energy E and angular momentum $\ell = 0$ experiences a potential which can be approximated as

$$V(r) = \begin{cases} 0 & (0 \le r \le a) \\ V_0 & (a < r < b), \\ 0 & (r \ge b) \end{cases}$$

where $V_0 > 0$. Determine the eigenfunctions of the energy for $E < V_0$. In the case that $\sqrt{2\mu(V_0 - E)/\hbar^2}(b-a) >> 1$, are there values of the energy for which the probability to find the particle outside the nucleus is negligible (*quasi-stationary levels*)?

References

1. E. Merzbacher, *Quantum Mechanics*, 3rd edn. (Wiley, New York, 1998)

2. A. Messiah, *Quantum Mechanics* (Dover publications, Mineola, 1958). Single-volume reprint of the John Wiley & Sons, New York, two-volume 1958 edition
3. K. Kurt Gottfried, T.M. Yan, *Quantum Mechanics: Fundamentals* (Springer, Berlin, 2004)
4. G. Baym, *Lectures in Quantum Mechanics*, Lecture Notes and Supplements in Physics (ABP, 1969)

Chapter 14
Systems of Particles

14.1 Systems of Many Degrees of Freedom

The quantum mechanical treatment of systems of more than one particle is essentially the same as the treatment of the system of one particle. We start with a classical analogue of the system, identifying its independent variables and writing down its Hamiltonian. Then, we quantize the system promoting its independent canonical variables to operators satisfying canonical commutation relations. In order to illustrate the procedure let's consider a general system of N *independent* particles, each associated with a set of variables

$$\mathbf{r}_1, \, \mathbf{p}_1, \, \mathbf{S}_1, \, \ldots, \, \mathbf{r}_2, \, \mathbf{p}_2, \, \mathbf{S}_2, \, \ldots\ldots, \, \mathbf{r}_a, \, \mathbf{p}_a, \, \mathbf{S}_a, \, \ldots \quad (a = 1, \, 2, \, \ldots, \, N).$$
(14.1)

By definition, the independence of two particles corresponds to the statement that all their respective variables commute

$$\left[\mathbf{r}_1, \, \mathbf{p}_2\right] = 0, \quad [\mathbf{S}_1, \, \mathbf{S}_2] = 0, \quad \ldots \, etc.$$
(14.2)

Thus, the commutation relations of the system variables are

$$\left[x_{i\,a}, \, \hat{p}_{b\,j}\right] = i\hbar\,\delta_{ab}\,\delta_{ij}$$

$$\left[\hat{S}_{a\,i}, \, \hat{S}_{b\,j}\right] = i\hbar\,\epsilon_{ijk}\delta_{ab}\hat{S}_{a\,k}$$
(14.3)

$$\left[x_{a\,i}, \, x_{b\,j}\right] = \left[\hat{p}_{a\,i}, \, \hat{p}_{b\,j}\right] = 0.$$

The indices $i, \, j, \, k, \, \ldots$ are spatial indices, while the indices $a, b = 1, 2, \ldots, N$ are indices denoting different particles. The Hamilton operator of the above system will be

$$\hat{H} = \frac{\hat{\mathbf{p}}_1^2}{2m_1} + \frac{\hat{\mathbf{p}}_2^2}{2m_2} + \cdots \frac{\hat{\mathbf{p}}_a^2}{2m_a} + \cdots + V(\mathbf{r}_1, \, \mathbf{r}_2, \, \ldots),$$
(14.4)

© Springer Nature Switzerland AG 2019
K. Tamvakis, *Basic Quantum Mechanics*, Undergraduate Texts in Physics,
https://doi.org/10.1007/978-3-030-22777-7_14

where V is a potential describing interactions of the particles among themselves or with an external source. V could in general be spin-dependent.

The state of the system is, at every time instant, described by the multiparticle wave function

$$\Psi(\mathbf{r}_1, \mathbf{r}_2, \ldots; t) \tag{14.5}$$

with the straightforward interpretation that the probability amplitude to find the particle-1 in the volume d^3r_1, the particle-2 in the volume d^3r_2, etc. is

$$d^3r_1 \, d^3r_2 \cdots d^3r_N \, |\Psi(\mathbf{r}_1, \mathbf{r}_2, \ldots)|^2 . \tag{14.6}$$

The corresponding normalization condition is

$$\int \int \cdots \int d^3r_1 \, d^3r_2 \cdots d^3r_N \, |\Psi(\mathbf{r}_1, \mathbf{r}_2, \ldots)|^2 = 1 . \tag{14.7}$$

The time-evolution of the system is controlled by the multiparticle Schroedinger equation

$$\hat{H}\Psi = i\hbar \frac{\partial \Psi}{\partial t} . \tag{14.8}$$

The states of the above multiparticle system $|\Psi\rangle$ span a Hilbert space [1, 2]

$$\mathcal{E} = \mathcal{E}_1 \otimes \mathcal{E}_2 \otimes \cdots \otimes \mathcal{E}_N, \tag{14.9}$$

which is the *tensor product* of the individual single-particle Hilbert spaces (see the Appendix). Each of the above single-particle operators, say $\mathcal{O}_1$, is generalized to an operator in $\mathcal{E}$ which acts as before in $\mathcal{E}_1$ but acts as the unit operator on each of the $\mathcal{E}_a$ with $a \neq 1$. Operators referring to the composite system as, for example, the total spin

$$\hat{\mathbf{S}} = \hat{\mathbf{S}}_1 + \hat{\mathbf{S}}_2 + \cdots + \hat{\mathbf{S}}_N , \tag{14.10}$$

act on the full tensor product space. The Hilbert space of the composite system can be built by tensor products of single-particle states. In the above example the tensor products of single-particle eigenstates

$$|S_1^2, S_{1z}\rangle \otimes |S_2^2, S_{2z}\rangle \times \otimes |S_N^2, S_{Nz}\rangle \tag{14.11}$$

can serve as a basis for the composite Hilbert space $\mathcal{E}$. Note that in a shorthand notation the $\otimes$ symbol will be often omitted and the tensor product will be simply denoted as $|a\rangle|b\rangle$, provided it is understood that $|a\rangle$ refers to $\mathcal{E}_1$ while $|b\rangle$ refers to $\mathcal{E}_2$. Also operator tensor products $\hat{S}_{1i} \otimes \hat{S}_{2j}$ are written just as $\hat{S}_{1i}\hat{S}_{2j}$ with the understanding that $\hat{\mathbf{S}}_1$ acts on $\mathcal{E}_1$ while $\hat{\mathbf{S}}_2$ acts on $\mathcal{S}_2$. Similarly, $\hat{\mathbf{S}}_1 + \hat{\mathbf{S}}_2$ is a shorthand in the place of the rather cumbersome $\hat{\mathbf{S}}_1 \otimes \hat{\mathbf{I}}_{N-1} + \hat{\mathbf{S}}_2 \otimes \hat{\mathbf{I}}_{N-2}$.

The multiparticle wave function (14.5) is obtained from the state vector of the multiparticle system $|\Psi\rangle$, expanded in the tensor product basis of position eigenvectors

$$|\mathbf{r}_1, \mathbf{r}_2, \ldots, \mathbf{r}_N\rangle = |\mathbf{r}_1\rangle \otimes |\mathbf{r}_2\rangle \cdots \otimes |\mathbf{r}_N\rangle \qquad (14.12)$$

as

$$\Psi(\mathbf{r}_1, \ldots, \mathbf{r}_N) = \langle \mathbf{r}_1, \mathbf{r}_2, \ldots, \mathbf{r}_N | \Psi \rangle. \qquad (14.13)$$

Orthonormality and completeness for the basis (14.12) follows directly from orthonormality and completeness of single particle states

$$\langle \mathbf{r}_1, \mathbf{r}_2, \ldots, \mathbf{r}_N | \mathbf{r}_1', \mathbf{r}_2', \ldots, \mathbf{r}_N' \rangle = \delta(\mathbf{r}_1 - \mathbf{r}_1') \ldots \delta(\mathbf{r}_N - \mathbf{r}_N') \qquad (14.14)$$

$$\int d^3 r_1 \int d^3 r_2 \ldots \int d^3 r_N |\mathbf{r}_1, \mathbf{r}_2, \ldots, \mathbf{r}_N \rangle \langle \mathbf{r}_1, \mathbf{r}_2, \ldots, \mathbf{r}_N | = \mathbf{I}. \qquad (14.15)$$

Only in the special case that the state vector itself is a tensor product of single-particle kets, only then, the wave function is a product of single-particle wave functions, e.g.,

$$|\Psi\rangle = |\psi_1\rangle \otimes \ldots |\psi_N\rangle \implies \Psi(\mathbf{r}_1, \mathbf{r}_2, \ldots, \mathbf{r}_N) = \psi_1(\mathbf{r}_1) \ldots \psi_N(\mathbf{r}_N), \qquad (14.16)$$

where $\psi_a(\mathbf{r}_a) = \langle \mathbf{r}_a | \psi_a \rangle$.

In this basis ($\{x\}$-*representation*) the one-particle operator $\hat{\mathbf{p}}_a$ acts on $|\mathbf{r}_1, \mathbf{r}_2, \ldots\rangle$ exactly as it would act on the single-particle state $|\mathbf{r}_a\rangle$. Therefore, we have again $\hat{\mathbf{p}}_a \to -i\hbar\nabla_a$. Thus, the Schroedinger equation

$$\hat{H}|\Psi\rangle = i\hbar\frac{d|\Psi\rangle}{dt} \qquad (14.17)$$

takes the form

$$\left\{ -\sum_{a=1}^{N} \frac{\hbar^2}{2m_a} \nabla_a^2 + V(\mathbf{r}_1, \ldots, \mathbf{r}_a) \right\} \Psi(\mathbf{r}_1, \ldots, \mathbf{r}_a) = i\hbar\frac{\partial}{\partial t}\Psi(\mathbf{r}_1, \ldots, \mathbf{r}_a). \qquad (14.18)$$

The so-called time-independent Schroedinger equation, corresponding to the energy eigenvalue problem $\hat{H}|\Psi_E\rangle = E|\Psi_E\rangle$, is

$$\left\{ -\sum_{a=1}^{N} \frac{\hbar^2}{2m_a} \nabla_a^2 + V(\mathbf{r}_1, \mathbf{r}_2, \ldots) \right\} \Psi_E(\mathbf{r}_1, \mathbf{r}_2, \ldots) = E\,\Psi_E(\mathbf{r}_1, \mathbf{r}_2, \ldots). \qquad (14.19)$$

In the case that the Hamiltonian is the sum of commuting parts (e.g., the free case of $V = 0$ or the case of external potential of the form $\sum_a V(\mathbf{r}_a)$)

$$\hat{H} = \sum_{a=1}^{N} \hat{H}_a(\mathbf{r}_a, \hat{\mathbf{p}}_a) \tag{14.20}$$

the eigenfunctions are products

$$\Psi_E(\mathbf{r}_1, \mathbf{r}_2, \ldots) = \psi_1(\mathbf{r}_1)\,\psi_2(\mathbf{r}_2)\ldots\psi_N(\mathbf{r}_N), \tag{14.21}$$

where

$$\hat{H}_a\,\psi_a(\mathbf{r}_a) = \epsilon_a\,\psi_a(\mathbf{r}_a) \tag{14.22}$$

and the eigenvalues are sums of the single-particle eigenvalues ϵ_a

$$E = \epsilon_1 + \epsilon_2 + \cdots + \epsilon_N. \tag{14.23}$$

14.2   The Two-Body Problem

Consider the system of two independent but in general interacting particles of masses m_1 and m_2. The *Center of Mass* of the system of the two particles is defined as the point

$$\mathbf{R} = \frac{m_1\mathbf{r}_1 + m_2\mathbf{r}_2}{m_1 + m_2}. \tag{14.24}$$

It is a very common situation that the potential depends on the distance between the two particles $|\mathbf{r}_1 - \mathbf{r}_2|$. Then, we may introduce also the relative position

$$\mathbf{r} = \mathbf{r}_1 - \mathbf{r}_2 \tag{14.25}$$

and make the change of coordinates

$$\mathbf{r}_1, \mathbf{r}_2 \implies \mathbf{R}, \mathbf{r}.$$

The pair $\mathbf{R}$, $\mathbf{r}$ are called *Center of Mass variables*. The inverse relations to (14.24) and (14.25) are

$$\mathbf{r}_1 = \mathbf{R} + \frac{m_2}{M}\mathbf{r}$$

$$\mathbf{r}_2 = \mathbf{R} - \frac{m_1}{M}\mathbf{r}, \tag{14.26}$$

where $M = m_1 + m_2$ is the total mass.

It is not difficult to see that the correct canonical momenta corresponding to the center of mass variables are

$$\mathbf{P} = \mathbf{p}_1 + \mathbf{p}_2$$

$$\mathbf{p} = \frac{m_2}{M}\mathbf{p}_1 - \frac{m_1}{M}\mathbf{p}_2. \tag{14.27}$$

This choice satisfies the fundamental commutation relations

$$[x_{ia}, \hat{p}_{jb}] = i\hbar\delta_{ij}\delta_{ab} \implies \begin{cases} \left[X_i, \hat{P}_j\right] = i\hbar\delta_{ij} \\[2mm] \left[x_i, \hat{p}_j\right] = i\hbar\delta_{ij} \\[2mm] \left[x_i, \hat{P}_j\right] = \left[X_i, \hat{p}_j\right] = 0. \end{cases} \tag{14.28}$$

The kinetic energy in terms of $\mathbf{P}$, $\mathbf{p}$ becomes

$$\frac{\mathbf{p}_1^2}{2m_1} + \frac{\mathbf{p}_2^2}{2m_2} = \frac{\mathbf{P}^2}{2M} + \frac{\mathbf{p}^2}{2\mu}, \tag{14.29}$$

where μ is the so-called *reduced mass*

$$\frac{1}{\mu} = \frac{1}{m_1} + \frac{1}{m_2} \implies \mu = \frac{m_1 m_2}{m_1 + m_2}. \tag{14.30}$$

Note that, if $m_2 \gg m_1$, the reduced mass is to a good approximation equal to the mass of the lightest of the two particles

$$\mu = \frac{m_1}{1 + \frac{m_1}{m_2}} \approx m_1 - \frac{m_1^2}{m_2} + \cdots.$$

Now, if the potential depends only on the distance between the particles

$$V(|\mathbf{r}_1 - \mathbf{r}_2|) = V(r)$$

the Hamiltonian separates into two commuting pieces

$$\hat{H} = \frac{\mathbf{P}^2}{2M} + \underbrace{\frac{\mathbf{p}^2}{2\mu} + V(r)}_{\hat{H}_{rel}}, \tag{14.31}$$

the second of which describes only the relative motion of the system. Following what has been said at the end of last section, the energy eigenfunctions will be products of a *"Center of Mass wave function"* and a *"relative motion wave function"*

$$\Psi_E(\mathbf{R}, \mathbf{r}) = \psi_{CM}(\mathbf{R})\,\psi_{rel}(\mathbf{r}), \tag{14.32}$$

where

$$\begin{cases} \frac{\mathbf{P}^2}{2M}\, \psi_{CM}(\mathbf{R}) = E_{CM}\, \psi_{CM}(\mathbf{R}) \\[2mm] \left\{ \frac{\mathbf{p}^2}{2\mu} + V(r) \right\} \psi_{rel}(\mathbf{r}) = \epsilon\, \psi_{rel}(\mathbf{r}) \end{cases} \qquad with \quad E = E_{CM} + \epsilon. \qquad (14.33)$$

We can actually proceed further and solve the center of mass equation which is free. We obtain

$$\psi_{CM}(\mathbf{R}) = \frac{e^{i\mathbf{K}\cdot\mathbf{R}}}{(2\pi)^{3/2}} \quad and \quad E_{CM} = \frac{\hbar K^2}{2M}. \qquad (14.34)$$

Therefore, the two particle problem has been reduced to the problem of one particle (the reduced mass particle) moving under the influence of a central potential.

14.3 Identical Particles

Consider a system of two particles, which are *identical*, i.e., have all their properties, like mass, electric charge, spin,... etc., the same (e.g., two electrons). Two such particles are not automatically *distinguishable*. If they are macroscopic particles obeying the laws of Classical Mechanics, their trajectories can be followed at any time and we can identify them, despite the fact that they are physically identical. Nevertheless, for microscopic particles like electrons their distinguishability presupposes the existence of an experiment that could tell them apart. If one were to follow along the lines of what is done in the above case of classical particles and follow the evolution of their respective wavepackets, he would have to face sooner or later the fact that due to spreading their wavepackets would overlap, something that would amount to the loss of identity. What lies behind that is, of cource, the uncertainty principle. In any case, we may conclude that *identical particles*[1] obeying the laws of Quantum Mechanics are not distinguishable.

For a system of two identical particles, putting aside the spin in order to simplify notation, we proceed to write down the corresponding energy eigenvalue problem

$$\hat{H}(\mathbf{r}_1, \hat{\mathbf{p}}_1; \mathbf{r}_2, \hat{\mathbf{p}}_2)\, \Psi_E(\mathbf{r}, \mathbf{r}') = E\, \Psi_E(\mathbf{r}, \mathbf{r}'). \qquad (14.35)$$

Nevertheless, we can also have

$$\hat{H}(\mathbf{r}_2, \hat{\mathbf{p}}_2; \mathbf{r}_1, \hat{\mathbf{p}}_1)\, \Psi_E(\mathbf{r}', \mathbf{r}) = E\, \Psi_E(\mathbf{r}', \mathbf{r}). \qquad (14.36)$$

But, since the particles are identical, the Hamiltonian will be invariant in the interchange of their variables, namely

---

[1]By definition particles that have all their properties, like mass, electric charge, spin,... etc., the same will be called *identical particles*.

$$\hat{H}(\mathbf{r}_1, \hat{\mathbf{p}}_1; \mathbf{r}_2, \hat{\mathbf{p}}_2) = \hat{H}(\mathbf{r}_2, \hat{\mathbf{p}}_2; \mathbf{r}_1, \hat{\mathbf{p}}_1). \tag{14.37}$$

Therefore, the above two statements of the eigenvalue problem can be written

$$\begin{cases} \hat{H}(\mathbf{r}_1, \hat{\mathbf{p}}_1; \mathbf{r}_2, \hat{\mathbf{p}}_2)\,\psi_E(\mathbf{r}, \mathbf{r}') = E\,\psi_E(\mathbf{r}, \mathbf{r}') \\ \hat{H}(\mathbf{r}_1, \hat{\mathbf{p}}_1; \mathbf{r}_2, \hat{\mathbf{p}}_2)\psi_E(\mathbf{r}', \mathbf{r}) = E\,\psi_E(\mathbf{r}', \mathbf{r}) \end{cases} \tag{14.38}$$

which means that any combination of $\Psi(\mathbf{r}, \mathbf{r}')$ and $\Psi(\mathbf{r}', \mathbf{r})$, e.g.,

$$C_1\,\psi_E(\mathbf{r}, \mathbf{r}') + C_2\,\psi_E(\mathbf{r}', \mathbf{r}) \tag{14.39}$$

is also an eigenstate of the same eigenvalue E. This phenomenon is called *exchange degeneracy*. We shall see shortly that only two types of linear combinations are allowed.

Since the two particles are identical, the probability density for one of them being at $\mathbf{r}$ and the other being at $\mathbf{r}'$ should be equal to the probability density with the two particles interchanged, namely

$$\mathcal{P}(\mathbf{r}, \mathbf{r}') = \mathcal{P}(\mathbf{r}', \mathbf{r}) \tag{14.40}$$

or

$$|\Psi(\mathbf{r}, \mathbf{r}')|^2 = |\Psi(\mathbf{r}', \mathbf{r})|^2. \tag{14.41}$$

As a result, the wave functions should be equal up to a phase

$$\Psi(\mathbf{r}, \mathbf{r}') = e^{i\phi}\,\Psi(\mathbf{r}', \mathbf{r}). \tag{14.42}$$

Interchanging the positions once more, we obtain

$$\Psi(\mathbf{r}', \mathbf{r}) = e^{i\phi}\,\Psi(\mathbf{r}, \mathbf{r}') = e^{2i\phi}\Psi(\mathbf{r}', \mathbf{r}). \tag{14.43}$$

This means that

$$e^{2i\phi} = 1 \implies \phi = 0, \pi. \tag{14.44}$$

Therefore, the effect of the interchange is

$$\Psi(\mathbf{r}', \mathbf{r}) = \pm\Psi(\mathbf{r}, \mathbf{r}'). \tag{14.45}$$

Thus, *the two-particle wave functions of two identical particles can only be symmetric or antisymmetric in the particle interchange*

$$\begin{aligned} \Psi_S(\mathbf{r}, \mathbf{r}') &= \Psi_S(\mathbf{r}', \mathbf{r}) \quad \textit{symmetric} \\ \Psi_A(\mathbf{r}, \mathbf{r}') &= -\Psi_A(\mathbf{r}', \mathbf{r}) \quad \textit{antisymmetric}. \end{aligned} \tag{14.46}$$

It can be shown that particles of half-integer spin $(1/2, 3/2, \ldots)$ always have anti-symmetric wave functions. Such particles (electrons, protons,... etc.) are called *fermions*. The wave function of two electrons will always be antisymmetric in the exchange of the two particles. Fermions obey *Fermi-Dirac Statistics*. Particles of integer spin $(0, 1, 2, \ldots)$ always have symmetric wave functions. Such particles (photons, pions, ρ-mesons,...) are called *bosons*. The wave function of two pions will always be symmetric in the interchange of the two particles. Bosons obey *Bose-Einstein Statistics*. This fact, supported by all experiments, can be proven if we go beyond nonrelativistic Quantum Mechanics. It can be shown in the framework of relativistic Quantum Field Theory, where it is known under the name of *"Spin and Statistics Theorem"*. Added to the set of axioms of Quantum Mechanics as the so-called *Symmetrization Postulate*, it holds that for any number of identical particles, e.g., the wave function of N fermions has to obey[2]

$$\Psi(1, 2, \ldots, i, \ldots, j, \ldots, N) = -\Psi(1, 2, \ldots, j, \ldots, i, \ldots, N), \quad (14.47)$$

while the wave function of N bosons has to obey

$$\Psi(1, 2, \ldots, i, \ldots, j, \ldots, N) = \Psi(1, 2, \ldots, j, \ldots, i, \ldots, N). \quad (14.48)$$

Note that the symmetric/antisymmetric nature of the wave function refers to all kinds of variables, spatial as well as spin.

The Pauli Principle. Consider a pair of identical fermions (e.g., two electrons). Their wave function will be antisymmetric satisfying

$$\Psi(\mathbf{r}_1, S_{z1}; \mathbf{r}_2, S_{z2}) = -\Psi(\mathbf{r}_2, S_{z2}; \mathbf{r}_1, S_{z1}). \quad (14.49)$$

If the two particles occupy the same space point and have the same spin, their wave function has to obey

$$\Psi(\mathbf{r}_1, S_{z1}; \mathbf{r}_1, S_{z1}) = -\Psi(\mathbf{r}_1, S_{z1}; \mathbf{r}_1, S_{z1}) \implies \Psi(\mathbf{r}_1, S_{z1}; \mathbf{r}_1, S_{z1}) = 0. \quad (14.50)$$

This means that the probability amplitude to find two electrons at the same point, having the same spin is zero. This is the so-called *Exclusion Principle* or *Pauli Principle* according to which *two identical fermions cannot be in the same state*. However, the Exclusion Principle allows two fermions to occupy the same point and have opposite spins.

Consider the system of two electrons and let's assume that to a good approximation the forces between them are spin-independent and, therefore, the Hamiltonian does not depend on spin. Then, the wave function is a product of a *spatial wave function* times a *spin wave function (spinor)*

$$\Psi(\mathbf{r}_1, \mathbf{S}_1; \mathbf{r}_2, \mathbf{S}_2) = \psi(\mathbf{r}_1, \mathbf{r}_2) \chi(\mathbf{S}_1, \mathbf{S}_2). \quad (14.51)$$

---

[2]See also [1–4].

The Pauli Principle demands that the product wave function is antisymmetric. This can be met if the spatial part is antisymmetric and the spinor part is symmetric or if the spatial part is symmetric and the spinor part antisymmetric. We already know that if we adopt for the spinor part an eigenfunction of the total spin it comes automatically either as symmetric or antisymmetric. Thus, for the spinor wave function we can take the *triplet* (symmetric)

$$
\begin{cases}
\chi_{1,-1} = \chi_\downarrow^{(1)} \chi_\downarrow^{(2)} \\[2mm]
\chi_{1,0} = \frac{1}{\sqrt{2}} \left(\chi_\uparrow^{(1)} \chi_\downarrow^{(2)} + \chi_\downarrow^{(1)} \chi_\uparrow^{(2)} \right) \\[2mm]
\chi_{1,1} = \chi_\uparrow^{(1)} \chi_\uparrow^{(2)}
\end{cases}
\tag{14.52}
$$

and the *singlet* (antisymmetric)

$$
\chi_{0,0} = \frac{1}{\sqrt{2}} \left(\chi_\uparrow^{(1)} \chi_\downarrow^{(2)} - \chi_\downarrow^{(1)} \chi_\uparrow^{(2)} \right).
\tag{14.53}
$$

Therefore, there are two possibilities for a totally antisymmetric wave function

$$
\begin{cases}
\psi_S(\mathbf{r}_1, \mathbf{r}_2) \chi_{0,0} \\[2mm]
\psi_A(\mathbf{r}_1, \mathbf{r}_2) \chi_{1,m}
\end{cases}
\tag{14.54}
$$

The first of these does not vanish when the electrons are at the same point because their spins are opposite. The second however gives zero for $\mathbf{r}_1 = \mathbf{r}_2$.

Let's assume now, as it is very often the case, that the two electrons interact through forces dependent only on their relative distance. Then, the potential is $V = V(|\mathbf{r}_1 - \mathbf{r}_2|)$. Going into center of mass coordinates

$$
\mathbf{R} = \frac{1}{2} (\mathbf{r}_1 + \mathbf{r}_2), \quad \mathbf{r} = \mathbf{r}_1 - \mathbf{r}_2,
\tag{14.55}
$$

the Hamiltonian of the system becomes ($M = 2m_e$, $\mu = m_e/2$)

$$
\hat{H} = \underbrace{\frac{\mathbf{P}^2}{2M}}_{H_{CM}} + \underbrace{\frac{\mathbf{p}^2}{2\mu} + V(r)}_{H_{rel}},
\tag{14.56}
$$

which is the sum of two commuting parts. Therefore, the spatial wave function is a product of the *center of mass wave function* $\psi_{CM}(\mathbf{R})$ and the *relative motion wave function* $\psi_{rel}(\mathbf{r})$

$$
\psi(\mathbf{r}_1, \mathbf{r}_2) = \psi_{CM}(\mathbf{R}) \, \psi_{rel}(\mathbf{r}),
\tag{14.57}
$$

where

$$\psi_{CM}(\mathbf{R}) = \frac{e^{i\mathbf{K}\cdot\mathbf{R}}}{(2\pi)^{3/2}}.$$

In center of mass coordinates the particle interchange operation is

$$\mathbf{r}_1 \to \mathbf{r}_2 \implies \begin{cases} \mathbf{R} \to \mathbf{R} \\ \\ \mathbf{r} \to -\mathbf{r} \end{cases} \tag{14.58}$$

Thus, it amounts to a spatial reflection (parity) of the relative coordinate. Note that $\hat{H}_{rel}$ commutes with the relative angular momentum $\mathbf{r} \times \mathbf{p}$. Thus, the $\hat{H}_{rel}$ eigenstates can be products of spherical harmonics and radial wave functions

$$\psi_{rel}(\mathbf{r}) = Y_{\ell m}(\theta, \phi)\, R_{E\ell}(r), \tag{14.59}$$

where r, θ, ϕ are the spherical coordinates corresponding to $\mathbf{r}$.

Summarizing, our possible energy eigenfunctions are

$$\begin{cases} \underbrace{\dfrac{e^{i\mathbf{K}\cdot\mathbf{R}}}{(2\pi)^{3/2}}\, Y_{\ell m}(\theta, \phi)\, R_{E\ell}(r)\, \chi_{00}}_{sym} \\ \\ \underbrace{\dfrac{e^{i\mathbf{K}\cdot\mathbf{R}}}{(2\pi)^{3/2}}\, Y_{\ell m}(\theta, \phi)\, R_{E\ell}(r)\, \chi_{1m}}_{antisym} \end{cases} \tag{14.60}$$

The relative coordinate parity operation in spherical coordinates amounts to

$$\theta, \phi \to \pi - \theta, \pi + \phi \implies Y_{\ell m}(\theta, \phi) \to (-1)^{\ell} Y_{\ell m}(\theta, \phi).$$

Therefore, we must have

$$\begin{cases} \underbrace{\dfrac{e^{i\mathbf{K}\cdot\mathbf{R}}}{(2\pi)^{3/2}}\, Y_{\ell m}(\theta, \phi)\, R_{E\ell}(r)\, \chi_{00}}_{\ell=0,2,\dots} \\ \\ \underbrace{\dfrac{e^{i\mathbf{K}\cdot\mathbf{R}}}{(2\pi)^{3/2}}\, Y_{\ell m}(\theta, \phi)\, R_{E\ell}(r)\, \chi_{1m}}_{\ell=1,3,\dots} \end{cases} \implies \begin{cases} \ell = 0, 2, \dots \ S = 0 \\ \\ \ell = 1, 3, \dots \ S = 1 \end{cases} \tag{14.61}$$

Thus, the system of two electrons interacting with forces depending only on the relative distance can exist in energy eigenstates of even values of the relative angular momentum and antiparallel electron spins ($S = 0$) or of odd values of the relative angular momentum and total spin $S = 1$.

Isolated electrons. The central new concept established in the present chapter is that of symmetrization or antisymmetrization of the wave function of systems of identical particles. This means that the wave function of *all* electrons has to be antisymmetrized. Does this mean that we cannot consider and study the system of one electron, forgetting about the rest of the electrons in the Universe, as we have been doing up to now? The answer is that it is possible to ignore all other electrons and study the one particle of interest if the overlap[3] of its wave function with the wave functions of the rest is negligible. To illustrate this, consider a system of two electrons. The system will be described with an antisymmetric wave function $\Psi(\mathbf{r}_1, \mathbf{r}_2) = -\Psi(\mathbf{r}_2, \mathbf{r}_1)$. We suppress the spinor variables for simplicity of notation. The probability density to find one electron anywhere would be

$$\mathcal{P}(\mathbf{r}) = \int d^3r_1 \, |\Psi(\mathbf{r}_1, \mathbf{r})|^2 + \int d^3r_2 \, |\Psi(\mathbf{r}, \mathbf{r}_2)|^2 . \tag{14.62}$$

Suppose now that one of the electrons is localized in the greater area of London, having the wave function ψ_L, while a second electron is localized in Paris, having the wave function ψ_P. The two electrons are far enough not to have any appreciable interaction. The wave function of the system will be an antisymmetric combination of products of one-particle wave functions

$$\Psi(\mathbf{r}_1, \mathbf{r}_2) = \frac{1}{\sqrt{2}} \left(\psi_L(\mathbf{r}_1)\psi_P(\mathbf{r}_2) - \psi_P(\mathbf{r}_1)\psi_L(\mathbf{r}_2) \right) . \tag{14.63}$$

Assuming that the overlap of the wave functions ψ_L and ψ_P is negligible, i.e.,

$$\psi_L(\mathbf{r}) \, \psi_P(\mathbf{r}) \sim 0 \quad \forall \mathbf{r} , \tag{14.64}$$

we may substitute (14.63) into (14.62) and obtain

$$\mathcal{P}(\mathbf{r}) \sim |\psi_L(\mathbf{r})|^2 + |\psi_P(\mathbf{r})|^2 . \tag{14.65}$$

For a point $\mathbf{r}$ in the London area, $\psi_P(\mathbf{r}) \approx 0$ and the second term in (14.65) can be neglected. The probability density will be just

$$\mathcal{P}(\mathbf{r}) \sim |\psi_L(\mathbf{r})|^2 . \tag{14.66}$$

This is the standard one-particle result, meaning that the existence of the second electron can be ignored.

Composite particles. Ordinary matter is organized in the form of atoms and molecules which are composite structures consisting of more than one particles.

---

[3] Two functions f and g have a negligible overlap if $f(\mathbf{r}) \, g(\mathbf{r}) \sim 0$ for every point $\mathbf{r}$.

What kind of wave functions should we be using for systems of identical atoms or other composite particles? Consider the case of two hydrogen atoms consisting each of one proton (nucleus) and one electron. Both protons and electrons are fermions of spin $1/2$. The wave function of the system of the two atoms will be (again, for simplicity we suppress the spin variables)

$$\Psi(\underbrace{\mathbf{r}_{e1},\ \mathbf{r}_{p1}}_{H_1};\ \underbrace{\mathbf{r}_{e2},\ \mathbf{r}_{p2}}_{H_2}).$$

Any interchange between the two electrons will generate a minus sign due to anti-symmetry. Any interchange between the two protons will also generate a minus sign because of antisymmetry. In contrast, any interchange between electron and proton does not dictate a change in sign, since electrons and protons are not identical particles. As a result, we may have

$$\Psi(\mathbf{r}_{e1},\ \mathbf{r}_{p1};\ \mathbf{r}_{e2},\ \mathbf{r}_{p2}) = -\Psi(\mathbf{r}_{e2},\ \mathbf{r}_{p1};\ \mathbf{r}_{e1},\ \mathbf{r}_{p2}) = \Psi(\mathbf{r}_{e2},\ \mathbf{r}_{p2};\ \mathbf{r}_{e1},\ \mathbf{r}_{p1})$$

or

$$\Psi(H_1;\ H_2) = \Psi(H_2;\ H_1). \tag{14.67}$$

This means that the two hydrogen atoms behave as bosons. This is in perfect agreement with the fact that the spin of the hydrogen atom is an integer. This is quite general dictating that composite particles are classified as bosons or fermions according to their total spin. Composite particles consisting of an even number of fermions are bosons, while composites consisting of an odd number of fermions are fermions.

Systems of noninteracting identical particles. Consider N identical particles which do not interact or, in any case, their interaction could be neglected. The Hamiltonian of the system is the sum of commuting single-particle Hamiltonians

$$\hat{H} = \sum_{i=1}^{N} \hat{H}_i \quad where \quad \hat{H}_i = \frac{\hat{\mathbf{p}}_i^2}{2m} + V(\mathbf{r}_i). \tag{14.68}$$

The identical one-particle Hamiltonians will have the same set of energy eigenvalues ϵ_a corresponding to one-particle eigenfunctions ψ_a.

An energy eigenfunction of the full system $\Psi(\mathbf{r}_1, \mathbf{r}_2, \ldots)$ will be a product of the single-particle eigenfunctions $\psi_a(\mathbf{r}_i)$, namely

$$\Psi(\mathbf{r}_1, \mathbf{r}_2, \ldots) = \psi_a(\mathbf{r}_1)\,\psi_b(\mathbf{r}_2)\ldots\psi_c(\mathbf{r}_N). \tag{14.69}$$

However, since the particles are identical this product must be either symmetric in the particle exchange or antisymmetric.

For N identical bosons we have

$$\Psi_S(\mathbf{r}_1, \mathbf{r}_2, \ldots) = \frac{1}{N!} \sum_P P \, \Psi(\mathbf{r}_1, \mathbf{r}_2, \ldots), \tag{14.70}$$

where P signifies any *permutation* of the particles involved. For N identical fermions we have

$$\Psi_A(\mathbf{r}_1, \mathbf{r}_2, \ldots) = \frac{1}{N!} \sum_P (-1)^P P \, \Psi(\mathbf{r}_1, \mathbf{r}_2, \ldots). \tag{14.71}$$

The factor $(-1)^P$ stands for a minus sign associated with every odd permutation.

To illustrate the above let's consider three identical fermions in the states ψ_a, ψ_b, ψ_c. There are six ways in which three particles can be accommodated in these three states corresponding to permutations of a starting configuration (a, b, c). These are

$$(a, b, c), (b, a, c), (c, b, a), (a, c, b), (b, c, a), (c, a, b).$$

These can be obtained from (a, b, c) by the permutations

$$P(ab), \; P(ac), \; P(bc), \; P(ab)P(ac), \; P(cb)P(ac).$$

The antisymmetrised wave function of the three fermions will be

$$\Psi(\mathbf{r}_1, \mathbf{r}_2, \mathbf{r}_3) = \frac{1}{6} \{ \psi_a(\mathbf{r}_1)\psi_b(\mathbf{r}_2)\psi_c(\mathbf{r}_3) - \psi_b(\mathbf{r}_1)\psi_a(\mathbf{r}_2)\psi_c(\mathbf{r}_3) - \psi_c(\mathbf{r}_1)\psi_b(\mathbf{r}_2)\psi_a(\mathbf{r}_3)$$
$$-\psi_a(\mathbf{r}_1)\psi_c(\mathbf{r}_2)\psi_b(\mathbf{r}_3) + \psi_b(\mathbf{r}_1)\psi_c(\mathbf{r}_2)\psi_a(\mathbf{r}_3) + \psi_c(\mathbf{r}_1)\psi_a(\mathbf{r}_2)\psi_b(\mathbf{r}_3) \}. \tag{14.72}$$

It can be expressed as the *Slater Determinant*

$$\Psi(\mathbf{r}_1, \mathbf{r}_2, \mathbf{r}_3) = \frac{1}{3!} \begin{vmatrix} \psi_a(\mathbf{r}_1) & \psi_a(\mathbf{r}_2) & \psi_a(\mathbf{r}_3) \\ \psi_b(\mathbf{r}_1) & \psi_b(\mathbf{r}_2) & \psi_b(\mathbf{r}_3) \\ \psi_c(\mathbf{r}_1) & \psi_c(\mathbf{r}_2) & \psi_c(\mathbf{r}_3) \end{vmatrix}. \tag{14.73}$$

The Slater determinant can be generalized to N fermions.

Example 14.1 Consider the system of two particles of the same mass m, electric charge q and spin $s = 1/2$. The system is subject to an external homogeneous magnetic field $\mathbf{B}$.

(a) Write the Hamiltonian in the Center of Mass System. The magnetic field is sufficiently weak to justify neglecting the corresponding quadratic term.

(b) Write down the Heisenberg equations for $\mathbf{R}$, $\mathbf{r}$, $\mathbf{P}$, $\mathbf{p}$ and solve the CM motion.

(c) Consider the case that the two particles are identical fermions that initially ($t = 0$) have a Gaussian spatial wave function Ψ and spin wave function χ.

$$\Psi(\mathbf{r}_1, \mathbf{r}_2, 0) = \left(\frac{\alpha}{\pi}\right)^{3/2} e^{-\frac{\alpha}{2}(r_1^2 + r_2^2)} e^{i(\mathbf{q}_1 \cdot \mathbf{r}_1 + \mathbf{q}_2 \cdot \mathbf{r}_2)}$$

$$\chi = \frac{1}{\sqrt{2}}\left(\chi_\uparrow^{(1)}\chi_\uparrow^{(2)} - \chi_\downarrow^{(1)}\chi_\downarrow^{(2)}\right). \tag{14.74}$$

Calculate the uncertainty in their center of mass position $(\Delta \mathbf{R}_{CM})_t^2$ at later times $t > 0$.

(a) The Hamiltonian of the system is

$$\hat{H} = \frac{1}{2m}\left(\mathbf{p}_1 - \frac{q}{c}\mathbf{A}(\mathbf{r}_1)\right)^2 + \frac{1}{2m}\left(\mathbf{p}_2 - \frac{q}{c}\mathbf{A}(\mathbf{r}_2)\right)^2 - \frac{qg}{2mc}\mathbf{B}\cdot(\mathbf{s}_1 + \mathbf{s}_2), \tag{14.75}$$

with $\mathbf{A} = \frac{1}{2}\mathbf{B} \times \mathbf{r}$, assuming that the Lande factor g is the same for both particles. Taking $\mathbf{B} = B\hat{z}$ we rewrite the Hamiltonian as

$$\hat{H} = \frac{1}{2m}\left(\mathbf{p}_1^2 + \mathbf{p}_2^2\right) - \frac{q}{2mc}\mathbf{B}\cdot(\mathbf{L}_1 + \mathbf{L}_2) + \frac{q^2B^2}{8mc}\left(x_1^2 + y_1^2 + x_2^2 + y_2^2\right)$$

$$+ \frac{q^2}{|\mathbf{r}_1 - \mathbf{r}|} - \frac{qgB}{2mc}(s_{1z} + s_{2z}). \tag{14.76}$$

Introducing the Center of Mass operators

$$\mathbf{P} = \mathbf{p}_1 + \mathbf{p}_2, \quad \mathbf{p} = \frac{1}{2}(\mathbf{p}_1 - \mathbf{p}_2)$$
$$\mathbf{R} = \frac{1}{2}(\mathbf{r}_1 + \mathbf{r}_2), \quad \mathbf{r} = \mathbf{r}_1 - \mathbf{r}_2 \tag{14.77}$$

we get the Hamiltonian as a sum of three commuting terms in the form

$$\hat{H} = \hat{H}_{CM} + \hat{H}_r + \hat{H}_{spin} \tag{14.78}$$

where

$$\hat{H}_{CM} = \frac{\mathbf{P}^2}{4m} - \frac{q}{2mc}\mathbf{B}\cdot\mathbf{L} + \frac{q^2B^2}{16mc}\left(X^2 + Y^2\right)$$
$$\hat{H}_r = \frac{\mathbf{p}^2}{m} - \frac{q}{2mc}\mathbf{B}\cdot\mathbf{L}_r + \frac{q^2B^2}{4mc}(x^2 + y^2) + \frac{q^2}{r} \tag{14.79}$$
$$\hat{H}_{spin} = -\frac{qgB}{2mc}S_z$$

and, where $\mathbf{L} = \mathbf{R} \times \mathbf{P}$ is the total orbital angular momentum, $\mathbf{L}_r = \mathbf{r} \times \mathbf{p}$ the relative one and $\mathbf{S} = \mathbf{s}_1 + \mathbf{s}_2$ the total spin.

(b) Setting $\omega = qB/2mc$ we have

$$\dot{\mathbf{P}} = -\omega\hat{z} \times \mathbf{P}, \quad \dot{\mathbf{R}} = \frac{\mathbf{P}}{2m} - \omega\hat{z} \times \mathbf{R} \tag{14.80}$$

and

$$\dot{\mathbf{p}} = -\nabla\left(\frac{q^2}{r}\right) - \omega\hat{z} \times \mathbf{p}, \quad \dot{\mathbf{r}} = 2\frac{\mathbf{p}}{m} - \omega\hat{z} \times \mathbf{r}. \tag{14.81}$$

The Center of Mass momentum operators are solved to be

$$
\begin{cases}
\hat{P}_x(t) = \hat{P}_x(0)\cos\omega t + \hat{P}_y(0)\sin\omega t \\
\hat{P}_y(t) = \hat{P}_y(0)\cos\omega t - \hat{P}_x(0)\sin\omega t \\
\hat{P}_z(t) = \hat{P}_z(0)
\end{cases}
\tag{14.82}
$$

Substituting the Center of Mass velocity equation into the momentum equation we obtain

$$
\ddot{\mathbf{R}} + 2\omega\hat{z}\times\dot{\mathbf{R}} + \omega^2\hat{z}\left(\hat{z}\cdot\mathbf{R}\right) - \omega^2\mathbf{R} = 0
\tag{14.83}
$$

$$
\begin{cases}
\ddot{X} - 2\omega\dot{Y} - \omega^2 X = 0 \\
\ddot{Y} + 2\omega\dot{X} - \omega^2 Y = 0 \\
\ddot{Z} = 0
\end{cases}
\implies
\begin{cases}
\hat{X}(t) = \hat{X}(0)\cos\omega t + \hat{Y}(0)\sin\omega t \\
\hat{Y}(t) = \hat{Y}(0)\cos\omega t - \hat{X}(0)\sin\omega t \\
\hat{Z}(t) = \hat{Z}(0) + \frac{\hat{P}_z(0)}{2m}t
\end{cases}
\tag{14.84}
$$

and

$$
\mathbf{R}(t) = \hat{x}\left(\hat{X}(0)\cos\omega t + \hat{Y}(0)\sin\omega t\right) + \hat{y}\left(\hat{Y}(0)\cos\omega t - \hat{X}(0)\sin\omega t\right) + \hat{z}\left(\hat{Z}(0) + \frac{\hat{P}_z(0)}{2m}t\right).
\tag{14.85}
$$

Note that

$$
\hat{R}^2(t) = \hat{R}^2(0) + \frac{t^2}{4m^2}\hat{P}_z^2(0) + \frac{t}{2m}\left(\hat{Z}(0)\hat{P}_z(0) + \hat{P}_z(0)\hat{Z}(0)\right).
\tag{14.86}
$$

(c) The initial spatial wave function can be written

$$
\Psi(\mathbf{R}, \mathbf{r}, 0) = \psi_{CM}(\mathbf{R}, 0)\,\psi_r(\mathbf{r}, 0) = \left(\left(\frac{2\alpha}{\pi}\right)^{3/4} e^{-\alpha R^2} e^{i\mathbf{q}_+\cdot\mathbf{R}}\right)\left(\left(\frac{\alpha}{2\pi}\right)^{3/4} e^{-\frac{\alpha}{4}r^2} e^{i\mathbf{q}_-\cdot\mathbf{r}}\right),
\tag{14.87}
$$

where $\mathbf{q}_+ = \mathbf{q}_1 + \mathbf{q}_2$ and $\mathbf{q}_- = \frac{1}{2}\left(\mathbf{q}_1 - \mathbf{q}_2\right)$. The initial CM expectation values are

$$
\begin{aligned}
\langle\mathbf{R}\rangle_0 &= \left(\frac{2\alpha}{\pi}\right)^{3/2}\int d^3R\, e^{-2\alpha R^2}\mathbf{R} = 0, \\
\langle\mathbf{P}\rangle_0 &= -i\hbar\left(\frac{2\alpha}{\pi}\right)^{3/2}\int d^3R\,(-2\alpha\mathbf{R} + i\mathbf{q}_+)\,e^{-2\alpha R^2} = \hbar\mathbf{q}_+
\end{aligned}
\tag{14.88}
$$

using parity. Therefore,

$$
\langle\mathbf{R}\rangle_t = \hat{z}\frac{\hbar(q_+)_z t}{2m}.
\tag{14.89}
$$

We also have

$$
\begin{aligned}
\langle R^2\rangle_0 &= \left(\frac{2\alpha}{\pi}\right)^{3/2}\int d^3R\, e^{-2\alpha R^2}R^2 = -\frac{1}{2}\left(\frac{2\alpha}{\pi}\right)^{3/2}\frac{\partial}{\partial\alpha}\int d^3R\, e^{-2\alpha R^2} = \frac{3}{4\alpha} \\
\langle\hat{P}_z^2\rangle_0 &= -\hbar^2\left(\frac{2\alpha}{\pi}\right)^{3/2}\int d^3R\, e^{-\alpha R^2}e^{-i\mathbf{q}_+\cdot\mathbf{R}}\frac{\partial^2}{\partial Z^2}e^{-\alpha R^2}e^{i\mathbf{q}_+\cdot\mathbf{R}} = \ldots = \hbar^2\left(\alpha + (q_+)_z^2\right) \\
\langle\hat{Z}\hat{P}_z + \hat{P}_z\hat{Z}\rangle &= -i\hbar - 2i\hbar\left(\frac{2\alpha}{\pi}\right)^{3/2}\int d^3R\, e^{-\alpha R^2}e^{-i\mathbf{q}_+\cdot\mathbf{R}}Z\frac{\partial}{\partial Z}e^{-\alpha R^2}e^{i\mathbf{q}_+\cdot\mathbf{R}} = \ldots = 0
\end{aligned}
\tag{14.90}
$$

Thus, finally, we have

$$
\langle\hat{R}^2\rangle_t = \frac{3}{4\alpha} + \frac{\hbar^2 t^2}{4m^2}\left(\alpha + (q_+)_z^2\right) \implies (\Delta\mathbf{R})_t^2 = \frac{3}{4\alpha}\left(1 + \frac{1}{3}\left(\frac{\hbar t\alpha}{m}\right)^2\right).
\tag{14.91}
$$

Example 14.2 Two identical fermions of spin $1/2$ interact in terms of a potential

$$V = \begin{cases} +\infty & (|\mathbf{r}_1 - \mathbf{r}_2| \geq a) \\ \\ 0 & (|\mathbf{r}_1 - \mathbf{r}_2| < a) \end{cases} \tag{14.92}$$

A measurement of the relative orbital angular momentum is performed with the outcomes $\ell = 1$ and $\ell = 2$ with 50% probability each. What is the state of the system prior to the measurement?

Applying the analysis of the system of two identical fermions in a potential $V(|\mathbf{r}_1 - \mathbf{r}_2|)$, we obtain that the relative motion is that of a particle of mass $m/2$ trapped in a spherical cavity of radius a with the allowed energies for each relative orbital angular momentum ℓ being given by the condition $j_\ell(ka) = 0$. Then, the two possible totally antisymmetric outcomes are $j_2(k_2 r)Y_{2,m}(\hat{r})\chi_{0,0}$ and $j_1(k_1 r)Y_{1,m'}(\hat{r})\chi_{1,m_s}$ and, since they are equiprobable, the state has to be

$$\frac{1}{\sqrt{2}}\left(j_2(k_2 r)Y_{2,m}(\hat{r})\chi_{0,0} + e^{i\beta} j_1(k_1 r)Y_{1,m'}(\hat{r})\chi_{1,m_s} \right), \tag{14.93}$$

up to a phase β. The wave numbers k_1 and k_2 are given by the conditions $j_1(k_1 a) = j_2(k_2 a) = 0$.

Problems and Exercises

14.1 Consider a system of N interacting particles with a Hamiltonian

$$\hat{H} = \sum_{a=1}^{N} \frac{\hat{p}_a^2}{2m_a} + \sum_a V_a(\mathbf{r}_a) + \frac{1}{2}\sum_{a \neq b} V_{ab}(\mathbf{r}_a, \mathbf{r}_b).$$

Consider the *one-particle current probability density* $\mathbf{j}_a = \frac{1}{m_a} Re\left\{ \Psi^* \hat{\mathbf{p}}_a \Psi \right\}$, where $\Psi(\mathbf{r}_1, \ldots, \mathbf{r}_N)$ is the system wave function. Introduce the *density and current density operators*

$$\hat{\rho}(\mathbf{r}) \equiv \sum_{a=1}^{N} \delta(\mathbf{r} - \mathbf{r}_a) \quad and \quad \hat{\mathbf{J}}(\mathbf{r}) \equiv \sum_{a=1}^{N} \frac{1}{2m_a}\left\{ \hat{\mathbf{p}}_a, \delta(\mathbf{r} - \mathbf{r}_a) \right\}$$

and prove that the quantities

$$\rho(\mathbf{r}, t) = \langle \Psi | \hat{\rho}(\mathbf{r}) | \Psi \rangle, \quad \mathbf{J}(\mathbf{r}, t) = \langle \Psi | \hat{\mathbf{J}}(\mathbf{r}) | \Psi \rangle$$

satisfy the continuity equation $\frac{\partial \rho}{\partial t} + \nabla \cdot \mathbf{J} = 0$.

14.2 Three identical bosons are trapped inside a one-dimensional infinite square well $[-L, L]$. Measurement of each boson energy has given the values

$$\frac{9\hbar^2\pi^2}{8mL^2}, \quad \frac{9\hbar^2\pi^2}{8mL^2}, \quad \frac{2\hbar^2\pi^2}{mL^2}.$$

Write down the wave function of the system.

14.3 Two electrons are in the same spin state and their total momentum vanishes. The two electrons are constrained to move in one dimension and interact through the potential

$$V(|x_1 - x_2|) = \begin{cases} -|V_0| & (|x_1 - x_2| \le a) \\ \\ 0 & (|x_1 - x_2| > a) \end{cases}$$

What is the lowest possible energy of the system?

14.4 (a) Consider two particles, each with orbital angular momentum $\ell = 1$, $m_\ell = 0$. Write down the possible values of the total orbital angular momentum and calculate the probability that a measurement will find each of these values. (b) Assume that the two particles are electrons and that their spatial wave functions are identical. Predict the total spin and total angular momentum of the system.

14.5 Two noninteracting identical particles of mass m are trapped in a one-dimensional infinite square well of width $2L$. The particles occupy the $n = 1$ and $n = 2$ energy eigenstates. Calculate the uncertainty $(\Delta x)_B^2 = \langle (x_1 - x_2)^2 \rangle$ in the case that the two particles are bosons of zero spin. Calculate the same quantity $(\Delta x)_D^2 = \langle (x_1 - x_2)^2 \rangle$ assuming that the two particles are distinguishable. What conclusions can you draw about effective forces between the particles from the sign difference $\Delta = (\Delta x)_B^2 - (\Delta x)_D^2$?

14.6 Consider the system of two particles of mass m each, kept at a constant relative distance R. The rotational motion of the system is described by the Hamiltonial $\hat{H} = \hat{L}^2/2I$, where I is the moment of inertia with respect to the center of mass. Which are the possible eigenvalues and eigenstates of the system if the two particles are (a) spin-1/2 fermions, (b) spin-0 bosons? or (c) distinguishable particles?

14.7 The ^{3}He is a nucleus consisting of two protons and one neutron, while ^{4}He is an *isotope* consisting of two protons and two neutrons. Protons and neutrons are spin 1/2 fermions. Ignore spatial degrees of freedom and possible interactions between these particles and write down the possible wave functions for ^{3}He and ^{4}He.

14.8 The spatial part of the wave function of two identical spin 1/2 fermions constrained to move in one dimension is

$$\psi(x_1, x_2) = N \cos(\pi(x_1 - x_2)/L) \sin(\pi(x_1 + x_2)/L).$$

What is the total spin and the full wave function?

14.9 Consider five identical particles of spin 3/2. The particles are not interacting among each other but are subject to isotropic harmonic oscillator forces of frequency ω. What is the energy of the ground state?

References

1. E. Merzbacher, *Quantum Mechanics*, 3rd edn. (Wiley, Hoboken, 1998)
2. A. Messiah, *Quantum Mechanics* (Dover publications, Mineola, 1958). Single-volume reprint of the John Wiley & Sons, New York, two-volume 1958 edition
3. G. Baym, *Lectures in Quantum Mechanics*, Lecture Notes and Supplements in Physics (ABP, 1969)
4. F. Levin, *An introduction to Quantum Theory* (Cambridge University Press, Cambridge, 2002)

Chapter 15
Atoms

Matter under normal conditions is organized in bound systems of particles, namely *atoms*. It was shown by Rutherford's experiments that atoms are composed of electrons bound by the electromagnetic field of a *nucleus*. It turns out that nuclei are themselves composites of protons and neutrons bound by another type of force, the nuclear force. Electromagnetic forces can also hold together larger bound systems of more than one nuclei, namely *molecules*. Atoms and molecules are quantum systems that wouldn't exist as stable structures in the framework of classical physics. In contrast to analogous classical systems like, for example, the solar system, that could exist at many possible sizes depending on initial conditions, atoms have well-defined size in their ground state, determined by the uncertainty principle.

15.1 The Hydrogen Atom

The Hydrogen atom is just the system of an electron and a proton, both fermions of spin 1/2, of very different mass and opposite charges ($e < 0$ for the electron and $-e > 0$ for the proton). The dominant force between electron and proton is the Coulomb attraction[1]

$$V(r) = -\frac{e^2}{r}, \qquad (15.1)$$

where r is the electron-proton relative distance. The static Coulomb attraction is only part of the forces of electromagnetic origin that act between two moving charged particles. Nevertheless, the extra dynamical electromagnetic contributions are subdominant and will be treated at a later stage as perturbative corrections. Since the Coulomb force depends only on the relative distance of elecron and proton, we know from our analysis of the two-body problem in the previous chapter that the dynamics can be reduced to the motion of the center of mass, which is trivial, and to the relative

---

[1]See also [1, 2].

© Springer Nature Switzerland AG 2019
K. Tamvakis, *Basic Quantum Mechanics*, Undergraduate Texts in Physics,
https://doi.org/10.1007/978-3-030-22777-7_15

motion which is equivalent to the motion of a particle of mass $\mu = m_e m_p/(m_e + m_p)$ (reduced mass) in the central potential (15.1). Nevertheless, due to the fact that $\mu = m_e/(1 + m_e/m_p) \approx m_e(1 + O(10^{-3}))$, the relevant Hamiltonian of the system will be to a very good approximation

$$\hat{H} = \frac{\hat{p}^2}{2m_e} - \frac{e^2}{r}. \tag{15.2}$$

Since the potential tends to zero at very large distances, being negative at all points, the energy spectrum will consist of a continuum with $E > 0$, corresponding to scattering states and possibly a set of discrete energy levels of negative energies $E < 0$ (bound states). We shall concentrate in solving the latter part of the energy eigenvalue problem, obtaining the atomic energy levels and eigenfunctions of the Hydrogen atom. The corresponding radial Schroedinger equation is

$$-\frac{\hbar^2}{2m_e}\frac{1}{r}\frac{d^2}{dr^2}rR_{E\ell}(r) + \frac{\hbar^2\ell(\ell+1)}{2m_e r^2}R_{E\ell}(r) - \frac{e^2}{r}R_{E\ell}(r) = E\,R_{E\ell}(r) \tag{15.3}$$

or the one-dimensional one is

$$-\frac{\hbar^2}{2m_e}\frac{d^2 u_{E\ell}}{dr^2} + \frac{\hbar^2\ell(\ell+1)}{2m_e r^2}u_{E\ell}(r) - \frac{e^2}{r}u_{E\ell}(r) = E\,u_{E\ell}(r) \tag{15.4}$$

in terms of $u = rR$. Then, introducing

$$E = -\frac{\hbar^2\kappa^2}{2m_e} \quad \text{and} \quad \lambda^2 = \frac{2m_e e^2}{\hbar^2\kappa}, \tag{15.5}$$

we can rewrite (15.4) in terms of the variable

$$\rho = \kappa r \tag{15.6}$$

as

$$\left\{\frac{d^2}{d\rho^2} - 1 + \frac{\lambda^2}{\rho} - \frac{\ell(\ell+1)}{\rho^2}\right\}u_{\kappa\ell}(\rho) = 0. \tag{15.7}$$

We have more conveniently labeled the one-dimensional eigenfunction as $u_{\kappa\ell}(\rho)$. At a great distance from the origin ($\rho \to \infty$) we may neglect the inverse powers in (15.7) and obtain an approximate solution

$$\lim_{\rho \to \infty}\{u_{\kappa\ell}(\rho)\} = e^{-\rho}. \tag{15.8}$$

Note that this is not any different than what we would expect from our analysis of finite range central potentials, although the Coulomb potential is of infinite range ($\lim_{r \to \infty}\{rV(r)\} \neq 0$).

Next we proceed to solve the differential equation (15.7) by the power series method. It is convenient to factor out the asymptotic behaviour (15.8) by introducing a radial function $v_{\kappa\ell}(\rho)$ as

$$u_{\kappa\ell}(\rho) = e^{-\rho} v_{\kappa\ell}(\rho). \tag{15.9}$$

The differential equation satisfied by $v_{\kappa\ell}$ is actually more complicated. It is

$$\left\{ \frac{d^2}{d\rho^2} - 2\frac{d}{d\rho} + \frac{\lambda^2}{\rho} - \frac{\ell(\ell+1)}{\rho^2} \right\} v_{\kappa\ell}(\rho) = 0. \tag{15.10}$$

We should not forget that the one-dimensional radial wave function has to satisfy the boundary condition $u_{\kappa\ell}(0) = 0$. If we assume now that in the vicinity of the origin

$$v_{\kappa\ell}(\rho) \sim \rho^{\alpha} \quad \text{with } \alpha > 0, \tag{15.11}$$

substituting in (15.10) we obtain

$$(\alpha(\alpha-1) - \ell(\ell+1))\rho^{\alpha-2} + O(\rho^{\alpha-1}) \sim 0 \quad \Longrightarrow \quad \alpha = \ell+1. \tag{15.12}$$

Therefore, we may write a trial power series solution as

$$v(\rho) = \sum_{\nu=0}^{\infty} C_\nu \rho^{\nu+\ell+1}. \tag{15.13}$$

Inserting this power series in (15.10) we get

$$\sum_{\nu=0}^{\infty} C_\nu ((\ell+\nu+1)(\ell+\nu) - \ell(\ell+1)) \rho^{\nu-1} + \sum_{\nu=0}^{\infty} C_\nu \left(\lambda^2 - 2(\ell+\nu+1)\right) \rho^\nu = 0. \tag{15.14}$$

Note that the first term of the first series vanishes and this sum really starts from $\nu = 1$. Changing the summation index to $\nu + 1$ we can rewrite (15.14) as

$$\sum_{\nu=0}^{\infty} \left[((\ell+\nu+2)(\ell+\nu+1) - \ell(\ell+1)) C_{\nu+1} + \left(\lambda^2 - 2(\ell+\nu+1)\right) C_\nu \right] \rho^\nu = 0,, \tag{15.15}$$

which, supposedly being true for every ρ, implies that every term of the series should vanish. However, this can occur only if the coefficients obbey the recursive relation

$$\frac{C_{\nu+1}}{C_\nu} = \frac{2(\ell+\nu+1) - \lambda^2}{[(\ell+\nu+2)(\ell+\nu+1) - \ell(\ell+1)]} \quad \text{for } \nu = 0, 1, 2, \ldots \tag{15.16}$$

For large ν values the coefficient ratio behaves as

$$\nu \to \infty \quad \Longrightarrow \quad \frac{C_{\nu+1}}{C_\nu} \sim \frac{2}{\nu} \;. \tag{15.17}$$

Notice that the power series for $e^{+2\rho}$ has exacly the above large ν recursive relation $\left(2^{\nu+1}/(n+1)!\right)/\left(2^\nu/n!\right) = 2/(n+1) \approx 2/n$. Such an asymptotic behaviour is unacceptable, since it overpowers the $e^{-\rho}$ factor of $u_{\kappa\ell}(\rho)$ and gives

$$u_{\kappa\ell}(\rho) \;=\; e^{-\rho}\, v_{\kappa\ell}(\rho) \;\sim\; e^{+\rho}\,,$$

which is not a square integrable wave function. The only way out is if the power series somehow terminates and $v_{\kappa\ell}(\rho)$ is just a polynomial. This is possible only if

$$\lambda^2 \;=\; 2(\nu + 1 + \ell) \tag{15.18}$$

for some ν. Then, $C_{\nu+1}$ and all subsequent coefficients would vanish and $v_{\kappa\ell}(\rho)$ would be a polynomial of order ν. Of course, this requires the special values of the parameter λ given by (15.18). Since, λ is related to the energy through its definition (15.5), these correspond to special values of the energy

$$E \;=\; -\frac{m_e\, e^4}{2\hbar^2 (\nu + 1 + \ell)^2}\,. \tag{15.19}$$

Thus, we obtain a discrete bound state spectrum labeled by the integers $\nu = 0, 1, 2, \ldots$ and $\ell = 0, 1, 2, \ldots$. We are free however to replace the power series index ν with the so-called *principal quantum number* n, defined as

$$n = \nu + \ell + 1 \quad \Longrightarrow \quad n = 1, 2, \ldots \tag{15.20}$$

that, since $n = 2(\nu + 1 + \ell) = \lambda^2 > 0$, takes positive integer values. In terms of it the energy levels of Hydrogen are

$$E_n \;=\; -\frac{m_e e^4}{2\hbar^2 n^2}\,. \tag{15.21}$$

Nevertheless, now ℓ is constrained by (15.20) to obtain values limited by n, namely $\ell = n - \nu - 1 = (n-1),\ (n-2),\ \ldots,\ 1,\ 0$ or

$$\ell = 0, 1, \ldots, n - 1 \;. \tag{15.22}$$

The fact that many states with different ℓ have the same energy is not explained by the spherical symmetry of the system. The spherical symmetry can only account for the $2\ell + 1$ states of different m having the same energy. The existence of the additional degeneracy means that the system possesses further symmetries beyond the spherical symmetry. We are going to return to this point in a later chapter. For the moment we may calculate that the overall degeneracy is

$$\sum_{\ell=0}^{n-1}(2\ell + 1) = n^2. \tag{15.23}$$

Thus, n^2 different energy eigenfunctions will correspond to a single eigenvalue of the energy $E_n = -e^4 m_e/2\hbar^2 n^2$.

It is remarkable that the energy of the ground state of the Hydrogen atom can be estimated by a simple argument based on the uncertainty principle. Something analogous was done in the case of the simple harmonic oscillator. The argument goes as follows. Let r_0 be the *radius* of the atom beyond which the probability to find the electron is negligible. The order of magnitude of the potential energy will be $-e^2/r_0$. The uncertainty principle puts a lower bound on the momentum $p \sim \hbar/r_0$ and, therefore, the kinetic energy will be at least $\hbar^2/2mr_0^2$. Putting these together into the total energy

$$E(r_0) = \frac{\hbar^2}{2m_e r_0^2} - \frac{e^2}{r_0}$$

we may ask which is the value r_0 for which E has a minimum. This amounts to

$$E'(r_0) = 0 \implies -\frac{\hbar^2}{m_e r_0^3} + \frac{e^2}{r_0^2} = 0 \implies r_0 = \frac{\hbar^2}{m_e e^2}$$

and gives the energy

$$E_0 = -\frac{m_e e^4}{2\hbar^2},$$

which is exactly the ground state energy obtained by solving the Schroedinger equation. This result cries out for the interpretation that the size of the atom is entirely a consequence of the uncertainty principle and its stability a consequence of the balancing between kinetic and potential energy.

At this point, it is usefull to introduce the so-called *Bohr radius* a_0 defined as

$$a_0 = \frac{\hbar^2}{m_e e^2}. \tag{15.24}$$

As we saw above in our estimate of the ground state energy through the uncertainty principle the combination $\hbar^2/m_e e^2$ gives the characteristic size of the atom. In terms of the Bohr radius, the energy levels are

$$E_n = -\frac{e^2}{2n^2 a_0}. \tag{15.25}$$

The numerical value of the Bohr radius is

$$a_0 = 0.52912\ldots \times 10^{-8}\,\text{cm} \tag{15.26}$$

and leads to a value for the Hydrogen ground state energy

$$|E_1| = \frac{e^2}{2a_0} = 13.606\ldots \text{ eV} \tag{15.27}$$

The excited energy levels are $E_n = -|E_1|/n^2$. Although this is not yet explained within the above framework of Coulomb interaction,[2] transitions between energy levels are possible provided the energy difference is emitted or absorbed in the form of electromagnetic radiation. The relation of the energy differences between neighboring levels and the corresponding radiation wavelength is

$$\Delta_n = E_{n+1} - E_n = \frac{(2n+1)|E_1|}{n^2(n+1)^2} \implies \lambda_n = \frac{hc}{\Delta_n}. \tag{15.28}$$

For Δ_1 this wavelength is

$$\lambda_1 = \frac{hc}{\Delta_1} = \frac{4hc}{3|E_1|} \approx 1200 \times 10^{-8} \text{ cm}$$

corresponding to the ultraviolet.

The Hydrogen atom energy eigenfunctions. Our next goal is to determine the radial wave functions $R_{n\ell}(\rho)$. From our analysis we have determined that they are of the form

$$R_{n\ell} \sim e^{-\rho} \rho^\ell \Lambda_{n\ell}(\rho) \quad \textbf{or} \quad v_{n\ell}(\rho) \sim \rho^{\ell+1} \Lambda_{n\ell}(\rho), \tag{15.29}$$

where $\Lambda_{n\ell}(\rho)$ is a polynomial of degree $n - \ell - 1$. Substituting this into (15.10) we obtain for $\Lambda_{n\ell}$ the equation

$$\frac{d^2\Lambda_{n\ell}}{d\rho^2} + 2\left(\frac{\ell+1}{\rho} - 1\right)\frac{d\Lambda_{n\ell}}{d\rho} + 2\frac{(n-\ell-1)}{\rho}\Lambda_{n\ell} = 0. \tag{15.30}$$

Comparing this to the *Laguerre differential equation*

$$\frac{d^2L_p^q}{dx^2} + \left(\frac{q+1}{x} - 1\right)\frac{dL_p^q}{dx} + \frac{p}{x}L_p^q = 0, \tag{15.31}$$

which has as solutions the classified *associated Laguerre polynomials* $L_p^q(x)$, we conclude that $\Lambda_{n\ell}(\rho)$ stands for $L_{n-\ell-1}^{2\ell+1}(2\rho)$. Thus, we obtain the radial energy eigenfunctions as

$$R_{n\ell}(\rho) = N_{n\ell}\, e^{-\rho} \rho^\ell\, L_{n-\ell-1}^{2\ell+1}(2\rho), \tag{15.32}$$

where $N_{n\ell}$ is a normalization factor to meet $\int_0^\infty dr\, r^2 R^2 = 1$. The associated Laguerre polynomials can be constructed from the ordinary Laguerre polynomials $L_s(x)$

---

[2]Radiation absorption and emmission phenomena will be discussed in a future chapter.

Table 15.1 Associated laguerre polynomials

$n = 1, \ \ell = 0$	$L_0^1(x) = 1$
$n = 2, \ \ell = 0$	$L_1^1(x) = 2 - x$
$n = 2, \ \ell = 1$	$L_0^3(x) = 1$
$n = 3, \ \ell = 0$	$L_2^1(x) = 3 - x$
$n = 3, \ \ell = 1$	$L_1^3(x) = 4 - x$
$n = 3, \ \ell = 2$	$L_0^5(x) = 1$

according to the formula

$$L_p^q(x) = (-1)^q \frac{d^q}{dx^q} L_{p+q}(x) \quad \text{where} \quad L_s(x) = \frac{1}{s!} e^x \frac{d^s}{dx^s} \left(e^{-x} x^s \right) . \tag{15.33}$$

Based on these definitions we can show that

$$L_0^q(x) = 1, \quad L_1^q(x) = 1 + q - x . \tag{15.34}$$

Some relevant associated Laguerre polynomials are shown in Table 15.1.[3]
The associated Laguerre polynomials satisfy the following normalization relation

$$\int_0^\infty dx \, e^{-x} \, x^{2(\ell+1)} \left(L_{n-\ell-1}^{2\ell+1}(x) \right)^2 = \frac{2n(n+\ell)!}{(n-\ell-1)!} . \tag{15.35}$$

The resulting Hydrogen energy eigenfunctions are

$$\psi_{n\ell m}(\mathbf{r}) = N_{n\ell} \, Y_{\ell m}(\Omega) \, e^{-\kappa r} \, (\kappa r)^\ell \, L_{n-\ell-1}^{2\ell+1}(2\kappa r) . \tag{15.36}$$

Note that κ has a dependence on n, given by

$$\kappa = \sqrt{\frac{2m_e |E|}{\hbar^2}} = \frac{m_e e^2}{\hbar^2 n} \quad \text{or} \quad \kappa = \frac{1}{n a_0} . \tag{15.37}$$

Thus, the general eigenfunction can be written

$$\psi_{n\ell m}(\mathbf{r}) = N_{n\ell} \, (n a_0)^{-\ell} \, Y_{\ell m}(\Omega) \, e^{-r/a_0 n} \, r^\ell \, L_{n-\ell-1}^{2\ell+1}(2r/a_0 n) . \tag{15.38}$$

Normalization determines the constant $N_{n\ell}$ to be

---

[3]Note that an alternative normalization is sometimes used in the literature in which the $1/s!$ factor in the generating formula for Laguerre polynomials is missing. Nevertheless, apart from a modification of the wave function normalization factor, we obtain ultimately the same Hydrogen energy eigenfunctions.

$$N_{n\ell} = a_0^{-3/2} \frac{2^{\ell+1}}{n^2} \sqrt{\frac{(n-\ell-1)!}{(n+\ell)!}} . \tag{15.39}$$

The first few radial wave functions are

$$R_{10}(r) = \frac{2}{a_0^{3/2}} e^{-r/a_0}$$

$$R_{20}(r) = \frac{1}{\sqrt{8a_0^3}} e^{-r/2a_0} \left(2 - \frac{r}{a_0}\right) \tag{15.40}$$

$$R_{21}(r) = \frac{1}{\sqrt{6a_0^3}} e^{-r/2a_0} \frac{r}{2a_0}$$

Some very usefull radial expectation values are the following:

$$\langle r \rangle_{n\ell} = \frac{a_0}{2} \left(3n^2 - \ell(\ell+1)\right)$$

$$\langle r^2 \rangle_{n\ell} = \frac{a_0^2}{2} n^2 \left(5n^2 + 1 - 3\ell(\ell+1)\right)$$

$$\langle r^{-1} \rangle_{n\ell} = \frac{1}{a_0 n^2} \tag{15.41}$$

$$\langle r^{-2} \rangle_{n\ell} = \frac{1}{a_0^2 n^3 \left(\ell+\frac{1}{2}\right)}$$

$$\langle r^{-3} \rangle_{n\ell} = \frac{1}{a_0^3 n^3 \ell(\ell+1)\left(\ell+\frac{1}{2}\right)}$$

where

$$\langle r^j \rangle = \int_0^\infty dr\, r^2\, r^j \left(R_{n\ell}(r)\right)^2 .$$

Finally, the complete expression for the energy eigenfunctions is

$$\psi_{n\ell m}(\mathbf{r}) = a_0^{-3/2} \frac{2^{\ell+1}}{n^2} \sqrt{\frac{(n-\ell-1)!}{(n+\ell)!}} \, Y_{\ell m}(\Omega) \, e^{-r/a_0 n} \, (r/na_0)^\ell \, L_{n-\ell-1}^{2\ell+1}(2r/a_0 n) \tag{15.42}$$

The first few eigenfunctions are given in Table 15.2.

Closing this section, it is useful for subsequent discussions on atomic structure to mention the established terminology. The Hydrogen eigenstates are classified according to the values of ℓ as s, p, d, f, g, ... corresponding to $\ell = 0, 1, 2, 3, 4, \ldots$. Thus, the state $n = 1$, $\ell = 0$ is referred to as $1s$, the state $n = 2$, $\ell = 0$ as $2s$, the state $n = 2$, $\ell = 1$ as $2p$. Analogously, the state $n = 3$, $\ell = 2$ is referred to as $3d$.

Example 15.1 Calculate the expectation value of the Hydrogen atom radius $\langle r \rangle$. What is the probability to find the electron at a point $r \geq \langle r \rangle$?

The expectation value of the Hydrogen atom radius in the state $\psi_{100}(r) = (\pi a_0^3)^{-1/2} e^{-r/a_0}$ is

Table 15.2 Hydrogen energy eigenfunctions

$\psi_{100}(r) = (\pi a_0^3)^{-1/2} e^{-r/a_0}$
$\psi_{200}(r) = (32\pi a_0^3)^{-1/2} e^{-r/2a_0} \left(2 - \frac{r}{a_0}\right)$
$\psi_{21-1}(\mathbf{r}) = (16\pi a_0^3)^{-1/2} e^{-i\phi} \sin\theta \, e^{-r/2a_0} \frac{r}{2a_0}$
$\psi_{210}(r) = (8\pi a_0^3)^{-1/2} \cos\theta \, e^{-r/2a_0} \frac{r}{2a_0}$
$\psi_{211}(\mathbf{r}) = -(16\pi a_0^3)^{-1/2} e^{i\phi} \sin\theta \, e^{-r/2a_0} \frac{r}{2a_0}$

$$\langle r \rangle = \int d^3r \, r \, \psi_{100}^2(r) = (\pi a_0)^{-1} \int d\Omega \int_0^\infty dr \, r^3 \, e^{-2r/a_0} =$$

$$\frac{a_0}{4} \int_0^\infty d\xi \, \xi^3 \, e^{-\xi} = \frac{a_0}{4} 3! = \frac{3}{2} a_0$$

The probability to find the electron at a point $r > 3a_0/2$ is

$$P = \int d\Omega \int_{3a_0/2}^\infty dr \, r^2 \, (\pi a_0)^{-1} e^{-2r/a_0} = \frac{1}{2} \int_3^\infty d\xi \, \xi^2 \, e^{-\xi} = \frac{17}{2e^3} \approx 0.423.$$

Example 15.2 Calculate the probability current density for a Hydrogen energy eigenstate and verify that, being stationary, it is divergenceless, namely that $\nabla \cdot \mathbf{J} = 0$.

By definition we have

$$\mathbf{J} = \frac{\hbar}{2m_e i} \left(\psi_{n\ell m}^* \nabla \psi_{n\ell m} - \psi_{n\ell m} \nabla \psi_{n\ell m}^*\right). \tag{15.43}$$

We note that $\psi_{n\ell m} = R_{n\ell}(r) Y_{\ell m}(\theta, \phi) = R_{n\ell}(r) Y_{\ell m}(\theta, 0) e^{im\phi}$, where $R_{n\ell}(r)$ and $Y_{\ell m}(\theta, 0)$ are real. Substituting the gradient in spherical coordinates, we obtain

$$\mathbf{J} = \cdots = \frac{\hbar}{2m_e i} (R_{n\ell}(r) Y_{\ell m}(\theta, 0))^2 \left(e^{-im\phi} \frac{\hat{\phi}}{r\sin\theta} \frac{\partial}{\partial\phi} e^{im\phi} - c.c.\right)$$

or

$$\mathbf{J} = \frac{\hbar m}{m_e r} (R_{n\ell}(r) Y_{\ell m}(\theta, 0))^2 \frac{\hat{\phi}}{\sin\theta}. \tag{15.44}$$

The divergence of this current density should vanish, since it corresponds to a stationary state. Indeed, we have

$$\nabla \cdot \mathbf{J} = \left(\hat{r}\frac{\partial}{\partial r} + \frac{\hat{\theta}}{r}\frac{\partial}{\partial\theta} + \frac{\hat{\phi}}{r\sin\theta}\frac{\partial}{\partial\phi}\right) \cdot \frac{\hbar m}{m_e r} (R_{n\ell}(r) Y_{\ell m}(\theta, 0))^2 \frac{\hat{\phi}}{\sin\theta} = \cdots$$

$$= \frac{\hbar m}{m_e r \sin\theta} \frac{\partial}{\partial\phi} \left(\frac{(R_{n\ell}(r) Y_{\ell m}(\theta, 0))^2}{r\sin\theta}\right) = 0. \tag{15.45}$$

Example 15.3 Consider the Hydrogen atom and

(a) Calculate the expectation value $\langle \frac{1}{r} \rangle_{n\ell m}$ in an energy eigenstate $\psi_{n\ell m}(\mathbf{r})$.[4]

---

[4]You may use the property of associated Laguerre polynomials $\int_0^\infty dx\, x^\alpha \, e^{-x} \left(L_n^\alpha(x)\right)^2 = \Gamma(n + \alpha + 1)/n!$.

(b) Determine the Hydrogen atom energy eigenvalues using the Virial Theorem.

(a) We have

$$\left\langle \frac{1}{r} \right\rangle_{n\ell m} = \mathcal{N}_{n\ell}^2 \int d\Omega \, |Y_{\ell m}(\Omega)|^2 \int_0^\infty dr \, r \, e^{-2r/na_0}(r/na_0)^{2\ell} \left(L_{n-\ell-1}^{2\ell+1}(2r/na_0) \right)^2$$

$$= \frac{1}{a_0 n^2} \frac{(n-\ell-1)!}{(n+\ell)!} \int_0^\infty dx \, x^{2\ell+1} e^{-x} \left(L_{n-\ell-1}^{2\ell+1}(x) \right)^2 = \frac{1}{a_0 n^2}. \qquad (15.46)$$

(b) The Virial Theorem states that $\langle T \rangle = \frac{1}{2}\langle \mathbf{r} \cdot \nabla V(r) \rangle$, where the expectation value refers to a stationary state. Thus, in the case of the Hydrogen atom Coulomb potential $V(r) = -e^2/r$ and for an energy eigenstate $|n\ell m\rangle$, we obtain

$$\langle T \rangle_{n\ell m} = \frac{e^2}{2} \left\langle \frac{1}{r} \right\rangle_{n\ell m} = \frac{e^2}{2a_0 n^2}. \qquad (15.47)$$

We also have

$$\langle V \rangle_{n\ell m} = -e^2 \left\langle \frac{1}{r} \right\rangle_{n\ell m} = -\frac{e^2}{a_0 n^2}. \qquad (15.48)$$

The energy eigenvalues will be

$$E_n = \langle T \rangle_{n\ell m} + \langle V \rangle_{n\ell m} = -\frac{e^2}{2a_0 n^2}. \qquad (15.49)$$

15.2 The Zeeman Effect

A charged particle possessing angular momentum has a magnetic dipole moment. This is true for the electron with the important addition of the magnetic dipole moment due to its spin. A homogeneous magnetic field **B** acting on a Hydrogen atom will induce interaction terms between the magnetic dipole moments of the electron and the magnetic field. The interaction terms in the Hamiltonian will be [2, 3][5]

$$\Delta \hat{H} = -\mathbf{m} \cdot \mathbf{B} - \mathbf{m}_s \cdot \mathbf{B}, \qquad (15.50)$$

where the corresponding magnetic dipole moments are

$$\mathbf{m} = \frac{e}{2m_e c}\mathbf{L}, \quad \mathbf{m}_s = \frac{ge}{2m_e c}\mathbf{S}. \qquad (15.51)$$

---

[5]In a subsequent chapter we are going to consider the general problem of a charged particle in a magnetic field. There, we will see that in addition to the interaction term linear in the magnetic field there is also an interaction term quadratic in the magnetic potential. Such a term turns out to be negligible in atoms subject to magnetic fields of moderate size. Nevertheless, it can become relevant in the presence of very strong magnetic fields, as the ones encountered in astrophysical environments.

The Lande factor g for the electron is to a very good approximation equal to 2. Without loss of generality, we can take the $\hat{z}$-axis to be along the direction of the magnetic field. Then, the above correction to the Hamilton operator is

$$\Delta\hat{H} = -\frac{eB}{2m_e c}\left(\hat{L}_z + 2\hat{S}_z\right). \tag{15.52}$$

The full Hamiltonian is

$$\hat{H} = \hat{H}_0 - \frac{eB}{2m_e c}\left(\hat{L}_z + 2\hat{S}_z\right) \quad \text{with} \quad \hat{H}_0 = \frac{\mathbf{p}^2}{2m_e} - \frac{e^2}{r}. \tag{15.53}$$

$\hat{H}_0$ is the *"unperturbed"* part with eigenvalues $E_n = -\frac{m_e e^4}{2n^2 \hbar^2}$. It is not difficult to see that the products

$$\psi_{n\ell m}(\mathbf{r})\,\chi \tag{15.54}$$

of the standard unperturbed Hydrogen energy eigenfunctions of $\hat{H}_0$ times a spinor χ that is an eigenstate of $\hat{S}_z$ are eigenfunctions of $\hat{H}$. Denoting by E'_n the corrected eigenvalues we have

$$\left(\hat{H}_0 - \frac{eB}{2m_e c}\left(\hat{L}_z + 2\hat{S}_z\right)\right)\psi_{n\ell m}(\mathbf{r})\,\chi = E'_n\,\psi_{n\ell m}(\mathbf{r})\,\chi. \tag{15.55}$$

Taking

$$\hat{S}_z\chi = \hbar m_s\,\chi = \pm\frac{\hbar}{2}\chi, \tag{15.56}$$

we obtain

$$\left(E_n - \frac{eB\hbar}{2m_e c}(m + 2m_s)\right)\psi_{n\ell m}(\mathbf{r})\,\chi = E'_n\,\psi_{n\ell m}(\mathbf{r})\,\chi \tag{15.57}$$

and the corrected eigenvalues

$$E'_{n\,m,\pm} = E_n - \frac{eB\hbar}{2m_e c}(m \pm 1). \tag{15.58}$$

for the eigenfunctions

$$\Psi_{n\ell m;\pm} = \psi_{n\ell m}(\mathbf{r})\,\chi_\pm. \tag{15.59}$$

As a result of the presence of the homogeneous magnetic field each of the Hydrogen energy levels splits into $2(2\ell + 1)$ levels. The $2n^2$-fold degeneracy of the Hydrogen spectrum is lifted due to the partial breaking of rotational invariance. These phenomena are known by the name *"Zeeman effect"*. For historical reasons, the part of the effect due to the electron spin bears the erroneous name *anomalous Zeeman effect*.

15.3 Fine Structure

Even in the absence of external fields, like the homogeneous magnetic field of the Zeeman effect, the Coulomb Hamiltonian $\hat{H}_0 = \mathbf{p}^2/2m_e - e^2/r$ is an approximation. The Coulomb potential describes the interaction of two static point charges, while in the case of the Hydrogen atom we have a moving electron, which in its rest frame *"observes"* a moving proton and should experience the intrinsic magnetic field generated by the moving proton electric current. This magnetic field is[6]

$$
\mathbf{B} = \frac{1}{c}\mathbf{v} \times \mathbf{E} \implies \mathbf{B} = \frac{1}{mc}\mathbf{p} \times \mathbf{E} = -\frac{e^2}{mc}\mathbf{p} \times \nabla\left(\frac{1}{r}\right). \tag{15.60}
$$

The effect of this field on the spin dipole magnetic moment of the electron generates the energy correction

$$
-\frac{e}{mc}\mathbf{S} \cdot \mathbf{B} = \frac{e^2}{m^2c^2}\mathbf{S} \cdot \left(\mathbf{p} \times \nabla\left(\frac{1}{r}\right)\right) = -\frac{e^2}{m^2c^2r^3}\mathbf{S} \cdot (\mathbf{p} \times \mathbf{r}) = \frac{e^2}{m^2c^2r^3}\mathbf{S} \cdot \mathbf{L}
$$

This derivation was based on the formula $\mathbf{B} = \frac{1}{c}\mathbf{v} \times \mathbf{E}$ which does not take into account the accelerated circular motion of the electron. A detailed derivation arrives at an analogous result with an extra $1/2$ factor. Thus, we have the so-called *LS-coupling* correction to the Hamiltonian [2–4]

$$
\Delta H_1 = \frac{e^2}{2m^2c^2}\frac{(\mathbf{S} \cdot \mathbf{L})}{r^3}. \tag{15.61}
$$

Our next step will be to determine the eigenstates and eigenvalues of the corrected Hamiltonian

$$
\hat{H} = \hat{H}_0 + \Delta H_1. \tag{15.62}
$$

Note however that the eigenfunctions of $\hat{H}_0$, namely $\psi_{n\ell m}\chi$, are not eigenfunctions of $\hat{H}$. The $\mathbf{L} \cdot \mathbf{S}$ operator appearing in (15.62) depends on the total angular momentum $\mathbf{J} = \mathbf{L} + \mathbf{S}$, i.e., the sum of orbital angular momentum of the electron and its spin. In fact, we have

$$
\mathbf{L} \cdot \mathbf{S} = \frac{1}{2}\left(J^2 - L^2 - S^2\right) = \frac{1}{2}\left(J^2 - L^2 - \frac{3\hbar^2}{4}\right). \tag{15.63}
$$

That means that we will have to resort to the basis of $\hat{J}^2$ eigenstates. Before we do that we shall digress briefly on the general method of computation of small corrections to the energy spectrum (*perturbation theory*).

---

[6]In the discussion of corrections to the Hydrogen atom Hamiltonian we may also include the fact that the proton is not immovable by simply replacing the electron mass with the reduced mass $m = m_e m_p/(m_e + m_p)$.

Digression on the computation of Small Corrections to the Energy Spectrum.
The problem of incorporation of *perturbative*, i.e., "small", corrections to the Hydrogen energy spectrum is addressed in the framework of Perturbation Theory, which will be the subject of a subsequent chapter. Nevertheless, the effect of the above small corrections to the Hydrogen Hamiltonian is simple enough to allow the first acquaintance with this method here.

Assume that we have a Hamiltonian $\hat{H} = \hat{H}_0 + \delta\hat{H}$ that is the sum of a standard part $\hat{H}_0$ (with known eigenstates and eigenvalues $\hat{H}_0|\psi_n^{(0)}\rangle = E_n^{(0)}|\psi_n^{(0)}\rangle$)) and a small perturbation $\delta\hat{H}$. We expect that the exact eigenstates $|\psi_n\rangle$ will only differ from $|\psi_n^{(0)}\rangle$ by a small part. Similarly for the eigenvalues

$$|\psi_n\rangle = |\psi_n^{(0)}\rangle + \delta|\psi_n\rangle \quad and \quad E_n = E_n^{(0)} + \delta E_n. \tag{15.64}$$

Substituting in the exact eigenvalue equation and neglecting $O(\delta^2)$ terms that are much smaller, we obtain

$$\hat{H}|\psi_n\rangle = E_n|\psi_n\rangle \implies \tag{15.65}$$

$$\left(\hat{H}_0 + \delta\hat{H}\right)\left(|\psi_n^{(0)}\rangle + \delta|\psi_n\rangle\right) = \left(E_n^{(0)} + \delta E_n\right)\left(|\psi_n^{(0)}\rangle + \delta|\psi_n\rangle\right) \tag{15.66}$$

or

$$\delta\hat{H}|\psi_n^{(0)}\rangle + \hat{H}_0\delta|\psi_n\rangle = E_n^{(0)}\delta|\psi_n\rangle + \delta E_n|\psi_n^{(0)}\rangle. \tag{15.67}$$

Taking the inner product with $\langle\psi_n^{(0)}|$ we obtain

$$\langle\psi_n^{(0)}|\delta\hat{H}|\psi_n^{(0)}\rangle + \langle\psi_n^{(0)}|\hat{H}_0\left(\delta|\psi_n\rangle\right) = E_n^{(0)}\langle\psi_n^{(0)}|\left(\delta|\psi_n\rangle\right) + \delta E_n \tag{15.68}$$

or

$$\delta E_n = \langle\psi_n^{(0)}|\delta\hat{H}|\psi_n^{(0)}\rangle. \tag{15.69}$$

Thus, to first order in the approximation the corrections to the energy eigenvalues are equal to the matrix elements of the *"perturbing"* part of the Hamiltonian in the unperturbed eigenstates.

Returning to the specific case of the $L \cdot S$ coupling correction $\Delta\hat{H}_1$, we note that the set of eigenfunctions $\psi_{n\ell m}\chi \sim R_{n\ell}Y_{\ell m}\chi$ are not the most suitable as the set of unperturbed eigenfunctions[7] $|\psi_n^{(0)}\rangle$, since they are not eigenstates of the operator $L \cdot S$. Instead, we may use as a basis the equivalent set of the eigenstates $|n;\ \ell,\ s;\ j,\ m_j\rangle$, i.e., $R_{n\ell}$ for the radial part and the L^2, S^2, J^2, J_z eigenstates $|\ell,\ s;\ j,\ m_j\rangle$ for the angular momentum and spin part. This last set we have considered before in the chapter on angular momentum addition. There we found that the allowed values for j are $\ell + 1/2$ and $\ell - 1/2$. Acting with $L \cdot S$ we obtain

---

[7] These are eigenstates of L^2, L_z, S^2, S_z and, of course, $\hat{H}_0$.

$$\begin{cases} \mathbf{L} \cdot \mathbf{S} | j = \ell + 1/2, \, m_j \rangle = \ell \frac{\hbar^2}{2} | j = \ell + 1/2, \, m_j \rangle \\ \mathbf{L} \cdot \mathbf{S} | j = \ell - 1/2, \, m_j \rangle = -(\ell + 1) \frac{\hbar^2}{2} | j = \ell - 1/2, \, m_j \rangle \end{cases} \quad (15.70)$$

Therefore, the corresponding matrix elements will be

$$\langle n; \ell, s; j, m_j | \Delta \hat{H}_1 | n; \ell, s; j, m_j \rangle = \frac{e^2 \hbar^2}{4m^2 c^2} \int_0^\infty dr \, r^2 \frac{R_{n\ell}^2(r)}{r^3} \begin{cases} \ell \\ -(\ell + 1) \end{cases}$$

Copying $\langle r^{-3} \rangle_{n\ell}$ from (15.41) we get

$$\langle n; \ell, s; j, m_j | \Delta \hat{H}_1 | n; \ell, s; j, m_j \rangle = \frac{me^8 / 4\hbar^4 c^2}{n^3 \ell (\ell + 1/2)(\ell + 1)} \begin{cases} \ell \\ -(\ell + 1) \end{cases} \quad (15.71)$$

Note that there seems to be a problem with this correction in the case of $\ell = 0$, however, this will go away as soon as we add an extra correction of the same significance as $\Delta \hat{H}_1$. This will be done immediately.

The correction $\Delta \hat{H}_1$ is a relativistic correction of order $O(c^{-2})$. Our analysis would be incorrect if we do not include *all* corrections of that order. So the relevant question is "*are there any other corrections of order* $O(c^{-2})$?". The answer is positive and comes from the kinetic part of the Hamiltonian. The nonrelativistic kinetic energy $p^2/2m$ should be replaced by the relativistic expression

$$\sqrt{c^2 \hat{p}^2 + m^2 c^4} \approx mc^2 \left(1 + \frac{\hat{p}^2}{2m^2 c^2} - \frac{\hat{p}^4}{8m^4 c^4} + \cdots \right) \quad (15.72)$$

The first term is just a constant that plays no role.[8] The first significant correction is

$$\Delta \hat{H}_2 = -\frac{\hat{p}^4}{8mc^2} . \quad (15.73)$$

Let's compute now the correction to the Hydrogen energy levels resulting from this term. As in the case of $\Delta \hat{H}_1$ we consider the basis eigenvectors $| n; \ell, s; j, m_j \rangle$. Thus, the corrections we look for will be

$$\langle n; \ell s j m_j | \Delta \hat{H}_2 | n; \ell s j m_j \rangle = -\frac{1}{2mc^2} \langle n; \ell s j m_j | \left(\frac{\hat{p}^2}{2m} \right)^2 | n; \ell s j m_j \rangle \quad (15.74)$$

or

---

[8] A constant in the energy can always be subtracted away and has no physical effect. Such a constant becomes relevant only in the framework of the theory of gravity (General Relativity).

$$\langle n; \ell s j m_j | \Delta \hat{H}_2 | n; \ell s j, m_j \rangle = -\frac{1}{2mc^2} \langle n; \ell s j m_j | \left(\hat{H}_0 + \frac{e^2}{r} \right)^2 | n; \ell s j m_j \rangle =$$

$$-\frac{1}{2mc^2} \langle n; \ell s j m_j | \left(E_n + \frac{e^2}{r} \right)^2 | n; \ell s j m_j \rangle = E_n^2 + 2e^2 E_n \langle r^{-1} \rangle_{n\ell} + e^4 \langle r^{-2} \rangle_{n\ell}$$

$$(15.75)$$

or, copying from (15.41) the expressions for $\langle r^{-1} \rangle_{n\ell}$ and $\langle r^{-2} \rangle_{n\ell}$,

$$\langle v; \ell, s; j, m_j | \Delta \hat{H}_2 | n; \ell, s; j, m_j \rangle = -\frac{me^8}{4\hbar^4 c^2} \left(\frac{2}{n^3(\ell + 1/2)} - \frac{3}{2n^4} \right) . \quad (15.76)$$

Adding the two corrections we obtain

$$\Delta E = \langle n; \ell, s; j, m_j | \left(\Delta \hat{H}_1 + \Delta \hat{H}_2 \right) | n; \ell, s; j, m_j \rangle$$

$$= \frac{me^8}{4\hbar^4 c^2} \left(\frac{3}{2n^4} - \frac{2}{n^3(\ell + 1/2)} + \frac{1}{n^3(\ell + 1/2)} \begin{Bmatrix} 1/(\ell + 1) \\ -1/\ell \end{Bmatrix} \right)$$

or

$$\Delta E = \frac{me^8}{4\hbar^4 c^2} \left(\frac{3}{2n^4} - \frac{2}{n^3(j + 1/2)} \right) \quad (15.77)$$

or, introducing the so-called *fine-structure constant* $\alpha = \frac{e^2}{\hbar c}$,

$$(\Delta E)_{nj} = -\alpha^2 \frac{|E_n|}{n} \left(\frac{1}{j + 1/2} - \frac{3}{4n} \right) , \quad (15.78)$$

where $E_n = -me^4/2n^2\hbar^2 = -e^2/2n^2 a_0$, the unperturbed energy eigenvalues. Note that, since $\alpha \approx 1/137$, this is a rather small correction, namely

$$\frac{\Delta E}{E} \sim (1/137)^2 \sim 0.53 \times 10^{-4} .$$

15.4 Hyperfine Structure

In addition to the corrections considered in the previous section, which are due to the relativistic motion of the electron, referred to by the name *fine structure* of the energy spectrum, there are additional even smaller corrections, referred to by the name *hyperfine structure*. Their origin is the magnetic field of the nucleus, arising from the nuclear magnetic dipole moment due to the proton spin

$$\mathbf{m}_p = -\frac{eg_p}{2m_p c}\mathbf{S}_p \,, \tag{15.79}$$

where $\mathbf{S}_p$ is the proton spin ($s_p = 1/2$). From standard Electrodynamics it is known that a magnetic dipole moment $\mathbf{m}_p$ will generate a magnetic field, the vector potential of which is given by

$$\mathbf{A}(\mathbf{r}) = \frac{\mathbf{m}_p \times \mathbf{r}}{4\pi r^3} \,. \tag{15.80}$$

The corresponding magnetic field is

$$B_i = (\nabla \times \mathbf{A})_i = -\frac{1}{4\pi}\left(m_{p_i}\nabla^2(1/r) - (\mathbf{m}_p \cdot \nabla)\nabla_i(1/r)\right). \tag{15.81}$$

Because of this magnetic field, there will be a correction to the Hamiltonian due to its coupling to the *electron magnetic dipole moment* $\mathbf{m}_e = \frac{e\mathbf{S}_e}{m_e c}$. We have added a subscript in the electron spin to distinguish it from the proton spin. Notice also the sign difference in the dipole moments due to the opposite charge of the proton ($-e > 0$). The correction is

$$\Delta \hat{H}_3 = -\mathbf{m}_e \cdot \mathbf{B} \,. \tag{15.82}$$

Substituting the expression of the magnetic moment, we obtain [2, 3]

$$\Delta \hat{H}_3 = -\frac{e^2 g_p}{8\pi m_p m_e c^2}\left((\mathbf{S}_e \cdot \mathbf{S}_p)\nabla^2(1/r) - (\mathbf{S}_p \cdot \nabla)(\mathbf{S}_e \cdot \nabla(1/r))\right). \tag{15.83}$$

We shall restrict ourselves to the correction of the ground state level

$$\Delta E = \langle n = 1, \ell = m_\ell = 0; S_{ez}, S_{pz}|\Delta \hat{H}_3|n = 1, \ell = m_\ell = 0; S_{ez}, S_{pz}\rangle \tag{15.84}$$

or

$$\Delta E = -\frac{e^2 g_p}{8\pi m_p m_e c^2}\chi^\dagger \mathcal{J}\chi \,, \tag{15.85}$$

where χ are the spinor wave functions and

$$\mathcal{J} = \int d^3r\,|\psi_{100}(r)|^2\left((\mathbf{S}_e \cdot \mathbf{S}_p)\nabla^2(1/r) - (\mathbf{S}_p \cdot \nabla)(\mathbf{S}_e \cdot \nabla(1/r))\right). \tag{15.86}$$

Note that, since $|\psi_{100}(r)|^2$ is spherically symmetric, we have

$$\int d^3r\,|\psi_{100}(r)|^2\nabla_i\nabla_j(1/r) = \frac{\delta_{ij}}{3}\int d^3r\,|\psi_{100}(r)|^2\nabla^2(1/r) \,. \tag{15.87}$$

Thus, we have

$$\mathcal{J} = \frac{2}{3}(\mathbf{S}_e \cdot \mathbf{S}_p) \,, \int d^3r\,|\psi_{100}(r)|^2\,\nabla^2(1/r) \tag{15.88}$$

Using the fact that $\nabla^2(1/r) = -4\pi\delta(\mathbf{r})$, we obtain

$$\mathcal{J} = -\frac{8\pi}{3}\left(\mathbf{S}_e \cdot \mathbf{S}_p\right)|\psi_{100}(0)|^2 \tag{15.89}$$

and

$$\Delta E = \frac{e^2 g_p}{3m_p m_e c^2}|\psi_{100}(0)|^2\, \chi^\dagger\left(\mathbf{S}_e \cdot \mathbf{S}_p\right)\chi. \tag{15.90}$$

The operator $\mathbf{S}_e \cdot \mathbf{S}_p$ can be written in terms of the total, i.e., nuclear and electronic, spin

$$\mathbf{S} = \mathbf{S}_e + \mathbf{S}_p \tag{15.91}$$

as

$$\mathbf{S}_e \cdot \mathbf{S}_p = \frac{1}{2}\left(\hat{S}^2 - \hat{S}_e^2 - \hat{S}_p^2\right) = \frac{1}{2}\left(\hat{S}^2 - \frac{3\hbar^2}{2}\right). \tag{15.92}$$

If we choose the spinors χ to be eigenstates of the total spin S, we obtain

$$\Delta E = \frac{e^2 g_p \hbar^2}{6m_p m_e c^2}|\psi_{100}(0)|^2\,(S(S+1) - 3/2). \tag{15.93}$$

For the two possible eigenvalues of S we have

$$\begin{aligned}
S = 1 &\implies \Delta E_{ortho} = \frac{g_p}{6\pi}\alpha^4\left(m_e c^2\right)\left(\frac{m_e}{m_p}\right)(1/2) \\
S = 0 &\implies \Delta E_{para} = \frac{g_p}{6\pi}\alpha^4\left(m_e c^2\right)\left(\frac{m_e}{m_p}\right)(-3/2)
\end{aligned} \tag{15.94}$$

The *ortho-Hydrogen* corresponds to electron-proton spins parallel, while the *para-Hydrogen* to spins antiparallel. Numerically this correction is[9]

$$\frac{\Delta E}{|E_1|} \sim \frac{\alpha^2}{6\pi} \times \frac{m_e}{m_p} \approx 10^{-6}.$$

The wavelength of radiation emmitted during transitions between these two levels is 10^9 times larger than the characteristic atomic wave length, being in the neighborhood of microwaves

$$\lambda \sim 21.1\,\mathrm{cm}.$$

---

[9] $2|E_1| = m_e e^4 \hbar^2 = (m_e c^2)\frac{e^4}{\hbar^2 c^2} = \alpha^2(m_e c^2)$.

15.5   Other Atoms

15.5.1   Hydrogenic Ions and Other Atoms

The analysis carried out in the preceding sections on the Hydrogen atom can be carried over without drastic changes to the case of single electron ions with a nucleus of $-\mathcal{Z}e$ charge, where $\mathcal{Z}$ corresponds to the number of protons making up the nucleus. The energy spectrum is obtained with the replacement

$$e^2 \rightarrow \mathcal{Z}\,e^2 \tag{15.95}$$

to be

$$E_n^{(\mathcal{Z})} = \mathcal{Z}^2\, E_n^{(\mathcal{Z}=1)} = -\frac{\mathcal{Z}^2 e^4}{2\hbar^2 n^2}. \tag{15.96}$$

The eigenfunctions, expressed in terms of the Bohr radius $a_0 = \hbar^2/m_e e^2$, are obtained with the single replacement

$$a_0 \rightarrow \frac{a_0}{\mathcal{Z}}. \tag{15.97}$$

The fine structure and hyperfine structure corrections, being both proportional to e^8, are just multiplied with $\mathcal{Z}^4$.

Neutral atoms beyond the Hydrogen atom require the addition of extra electrons. As a consequence, the dynamical problem is not any more a two-body problem. Furthermore, the Pauli Exclusion Principle has to be taken into account, limiting the available possibilities and leading us to specific options. For example, if for instance we assume that the hydrogen energy levels are not significantly modified by the presence of a second electron, the ground state can only be occupied by two electrons of opposite spin. This would be the case of the simplest *element* beyond Hydrogen, provided that the nuclear charge is doubled by the addition of a second proton. This way of building up the atoms, is, of course, a crude approximation, since the hydrogen energy levels are modified by the addition of the extra electrons and their mutual repulsion. A quantitative study of chemical elements requires the systematic use of approximation methods.

15.5.2 The Helium Atom

The Helium atom[10] has a nucleus[11] of *atomic number*, i.e., postive charge $\mathcal{Z} = 2$, and two electrons. In addition to the attraction of the electrons by the nucleus, the potential includes the mutual repulsion of the two electrons. Ignoring to a first approximation the motion of the nucleus, the Hamiltonian is

$$H = \frac{p_1^2}{2m_e} + \frac{p_2^2}{2m_e} - \frac{\mathcal{Z}e^2}{r_1} - \frac{\mathcal{Z}e^2}{r_2} + \frac{e^2}{|\mathbf{r}_1 - \mathbf{r}_2|}. \tag{15.98}$$

We have also ignored relativistic corrections and $\mathbf{L} \cdot \mathbf{S}$ coupling phenomena analogous to the hydrogen fine structure, phenomena analogous to the Thomas precession related to the relative motion of the two electrons and, finally, hyperfine structure corrections.

The Hamilton operator (15.98) is made of a *solvable* piece consisting of two independent hydrogen-type Hamiltonians and the repulsion term

$$\hat{H} = \sum_{i=1,2} \hat{H}_0^{(i)} + \hat{V} \quad with \quad \begin{cases} \hat{H}_0^{(i)} = \frac{p_i^2}{2m_e} - \frac{\mathcal{Z}e^2}{r_i} \\ \hat{V} = \frac{e^2}{|\mathbf{r}_1 - \mathbf{r}_2|}. \end{cases} \tag{15.99}$$

The repulsion term cannot be treated as a perturbation, since it is not a priori small. Nevertheless, we may analyze the features of the approximate solution that we obtain if we ignore it. In this case, we have just a pair of two mutually noninteracting electrons. The corresponding energy eigenfunctions will be products of single electron ones. Their spatial part will be

$$\psi_{n_1 \ell_1 m_1}(\mathbf{r}_1) \, \psi_{n_2 \ell_2 m_2}(\mathbf{r}_2). \tag{15.100}$$

The energy eigenvalues will be sums

$$E = E_{n_1} + E_{n_2} = -\frac{\mathcal{Z}^2 e^2}{2a_0} \left(\frac{1}{n_1^2} + \frac{1}{n_2^2} \right). \tag{15.101}$$

With n_1 and n_2 we have denoted the principal hydrogen quantum number for each electron, taking up the standard positive integer values.

Since, the two electrons are indentical particles, their total wave function should be antisymmetric. The lowest energy eigenvalue (ground state), corresponding to $n_1 = n_2 = 1$ and $\ell_1 = \ell_2 = 0$, has a symmetric spatial part $\psi_{100}(r_1)\psi_{100}(r_2)$. Thus, we must necessarily have an antisymmetric spinorial part. Recall that the antisymmetric

---

[10]See also [2, 3].

[11]Helium exists in the form of two *isotopes*, namely as ^{3}He, with a nucleus of two protons and a neutron, and as ^{4}He, with a nucleus of two protons and two neutrons. The two types of nuclei have drastically different properties due to the fact that the former is a fermion, while the latter a boson.

combination of two spins is the so-called *singlet* with total spin $S = S_z = 0$

$$\chi_{0,0} = \frac{1}{\sqrt{2}} \left(\chi_\uparrow^{(1)} \chi_\downarrow^{(2)} - \chi_\downarrow^{(1)} \chi_\uparrow^{(2)} \right) . \tag{15.102}$$

Therefore, in this approximation of mutually non-interacting electrons, the total ground state wave function will be

$$\Psi_0(1, 2) = \psi_{100}(r_1) \psi_{100}(r_2) \frac{1}{\sqrt{2}} \left(\chi_\uparrow^{(1)} \chi_\downarrow^{(2)} - \chi_\downarrow^{(1)} \chi_\uparrow^{(2)} \right) , \tag{15.103}$$

where

$$\psi_{100}(r) = \left(\frac{8}{\pi a_0^3} \right)^{1/2} e^{-\frac{2r}{a_0}} . \tag{15.104}$$

Within this approximation, the state in which one of the electrons has $n = 1$, $\ell = 0$, while the other has a higher n, corresponds to a series of excited levels. The total wave function for such a state will be either

$$\frac{1}{\sqrt{2}} \left(\psi_{100}(r_1) \psi_{n\ell m}(\mathbf{r}_2) + \psi_{100}(r_2) \psi_{n\ell m}(\mathbf{r}_1) \right) \chi_{0,0} \tag{15.105}$$

or

$$\frac{1}{\sqrt{2}} \left(\psi_{100}(r_1) \psi_{n\ell m}(\mathbf{r}_2) - \psi_{100}(r_2) \psi_{n\ell m}(\mathbf{r}_1) \right) \chi_{1,m_s} \tag{15.106}$$

where χ_{1,m_s} stands for the triplet spinor wave function of the total spin ($m_s = 0, \pm 1$).

In the framework of this approximation the energy of the ground state is $E_{n_1 = n_2 = 1} = -8|E_1| = -108.8\,\text{eV}$, where $E_1 = -13.6\,\text{eV}$ stands for the Hydrogen ground state energy. The next excited energy level corresponds to $E_{n_1 = 1, n_2 = 2} = -4|E_1| = -54.4\,\text{eV}$. Note however that the *ionization energy* corresponds to the difference between $E_{11} = -8|E_1|$ and the energy of an one-electron-ion $-4|E_1|$, i.e., $E_i = -4|E_1| - (-8|E_1|) = 4|E_1| = 54.4\,\text{eV}$. Therefore, the continuum starts at $E \geq E_{11} + E_i = -54.4\,\text{eV}$ and the second *would-be excited state* ($E_{22} = -2|E_1| = -27.2\,\text{eV}$) is part of the continuum.

Next, let's consider the repulsion term and let's treat it as if it were a perturbation. We may compute the correction to the energy of the ground state following our recipe for the computation of small corrections to the energy spectrum of Sect. 15.3. According to that the correction to the ground state energy in the lowest approximation will be

$$\Delta E = \int d^3 r_1 \int d^3 r_2 \, |\Psi_0(r_1, r_2)|^2 \frac{e^2}{|\mathbf{r}_1 - \mathbf{r}_2|} \tag{15.107}$$

or

$$\Delta E = e^2 \left(\frac{8}{\pi a_0^3}\right)^2 \int_0^\infty dr_1 r_1^2 e^{-\frac{2Zr_1}{a_0}} \int_0^\infty dr_2 r_2^2 e^{-\frac{2Zr_2}{a_0}} \int\int \frac{d\Omega_1 d\Omega_2}{|\mathbf{r}_1 - \mathbf{r}_2|} . \quad (15.108)$$

Without loss of generality we may choose $\hat{z}_2$ along the direction of $\mathbf{r}_1$ and have

$$\int \frac{d\Omega_2}{|\mathbf{r}_1 - \mathbf{r}_2|} = \int_0^{2\pi} d\phi_2 \int_{-1}^1 \frac{d\cos\theta_2}{\sqrt{r_1^2 + r_2^2 - 2r_1 r_2 \cos\theta_2}} = \frac{2\pi}{r_1 r_2} (r_1 + r_2 - |r_1 - r_2|) .$$

The remaining angular integration is trivial giving just 4π. Thus, we have

$$\Delta E = 8 e^2 \left(\frac{8}{a_0^3}\right)^2 \int_0^\infty dr_1 r_1 e^{-\frac{4r_1}{a_0}} \int_0^\infty dr_2 r_2 e^{-\frac{4r_2}{a_0}} (r_1 + r_2 - |r_1 - r_2|) .$$

$$(15.109)$$

Performing the tedious but, otherwise, straightforward integrations, we obtain

$$\Delta E = \frac{5}{4} \frac{e^2}{a_0} \sim 34 \, \text{eV}. \quad (15.110)$$

Including this correction, we obtain for the ground state energy

$$E \sim -74.8 \, \text{eV}, \quad (15.111)$$

which is not very different from the actual measured ground state energy of $-78.975 \, \text{eV}$. A better estimate can be obtained employing the non-perturbative variational method which will be analyzed in a subsequent chapter on systematic approximation methods.

Problems and Exercises

15.1 Verify that for the Hydrogen atom

$$E_n = -\langle n, \ell, m | \frac{\hat{p}^2}{2m_e} | n, \ell, m \rangle = -\frac{1}{2} \langle n, \ell, m | \frac{e^2}{r} | n, \ell, m \rangle .$$

15.2 Calculate the expectation values

$$\langle \mathbf{r} \rangle, \ \langle r \rangle, \ \langle r^2 \rangle, \ \langle r^{-1} \rangle, \ \langle r^{-2} \rangle$$

for the ground state of the Hydrogen atom.

15.3 For a Hydrogen atom in the state

$$\frac{1}{6}\left[4|1,0,0\rangle + 3|2,1,1\rangle - |2,1,0\rangle + \sqrt{10}\,|2,1,-1\rangle\right]$$

calculate the expectation value of the energy, the expectation value of $\hat{L}^2$ and the expectation value of $\hat{L}_z$.

15.4 Consider a Hydrogen atom in its ground state. Calculate the probability to find the electron at a distance larger than the Bohr radius a_0.

15.5 For the ground state of the Hydrogen atom calculate the uncertainties (Δx) and (Δp_x) and verify the Heisenberg inequality.

15.6 A Hydrogen atom is in the state

$$|\psi\rangle = N\left[|1,0,0\rangle + (1+i)|2,1,1\rangle + i|3,2,-1\rangle\right].$$

Calculate the constant N. Calculate the expectation value of the energy. Calculate the uncertainty (ΔL_z). Assume that the atom is in this state at $t = 0$, obtain the time-evolved state at $t > 0$ and calculate the probability to find the atom in $|\psi\rangle$ at a time $t > 0$.

15.7 Consider a Hydrogen atom in the presence of a uniform time-independent magnetic field directed along the $\hat{z}$-axis. Assume that at $t = 0$ the particle is in a state $|n, \ell, 1/2; j, m\rangle$ with $j = \ell + 1/2$. Find the probability that at time $t > 0$ the particle is in a state $|n', \ell', 1/2; j', m'\rangle$ with j' equal to either $\ell + 1/2$ or $\ell - 1/2$.

15.8 Consider a Hydrogen atom subject simultaneously to a homogeneous magnetic and a homogeneous electric field, generating the extra interaction terms $\hat{H}_B = -\frac{e}{2m_e c}\mathbf{B} \cdot (\mathbf{L} + 2\mathbf{S})$ and $\hat{H}_E = -e\mathbf{r} \cdot \mathbf{E}$. Show that

$$\left|\langle n, \ell, m, m_s|\left(\hat{H}_B + \hat{H}_E\right)|n', \ell', m', m'_s\rangle\right|^2 =$$

$$\left|\langle n, \ell, m, m_s|\hat{H}_B|n', \ell', m', m'_s\rangle\right|^2 + \left|\langle n, \ell, m, m_s|\hat{H}_E|n', \ell', m', m'_s\rangle\right|^2$$

and that one of these two terms will be zero.

15.9 Consider the Hydrogen energy eigenfunctions. Explain why the radial wave functions $R_{n\ell}(r)$ can always be chosen to be real. Show that $\psi_{n\ell m=0}$ are also real.

(a) Consider the state $|\psi\rangle = a|2,0,0\rangle + b|2,1,0\rangle$, with a, b real coefficients, and calculate the expectation values $\langle\psi|\mathbf{r}|\psi\rangle$, $\langle\psi|\mathbf{p}|\psi\rangle$. Do the same for the nonstationary state $|\psi'\rangle = a|1,0,0\rangle + b|2,1,0\rangle$. Note and explain the difference between the expectation values of the position and those of the momentum.

(b) Assume that at time $t = 0$ the Hydrogen atom is in the above state $|\Psi(0)\rangle = |\psi'\rangle$. Calculate the expectation values of $\mathbf{r}$ and $\mathbf{p}$ at a time $t > 0$.

15.10 Consider the *"one-dimensional"* radial equation for the Hydrogen atom and make the transformation of variables

$$r = Cr'^2, \quad u(r) = \sqrt{r'}\, u'(r')\,.$$

With the appropriate choice of the constant C, show that the equation for $u'(r')$ is that of the isotropic oscillator. Exhibit the relation for the energy eigenvalues and the radial quantum numbers for the two systems.

References

1. D. Griffiths, *Introduction to Quantum Mechanics*, 2nd edn. (Cambridge University Press, Cambridge, 2017)
2. E. Merzbacher, *Quantum Mechanics*, 3rd edn. (Wiley, Hoboken, 1998)
3. G. Baym, *Lectures in Quantum Mechanics*, Lecture Notes and Supplements in Physics (ABP, 1969)
4. F. Levin, *An Introduction to Quantum Theory* (Cambridge University Press, Cambridge, 2002)

Chapter 16
Molecules

16.1 Born–Oppenheimer

Molecules are composite structures made up of electrons and more than one nuclei. The simplest molecule is the *Hydrogen ion* H_2^+ made up of two Hydrogen nuclei (protons) and a single electron. As in the case of atoms beyond the Hydrogen atom the molecule energy eigenvalue problem corresponds to a very complicated system of Schroedinger equations that can only be treated in the framework of approximation methods. The active degrees of freedom in molecules are clearly more than those of atoms. For example, H_2^+ has six degrees of freedom (three for the relative distance of the two protons and three for the electron position). A simplifying fact is that the nuclei are much heavier than electrons ($m_e/M_N \sim 10^{-3}$) and, to a leading approximation, the motion of the nuclei could be ignored. In this approximation, the relative nuclei distances will appear as parameters in the effective Hamiltonian that describes the motion of the electrons. This picture can be corrected by taking into account the slow nuclear motion.

We may begin by writing down the exact Hamilton operator for a general molecular system

$$\hat{H} = \hat{T}_e + \hat{T}_N + \hat{V}_{ee} + \hat{V}_{eN} + \hat{V}_{NN}. \qquad (16.1)$$

The appearing terms are

$$\hat{T}_e = \sum_j \frac{p_j^2}{2m_e} \qquad \textit{Kinetic Energy of Electrons}$$

$$\hat{T}_N = \sum_\alpha \frac{P_\alpha^2}{2M_\alpha} \qquad \textit{Kinetic Energy of Nuclei}$$

$$V_{ee} = \sum_{i>j} \frac{e^2}{|\mathbf{r}_i - \mathbf{r}_j|} \qquad \textit{Electron Coulomb Repulsion}$$

© Springer Nature Switzerland AG 2019
K. Tamvakis, *Basic Quantum Mechanics*, Undergraduate Texts in Physics,
https://doi.org/10.1007/978-3-030-22777-7_16

$$V_{eN} = -\sum_{\alpha}\sum_{j}\frac{Z_\alpha e^2}{|\mathbf{R}_\alpha - \mathbf{r}_j|} \qquad \textit{Electron-Nuclei Coulomb Attraction}$$

$$V_{NN} = \sum_{\alpha,\beta}\frac{Z_\alpha Z_\beta e^2}{|\mathbf{R}_\alpha - \mathbf{R}_\beta|} \qquad \textit{Nuclei Coulomb Repulsion}$$

The nuclear kinetic energy due to the large nuclear masses can be considered as a *perturbation*. This is the so-called *Born–Oppenheimer Approximation Method* [1, 2]. In this approach, an expansion can be set up characterized by the small parameter m_e/M_N. The essence of the Born–Oppenheimer method is that, since the nuclei are more massive and, therefore, slow, their wave functions are more localized than those of the electrons. Thus, in the leading approximation, where their kinetic energy could be neglected, their positions can be treated as fixed parameters and the molecular dynamics evolve entirely in terms of the electrons. At a higher level of precision we may solve for the nuclei dynamics.

The exact Hamiltonian can be separated into *"nuclear"* and *"electronic"* parts[1] as

$$\hat{H} = \hat{H}_N + \hat{H}_e, \tag{16.2}$$

where

$$\hat{H}_e = \frac{-\hbar^2}{2m_e}\sum_j \nabla_j^2 + \sum_{i>j}\frac{e^2}{|\mathbf{r}_i - \mathbf{r}_j|} - \sum_{\alpha,j}\frac{Z_\alpha e^2}{|\mathbf{R}_\alpha - \mathbf{r}_j|} + \sum_{\alpha,\beta}\frac{Z_\alpha Z_\beta e^2}{|\mathbf{R}_\alpha - \mathbf{R}_\beta|} \tag{16.3}$$

$$\hat{H}_N = -\sum_\alpha \frac{\hbar^2}{2M_\alpha}\nabla_{R_\alpha}^2$$

The Schroedinger equation for the electronic Hamiltonian is

$$\left\{\frac{-\hbar^2}{2m_e}\sum_j \nabla_j^2 + \sum_{i>j}\frac{e^2}{|\mathbf{r}_i - \mathbf{r}_j|} - \sum_{\alpha,j}\frac{Z_\alpha e^2}{|\mathbf{R}_\alpha - \mathbf{r}_j|}\right\}\psi_n(\mathbf{r}, \mathbf{R}) =$$

$$\left(\epsilon_n(\mathbf{R}) - \sum_{\alpha,\beta}\frac{Z_\alpha Z_\beta e^2}{|\mathbf{R}_\alpha - \mathbf{R}_\beta|}\right)\psi_n(\mathbf{r}, \mathbf{R}). \tag{16.4}$$

The eigenfunctions of the electronic Hamiltonian $\psi_n(\ldots \mathbf{r}_j \ldots, \ldots \mathbf{R}_\alpha \ldots)$ are a complete orthonormal basis in terms of which we may expand any molecular wave function. Specifically, the *exact eigenstates of the full Hamiltonian* (16.1), defined by

$$\hat{H}\,\Psi(\mathbf{r}, \mathbf{R}) = E\,\Psi(\mathbf{r}, \mathbf{R}) \tag{16.5}$$

---

[1] It is understood that this is not a very precise terminology, since the "electronic" part contains the mutual repulsion term of the nuclei. Nevertheless, this separation embodies the fact that all dynamical effects of the nuclei are generated by the "nuclear" part.

can always be expanded in terms of the ψ_n's as

$$\Psi(\mathbf{r},\ \mathbf{R}) = \sum_n \chi_n(\mathbf{R})\, \psi_n(\mathbf{r},\mathbf{R}). \qquad (16.6)$$

Here and in what follows $\mathbf{r}$ and $\mathbf{R}$ stand for the set of position operators of all electrons and all nuclei. Substituting (16.6) into the full Schroedinger equation and using (16.1) we obtain

$$\sum_n \left(\epsilon_n(\mathbf{R}) + \hat{T}_N\right) \chi_n(\mathbf{R})\, \psi_n(\mathbf{r},\mathbf{R}) = E \sum_n \chi_n(\mathbf{R})\, \psi_n(\mathbf{r},\mathbf{R}). \qquad (16.7)$$

Multiplying with $\psi_{n'}^*(\mathbf{r},\mathbf{R})$ and integrating with respect to the electron positions $\mathbf{r}$, we obtain

$$(E - \epsilon_{n'}(\mathbf{R}))\, \chi_{n'}(\mathbf{R}) = -\sum_n \int d^3r\, \psi_{n'}^*(\mathbf{r},\mathbf{R}) \sum_\alpha \frac{\hbar^2}{2M_\alpha} \nabla_{R_\alpha}^2 \left(\chi_n(\mathbf{R})\psi_n(\mathbf{r},\mathbf{R})\right).$$

$$(16.8)$$

The integration $\int d^3r$ stands symbolically for the multiple integration of all electron positions. In order to proceed we note the identity

$$\nabla_R^2(\chi\psi) = (\nabla_R^2\chi)\psi + \chi(\nabla_R^2\psi) + 2(\nabla_R\chi)\cdot(\nabla_R\psi) \qquad (16.9)$$

in terms of which the right-hand side of (16.8) becomes

$$-\sum_\alpha \frac{\hbar^2}{2M_\alpha} \sum_n \int d^3r\, \psi_{n'}^* \left((\nabla_{R_\alpha}^2\chi)\psi + \phi(\nabla_{R_\alpha}^2\psi) + 2(\nabla_{R_\alpha}\chi)\cdot(\nabla_{R_\alpha}\psi)\right)$$

$$= -\sum_\alpha \frac{\hbar^2}{2M_\alpha}\nabla_{R_\alpha}^2 \chi_{n'}(\mathbf{R}) + \Delta_{n'}, \qquad (16.10)$$

where Δ_n stands for

$$\Delta_{n'} \equiv -\sum_\alpha \frac{\hbar^2}{2M_\alpha} \sum_n \int d^3r\, \psi_{n'}^* \left(\chi(\nabla_R^2\psi) + 2(\nabla_R\chi)\cdot(\nabla_R\psi)\right). \qquad (16.11)$$

We finally arrive at a modified nuclear Schroedinger equation

$$\left(\hat{T}_N + \epsilon_{n'}(\mathbf{R})\right) \chi_{n'}(\mathbf{R}) = E\, \chi_{n'}(\mathbf{R}) - \Delta_{n'}. \qquad (16.12)$$

Thus, the molecule eigenvalue problem has been finally put in the form of the following set of equations

$$\left\{\frac{-\hbar^2}{2m_e}\sum_j \nabla_j^2 + \sum_{i>j}\frac{e^2}{|\mathbf{r}_i-\mathbf{r}_j|} - \sum_{\alpha,j}\frac{Z_\alpha e^2}{|\mathbf{R}_\alpha-\mathbf{r}_j|}\right\}\psi_n(\mathbf{r},\mathbf{R}) =$$

$$\left(\epsilon_n(\mathbf{R}) - \sum_{\alpha,\beta}\frac{Z_\alpha Z_\beta e^2}{|\mathbf{R}_\alpha-\mathbf{R}_\beta|}\right)\psi_n(\mathbf{r},\mathbf{R})$$

$$\left(\hat{T}_N + \epsilon_n(\mathbf{R})\right)\chi_n(\mathbf{R}) = E\,\chi_n(\mathbf{R}) - \Delta_n \qquad (16.13)$$

$$\Psi(\mathbf{r},\mathbf{R}) = \sum_n \chi_n(\mathbf{R})\,\psi_n(\mathbf{r},\mathbf{R})$$

$$\Delta_{n'} \equiv -\sum_\alpha \frac{\hbar^2}{2M_\alpha}\sum_n \int d^3r\,\psi_{n'}^*\left(\chi_n(\nabla_R^2\psi_n) + 2(\nabla_R\chi_n)\cdot(\nabla_R\psi_n)\right).$$

The term Δ_n is naturally of order m_e/M_N or smaller and to the leading Born–Oppenheimer approximation it may be ignored. We may also assume that $\psi_n(\mathbf{r},\mathbf{R})$ do not have a strong dependence on the nuclei positions, i.e. $\nabla_{R_\alpha}\psi_n \approx 0$, at least near the minimum of the electron energy eigenvalues

$$\nabla_R\{\epsilon_n(\mathbf{R})\}|_{\mathbf{R}_0} = 0. \qquad (16.14)$$

Thus, ignoring the Δ_n's we can write the Schroedinger equation for the nuclei wave functions as

$$\left(\hat{T}_N + \epsilon_n(\mathbf{R})\right)\chi_n(\mathbf{R}) = E\,\chi_n(\mathbf{R}). \qquad (16.15)$$

Furthermore, expanding the electron energy eigenvalues

$$\epsilon_n(\mathbf{R}) \approx \epsilon_n(\mathbf{R}_0) + \frac{1}{2}\sum_\alpha (X_{ai}-X_{ai}^{(0)})(X_{aj}-X_{aj}^{(0)})\left(\frac{\partial^2\epsilon_n}{\partial X_{ia}\partial X_{aj}}\right)_{\mathbf{R}_0} + \cdots \qquad (16.16)$$

we may conclude that, within these approximations, the nuclei move under the influence of harmonic forces. Estimating the characteristic order of magnitude of electronic energies from the uncertainty principle

$$\epsilon \sim \frac{(\hbar/R)^2}{2m_e}, \qquad (16.17)$$

we may conclude from (16.16) that the *vibrational energies* of molecules will be of the order of

$$M\omega^2 = \frac{\partial^2\epsilon}{\partial R^2} \sim \frac{\hbar^2}{m_e R^4}, \qquad (16.18)$$

leading to frequencies

$$\omega \sim \left(\frac{m_e}{M}\right)^{1/2} \frac{\hbar^2}{m_e R^2} \, . \tag{16.19}$$

The ratio of vibrational energies to electronic energies will be

$$\frac{\epsilon_{osc}}{\epsilon} \sim \frac{\hbar\omega}{\epsilon} \sim \left(\frac{m_e}{M}\right)^{1/2} \sim O(10^{-1}) \, . \tag{16.20}$$

Finally, the order of magnitude of *molecular rotational energies* will be

$$\epsilon_R \sim \frac{\ell(\ell+1)\hbar^2}{2I} \sim \frac{\hbar^2}{M R^2} \tag{16.21}$$

and

$$\frac{\epsilon_R}{\epsilon} \sim \frac{m_e}{M} \sim O(10^{-3}) \, . \tag{16.22}$$

Thus, molecules display a triple hierarchy in their spectrum

$$\epsilon_R \ll \epsilon_{osc} \ll \epsilon \tag{16.23}$$

with corresponding wavelengths in the microwave (rotational), infrared (vibrational), and ultraviolet (electronic) regions.

16.2 The Hydrogen Ion H_2^+

The simplest example of the application of the Born–Oppenheimer procedure is on the formation of the Hydrogen ion H_2^+ consisting of two protons and a single electron [1]. The exact Hamiltonian is

$$\hat{H} = \frac{P_1^2}{2M_p} + \frac{P_2^2}{2M_p} + \frac{p^2}{2m_e} - \frac{e^2}{|\mathbf{r} - \mathbf{R}_1|} - \frac{e^2}{|\mathbf{r} - \mathbf{R}_2|} + \frac{e^2}{|\mathbf{R}_1 - \mathbf{R}_2|} \, . \tag{16.24}$$

Applying the Born–Oppenheimer procedure we write down the Schroedinger equation for the electronic part

$$\left\{ -\frac{\hbar^2}{2m_e} \nabla^2 - \frac{e^2}{|\mathbf{r} - \mathbf{R}_1|} - \frac{e^2}{|\mathbf{r} - \mathbf{R}_2|} + \frac{e^2}{|\mathbf{R}_1 - \mathbf{R}_2|} \right\} \psi_n(\mathbf{r}, \mathbf{R}) = \varepsilon_n(\mathbf{R}) \, \psi_n(\mathbf{r}, \mathbf{R}) \tag{16.25}$$

while determining the proton positions from the minimization equation

$$\frac{\partial \varepsilon_n}{\partial \mathbf{R}} = 0 \, . \tag{16.26}$$

Within the lowest order Born–Oppenheimer approximation, where the nuclei positions are parameters, we may take the middle of the proton distance as the origin of the coordinate system and simplify the potential as

$$-\frac{e^2}{|\mathbf{r} - \mathbf{R}/2|} - \frac{e^2}{|\mathbf{r} + \mathbf{R}/2|} + \frac{e^2}{R}.$$

(16.27)

Then, the electronic Schroedinger equation is

$$\left\{ -\frac{\hbar^2}{2m_e}\nabla^2 - \frac{e^2}{|\mathbf{r} - \mathbf{R}/2|} - \frac{e^2}{|\mathbf{r} + \mathbf{R}/2|} + \frac{e^2}{R} \right\} \psi_n(\mathbf{r},\ \mathbf{R}) = \varepsilon_n(\mathbf{R})\, \psi_n(\mathbf{r},\ \mathbf{R}).$$

(16.28)

Although this is the lowest order approximation in the Born–Oppenheimer expansion procedure, it is still a rather complicated problem. The way to proceed is to introduce an ansatz for the ground state, calculate the corresponding ground state energy and apply the minimization (16.26). This would be a simple application of the *Variational Approximation Method* to be described in a systematic way in a subsequent chapter.

As an ansatz we may adopt a wave function that represents the superposition of two extreme situations, namely to the cases in which the electron is attached to either one of the nuclei, being in the ground state, while ignoring the existence of the other nucleus. Since the Hamiltonian is parity invariant, we may choose the adopted ansatz to have a definite parity being even or odd. Therefore, we introduce the trial wave function

$$\psi_\pm(\mathbf{r}) = N_\pm\left(\psi_1(\mathbf{r} - \mathbf{R}/2) \pm \psi_1(\mathbf{r} + \mathbf{R}/2) \right),$$

(16.29)

where

$$\psi_1(r) = \frac{e^{-r/a_0}}{\sqrt{\pi a_0^3}}.$$

(16.30)

More explicitly, we have

$$\psi_\pm(\mathbf{r}) = \frac{N_\pm}{\sqrt{\pi a_0^3}}\left(e^{-|\mathbf{r}-\mathbf{R}/2|/a_0} \pm e^{-|\mathbf{r}+\mathbf{R}/2|/a_0} \right).$$

(16.31)

The normalization factor can be calculated to be

$$N_\pm = \frac{1}{\sqrt{2(1 \pm S)}} \quad with \quad S = \left(1 + \frac{R}{a_0} + \frac{R^2}{3a_0^2} \right) e^{-R/a_0}.$$

(16.32)

The expectation value of the electronic energy in this state will be

$$\varepsilon_n\pm(\mathbf{R}) = \int d^3r\, \psi_\pm^*(\mathbf{r})\left(-\frac{\hbar^2}{2m_e}\nabla^2 - \frac{e^2}{|\mathbf{r} - \mathbf{R}/2|} - \frac{e^2}{|\mathbf{r} + \mathbf{R}/2|} + \frac{e^2}{R} \right)\psi_\pm(\mathbf{r}).$$

(16.33)

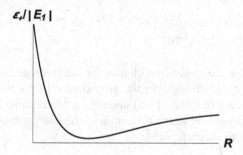

Fig. 16.1 Plot of $\varepsilon_+/|E_1|$

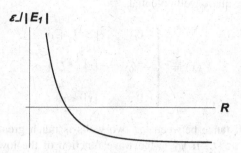

Fig. 16.2 Plot of $\varepsilon_-/|E_1|$

Substituting our ansatz, we obtain after some calculation

$$\varepsilon_\pm(R) = 2N_\pm^2 \left\{ E_1 + \frac{e^2}{R}e^{-2R/a_0}\left(1 + \frac{R}{a_0}\right) \pm \left(E_1 + \frac{e^2}{R}\right)S \mp \frac{e^2}{a_0}\left(1 + \frac{R}{a_0}\right)e^{-R/a_0} \right\}$$
$$(16.34)$$

where $E_1 = -e^2/2a_0$ is the Hydrogen atom ground state energy. Plotting $\varepsilon_\pm/E_1$ in Figs. 16.1 and 16.2, we see that only the even branch possesses a minimum. Thus, the ground state wave function of the H_2^+ ion is approximated by $\psi_+(\mathbf{r})$ and the estimated ground state energy is $\varepsilon_+ \approx -1.76\,\text{eV}$, while the internuclear size of the ion is at $R_0 \approx 1.3\,\text{Å}$. The experimental values are not very far, being $\varepsilon = -2.8\,\text{eV}$ and $R = 1.08\,\text{Å}$.

Problems and Exercises

16.1 Calculate the expectation value of the energy of the Hydrogen ion $\epsilon_\pm$ for the trial wave function

$$\psi_{\pm}(\mathbf{r}) = \frac{N_{\pm}}{\sqrt{\pi a_0^3}} \left(e^{-|\mathbf{r}-\mathbf{R}/2|/a_0} \pm e^{-|\mathbf{r}+\mathbf{R}/2|/a_0} \right) .$$

16.2 Write down the electronic Hamiltonian for the Hydrogen molecule. Assume that an acceptable approximation for the ground state of the molecule is a properly symmetrized wave function of two mutually noninteracting electrons, each in a Hydrogen ground state. Find the correction to the energy in the presence of an external homogeneous magnetic field.

16.3 *A one-dimensional molecule.* Consider a particle moving in one dimension and subject to a double square well potential

$$V(x) = \begin{cases} 0 & |x| > L+a \\ -V_0 & a < |x| < L+a \\ 0 & |x| < a \end{cases}$$

Assume that the distance between the two wells is much greater than the width of each well, i.e., $a \gg L$. If ψ_1 is the wave function of the lowest energy bound state localized in the left well and ψ_2 is the wave function of the lowest energy bound state localized in the right well, with corresponding energies $E_1 = E_2 = E_0$, calculate the expectation value of the energy for each of the trial wave functions $\psi = \frac{1}{\sqrt{2}} (\psi_1 \pm \psi_2)$, in terms of the integral $A = \int dx \, \psi_1 \hat{H} \psi_2$. You may assume that the overlap of ψ_1 and ψ_2 is very small, so that $A \ll E_0$. Which of the two choices corresponds to the ground state of the system?

References

1. G. Baym, *Lectures in Quantum Mechanics*, Lecture Notes and Supplements in Physics (ABP, 1969)
2. E. Merzbacher, *Quantum Mechanics*, 3rd edn. (Wiley, Hoboken, 1998)

Chapter 17
Particle Interactions with EM Fields

17.1 General Considerations

We have already discussed simple cases of the motion of charged particles under the influence of static electric and magnetic fields, as is the case of Coulomb force or the case of the magnetic dipole moment coupling to a homogeneous magnetic field. In these cases, the electromagnetic fields were treated classically while the particles were quantized. Such a framework can only be approximate, since ultimately the electromagnetic field has to be quantized as well. Despite that, it turns out it is a reliable approximation within the realm of non-relativistic Quantum Mechanics. Such is the case of the Hydrogen atom with characteristic energies of $O(10\,\mathrm{eV})$, much smaller than energies $m_e c^2 \sim O(10^5\,\mathrm{eV})$. In contrast, the explanation of phenomena outside this realm, i.e., with characteristic energy scales of $O(mc^2)$, can be achieved only in the framework of a theory with an infinite number of degrees of freedom, namely a (relativistic) Quantum Field Theory. In such a theoretical framework both the matter particles (electrons, protons, etc.) and electromagnetic radiation (photons) are treated quantum mechanically. In the present chapter, we shall remain within the framework of one or more quantized particles (electrons or others) interacting with a classical electromagnetic field, leaving the case of the fully quantized system of particles and photons for a later chapter.

Let's begin considering the classical system of a particle of charge e interacting with an electromagnetic field. Recall that the electric field $\mathbf{E}(\mathbf{r}, t)$ and the magnetic field $\mathbf{B}(\mathbf{r}, t)$ can be expressed in terms of the *electromagnetic potentials* $\phi(\mathbf{r}, t)$ (scalar potential) and $\mathbf{A}(\mathbf{r}, t)$ (vector potential) as[1]

---

[1] Electric and magnetic fields satisft *Maxwell's equations*

$$\nabla \cdot \mathbf{E} = 4\pi\rho_e, \quad \nabla \times \mathbf{E} = -\frac{1}{c}\frac{\partial \mathbf{B}}{\partial t}$$

$$\nabla \cdot \mathbf{B} = 0, \quad \nabla \times \mathbf{E} = \frac{4\pi}{c}\mathbf{J}_e + \frac{1}{c}\frac{\partial \mathbf{E}}{\partial t}$$

The pair of homogeneous Maxwell's equations are *"solved"* by the potentials. The inhomogeneous ones imply local charge conservation expressed by the continuity equation $\nabla \cdot \mathbf{J}_e + \frac{\partial \rho_e}{\partial t} = 0$.

© Springer Nature Switzerland AG 2019
K. Tamvakis, *Basic Quantum Mechanics*, Undergraduate Texts in Physics,
https://doi.org/10.1007/978-3-030-22777-7_17

$$\mathbf{E} = -\nabla\phi - \frac{1}{c}\frac{\partial\mathbf{A}}{\partial t}, \quad \mathbf{B} = \nabla\times\mathbf{A}. \tag{17.1}$$

The electric and magnetic fields are invariant under *gauge transformations* of the potentials

$$\phi \to \phi - \frac{1}{c}\frac{\partial\Lambda}{\partial t}, \quad \mathbf{A} \to \mathbf{A} + \nabla\Lambda \tag{17.2}$$

in terms of an arbitrary function Λ. The classical equation of motion for the particle is simply Newton's equation with the *Lorentz force* in the right-hand side

$$m\frac{d^2\mathbf{r}}{dt^2} = e\left(\mathbf{E} + \frac{1}{c}\mathbf{v}\times\mathbf{B}\right) \tag{17.3}$$

c stands for the velocity of light.[2] It is not difficult to show that the Lorentz force law (17.3) emerges from a Lagrangian

$$L = \frac{1}{2}mv^2 + \frac{e}{c}\mathbf{v}\cdot\mathbf{A} - e\phi. \tag{17.4}$$

Indeed, we have

$$\frac{d}{dt}\left(\frac{\partial L}{\partial v_i}\right) = m\frac{dv_i}{dt} + \frac{e}{c}\frac{dA_i}{dt} = m\frac{d^2x_i}{dt^2} + \frac{e}{c}\frac{\partial A_i}{\partial t} + \frac{e}{c}\frac{dx_j}{dt}\frac{\partial A_i}{\partial x_j}$$

$$\frac{\partial L}{\partial x_i} = \frac{e}{c}v_j\frac{\partial A_j}{\partial x_i} - e\frac{\partial\phi}{\partial x_i}$$

Therefore, Lagrange's equations read

$$\frac{d}{dt}\left(\frac{\partial L}{\partial v_i}\right) - \frac{\partial L}{\partial x_i} = 0 \implies \begin{cases} m\frac{d^2x_i}{dt^2} = -e\nabla_i\phi + \frac{e}{c}\frac{\partial A_i}{\partial t} + \frac{e}{c}\left(v_j\nabla_j A_i - v_j\nabla_i A_j\right) \\ \text{or} \\ m\frac{d^2x_i}{dt^2} = eE_i + \frac{e}{c}\left(\mathbf{v}\times\mathbf{B}\right)_i \end{cases}$$

The Lagrangian (17.4) is invariant under the above considered *gauge transformations*. This goes as follows

$$L \to L + \frac{e}{c}\mathbf{v}\cdot\nabla\Lambda + \frac{e}{c}\frac{\partial\Lambda}{\partial t} = L + \frac{e}{c}\frac{d\Lambda}{dt}. \tag{17.5}$$

---

[2]In the system of units used fields and charges are related to those in the MKSA system as

$$q' = q\sqrt{4\pi\epsilon_0}, \ E' = E/\sqrt{4\pi\epsilon_0}, \ B' = B\sqrt{\mu_0/4\pi}$$

and $c = 1/\sqrt{\mu_0\epsilon_0}$.

A total time derivative has no contribution to the Action or to the equations of motion and, therefore, the Lagrangian is invariant.

Dynamics can equally well be described in terms of the Hamiltonian function of momenta and positions. The *canonical momentum* of the particle is defined in terms of the Lagrangian as

$$\mathbf{p} = \frac{\partial L}{\partial \mathbf{v}} = m\mathbf{v} + \frac{e}{c}\mathbf{A}. \tag{17.6}$$

It is important to note that the momentum of a charged particle in a magnetic field is not just its velocity times its mass. An extra *magnetic momentum* $\mathbf{p}_m = \frac{e}{c}\mathbf{A}$ is added by the presence of the magnetic field. The Hamiltonian can now be obtained from its standard definition

$$H(\mathbf{r}, \mathbf{p}) = \mathbf{p} \cdot \mathbf{v} - L = \frac{1}{m}\mathbf{p} \cdot \left(\mathbf{p} - \frac{e}{c}\mathbf{A}\right) - \frac{1}{2m}\left(\mathbf{p} - \frac{e}{c}\mathbf{A}\right)^2 - \frac{e}{mc}\left(\mathbf{p} - \frac{e}{c}\mathbf{A}\right) \cdot \mathbf{A} + e\phi \tag{17.7}$$

or

$$H(\mathbf{r}, \mathbf{p}) = \frac{1}{2m}\left(\mathbf{p} - \frac{e}{c}\mathbf{A}\right)^2 + e\phi. \tag{17.8}$$

Quantization proceeds in a straightforward fashion by the standard replacement of the classical physical variables with operators

$$x_j \rightarrow \hat{x}_j = x_j, \quad p_j \rightarrow \hat{p}_j = -i\hbar\nabla_j \tag{17.9}$$

obeying canonical commutation relations

$$[x_i, \hat{p}_j] = i\hbar\delta_{ij}, \quad [x_i, x_j] = [\hat{p}_i, \hat{p}_j] = 0. \tag{17.10}$$

However, note that the velocity operators $\mathbf{v} = (\mathbf{p} - e\mathbf{A}/c)/m$ obey nontrivial commutation relations with different components having a nonzero commutator

$$\left[\hat{v}_i, \hat{v}_j\right] = -\frac{i\hbar e}{cm^2}\left(\nabla_i A_j - \nabla_j A_i\right) = -\frac{i\hbar e}{cm^2}\epsilon_{ijk}\left(\nabla \times \mathbf{A}\right)_k \tag{17.11}$$

or

$$\left[\hat{v}_i, \hat{v}_j\right] = -\frac{i\hbar e}{cm^2}\epsilon_{ijk}B_k. \tag{17.12}$$

The Schroedinger equation is

$$\left\{\frac{1}{2m}\left(-i\hbar\nabla - \frac{e}{c}\mathbf{A}\right)^2 + e\phi\right\}\psi = i\hbar\frac{\partial\psi}{\partial t}. \tag{17.13}$$

It is very important to note that, in contrast to the classical case, the quantum dynamics of a charged particle is not determined by the electric and magnetic fields alone but depends directly on the electromagnetic potentials. Another important point is that the Schroedinger equation is invariant under a gauge transformation provided that the wave function changes as

$$\psi \rightarrow e^{\frac{ie}{\hbar c}\Lambda}\psi.$$ (17.14)

We'll come back to this point in a subsequent section.

Local conservation of probability can be formulated in terms of the probability density $|\psi|^2$ and a probability current density suitably defined. The correct gauge-invariant probability current can be deduced recalling that the corresponding current in the absence of electromagnetic fields could also be written as

$$\mathbf{J} = \frac{\hbar}{2mi}\left(\psi^*\nabla\psi - \psi\nabla\psi^*\right) == \frac{1}{m}Re\left(\psi^*\mathbf{p}\psi\right).$$

Generalizing this expression to

$$\mathbf{J} = \frac{1}{m}Re\left(\psi^*\left(\mathbf{p} - \frac{e}{c}\mathbf{A}\right)\psi\right)$$ (17.15)

we obtain

$$\mathbf{J} = \frac{\hbar}{2mi}\left(\psi^*\nabla\psi - \psi\nabla\psi^* - \frac{ie}{c\hbar}\mathbf{A}|\psi|^2\right).$$ (17.16)

In terms of this current, we have local conservation of probability through a gauge-invariant continuity equation

$$\nabla\cdot\mathbf{J} + \frac{\partial|\psi|^2}{\partial t} = 0.$$ (17.17)

Example 17.1 Consider a particle of charge $e < 0$ in a homogeneous magnetic field $\mathbf{B}$ and calculate the expectation values $\langle\mathbf{r}\rangle_t$ at any time t in an arbitrary normalizable state. Take as a vector potential the symmetric choice $\mathbf{A} = \frac{B}{2}\left(x\hat{y} - y\hat{x}\right)$.

Starting with the Hamiltonian in the form

$$\hat{H} = \frac{m}{2}v^2 \quad with \quad \mathbf{v} = \frac{1}{m}\left(\mathbf{p} - \frac{e}{c}\mathbf{A}\right).$$ (17.18)

From Heisenberg's equation, we obtain

$$\frac{d\hat{v}_i}{dt} = \frac{i}{\hbar}\left[\hat{H}, \hat{v}_i\right] = \cdots = -\frac{eB}{mc}\epsilon_{ij3}\hat{v}_j$$ (17.19)

or

$$\frac{d\hat{v}_x}{dt} = \omega\,\hat{v}_y, \quad \frac{d\hat{v}_y}{dt} = -\omega\,\hat{v}_x, \quad \frac{d\hat{v}_z}{dt} = 0,$$ (17.20)

where $\omega \equiv -\frac{eB}{mc}$. Note that these equations are linear and, therefore, will also be satisfied by the corresponding expectation values. The longitudinal motion is solved trivially to give

$$\hat{v}_z(t) = \hat{v}_z(0) \implies z(t) = z(0) + \hat{v}_z(0)t$$ (17.21)

or

$$\langle z \rangle_t = \langle z \rangle_0 + \langle \hat{p}_z \rangle_0 \frac{t}{m}. \tag{17.22}$$

Diferentiating the transverse equations once more we decouple them and obtain

$$\frac{d^2 \hat{v}_x}{dt^2} = -\omega^2 \, \hat{v}_x$$
$$\frac{d^2 \hat{v}_y}{dt^2} = -\omega^2 \, \hat{v}_y \tag{17.23}$$

Their solutions are

$$\hat{v}_x(t) = \hat{v}_x(0) \cos(\omega t) + C_x \, \sin(\omega t)$$
$$\hat{v}_y(t) = \hat{v}_y(0) \cos(\omega t) + C_y \, \sin(\omega t) \tag{17.24}$$

The constants C_x, C_y are obtained by substituting these solutions in the original coupled equations. Thus, we obtain

$$\hat{v}_x(t) = \hat{v}_x(0) \cos(\omega t) + \hat{v}_y(0) \, \sin(\omega t)$$
$$\hat{v}_y(t) = \hat{v}_y(0) \cos(\omega t) - \hat{v}_x(0) \, \sin(\omega t) \tag{17.25}$$

or equivalently

$$\hat{v}_x(t) = \tfrac{1}{m} \left(\hat{p}_x(0) + \tfrac{eB}{2c} y(0) \right) \cos(\omega t) + \tfrac{1}{m} \left(\hat{p}_y(0) - \tfrac{eB}{2c} x(0) \right) \sin(\omega t)$$
$$\hat{v}_y(t) = \tfrac{1}{m} \left(\hat{p}_y(0) - \tfrac{eB}{2c} x(0) \right) \cos(\omega t) - \tfrac{1}{m} \left(\hat{p}_x(0) + \tfrac{eB}{2c} y(0) \right) \sin(\omega t) \tag{17.26}$$

Integrating we obtain

$$x(t) = x(0) + \tfrac{1}{m\omega} \left(\hat{p}_x(0) + \tfrac{eB}{2c} y(0) \right) \sin(\omega t) - \tfrac{1}{m\omega} \left(\hat{p}_y(0) - \tfrac{eB}{2c} x(0) \right) (\cos(\omega t) - 1)$$
$$y(t) = y(0) + \tfrac{1}{m\omega} \left(\hat{p}_y(0) - \tfrac{eB}{2c} x(0) \right) \sin(\omega t) + \tfrac{1}{m\omega} \left(\hat{p}_x(0) + \tfrac{eB}{2c} y(0) \right) (\cos(\omega t) - 1) \tag{17.27}$$

Thus, we finally arrive at the expectation values

$$\langle x \rangle_t = \langle x \rangle_0 + \tfrac{1}{m\omega} \left(\langle p_x \rangle_0 - \tfrac{m\omega}{2} \langle y \rangle_0 \right) \sin(\omega t) - \tfrac{1}{m\omega} \left(\langle p_y \rangle_0 + \tfrac{m\omega}{2} \langle x \rangle_0 \right) (\cos(\omega t) - 1)$$

$$\langle y \rangle_t = \langle y \rangle_0 + \tfrac{1}{m\omega} \left(\langle p_y \rangle_0 + \tfrac{m\omega}{2} \langle x \rangle_0 \right) \sin(\omega t) + \tfrac{1}{m\omega} \left(\langle p_x \rangle_0 - \tfrac{m\omega}{2} \langle y \rangle_0 \right) (\cos(\omega t) - 1) \tag{17.28}$$

The motion described by these expectation values is a circle in the transverse plane and coincides with the motion of the classical particle. Combining this with the longitudinal motion we obtain a circular helix with its principal axis along the magnetic field. These equations simplify in case our initial state has a definite parity and $\langle \mathbf{r} \rangle_0 = 0$. Then, we have

$$\left(\langle x \rangle_t - \frac{\langle p_y \rangle_0}{m\omega} \right)^2 + \left(\langle y \rangle_t + \frac{\langle p_x \rangle_0}{m\omega} \right)^2 = \frac{1}{(m\omega)^2} \left(\langle p_x \rangle_0^2 + \langle p_y \rangle_0^2 \right) \tag{17.29}$$

and $\langle z \rangle_t = \frac{\langle p_z \rangle_0}{m} t$. The transverse circle has its center at the point $\left(\frac{\langle p_y \rangle_0}{m\omega}, -\frac{\langle p_x \rangle_0}{m\omega} \right)$ and has a radius $\left(\langle p_x \rangle_0^2 + \langle p_y \rangle_0^2 \right)^{1/2} / m\omega$.

17.2   Landau Levels

In this section, we shall consider the system of a charged particle in a homogeneous, i.e., constant in space and time, magnetic field **B** and solve the energy eigenvalue problem obtaining the energy eigenvalues and eigenfunctions [1, 2]. Since there is no electric field present the corresponding scalar potential appearing in the Hamiltonian (17.8) will be set to zero. There is a relative freedom[3] in choosing the vector potential **A** that yields the magnetic field **B**. A suitable choice is

$$\mathbf{A} = \frac{1}{2}\mathbf{B} \times \mathbf{r}. \tag{17.30}$$

Taking $\mathbf{B} = \hat{z}B$ corresponds to

$$\mathbf{A} = \frac{1}{2}B\left(x\hat{y} - y\hat{x}\right). \tag{17.31}$$

Substituting into the Hamiltonian we obtain

$$\hat{H} = \frac{1}{2m}\left(\left(\hat{p}_x + \frac{eB}{2c}y\right)^2 + \left(\hat{p}_y - \frac{eB}{2c}x\right)^2 + \hat{p}_z^2\right). \tag{17.32}$$

The Hamiltonian is the sum of two commuting parts, describing the motion in the (x, y) plane on the one hand and the (free) motion along the z-direction on the other

$$\hat{H} = \hat{H}_\perp + \hat{H}_\| \implies \begin{cases} \hat{H}_\perp = \frac{1}{2m}\left(\hat{p}_x + \frac{eB}{2c}y\right)^2 + \frac{1}{2m}\left(\hat{p}_y - \frac{eB}{2c}x\right)^2 \\ \hat{H}_\| = \frac{\hat{p}_z^2}{2m} \end{cases}$$
$$\tag{17.33}$$

The eigenfunctions will be products

$$\psi_E(\mathbf{r}) = \psi^{(\perp)}(x, y)\frac{e^{ikz}}{\sqrt{2\pi}} \tag{17.34}$$

while the eigenvalues will be sums of the eigenvalues of the *"transverse"* part $E_\perp$ plus the *"longitudinal"* energy $\hbar^2 k^2/2m$

$$E = E_\perp + \frac{\hbar^2 k^2}{2m}. \tag{17.35}$$

In what follows we proceed to solve the transverse eigenvalue problem ignoring the longitudinal free motion or, equivalently, restricting ourselves to the $k = 0$ case. The structure of the transverse Hamiltonian motivates the introduction of the hermitian operators

---

[3]i.e., freedom under gauge transformations.

$$\hat{Q} = \frac{1}{\sqrt{2}} \left(y + \frac{2c}{eB} \hat{p}_x \right), \quad \hat{P} = \frac{1}{\sqrt{2}} \left(\hat{p}_y - x \frac{eB}{2c} \right), \tag{17.36}$$

which satisfy a canonical commutation relation

$$\left[\hat{Q}, \hat{P} \right] = i\hbar. \tag{17.37}$$

Furthermore, the transverse Hamiltonian takes the form

$$\hat{H}_\perp = \frac{\hat{P}^2}{m} + m \left(\frac{eB}{2mc} \right)^2 \hat{Q}^2 \tag{17.38}$$

or, introducing the parameter with dimensions of frequency

$$\omega = \frac{|e|B}{2mc} \tag{17.39}$$

we have

$$\hat{H}_\perp = 2 \left(\frac{\hat{P}^2}{2m} + \frac{1}{2} m\omega^2 \hat{Q}^2 \right). \tag{17.40}$$

This is twice the Hamiltonian of a simple harmonic oscillator. We may introduce creation/annihilation operators in the standard fashion as

$$\hat{a} = \sqrt{\frac{m\omega}{2\hbar}} \hat{Q} + \frac{i}{\sqrt{2m\hbar\omega}} \hat{P}, \quad \hat{a}^\dagger = \sqrt{\frac{m\omega}{2\hbar}} \hat{Q} - \frac{i}{\sqrt{2m\hbar\omega}} \hat{P} \tag{17.41}$$

obeying the standard commutation relation

$$\left[\hat{a}, \hat{a}^\dagger \right] = 1. \tag{17.42}$$

In terms of them the transverse Hamiltonian is

$$\hat{H}_\perp = 2\hbar\omega \left(\hat{a}^\dagger \hat{a} + \frac{1}{2} \right). \tag{17.43}$$

Its eigenvalues are twice the standard harmonic oscillator eigenvalues

$$E_{\perp n} = \hbar\omega (2n + 1) \quad \text{with} \quad n = 0, 1, 2, \ldots \tag{17.44}$$

Thus, the spectrum of the transverse Hamiltonian is the above discrete set of energy levels, the so-called *Landau levels*.

At this point, it is important to note that the Hamiltonian commutes with the angular momentum along the z-direction. In fact, we can write

$$\hat{H} = \frac{\mathbf{p}^2}{2m} - \frac{eB}{2mc}\hat{L}_z + \frac{e^2 B^2}{8mc^2}(x^2 + y^2) \qquad (17.45)$$

and, therefore,

$$\left[\hat{H}, \hat{L}_z\right] = \left[\hat{H}_\perp, \hat{L}_z\right] = 0. \qquad (17.46)$$

Thus, symbolizing the common eigenstates of the transverse Hamiltonian and angular momentum as $|n, M\rangle$, the ground state $|0, M\rangle$ is defined as

$$\hat{a}|0, M\rangle = \frac{\eta}{\sqrt{4\hbar m\omega}}\left\{\hat{p}_x + i\eta\,\hat{p}_y - im\omega\,(x + i\eta\,y)\right\}|0, M\rangle = 0, \qquad (17.47)$$

where η is the sign of the charge e. In the $\{x, y\}$-reprsentation, this reads

$$\frac{\eta}{\sqrt{4\hbar m\omega}}\left\{-i\hbar\left(\nabla_x + i\eta\,\nabla_y\right) - im\omega\,(x + i\eta\,y)\right\}\psi_{0,M}^{(\perp)}(x, y) = 0. \qquad (17.48)$$

At this point, it is convenient to consider separately the cases of a particle with positive or negative charge.

A particle of positive charge. For $\eta = +1$ the last equation is

$$\frac{1}{\sqrt{4\hbar m\omega}}\left\{-i\hbar\left(\nabla_x + i\,\nabla_y\right) - im\omega\,(x + i\,y)\right\}\psi_{0,M}^{(\perp)}(x, y) = 0. \qquad (17.49)$$

Introducing polar coordinates this is rewritten as

$$\frac{e^{i\phi}}{\sqrt{4m\hbar\omega}}\left[-i\hbar\left(\frac{\partial}{\partial\rho} + \frac{i}{\rho}\frac{\partial}{\partial\phi}\right) - im\omega\,\rho\right]\psi_{0,M}^{(\perp)}(\rho, \phi) = 0. \qquad (17.50)$$

However, since $\psi_{0,M}^{(\perp)}$ is an eigenfunction of $\hat{L}_z$,

$$-i\hbar\frac{\partial\psi_{0,M}^{(\perp)}}{\partial\phi} = \hbar M\psi_{0,M}^{(\perp)} \implies \psi_{0,M}^{(\perp)}(\rho, \phi) = \psi_{0,M}^{(\perp)}(\rho, 0)\,e^{iM\phi}. \qquad (17.51)$$

Therefore, we have

$$\frac{-i\hbar e^{i(M+1)\phi}}{\sqrt{4\hbar m\omega}}\left[\frac{d}{d\rho} - \frac{M}{\rho} + \frac{m\omega}{\hbar}\rho\right]\psi_{0,M}^{(\perp)}(\rho, 0) = 0, \qquad (17.52)$$

which, after integrating, leads to the solution

$$\psi_{0,M}^{(\perp)}(\rho, \phi) = \mathcal{N}\,\rho^M\,e^{-\frac{m\omega}{2\hbar}\rho^2}\,e^{iM\phi}. \qquad (17.53)$$

The normalization factor can be calculated to be $\mathcal{N} = \sqrt{(m\omega/\hbar)^{M+1}/\pi M!}$. The absence of a singularity at the origin requires $M \geq 0$.

Excited energy eigenstates can be created from the ground state acting on it with $\hat{a}^{\dagger}$. The first excited state $\hat{a}^{\dagger}|0, M\rangle$ of energy $3\hbar\omega$ corresponds to the eigenfunction

$$\psi_1^{(\perp)}(\rho, \phi) = \tilde{\mathcal{N}} \, e^{i(M-1)\phi} \left(\frac{M}{\rho} - \frac{m\omega}{\hbar}\rho \right) \rho^M e^{-\frac{m\omega}{2\hbar}\rho^2} . \tag{17.54}$$

Note that this is an eigenfunction of angular momentum $\hbar(M-1)$. Thus, the ascending energy eigenfunctions, defined by repeated n applications of the creation operator $\hat{a}^{\dagger}$, correspond to eigenstates of angular momentum $\hbar(M-n)$

$$\frac{(\hat{a}^{\dagger})^n}{\sqrt{n!}}|0, M\rangle = |n, M-n\rangle . \tag{17.55}$$

This is a consequence of the commutation rules

$$\left[\hat{a}^{\dagger}, \hat{L}_z \right] = \hbar\hat{a}^{\dagger}, \quad \left[(\hat{a}^{\dagger})^2, \hat{L}_z \right] = 2\hbar \, (\hat{a}^{\dagger})^2, \dots \tag{17.56}$$

or generally

$$\left[(\hat{a}^{\dagger})^n, \hat{L}_z \right] = \hbar n \, (\hat{a}^{\dagger})^n . \tag{17.57}$$

The obvious interpretation of Eq. (17.55) is that $\hat{a}^{\dagger}$ is a *anhillation operator for* $\hat{L}_z$, *although it is a creation operator for* $\hat{H}_{\perp}$. Each energy level corresponding to n can have angular momentum $-n, -n+1, \dots, +\infty$. The energy spectrum is infinitely degenerate.

A particle of negative charge. For $\eta = -1$ the ground state is defined by

$$-\frac{1}{\sqrt{4\hbar m\omega}} \left\{ -i\hbar \left(\nabla_x - i\,\nabla_y \right) - im\omega \, (x - i\,y) \right\} \psi_{0,M}^{(\perp)}(x, y) = 0 \tag{17.58}$$

or, introducing polar coordinates,

$$\frac{i\hbar}{\sqrt{4\hbar m\omega}} e^{-i\phi} \left(\frac{\partial}{\partial\rho} - \frac{i}{\rho}\frac{\partial}{\partial\phi} + \frac{m\omega}{\hbar}\rho \right) \psi_{0,M}^{(\perp)}(\rho, \phi) = 0 \tag{17.59}$$

or

$$\frac{i\hbar}{\sqrt{4\hbar m\omega}} e^{i(M-1)\phi} \left(\frac{d}{d\rho} + \frac{M}{\rho} + \frac{m\omega}{\hbar}\rho \right) \psi_{0,M}^{(\perp)}(\rho, 0) = 0 . \tag{17.60}$$

This is satisfied by

$$\psi_{0,M}^{(\perp)}(\rho, \phi) = \mathcal{N} \, \rho^{-M} e^{-\frac{m\omega}{2\hbar}\rho^2} e^{iM\phi} . \tag{17.61}$$

The absence of a singularity at the origin requires $M \le 0$. The first excited energy eigenstate is $|1\rangle = \hat{a}^{\dagger}|0, M\rangle$ and corresponds to the wave function

$$\psi_1^{(\perp)}(\rho, \phi) = \mathcal{N}' e^{i(M+1)\phi} \left(\frac{M}{\rho} + \frac{m\omega}{\hbar}\rho \right) \rho^{-M} e^{-\frac{m\omega}{2\hbar}\rho^2}. \qquad (17.62)$$

This is an eigenstate of angular momentum $\hbar(M + 1)$. The commutation relations

$$\left[(\hat{a}^\dagger)^n, \hat{L}_z \right] = -n\hbar (\hat{a}^\dagger)^n \qquad (17.63)$$

imply now that the states $(\hat{a}^\dagger)^n |0, M\rangle$ are $|n, M + n\rangle$ angular momentum eigenstates

$$\frac{(\hat{a}^\dagger)^n}{\sqrt{n!}} |0, M\rangle = |n, M + n\rangle. \qquad (17.64)$$

Thus, each energy level corresponding to n can have angular momentum quantum number values n, $n - 1$, $n - 2$, $\ldots - \infty$. Again the energy spectrum is infinitely degenerate.

The above picture of charged particles under the influence of a magnetic field is incomplete for electrons or fermions in general since we have not included the spin interaction part in the Hamiltonian

$$-\mathbf{m}_s \cdot \mathbf{B} = -\frac{egB}{2mc} S_z \qquad (17.65)$$

or, specifically for electrons in a homogeneous magnetic field, just $2\omega S_z$ in terms of the ω-definition (17.39). The energy eigenfunctions will be just products of the eigenfunctions obtained above times the S_z spin-up and spin-down eigenstates

$$\psi_E(\mathbf{r}) \chi = \psi_E(\mathbf{r}) \begin{cases} \chi_\uparrow \\ \\ \chi_\downarrow \end{cases} \qquad (17.66)$$

The eigenvalues will be shifted to

$$E_{\perp n} = \hbar\omega (2n + 1) \pm \hbar\omega. \qquad (17.67)$$

With the addition of spin, the system acquires an additional very interesting symmetry. The spin part of the Hamiltonian can be written

$$-\frac{egB}{2mc} \hat{S}_z = 2\hbar\omega \left(\hat{b}^\dagger \hat{b} - \frac{1}{2} \right), \qquad (17.68)$$

where

$$\hat{b} = \frac{1}{\hbar} \hat{S}_- = \frac{1}{\hbar} \left(\hat{S}_x - i \hat{S}_y \right). \qquad (17.69)$$

The operators $\hat{b}$, $\hat{b}^\dagger$ have the following properties

$$\hat{b}^2 = (\hat{b}^\dagger)^2 = 0 \quad and \quad \left\{\hat{b}, \hat{b}^\dagger\right\} = 1. \tag{17.70}$$

The operators $\hat{b}$, $\hat{b}^\dagger$ are annihilation and creation operators of $\bar{N} = \hat{b}^\dagger \hat{b}$ quanta *with the crucial difference that the only eigenvalues of $\bar{N}$ are 0 and 1*. Indeed, we have

$$\hat{b}|\dots,0\rangle = 0 \quad \hat{b}^\dagger|\dots,0\rangle = |\dots,1\rangle \tag{17.71}$$

and

$$(\hat{b}^\dagger)^2|\dots,1\rangle = \hat{b}^2|\dots,1\rangle = 0. \tag{17.72}$$

The quanta created by $\hat{b}^\dagger$ satisfy a kind of *Pauli principle*, since no states exist with more than one. The operators $\hat{b}$, $\hat{b}^\dagger$ commute with all operators that do not depend on the spin, e.g., $[\hat{b}, \hat{a}] = [\hat{b}, \hat{a}^\dagger] = 0$. With the addition of spin, the full transverse Hamiltonian can be written as

$$\hat{H}_\perp = 2\hbar\omega \left(\hat{a}^\dagger\hat{a} + \hat{b}^\dagger\hat{b}\right). \tag{17.73}$$

The energy eigenfunctions $\psi_{nM}^{(\perp)}$ found above, corresponding to the states $|n, M\rangle$, should also be labeled with the $\bar{N} = \hat{b}^\dagger\hat{b}$ quantum number $\bar{n}$ as $|n, \bar{n}, M\rangle$. Note that $\bar{n} = 0, 1$, although $n = 0, 1, \dots, \infty$.

Introduce now the operator

$$\hat{Q} \equiv \sqrt{2\hbar\omega}\,\hat{a}\hat{b}^\dagger. \tag{17.74}$$

Note that this operator commutes with the Hamiltonian $\hat{H}_\perp$

$$\left[\hat{H}_\perp, \hat{Q}\right] = 0. \tag{17.75}$$

This operator has the property to remove *"type-a"* quanta and add *"type-b"* ones. Starting from the commutators

$$\left[\hat{N}, \hat{Q}\right] = -\hat{Q}, \quad \left[\hat{\bar{N}}, \hat{Q}\right] = \hat{Q} \tag{17.76}$$

we obtain the relations

$$\hat{N}\hat{Q}|n, \bar{n}\rangle = (n-1)\hat{Q}|n, \bar{n}\rangle, \quad \hat{\bar{N}}\hat{Q}|n, \bar{n}\rangle = (n+1)\hat{Q}|n, \bar{n}\rangle, \tag{17.77}$$

from which it is clear that

$$\hat{Q}|n, \bar{n}\rangle \propto |n-1, \bar{n}+1\rangle. \tag{17.78}$$

Note that the operators $\hat{Q}$ and $\hat{Q}^\dagger$ have vanishing squares and that their anticommutator is equal to $\hat{H}_\perp$

$$\left\{\hat{Q}, \hat{Q}\right\} = \left\{\hat{Q}^\dagger, \hat{Q}^\dagger\right\} = 0 \tag{17.79}$$

$$\left\{\hat{Q}, \hat{Q}^\dagger\right\} = \hat{H}_\perp. \tag{17.80}$$

The fact that the operator $\hat{Q}$ commutes with the Hamiltonian means that it is a constant of the motion and it corresponds to a symmetry. This symmetry should be manifest in the energy spectrum. Indeed, the state $|n, \bar{n}\rangle$ has the same energy with the state $|n - 1, \bar{n} + 1\rangle$, meaning that the removal of a *"type-a"* quantum is equivalent to the addition of a *"type-b"* quantum. Since, there is no limitation on the number of n-quanta, we may loosely refer to these quanta as *bosons*, while referring to the $\bar{n}$-quanta as *fermions*. The existing symmetry, which may be called *supersymmetry*, amounts to the fact that a state $|0, 1\rangle$, with no *bosons* and one *fermion* is degenerate with a state $|1, 0\rangle$ of one boson and no fermion.

17.3   The Bohm–Aharonov Effect

Consider the setup shown in Fig. 17.1. A solenoid confining magnetic flux is placed between the slits of a two-slit experiment. Although there is no magnetic field in the outside region, there is nonzero vector potential (being just a gradient $\mathbf{A} = \nabla \Lambda$), which however has to be continuously matched with the non-trivial vector potential inside the solenoid.

The Schroedinger equation in the outside region, where $\mathbf{A} = \nabla \Lambda$, is

$$\left\{\frac{1}{2m}\left(-i\hbar\nabla - \frac{e}{c}\mathbf{A}\right)^2 + V(\mathbf{r})\right\}\psi(\mathbf{r}) = E\psi(\mathbf{r}) \tag{17.81}$$

$V(\mathbf{r})$ is any possibly existing potential. This equation can be rewritten as

$$\left\{-\frac{\hbar^2}{2m}\nabla^2 + V(\mathbf{r})\right\}\psi_A(\mathbf{r}) = E\psi_A(\mathbf{r}) \tag{17.82}$$

in terms of

$$\psi_A(\mathbf{r}) = e^{\frac{ie}{\hbar c}\Lambda}\psi(\mathbf{r}) = e^{\frac{ie}{\hbar c}\int^{\mathbf{r}} d\mathbf{r}\cdot\mathbf{A}}\psi(\mathbf{r}). \tag{17.83}$$

Note that the integral in the exponential is *path-dependent*. The particles arriving on the final screen have to be described with a superposition of wave functions corresponding to the two distinct paths (above and below the solenoid), namely

$$\psi(\mathbf{r}) = \psi_A^{(I)}(\mathbf{r}) + \psi_A^{(II)}(\mathbf{r}) = e^{\frac{ie}{\hbar c}\int_I^{\mathbf{r}} d\mathbf{r}\cdot\mathbf{A}}\psi^{(I)}(\mathbf{r}) + e^{\frac{ie}{\hbar c}\int_{II}^{\mathbf{r}} d\mathbf{r}\cdot\mathbf{A}}\psi^{(II)}(\mathbf{r}) \tag{17.84}$$

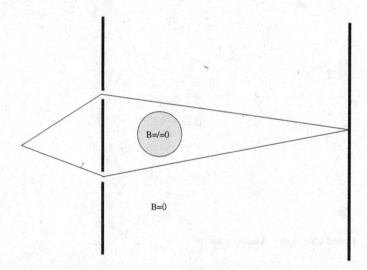

Fig. 17.1 The two-slit Bohm–Aharonov experimental setup

or

$$\psi(\mathbf{r}) = e^{\frac{ie}{\hbar c} \int_I^\mathbf{r} d\mathbf{r}\cdot\mathbf{A}} \left(\psi^{(I)}(\mathbf{r}) + e^{\frac{ie}{\hbar c} \oint d\mathbf{r}\cdot\mathbf{A}} \psi^{(II)}(\mathbf{r}) \right) . \qquad (17.85)$$

The line integral on the closed loop is related via Stokes theorem to the surface integral enclosing the magnetic field

$$\oint d\mathbf{r} \cdot \mathbf{A} = \int_S d\mathbf{s} \cdot (\nabla \times \mathbf{A}) = \int_S d\mathbf{S} \cdot \mathbf{B} = \Phi \qquad (17.86)$$

which is the magnetic flux passing through the solenoid. Thus, we have

$$|\psi(\mathbf{r})|^2 = \left| \psi^{(I)}(\mathbf{r}) + e^{\frac{ie}{\hbar c}\Phi} \psi^{(II)}(\mathbf{r}) \right|^2 \qquad (17.87)$$

and there is a measurable phase factor between the two wave function terms $e^{\frac{ie}{\hbar c}\Phi}$ that will create an observable interference pattern. This is the *Bohm–Aharonov effect* [1–4]. This phenomenon implies that in contrast to classical physics, in quantum mechanics the electromagnetic potentials have an imprint on physics.

There is a related setup, shown very schematically in Fig. 17.2, where the same phenomena have other very interesting conclusions. A superconducting fluid subject to a magnetic field displays the so-called *Meisner effect* according to which the *superconducting phase* expels the magnetic field so that the magnetic flux is confined in bounded regions of *normal phase* like the central region shown in the figure.

In contrast to the two-slit Bohm–Aharonov setup, where the electrons correspond to scattering states, the charged particles of the superconducting phase ("Cooper pairs" of charge $q = 2e$ composed of two electrons) circulating around the region

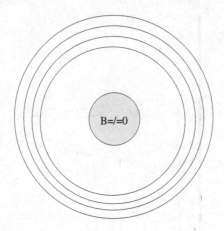

Fig. 17.2 Bound-state Bohm–Aharonov setup

that confines the magnetic flux, have wave functions that wrap around the flux tubes. Each circle endows them with a phase factor $e^{\frac{iq}{\hbar c}\Phi}$. However, single-valuedness of the wave function demands that this phase is an integer multiple of 2π, namely

$$\Phi = nhc/q \, . \tag{17.88}$$

Therefore, the superconducting versus normal phase regions are adjusted so that the area supporting the magnetic flux is such that it satisfies the above flux quantization condition. The effect has been observed experimentally. In addition, it has been verified that the charge appearing in the above formula is $2e$ justifying that indeed the superconductivity carriers are Cooper pairs.

Example 17.2 (**A Bohm–Aharonov Toy Model**) Consider a particle circulating in a circle of radius R, while in a concentric region of radius $a < R$ there is a homogeneous magnetic field perpendicular to the plane of motion of the particle. Solve the energy eigenvalue problem.

First let's consider the corresponding energy eigenvalue problem in the absence of the magnetic field, which is a compact version of free one-dimensional motion with the position variable replaced by the polar angle $0 \le \phi \le 2\pi$ according to $\mathbf{r} = R\left(\cos\phi\,\hat{x} + \sin\phi\,\hat{y}\right)$. The momentum is replaced by $\mathbf{p} = -i\hbar\nabla = -\frac{i\hbar}{R}\hat{\phi}\frac{d}{d\phi}$. The Schroedinger equation is solved by plane wave-type of eigenfunctions

$$\frac{1}{2m}\left(-\frac{i\hbar}{R}\frac{d}{d\phi}\right)^2\Psi(\phi) = \frac{\hbar^2 k^2}{2m}\Psi(\phi) \implies \Psi(\phi) = \frac{e^{ikR\phi}}{\sqrt{2\pi R}} \, . \tag{17.89}$$

Note however that, because of the required single-valuedness of the wave function, we must have

$$\Psi(\phi + 2\pi n) = \Psi(\phi) \implies k = \frac{n}{R} \quad (n = \pm 1, \, \pm 2, \, \dots) \, . \tag{17.90}$$

As a result, the energy eigenvalues will be

$$E_n = \frac{\hbar^2 n^2}{2mR^2} \quad (n = \pm 1, \pm 2, \ldots). \tag{17.91}$$

Consider now that a cylindrical region of radius $a < R$ carries a homogeneous magnetic field $\mathbf{B} = \hat{z}B$ perpendicular to the plane of motion of the particle, which is also assumed to carry electric charge e. A continuous vector potential that can represent the constant magnetic field is

$$\mathbf{A} = \begin{cases} \frac{1}{2}(\mathbf{B} \times \mathbf{r}) = \frac{B}{2}(-y\hat{x} + x\hat{y}) = \frac{B}{2}\rho\hat{\phi} & (\rho \leq a) \\ \frac{B}{2}\frac{a^2}{\rho^2}(-y\hat{x} + x\hat{y}) = \frac{B}{2}\frac{a^2}{\rho}\hat{\phi} & (\rho \geq a) \end{cases} \tag{17.92}$$

This $\mathbf{A}$ leads to the magnetic field

$$\mathbf{B} = \nabla \times \mathbf{A} = \begin{cases} \hat{z}B & (\rho \leq a) \\ 0 & (\rho > a) \end{cases} \tag{17.93}$$

Note that the vector potential in the outside region is just a gradient

$$\mathbf{A}(\rho \geq a) = \nabla\Lambda \quad \text{with} \quad \Lambda = \frac{B}{2}a^2\phi = \frac{B}{2}a^2 \arctan(y/x) \tag{17.94}$$

and gives a vanishing magnetic field.

Now the Schroedinger equation for a charged particle orbiting in a circle of radius $R > a$ will be

$$\frac{1}{2m}\left(-\frac{i\hbar}{R}\frac{d}{d\phi} - \frac{eBa^2}{2cR}\right)^2 \Psi(\phi) = E\Psi(\phi). \tag{17.95}$$

Substituting $\Psi(\phi) \propto e^{ikR\phi}$ we obtain

$$E = \frac{1}{2m}\left(\hbar k - \frac{eBa^2}{2cR}\right)^2. \tag{17.96}$$

Again, single-valuedness dictates that $k = n/R$. Finally, we may write the above correction to the energy eigenvalues in terms of the *magnetic flux* $\Phi = \pi a^2 B$ as

$$E_n = \frac{\hbar^2}{2mR^2}\left(n - \frac{e\Phi}{2\pi c\hbar}\right)^2. \tag{17.97}$$

Problems and Exercises

17.1 Consider a gauge transformation $\mathbf{A} \to \mathbf{A}' = \mathbf{A} + \nabla\Lambda$ and show that $e^{\frac{ie}{\hbar c}\Lambda}\left(\mathbf{p} - \frac{e}{c}\mathbf{A}\right)e^{-\frac{ie}{\hbar c}\Lambda} = \mathbf{p} - \frac{e}{c}\mathbf{A}'$. Then, show that the Schroedinger equation is gauge-transformed to

$$\frac{1}{2m}\left(\mathbf{p} - \frac{e}{c}\mathbf{A}'\right)^2 \psi'(\mathbf{r}) = E\psi'(\mathbf{r}) \quad \text{with} \quad \psi'(\mathbf{r}) = e^{\frac{ie}{\hbar c}\Lambda}\psi(\mathbf{r}).$$

17.2 For a homogeneous magnetic field $\mathbf{B} = \hat{z} B$ choose the asymmetric vector potential $\mathbf{A} = Bx\hat{y}$ and solve the energy eigenvalue problem.

17.3 Consider a charged particle in a homogeneous magnetic field. Show that a translation of the state $\hat{T}(a)|\psi\rangle = e^{\frac{i}{\hbar}\mathbf{a}\cdot\mathbf{p}}|\psi\rangle$ is equivalent to a gauge transformation.

17.4 A particle of mass m and electric charge e is subject to an external electric field $\mathbf{E}$. The Schroedinger equation is

$$\left(-\frac{\hbar^2}{2m}\nabla^2 + e\phi\right)\psi = i\hbar\dot{\psi}$$

where $\phi(\mathbf{r}, t)$ is the scalar potential corresponding to the electric field. Consider now a gauge transformation $\phi \rightarrow \phi + \frac{1}{c}\dot{\Lambda}$. Show that the following Schroedinger equation

$$\left\{\frac{1}{2m}\left(-i\hbar\nabla - \frac{e}{c}\nabla\Lambda\right)^2 + e\phi - \frac{e}{c}\dot{\Lambda}\right\}\psi_\Lambda = i\hbar\dot{\psi}_\Lambda$$

is satisfied by $\psi_\Lambda = e^{\frac{ie}{\hbar c}\Lambda}\psi$.

17.5 Two identical fermions of spin $1/2$ and electric charge q are subject to an external strong homogeneous magnetic field. The two particles have also a mutual interaction expressed through a potential that depends on the relative distance of the two particles. Write down the Hamiltonian in center of mass coordinates and determine the eigenvalues corresponding to the center of mass part of the Hamiltonian. If we neglect the mutual interaction, what are the eigenvalues of the energy? In the latter case, what is the ground state wave function of the system?

17.6 A beam of neutrons (neutral spin $1/2$ fermions possessing a magnetic moment due to their spin) splits in two halves, one of them passing through a region of homogeneous magnetic field $\mathbf{B}$. Ignore spatial degrees of freedom and assume that the Hamiltonian of the neutron system is $\hat{H} = E_0 - 2\frac{\mu_N}{\hbar}\mathbf{S}\cdot\mathbf{B}$, where μ_N is the neutron magnetic moment. If the initial state of the beam is $|\psi(0)\rangle = |\pm\rangle$, and the two branches interfere at time T, write down the evolved state $|\psi(T)\rangle$ and the corresponding probability amplitude, showing a sinusoidal variation as a function of the magnetic field. Find the difference in B required to produce two successive maxima of the amplitude.

17.7 Consider a charged particle subject to a strong homogeneous magnetic field. Ignore the part of the Hamiltonian associated with motion parallel to the magnetic field. Consider the ground state of the transverse part of the Hamiltonian and calculate for this state the uncertainty in the radius $\rho = \sqrt{x^2 + y^2}$.

17.8 Consider a particle of mass m and electric charge $q > 0$ subject to a homogeneous magnetic field $\mathbf{B}$ and simultaneously to a homogeneous electric field $\mathbf{E}$. Solve the Heisenberg equations and determine the velocity $\mathbf{v}(t) = \left(\mathbf{p}(t) - \frac{q}{c}\mathbf{A}(\mathbf{r}(t))\right)/m$, where $\mathbf{A} = \frac{1}{2}\mathbf{B} \times \mathbf{r}$ is the vector potential related to the magnetic field.

References

1. C. Cohen-Tannoudji, B. Diu, F. Laloe, *Quantum mechanics*, vols I and II, 1st edn. Wiley-VCH (1977)
2. F. Levin, *An introduction to quantum theory* (Cambridge University Press, Cambridge, 2002)
3. G. Baym, *Lectures in quantum mechanics* (Lecture notes and supplements in physics, ABP, 1969)
4. D. Griffiths, *Introduction to quantum mechanics*, 2nd edn. (Cambridge University Press, Cambridge, 2017)

Chapter 18
Approximation Methods

18.1 General Considerations

Only a few physical systems correspond to mathematical equations that can be solved analytically. Therefore, the development of *approximation methods* has been very important since the very early days of Quantum Mechanics. Approximation methods can be roughly separated into two broad categories, namely, *perturbative methods* and *non-perturbative* ones. The applicability of perturbative approximation methods requires the existence of a small parameter in the problem at hand. Denoting this small parameter by λ, such a perturbative approximation method can be applied if we know the solution of the problem for $\lambda = 0$. From a physical point of view, we might say that the method has application on problems which are infinitesimally close to a solved problem. For example, in the case of the Zeeman effect, as long as the magnetic field is *small*, the correction to the Hamiltonian $\frac{e}{2mc}(\mathbf{L} + 2\mathbf{S}) \cdot \mathbf{B}$ can be treated as a small perturbation. The term *small* has to be quantified. To make the statement *"small correction to the energy"* meaningful requires comparison to the characteristic energy of the solved unperturbed problem, i.e., in this particular case $e\hbar B/mc \ll me^4/\hbar^2$. Note, however, that the smallness of a parameter does not guarantee the success of the perturbative method. For example, in the case of an anharmonic perturbation λx^4 to the simple harmonic oscillator, there will always be distances x for which such a term will stop being small, no matter how small λ is. Furthermore, there are aspects of the solution to a physical problem that cannot be obtained by perturbative methods. A characteristic very simple example is the case of bound states of the one-dimensional square well. We know that at least one bound state exists no matter how shallow is the well. On the other hand, using the depth of the well as our small parameter, it is a fact that we will never obtain a bound state solution starting from the free-particle solution and treating the potential well as perturbation. Nevertheless, as a general rule, the success of the perturbative method should always be judged *a posteriori*, i.e., after numerical results have been obtained and compared with measured values.

© Springer Nature Switzerland AG 2019

K. Tamvakis, *Basic Quantum Mechanics*, Undergraduate Texts in Physics,
https://doi.org/10.1007/978-3-030-22777-7_18

The common characteristic of *non-perturbative* approximation methods is the fact that a small parameter is not available for the given problem at hand. Then, a way to proceed is to consider a set of trial candidate solutions characterized by one or more free parameters and fix the values of these parameters in order to obtain agreement with the known (experimental) data for physical observables. A guiding principle for the choice of trial solutions is that they should embody as much as possible the basic qualitative physical characteristics expected. For example, the central idea of the *Variational Method* is to consider a trial wave function, parametrized suitably in terms of a set of parameters, calculate the corresponding expectation value of the energy, and then minimize it with respect to the parameters, estimating in this way the ground state energy of the system.

18.2   The WKB Method

Consider a classical particle moving in one dimension under the influence of a potential $V(x)$. The expression for its conserved energy $E = \frac{p^2}{2m} + V(x)$ can be solved for the momentum as

$$p(x) = \sqrt{2m\left(E - V(x)\right)} \ . \tag{18.1}$$

Note that classically only $E > V(x)$ is allowed. Whenever the particle reaches a point at which the energy equals the potential, or, equivalently, its kinetic energy vanishes, it turns around and moves backward. The situation for a quantum particle, of course, is quite different. Assume that in a *"small enough"* region around a point we can approximate the potential with a constant. Then, the wave function will be approximately in the form of a plane wave

$$\psi(x) \approx e^{\frac{i}{\hbar}p(x)x}\,, \tag{18.2}$$

where $p(x)$ is the expression (18.1). Note, however, that for the points $E \leq V(x)$, where the wave function unavoidably extends, $p(x)$ becomes imaginary and the wave function has a decaying behavior as expected.

The WKB method [1–3][1] is essentially a systematic quantitative application of the above ideas. We start writing the Schroedinger equation

$$\frac{d^2\psi}{dx^2} + \frac{2m}{\hbar^2}\left(E - V(x)\right)\psi = 0 \tag{18.3}$$

and search for solutions of the form

$$\psi(x) = e^{\frac{i}{\hbar}W(x)}\,. \tag{18.4}$$

---

[1] The name is an acronym of the names of Wenzel, Kramers, and Brillouin.

Substituting this expression into the Schroedinger equation we obtain

$$i\hbar \frac{d^2 W}{dx^2} - \left(\frac{dW}{dx}\right)^2 + p^2 = 0, \tag{18.5}$$

where $p(x)$ is that of Eq. (18.1). We may then set up a *small $\hbar$ or semiclassical expansion* as

$$W(x) = W_0(x) + \hbar W_1(x) + \hbar^2 W_2(x) + \dots \tag{18.6}$$

Substituting into (18.5) we obtain the *order by order "solution"*

$$p^2(x) - \left(W_0'(x)\right)^2 + \hbar \left(i W_0''(x) - 2W_0'(x)W_1'(x)\right) + O(\hbar^2) = 0. \tag{18.7}$$

To lowest order we get

$$W_0'(x) = \pm p(x) \implies W_0(x) = \pm \int^x dx\, p(x). \tag{18.8}$$

The next order gives

$$W_1'(x) = \frac{i}{2}\frac{W_0''(x)}{W_0'(x)} = \frac{i}{2}\frac{p'(x)}{p(x)} \implies W_1(x) = \frac{i}{2}\ln p(x) + const. \tag{18.9}$$

Thus, the $O(\hbar)$ WKB solution is

$$\psi(x) \approx \frac{e^{\pm \frac{i}{\hbar}\int^x dx\, p(x)}}{\sqrt{p(x)}}. \tag{18.10}$$

The validity of the expansion rests on the condition

$$i\hbar W''(x) \ll (W')^2$$

or

$$i\hbar\, p'(x) \ll (p(x))^2. \tag{18.11}$$

This is usually expressed in terms of the *De Broglie wavelength* $\lambda = h/p$ as

$$\left|\frac{\lambda'(x)}{2\pi}\right| \ll 1. \tag{18.12}$$

This condition can be rephrased into $\lambda(x)|V'(x)| \ll p^2(x)/2m$, which states that the change of the potential over distances comparable to λ should be much smaller that the kinetic energy.

It is clear that the WKB approximation can be valid in the $E \gg V(x)$ and the $E \ll V(x)$ regions but it is never valid at the *turning point* $p(x_0) = 0$ where $E = V(x_0)$. This problem is dealt with by *matching* the WKB solution with a different solution obtained near the turning point. For the latter, we turn to the Schroedinger equation and expand the potential around the turning point keeping the lowest order linear term

$$\psi''(x) \approx \frac{2m}{\hbar^2} \left((x - x_0) V'(x_0) + \cdots \right) \psi(x) . \tag{18.13}$$

Considering the case $V'(x_0) > 0$, we define

$$z = (x - x_0) \left(\frac{2m V'(x_0)}{\hbar^2} \right)^{1/3} . \tag{18.14}$$

Then, the Schroedinger equation becomes

$$\frac{d^2 \psi(z)}{dz^2} = z \, \psi(z) . \tag{18.15}$$

This is a well-known differential equation, the *Airy equation*, having as solutions the *Airy functions*

$$Ai(z) = \frac{1}{\pi} \int_0^z dt \, \cos \left(\frac{t^3}{\pi} + zt \right) , \quad Bi(z) = \frac{1}{\pi} \int_0^z dt \left(\sin \left(\frac{t^3}{\pi} + zt \right) + e^{-\frac{t^3}{\pi} + zt} \right) . \tag{18.16}$$

Thus, we have

$$\psi(z) = \alpha \, Ai(z) + \beta \, Bi(z) . \tag{18.17}$$

The asymptotic behavior of the Airy functions is

$$Ai(z) \sim \begin{cases} \frac{z^{-1/4}}{2\sqrt{\pi}} e^{-\frac{2}{3} z^{3/2}} & (z \gg 0) \\ \frac{(-z)^{-1/4}}{\sqrt{\pi}} \cos \left(-\frac{2}{3}(-z)^{3/2} - \frac{\pi}{4} \right) & (z \ll 0) \end{cases} \tag{18.18}$$

and

$$Bi(z) \sim \begin{cases} \frac{z^{-1/4}}{\sqrt{\pi}} e^{\frac{2}{3} z^{3/2}} & (z \gg 0) \\ -\frac{(-z)^{-1/4}}{\sqrt{\pi}} \sin \left(-\frac{2}{3}(-z)^{3/2} - \frac{\pi}{4} \right) & (z \ll 0) \end{cases} \tag{18.19}$$

$Ai(z)$ has a decaying behavior beyond the turning point, while it oscillates before it reaches it. This is the physically expected behavior, since $z > 0$ corresponds to $E < V(x)$ and $z < 0$ to $E > V(x)$. In contrast, $Bi(z)$ blows up away from the turning point and, if the classically forbidden region continues to infinity, it is physically unacceptable.

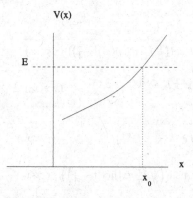 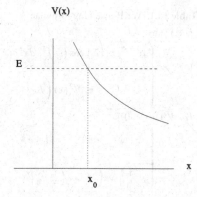

Fig. 18.1 Turning points

Let's proceed now to see how *matching* works in the case of a potential with an everywhere positive slope shown in the left-hand plot of Fig. 18.1.

In the $z \ll 0$ region, we have

$$\frac{2m}{\hbar^2}(V(x) - E) = \frac{2m}{\hbar^2}(x - x_0)V'(x_0) = z\left(\frac{2mV'(x_0)}{\hbar^2}\right)^{2/3}$$

or

$$-z = \frac{2m}{\hbar^2}(E - V(x))\left(\frac{\hbar^2}{2mV'(x_0)}\right)^{2/3}. \tag{18.20}$$

Note that

$$\frac{2}{3}(-z)^{3/2} = -\int_0^z dz'(-z')^{1/2} = \frac{1}{\hbar}\int_{x_0}^x dx'\sqrt{2m(E - V(x'))}. \tag{18.21}$$

Thus, if we take the wave function in the classically forbidden region $x \gg x_0$ (i.e., $z \gg 0$) to correspond to the decreasing exponential

$$\psi(x) \sim \frac{1}{\sqrt{\pi}}\frac{(2m\hbar V'(x_0))^{1/6}}{(2m(V(x) - E))^{1/4}}e^{-\frac{1}{\hbar}\int_{x_0}^x dx'\sqrt{2m(V(x')-E)}}, \tag{18.22}$$

this will be matched in the region $x \ll x_0$ with the wave function

$$\psi(x) \sim \frac{2}{\sqrt{\pi}}\frac{(2m\hbar V'(x_0))^{1/6}}{(2m(E - V(x)))^{1/4}}\cos\left(\frac{1}{\hbar}\int_x^{x_0} dx'\sqrt{2m(E - V(x'))} - \frac{\pi}{4}\right). \tag{18.23}$$

General matching formulae can be written as shown in Table 18.1, where

$$p(x) = \sqrt{2m(E - V(x))} \quad and \quad \pi(x) = \sqrt{2m(V(x) - E)}.$$

Table 18.1 WKB matching formulae

For $V'(x_0) > 0$

$$\psi(x) \sim \begin{cases} p^{-1/2}(x)\left(2A\cos\left(\frac{1}{\hbar}\int_x^{x_0} dx'\, p(x') - \frac{\pi}{4}\right) - B\sin\left(\frac{1}{\hbar}\int_x^{x_0} dx'\, p(x') - \frac{\pi}{4}\right)\right) & (x \ll x_0) \\[2ex] \pi^{-1/2}(x)\left(B e^{\frac{1}{\hbar}\int_{x_0}^x dx'\, \pi(x')} + A e^{-\frac{1}{\hbar}\int_{x_0}^x dx'\, \pi(x')}\right) & (x \gg x_0) \end{cases}$$

For $V'(x_0) < 0$

$$\psi(x) \sim \begin{cases} \pi^{-1/2}(x)\left(A e^{\frac{1}{\hbar}\int_x^{x_0} dx'\, \pi(x')} + B e^{-\frac{1}{\hbar}\int_x^{x_0} dx'\, \pi(x')}\right) & (x \ll x_0) \\[2ex] p^{-1/2}(x)\left(2B\cos\left(\frac{1}{\hbar}\int_{x_0}^x dx'\, p(x') - \frac{\pi}{4}\right) - A\sin\left(\frac{1}{\hbar}\int_{x_0}^x dx'\, p(x') - \frac{\pi}{4}\right)\right) & (x \gg x_0) \end{cases}$$

18.2.1   Bound States and WKB

Consider the potential shown in Fig. 18.2.

For a given energy E, the turning points are at $x = a$ and $x = b$. In the region $x \ll a$ the WKB wave function is

$$\psi(x) \sim \frac{1}{\sqrt{\pi(x)}}\, e^{-\frac{1}{\hbar}\int_x^a dx'\, \pi(x')}, \tag{18.24}$$

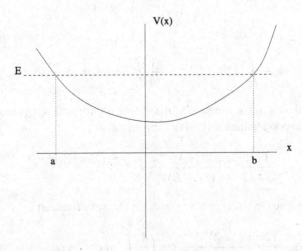

Fig. 18.2 Bound states in WKB

Similarly, in the opposite asymptotic region $x \gg b$ the WKB wave function will be

$$\psi(x) \sim \frac{1}{\sqrt{\pi(x)}} \, e^{-\frac{1}{\hbar} \int_b^x dx' \pi(x')} , \tag{18.25}$$

where $\pi(x) = \sqrt{2m \, (V(x) - E)}$. In the intermediate region far away from the turning points the WKB approximate wave function will be

$$\psi(x) \sim \frac{2}{\sqrt{p(x)}} \, \cos \left(\frac{1}{\hbar} \int_x^b dx' \, p(x') - \frac{\pi}{4} \right)$$

or

$$\psi(x) \sim \frac{2}{\sqrt{p(x)}} \, \cos \left(\frac{1}{\hbar} \int_a^x dx' \, p(x') - \frac{\pi}{4} \right) . \tag{18.26}$$

The last two expressions must agree. This implies that

$$\cos \left(\frac{1}{\hbar} \int_x^b dx' \, p(x') - \frac{\pi}{4} \right) = \pm \cos \left(\frac{1}{\hbar} \int_a^x dx' \, p(x') - \frac{\pi}{4} \right)$$

or

$$\cos \left(\frac{1}{\hbar} \int_x^b dx' \, p(x') - \frac{\pi}{4} \right) = \pm \cos \left(-\frac{1}{\hbar} \int_a^x dx' \, p(x') + \frac{\pi}{4} \right)$$

or

$$\int_a^b dx' \sqrt{2m \, (E - V(x'))} = \left(n + \frac{1}{2} \right) \hbar \pi , \tag{18.27}$$

where $n = 0, 1, 2, \ldots$.

If we apply this condition to the case of the simple harmonic oscillator $V(x) = m\omega^2 x^2 / 2$, we get

$$\int_{-\sqrt{2E/m\omega^2}}^{+\sqrt{2E/m\omega^2}} dx' \sqrt{2mE - (m\omega x')^2} = E\pi/\omega , \tag{18.28}$$

which leads to the standard oscillator spectrum $E_n = (n + 1/2)\hbar\omega$. Of course, this is a happy coincidence. In most cases, the result will be an approximation that improves with increasing n, i.e., when we approach the large quantum number or semiclassical limit.

18.2.2 Tunneling in WKB

Let's apply now the WKB method to calculate the transmission rate through a potential barrier as the one shown in Fig. 18.3. Assuming that the WKB approximation holds, no special assumption will be made for the exact shape of the potential.

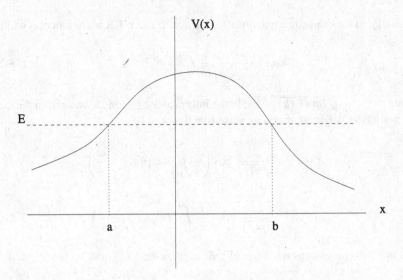

Fig. 18.3 Barrier penetration in WKB

We start with the outgoing wave function in the region $x \gg b$

$$\psi(x \gg b) \sim \frac{F\,e^{-i\pi/4}}{\sqrt{p(x)}} e^{\frac{i}{\hbar} \int_b^x dx'\, p(x')}, \qquad (18.29)$$

which can be written as

$$\psi(x \gg b) = \frac{F}{\sqrt{p(x)}} \left(\cos\left(\frac{1}{\hbar} \int_b^x dx'\, p(x') - \frac{\pi}{4} \right) + i \sin\left(\frac{1}{\hbar} \int_b^x dx'\, p(x') - \frac{\pi}{4} \right) \right). \qquad (18.30)$$

From the connection formulae for a downward slope, we connect this with the dominant part of the wave function in the classically inaccessible region

$$\psi(a \ll x \ll b) \sim -i \frac{F}{\sqrt{\pi(x)}} e^{\frac{1}{\hbar} \int_x^b dx'\, \pi(x')} = -i \frac{F e^{\frac{1}{\hbar} \int_a^b dx'\, \pi(x')}}{\sqrt{\pi(x)}} e^{-\frac{1}{\hbar} \int_a^x dx'\, \pi(x')}. \qquad (18.31)$$

This is matched at the upward slope as

$$\psi(x \ll a) \sim -2i \frac{F e^{\frac{1}{\hbar} \int_a^b dx'\, \pi(x')}}{\sqrt{p(x)}} \cos\left(\frac{1}{\hbar} \int_x^a dx'\, p(x') - \frac{\pi}{4} \right). \qquad (18.32)$$

The ingoing part of this wave function is

$$\psi_{inc}(x) \approx -i \frac{F e^{\frac{1}{\hbar} \int_a^b dx'\, \pi(x')}}{\sqrt{p(x)}} e^{\frac{i}{\hbar} \int_a^x dx'\, p(x')}. \qquad (18.33)$$

Since the transmitted wave function is

$$\psi_{trans}(x) \approx \frac{F e^{-i\pi/4}}{\sqrt{p(x)}} e^{\frac{i}{\hbar} \int_b^x dx'\, p(x')}, \tag{18.34}$$

the *transmission coefficient* will be

$$\mathcal{T} = \frac{v_{trans} |\psi_{trans}|^2}{v_{inc} |\psi_{inc}|^2} \approx e^{-\frac{2}{\hbar} \int_a^b dx\, \sqrt{2m(V(x)-E)}}. \tag{18.35}$$

Example 18.1 Consider the potential

$$V(x) = \begin{cases} 0 & (x \le 0) \\ V_1\left(1 - \frac{x}{b}\right) & (0 < x \le b) \\ 0 & (x > b). \end{cases}$$

Calculate the transmission coefficient for a particle of energy $E < V_1$ incident from the left in the WKB approximation.

The turning points are at $x = 0$ and at $x = a$, where $E = V(a)$ or $a = b\left(1 - \frac{E}{V_1}\right)$. Starting with the transmitted wave function

$$\psi(x \gg a) \sim \frac{F e^{-i\pi/4}}{\sqrt{p(x)}} e^{\frac{i}{\hbar} \int_a^x dx'\, p(x')}, \tag{18.36}$$

we connect it with the intermediate wave function

$$\psi(0 < x \ll a) \sim -i \frac{F}{\sqrt{\pi(x)}} e^{\frac{1}{\hbar} \int_0^a dx'\, \pi(x')} e^{-\frac{1}{\hbar} \int_a^x dx'\, \pi(x')}. \tag{18.37}$$

Next, we apply continuity at the point $x = 0$ between the exact wave function[2] ($E = \hbar^2 k^2 / 2m$)

$$\psi(x < 0) = \frac{A}{\sqrt{\hbar k}} \sin(kx + \varphi)' \tag{18.38}$$

and the approximate WKB wave function obtained above

$$\psi(x \sim 0) \approx -i \frac{F}{\sqrt{\pi(0)}} e^{\frac{1}{\hbar} \int_0^a dx'\, \pi(x')} e^{-\frac{1}{\hbar} \int_a^x dx'\, \pi(x')}, \tag{18.39}$$

where $\pi(0) = \sqrt{2m(V_1 - E)}$. We get

$$\frac{A}{\sqrt{\hbar k}} \sin\varphi = -i \frac{F}{\sqrt{\pi(0)}} e^{\frac{1}{\hbar} \int_0^a dx'\, \pi(x')}, \qquad \frac{A}{\sqrt{\hbar k}} k \cos\varphi = \frac{i}{\hbar} F \sqrt{\pi(0)} e^{\frac{1}{\hbar} \int_0^a dx'\, \pi(x')}. \tag{18.40}$$

---

[2] We have parametrized the exact wave function in the WKB fashion.

From these, we obtain

$$\tan \varphi = -1/\sqrt{V_1/E - 1} \quad \Longrightarrow \quad \sin \varphi = -\sqrt{\frac{E}{V_1}} \tag{18.41}$$

and

$$F = -i A \sqrt{\left(1 - \frac{E}{V_1}\right)} e^{-\frac{1}{\hbar} \int_0^a dx \sqrt{2m(V(x)-E)}} . \tag{18.42}$$

Therefore, the transmission coefficient $T = J_{tr}/J_{inc} = \left(|\psi_{tr}|^2 v_{tr}\right) / \left(|\psi_{inc}|^2 v_{inc}\right)$ will be[3]

$$T = \frac{|Fe^{-i\pi/4}/\sqrt{\hbar k}|^2 \hbar k}{|Ae^{i\varphi}/2i\sqrt{\hbar k}|^2 \hbar k} = 4\left(1 - \frac{E}{V_1}\right) e^{-\frac{2}{\hbar} \int_0^a dx \sqrt{2m(V(x)-E)}} \tag{18.43}$$

or

$$T = 4\left(1 - E/V_1\right) \exp\left(-\left(32m V_1 a^3/9\hbar^2 b\right)^{1/2}\right) . \tag{18.44}$$

18.3   The Adiabatic Approximation

Consider a quantum system characterized by a set of parameters. Imagine a *gradual change* of these parameters. Such a process is characterized as *adiabatic*[4] if the characteristic time of change of the parameters T_{ext} (*external timescale*) is much larger than the characteristic timescale of change of the system T_{int} (*internal timescale*). For example, for the system of a particle bound in a square well, consider a slow change of the depth of the potential at a rate $\dot{V}_0/V_0$. Such a process can be characterized adiabatic if the characteristic time of change $T_{ext} \sim V_0/\dot{V}_0$ is much larger than the characteristic timescale of the system given by the period of stationary states $T_{int} \sim \hbar/\Delta E \sim O(\hbar/V_0)$.

Let's assume that the Hamiltonian of the system depends on the set of parameters $\{\alpha\} = \{\alpha_1, \alpha_2, \dots\}$. The eigenstates and eigenvalues will also be α-dependent

$$\hat{H}(\alpha) |E_n(\alpha)\rangle = E_n(\alpha) |E_n(\alpha)\rangle . \tag{18.45}$$

Consider now an adiabatic change in these parameters

$$\alpha_j \rightarrow \alpha_j(t) . \tag{18.46}$$

Adiabaticity corresponds to the assumption that the rate of parameter change $\dot{\alpha}_j/\alpha_j$ is much smaller than the characteristic rate of change in the system defined by the energy splittings $(E_n - E_{n'})/\hbar$.

The general solution to the Schroedinger equation

$$\hat{H} |\psi(t)\rangle = i\hbar \frac{d}{dt} |\psi(t)\rangle \tag{18.47}$$

---

[3] $p(x \gg a) = \hbar k$
[4] See also Griffiths [3].

can be written as an expansion in $\{|E_n(\alpha)\rangle\}$ according to

$$|\psi(t)\rangle = \sum_n C_n(t)\, e^{i\varphi_n(t)}\, |E_n(\alpha)\rangle. \tag{18.48}$$

In the absence of any parameter change, the C_n's are time-independent and the phases φ_n are the standard $-E_n t/\hbar$.

We may adopt the following Ansatz for the φ_n's:

$$\varphi_n(t) = -\frac{1}{\hbar}\int_0^t dt'\, E_n(\alpha(t')) \tag{18.49}$$

and proceed to determine the C_n's by substituting it into the Schroedinger equation. We obtain

$$\sum_n \left\{ \frac{dC_n}{dt}\, e^{i\varphi_n}\, |E_n\rangle + C_n\, e^{i\varphi_n}\, \frac{d}{dt}|E_n\rangle \right\} = 0. \tag{18.50}$$

Taking an inner product with one of the states, say $|E_m\rangle$, we get

$$\frac{dC_m}{dt} = -\sum_n C_n e^{i(\varphi_n-\varphi_m)}\langle E_m|\frac{d}{dt}|E_n\rangle = -\sum_n \sum_j C_n e^{i(\varphi_n-\varphi_m)}\dot{\alpha}_j\langle E_m|\frac{\partial}{\partial\alpha_j}|E_n\rangle. \tag{18.51}$$

The right-hand side can be separated as

$$-C_m \sum_j \dot{\alpha}_j\langle E_m|\frac{\partial}{\partial\alpha_j}|E_m\rangle - \sum_{n\neq m}\sum_j C_n e^{i(\varphi_n-\varphi_m)}\dot{\alpha}_j\langle E_m|\frac{\partial}{\partial\alpha_j}|E_n\rangle \tag{18.52}$$

or, introducing

$$\mathcal{A}_j^{(m)} = -i\langle E_m|\frac{\partial}{\partial\alpha_j}|E_m\rangle, \tag{18.53}$$

as

$$-iC_m \sum_j \dot{\alpha}_j \mathcal{A}_j^{(m)} - \sum_{n\neq m}\sum_j C_n e^{i(\varphi_n-\varphi_m)}\dot{\alpha}_j\langle E_m|\frac{\partial}{\partial\alpha_j}|E_n\rangle. \tag{18.54}$$

Thus, the equation for the coefficients C_n's is

$$\frac{dC_m}{dt} = -iC_m \sum_j \dot{\alpha}_j \mathcal{A}_j^{(m)} - \sum_{n\neq m} C_n e^{i(\varphi_n-\varphi_m)}\sum_j \dot{\alpha}_j\langle E_m|\frac{\partial}{\partial\alpha_j}|E_n\rangle. \tag{18.55}$$

Note that, up to now, we have not made any use of the adiabatic nature of the change and the last equation is exact. Going back to the statement of the eigenvalue problem $\hat{H}|E_n\rangle = E_n|E_n\rangle$ and taking a derivative with respect to a parameter α_k, we obtain

$$\frac{\partial \hat{H}}{\partial \alpha_k}|E_n\rangle + \hat{H}\frac{\partial}{\partial \alpha_k}|E_n\rangle = \frac{\partial E_n}{\partial \alpha_k}|E_n\rangle + E_n\frac{\partial}{\partial \alpha_k}|E_n\rangle. \tag{18.56}$$

Multiplying with an eigenstate $|E_m\rangle$ for $m \neq n$, we obtain

$$\langle E_m|\frac{\partial \hat{H}}{\partial \alpha_k}|E_n\rangle + E_m\langle E_m|\frac{\partial}{\partial \alpha_k}|E_n\rangle = E_n\langle E_m|\frac{\partial}{\partial \alpha_k}|E_n\rangle \tag{18.57}$$

or

$$(E_m - E_n)\langle E_m|\frac{\partial}{\partial \alpha_k}|E_n\rangle = -\langle E_m|\frac{\partial \hat{H}}{\partial \alpha_k}|E_n\rangle. \tag{18.58}$$

This implies

$$\sum_j \dot{\alpha}_j\langle E_m|\frac{\partial}{\partial \alpha_j}|E_n\rangle = -\sum_j \langle E_m|\frac{\partial \hat{H}}{\partial \alpha_j}|E_n\rangle\frac{\dot{\alpha}_j}{(E_m - E_n)}. \tag{18.59}$$

For an adiabatic process, the right-hand side is by hypothesis a rather small number. Therefore, the second term in (18.55) can be ignored. Then, we have

$$\frac{dC_m}{dt} = -iC_m\sum_j \dot{\alpha}_j\mathcal{A}_j^{(m)} \implies C_n(t) = C_n(0)\,e^{-i\sum_j \int_0^t dt\,\dot{\alpha}_j\mathcal{A}_j^{(n)}}. \tag{18.60}$$

If we assume that we start at $t = 0$ in an eigenstate $|E_0\rangle$ (i.e., $C_n = \delta_{n0}$), then

$$C_n(t) = \delta_{n0}\,e^{-i\int d\alpha\cdot\mathcal{A}^{(0)}}. \tag{18.61}$$

Thus, *the system remains in the state $|E_0\rangle$ as we vary the α's.* This is the *Adiabatic Theorem*.

Summarizing, the system during the adiabatic change will stay in the same state, which, however, acquires a phase

$$|\psi(t)\rangle = e^{i\left(\varphi_0(t) - \int d\alpha_j\mathcal{A}_j^{(0)}\right)}|E_0\rangle \tag{18.62}$$

with

$$\varphi_0(t) = -\frac{1}{\hbar}\int_0^t dt\,E_0(\alpha(t)) \quad and \quad \mathcal{A}_j^{(0)} = -i\langle E_0|\frac{\partial}{\partial \alpha_j}|E_0\rangle. \tag{18.63}$$

The phase acquired by the eigenstate consists of two parts. In contrast to the first part ϕ_n (*dynamical phase*) that depends on the total time of change, the second part (*geometric phase*) does not. In fact, for an adiabatic change that returns to the initial parameter values, it is

$$\varphi_B = \oint d\alpha \cdot \mathcal{A} \tag{18.64}$$

and depends only on the path taken in parameter space. This is the so-called *Berry phase*, related nontrivially to the topology of parameter space.

Note that (18.61) is only a lowest order result and if we wanted to go beyond this approximation we should have kept the exact Eq. (18.55). This can be transformed to an equivalent integral equation going back to its initial form

$$\frac{dC_m}{dt} = -\sum_n C_n(t) e^{i(\varphi_n(t) - \varphi_m(t))} \sum_j \dot{\alpha}_j(t) \langle E_m | \frac{\partial}{\partial \alpha_j} | E_n \rangle$$

and integrating

$$C_m(t) = C_m(0) - \sum_n \int_0^t dt' C_n(t') e^{i(\varphi_n(t') - \varphi_m(t'))} \sum_j \dot{\alpha}_j(t') \langle E_m | \frac{\partial}{\partial \alpha_j} | E_n \rangle. \tag{18.65}$$

A first approximation to this equation consists in replacing the coefficients in the integral with their initial values

$$C_m(t) = C_m(0) - \sum_n C_n(0) \sum_j \int d\alpha_j \, e^{i(\varphi_n - \varphi_m)} \langle E_m | \frac{\partial}{\partial \alpha_j} | E_n \rangle. \tag{18.66}$$

This line of approximations can be continued in an iterative way and, thus, define a perturbative approach. For example, the second-order approximation would be to substitute (18.66) for the coefficients inside the integral of (18.65).

Example 18.2 Consider a two-state system characterized by the Hamiltonian matrix

$$\mathcal{H} = \begin{pmatrix} \epsilon & \eta \, e^{i\delta} \\ \eta \, e^{-i\delta} & -\epsilon \end{pmatrix}, \tag{18.67}$$

where ϵ, η are constant real parameters, while the real parameter $\delta(t)$ varies with time. We shall assume that $\eta \ll \epsilon$. Determine the evolution of the state of the system assuming the adiabatic approximation.

The energy eigenvalues are

$$E_\pm = \pm\sqrt{\epsilon^2 + \eta^2} \approx \pm\left(\epsilon + \frac{\eta^2}{2\epsilon}\right) \tag{18.68}$$

and do not depend on $\delta(t)$. The corresponding eigenvectors are

$$\chi_+ \approx \begin{pmatrix} 1 \\ \frac{\eta}{2\epsilon} e^{-i\delta} \end{pmatrix}, \quad \chi_- \approx \begin{pmatrix} \frac{\eta}{2\epsilon} \\ -e^{-i\delta} \end{pmatrix}. \tag{18.69}$$

The state of the system can be written as

$$\psi(t) = C_+(t)\,e^{i\phi_+(t)}\,\chi_+ + C_-(t)\,e^{i\phi_-(t)}\,\chi_-\,, \tag{18.70}$$

where the dynamical phases are just

$$\phi_\pm(t) = -\frac{1}{\hbar}\int_0^t dt'\,E_\pm = \mp E_\pm\frac{t}{\hbar}\,. \tag{18.71}$$

The coefficients satisfy the differential equations

$$\dot{C}_+ = -iC_+\dot{\delta}\mathcal{A}^{(+)} - C_-e^{i(\phi_--\phi_+)}\,\dot{\delta}\chi_+^\dagger\frac{\partial}{\partial\delta}\chi_- \\ \dot{C}_- = -iC_-\dot{\delta}\mathcal{A}^{(-)} - C_+e^{i(\phi_+-\phi_-)}\,\dot{\delta}\chi_-^\dagger\frac{\partial}{\partial\delta}\chi_+ \tag{18.72}$$

where

$$\mathcal{A}^{(+)} = -i\chi_+^\dagger\frac{\delta}{\delta}\chi_+ \approx -i\frac{\eta^2}{4\epsilon^2}, \quad \mathcal{A}^{(-)} = -i\chi_-^\dagger\frac{\delta}{\delta}\chi_- = -i\,. \tag{18.73}$$

We also have

$$\chi_+^\dagger\frac{\partial}{\partial\delta}\chi_- = \chi_-^\dagger\frac{\partial}{\partial\delta}\chi_+ \approx \frac{i\eta}{2\epsilon}\,. \tag{18.74}$$

Therefore, in the adiabatic approximation, we shall have

$$\dot{C}_+ \approx 0, \quad \dot{C}_- \approx i\dot{\delta}\,C_-\,,$$

giving

$$C_\pm(t) = C_\pm(0)\,e^{i\gamma_\pm} \text{ with } \gamma_+ = 0, \ \gamma_- = \delta(t) - \delta(0)\,. \tag{18.75}$$

Example 18.3 A simple harmonic oscillator is subject to a time-dependent external force $\alpha(t)$. Assuming that initially ($t = 0$) the system is in an eigenstate of the unperturbed harmonic oscillator ($\alpha(0) = 0$), find the probability to make a transition to another eigenstate at times $t > 0$.

The Hamiltonian of the system is

$$\hat{H}(\alpha) = \frac{\hat{p}^2}{2m} + \frac{1}{2}m\omega^2\hat{x}^2 - \alpha\hat{x} \tag{18.76}$$

and can be written as

$$\hat{H}(\alpha) = \frac{\hat{p}^2}{2m} + \frac{1}{2}m\omega^2\left(\hat{x} - \frac{\alpha}{m\omega^2}\right)^2 - \frac{\alpha^2}{2m\omega^2}$$

or

$$\hat{H}(\alpha) = e^{-i\frac{\alpha\hat{p}}{\hbar m\omega^2}}\hat{H}(0)e^{i\frac{\alpha\hat{p}}{\hbar m\omega^2}} - \frac{\alpha^2}{2m\omega^2}\,. \tag{18.77}$$

Then, it is clear that the eigenvalue problem

$$\hat{H}(\alpha)|\psi_n(\alpha)\rangle = E_n(\alpha)|\psi_n(\alpha)\rangle \tag{18.78}$$

has the solution

$$|\psi_n(\alpha)\rangle = e^{-i\frac{\alpha\hat{p}}{\hbar m\omega^2}}|\psi_n\rangle$$

$$E_n(\alpha) = E_n - \frac{\alpha^2}{2m\omega^2}\,, \tag{18.79}$$

where $|\psi_n\rangle$ and $E_n = \hbar\omega(n + 1/2)$ are the standard harmonic oscillator eigenstates and eigenvalues for $\alpha = 0$.

A general state $|\psi(t)\rangle$ can be written

$$|\psi(t)\rangle = \sum_n C_n(t) e^{i\varphi_n(\alpha)} |\psi_n(\alpha)\rangle, \tag{18.80}$$

where

$$\varphi_n(\alpha) = -\frac{1}{\hbar} \int_0^t dt' E_n(\alpha(t')) = -\omega t \left(n + \frac{1}{2}\right) + \frac{1}{2\hbar m\omega^2} \int_0^t dt' \alpha^2(t'), \tag{18.81}$$

and the coefficients $C_n(t)$ obey the equation

$$\frac{dC_n}{dt} = -\dot{\alpha}(t) \sum_{n'} C_{n'}(t) e^{i(\varphi_{n'} - \varphi_n)} \langle \psi_n(\alpha)| \frac{\partial}{\partial \alpha} |\psi_{n'}(\alpha)\rangle. \tag{18.82}$$

The matrix element appearing in this formula is

$$\langle \psi_n(\alpha)| \frac{\partial}{\partial \alpha} |\psi_{n'}(\alpha)\rangle = \int_{-\infty}^{+\infty} dx\, \psi_n \left(x - \frac{\alpha}{m\omega^2}\right) \frac{\partial}{\partial \alpha} \psi_n^* \left(x - \frac{\alpha}{m\omega^2}\right)$$

$$= -m\omega^2 \int_{-\infty}^{+\infty} dx\, \psi_n^*(x) \frac{\partial}{\partial x} \psi_{n'}(x) = \frac{i}{\hbar} \langle \psi_n| \hat{p} |\psi_{n'}\rangle = \sqrt{\frac{m\omega}{2\hbar}} \langle \psi_n| \left(\hat{a} - \hat{a}^\dagger\right) |\psi_{n'}\rangle$$

$$= \sqrt{\frac{m\omega}{2\hbar}} \left(\sqrt{n'}\, \delta_{n,n'-1} - \sqrt{n'+1}\, \delta_{n,n'+1}\right). \tag{18.83}$$

Note that only transitions to neighboring levels are allowed. Therefore, we obtain

$$\frac{dC_n(t)}{dt} = -\dot{\alpha} \sqrt{\frac{m\omega}{2\hbar}} \left(C_{n-1}(t)\sqrt{n-1}\, e^{i(\varphi_{n-1} - \varphi_n)} - C_{n+1}(t)\sqrt{n+1}\, e^{i(\varphi_{n+1} - \varphi_n)}\right)$$

$$= -\dot{\alpha} \sqrt{\frac{m\omega}{2\hbar}} \left(C_{n-1}(t)\sqrt{n-1}\, e^{i\omega t} - C_{n+1}(t)\sqrt{n+1}\, e^{-i\omega t}\right). \tag{18.84}$$

Integrating we obtain

$$C_n(t) = C_n(0) - \sqrt{\frac{m\omega}{2\hbar}} \left(\,\ldots\,\right)$$

with

$$\cdots = \sqrt{n-1} \int_0^t dt' \dot{\alpha}(t') C_{n-1}(t') e^{i\omega t'} - \sqrt{n+1} \int_0^t dt' \dot{\alpha}(t') C_{n+1}(t') e^{-i\omega t'}. \tag{18.85}$$

Assuming that we have an adiabatic rate of change and $\dot{\alpha}$ is small (i.e., $\dot{\alpha}/\alpha \ll \omega$), we may approximate the coefficients inside the integral with their initial values and have a first-order result

$$C_n(t) \approx C_n(0) - \sqrt{\frac{m\omega}{2\hbar}} \left(\sqrt{n-1}\, C_{n-1}(0) D^*(t) - \sqrt{n+1}\, C_{n+1}(0) D(t)\right), \tag{18.86}$$

where $D \equiv \int_0^t dt' \dot{\alpha}(t') e^{i\omega t'}$. If $\alpha(0) = 0$ and initially the system occupies an eigenstate $|\psi_\ell\rangle$ of the standard harmonic oscillator, $C_n(0) = \delta_{n\ell}$ and

$$C_n(t) \approx \delta_{n\ell} - \sqrt{\frac{m\ell\omega}{2\hbar}} \left(\delta_{n,\ell+1} \int_0^t dt' \dot{\alpha}(t') e^{i\omega t'} - \delta_{n,\ell+1} \int_0^t dt' \dot{\alpha}(t') e^{-i\omega t'}\right), \tag{18.87}$$

which means that transition is possible only to the immediately neighboring eigenstates with probability

$$\mathcal{P}_{n \to n\pm1}(t) = (n \pm 1) \frac{m\omega}{2\hbar} \left| \int_0^t dt' \dot{\alpha}(t') e^{i\omega t'} \right|^2. \tag{18.88}$$

18.3.1 The Berry Phase

As we saw above, for an adiabatic process, the expansion coefficients in (18.48) change only by acquiring a phase

$$C_n(t) = C_n(0) e^{-i \sum_j \int_0^t dt' \dot{\alpha}_j(t') \mathcal{A}_j^{(n)}}. \qquad (18.89)$$

As a result, if the system is initially in an energy eigenstate $|\psi_n\rangle$, its state at a later time will be

$$|\psi_n(t)\rangle = e^{i\varphi_n} e^{i\gamma_n} |\psi_n\rangle, \qquad (18.90)$$

where $\varphi_n = -\frac{1}{\hbar} \int_0^t dt' E_n(\alpha(t'))$ is a *dynamical phase* and $\gamma_n = -\sum_j \int_0^t dt' \dot{\alpha}_j(t') \mathcal{A}_j^{(n)}$ is a *geometrical phase*. The quantity $\mathcal{A}^{(n)} = -i \langle \psi_n(\alpha)| \frac{\partial}{\partial \alpha} |\psi_n(\alpha)\rangle$ is called the *Berry connection*. Assume now that the system makes a closed curve in parameter space returning to the same values of the parameters $\alpha_j(0) = \alpha_j(t)$. Then, the geometric phase will be

$$\gamma_n = -\oint d\alpha_j \, \mathcal{A}_j^{(n)}. \qquad (18.91)$$

This is the *Berry phase* [3]. The value of this phase will depend on the closed path taken in parameter space and will not be in general zero. It is not difficult to see that the Berry phase vanishes if the eigenfunctions $\psi_n(x, \alpha)$ are real. We have

$$\gamma_n = i \oint d\alpha \int dx \, \psi_n(\alpha, x) \frac{\partial}{\partial \alpha} \psi_n(\alpha, x) = \frac{i}{2} \oint d\alpha \frac{\partial}{\partial \alpha} \int dx \, \psi_n^2(\alpha) = 0.$$

Another question that could be asked is whether γ_n is real. This is crucial because an imaginary part in γ_n would signal a violation of the conservation of probability, since the norm of the state (18.90) has to be conserved. Starting from

$$0 = \frac{\partial}{\partial \alpha} \langle \psi_n(\alpha)|\psi_n(\alpha)\rangle = \left(\frac{\partial}{\partial \alpha} \langle \psi_n(\alpha)|\right) |\psi_n(\alpha)\rangle + \langle \psi_n(\alpha)| \left(\frac{\partial}{\partial \alpha}|\psi_n(\alpha)\rangle\right)$$

we have

$$\gamma_n^* = -i \oint d\alpha \left(\langle \psi_n(\alpha)| \frac{\partial}{\partial \alpha} |\psi_n(\alpha)\rangle\right)^* = -i \oint d\alpha \frac{\partial}{\partial \alpha} (\langle \psi_n(\alpha)|) |\psi_n(\alpha)\rangle$$

$$= i \oint d\alpha \langle \psi_n(\alpha)| \frac{\partial}{\partial \alpha} (|\psi_n(\alpha)\rangle) = \gamma_n,$$

which proves its reality.

Example 18.4 As an example of a system that develops a nontrivial Berry phase during an adiabatic process, we may consider the system of a particle (an electron) in a homogeneous magnetic field. Ignoring the particle's spatial motion, the Hamiltonian is $\hat{H} = -\frac{e}{m_e c} \mathbf{B} \cdot \mathbf{S}$.

Parameterizing the direction of the magnetic field in terms of a *polar angle θ and an azimuthal angle parameter ϕ*, the Hamiltonian in the space of spin eigenstates corresponds to the matrix

$$\mathcal{H} = \hbar\omega \begin{pmatrix} \cos\theta & e^{-i\phi}\sin\theta \\ e^{i\phi}\sin\theta & -\cos\theta \end{pmatrix}, \tag{18.92}$$

where $\omega = |e|B/2m_e c$. The energy eigenvalues are $E_\pm = \pm\hbar\omega$, the corresponding eigenstates being

$$\chi_+ = \begin{pmatrix} \cos(\theta/2) \\ e^{i\phi}\sin(\theta/2) \end{pmatrix} \qquad \chi_- = \begin{pmatrix} \sin(\theta/2) \\ -e^{i\phi}\cos(\theta/2) \end{pmatrix}. \tag{18.93}$$

Assume now that initially the system is in an energy eigenstate, say χ_+, and the direction of the magnetic field changes slowly with time, i.e., $\theta(t)$, $\phi(t)$ vary, in an adiabatic fashion ($\dot{\theta}, \dot{\phi} \ll \omega$). The change is such that the magnetic field returns to its initial direction tracing a circular path in parameter space as shown in Fig. 18.4. Here the parameter space coincides with the direction in ordinary space. The corresponding Berry phase will depend on the path followed.

Starting from the expression for the Berry phase in terms of the Berry connection

$$\mathbf{A} = -i\chi_+^\dagger \nabla \chi_+, \tag{18.94}$$

we have

$$\gamma = -\oint d\mathbf{r} \cdot \mathbf{A} = -\int d\mathbf{S} \cdot (\nabla \times \mathbf{A}), \tag{18.95}$$

where the surface integral is over a segment of a spherical surface defined by the closed loop in parameter space. We have

$$\mathbf{A} = -\frac{i}{r}\chi_+^\dagger \left(\hat{\theta}\frac{\partial}{\partial\theta} + \frac{\hat{\phi}}{\sin\theta}\frac{\partial}{\partial\phi} \right) \chi_+ = \hat{\phi}\frac{\sin^2(\theta/2)}{r\sin\theta} \tag{18.96}$$

and

$$\nabla \times \mathbf{A} = \frac{1}{r^2}\left(\hat{\theta}\frac{\partial}{\partial\theta} + \frac{\hat{\phi}}{\sin\theta}\frac{\partial}{\partial\phi} \right) \times \left(\hat{\phi}\frac{\sin^2(\theta/2)}{\sin\theta} \right)$$

Fig. 18.4 Adiabatic rotation of the direction of the magnetic field

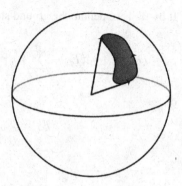

$$= \frac{\hat{r}}{r^2} \frac{\partial}{\partial \theta} \left(\frac{\sin^2(\theta/2)}{\sin \theta} \right) + \frac{\sin^2(\theta/2)}{r^2 \sin^2 \theta} \left(\hat{\phi} \times \frac{\partial \hat{\phi}}{\partial \phi} \right) =$$

$$\frac{\hat{r}}{r^2} \frac{\partial}{\partial \theta} \left(\frac{\sin^2(\theta/2)}{\sin \theta} \right) + \frac{\sin^2(\theta/2)}{r^2 \sin^2 \theta} \left(\sin \theta \hat{\theta} + \cos \theta \hat{r} \right) = \frac{\hat{r}}{2r^2} + \frac{\sin^2(\theta/2)}{r^2 \sin \theta} \hat{\theta} . \tag{18.97}$$

Therefore, we obtain

$$\gamma = - \int d\Omega \, r^2 \hat{r} \cdot \left(\frac{\hat{r}}{2r^2} \right) = -\frac{1}{2} \int d\Omega \implies \gamma = -\frac{1}{2} \Omega , \tag{18.98}$$

where Ω is the total solid angle traced by the circular path of the magnetic field direction.

18.4   The Variational Method

The central idea of the Variational Method [1, 3–5] is to parametrize the possible solutions to a given problem in terms of a set of suitable parameters and seek the actual solution as the one that minimizes a physical quantity (e.g., the energy). The method is particularly useful in placing an upper bound on the ground state energy of a system.

Consider a given quantum system and its energy eigenvalue problem

$$\hat{H} |\psi_n\rangle = E_n |\psi_n\rangle , \tag{18.99}$$

assuming that the system has a discrete energy spectrum. For a given normalized state

$$|\psi\rangle = \sum_{n=0}^{\infty} \psi_n |\psi_n\rangle \quad \left(\sum_{n=0}^{\infty} |\psi_n|^2 = 1 \right) , \tag{18.100}$$

we consider the expectation value of the energy

$$\langle E \rangle = \sum_{n,n'}^{\infty} \psi_n^* \psi_{n'} \langle \psi_n | \hat{H} | \psi_{n'} \rangle = \sum_{n=0}^{\infty} |\psi_n|^2 E_n . \tag{18.101}$$

If by $|\psi_0\rangle$ we denote the ground state, we can write

$$\langle E \rangle = |\psi_0|^2 E_0 + \sum_{n \geq 1}^{\infty} |\psi_n|^2 E_n = \left(1 - \sum_{n \geq 1}^{\infty} |\psi_n|^2 \right) E_0 + \sum_{n \geq 1}^{\infty} |\psi_n|^2 E_n$$

or

$$\langle E \rangle = E_0 + \sum_{n \geq 1}^{\infty} |\psi_n|^2 (E_n - E_0) . \tag{18.102}$$

Since, by hypothesis, E_0 is the ground state energy and $E_n > E_0$ for $\forall n \geq 1$, we must have

$$\langle E \rangle \geq E_0 . \tag{18.103}$$

Therefore, *the expectation value of the energy in any normalized state $|\psi\rangle$ obeys the inequality*

$$\langle E \rangle = \langle \psi | \hat{H} | \psi \rangle \geq E_0 , \tag{18.104}$$

where E_0 is the ground state energy.

Since the state $|\psi\rangle$ is an arbitrary state, we may consider a family of normalizable states depending on a set of parameters α_1, α_2, ... The expectation value of the energy for any such state will depend on these parameters

$$\mathcal{E}(\alpha) = \frac{\langle \psi(\alpha) | \hat{H} | \psi(\alpha) \rangle}{\langle \psi(\alpha) | \psi(\alpha) \rangle} . \tag{18.105}$$

According to what was shown above, we must have

$$\mathcal{E}(\alpha) \geq E_0 . \tag{18.106}$$

We may derive an upper bound for the ground state energy by minimizing the quantity $\mathcal{E}(\alpha)$ with respect to the parameters α, obtaining

$$E_0 \leq \mathcal{E}(\bar{\alpha}) , \tag{18.107}$$

where $\mathcal{E}(\bar{\alpha})$ will be the minimum of $\mathcal{E}(\alpha)$, defined by the condition for an extremum

$$\left. \frac{\partial \mathcal{E}(\alpha)}{\partial \alpha_j} \right|_{\bar{\alpha}} = 0 \tag{18.108}$$

and the requirement for a minimum, corresponding to the matrix $\left. \frac{\partial^2 \mathcal{E}}{\partial \alpha_i \partial \alpha_j} \right|_{\bar{\alpha}}$ having positive eigenvalues.

As a simple example of the above we may consider the system of a particle bound in the one-dimensional potential

$$V(x) = g^2 |x| . \tag{18.109}$$

Since we are seeking to determine its ground state energy, we note that for small values of x the potential is not very different from a harmonic oscillator potential and we may adopt for a trial wave function a Gaussian

$$\psi(x, \alpha) = N(\alpha) e^{-\frac{\alpha}{2} x^2} , \tag{18.110}$$

parametrized in terms of a parameter α that determines its width. The coefficient $N(\alpha)$ is fixed by normalization to be $N(\alpha) = (\alpha/\pi)^{1/4}$. Introducing this trial function into the expression for the expectation value of the energy, we obtain

$$\mathcal{E}(\alpha) = \frac{\hbar^2 \alpha}{4m} + \frac{g^2}{\sqrt{\pi \alpha}} . \tag{18.111}$$

The minimum occurs at

$$\bar{\alpha} = \left(\frac{2g^2 m}{\hbar^2 \sqrt{\pi}} \right)^{2/3} \implies \mathcal{E}(\bar{a}) = \frac{3}{2} \left(\frac{\hbar^2 g^4}{2\pi m} \right)^{1/3} . \tag{18.112}$$

Thus, we may conclude that

$$E_0 \le \frac{3}{2} \left(\frac{\hbar^2 g^4}{2\pi m} \right)^{1/3} . \tag{18.113}$$

Example 18.5 Consider a particle in the potential $V(x) = g^2 |x|$. Using the variational method estimate an upper bound for the first excited state energy.

The ground state energy for this potential was estimated above to be $E_0 \le \frac{3}{2} \left(\frac{\hbar^2 g^4}{2\pi m} \right)^{1/3}$ using the trial wave function $\psi(x, \alpha) = (\alpha/\pi)^{1/4} e^{-\frac{\alpha}{2} x^2}$. The parity invariance of the Hamiltonian implies that the energy eigenstates will be parity eigenstates. The ground state will be an even function ($\psi_0(-x) = \psi_0(x)$), while the first excited state will be an odd function ($\psi_1(-x) = -\psi(x)$). The expectation value of the energy in any normalized state $|\psi\rangle$ will be

$$\langle E \rangle = \frac{\sum_{n=0}^{\infty} |\langle \psi | \psi_n \rangle|^2 E_n}{\sum_{n=0}^{\infty} |\langle \psi | \psi_n \rangle|^2} . \tag{18.114}$$

For an odd wave function $\psi(-x) = -\psi(x)$ we have $\langle \psi | \psi_0 \rangle = 0$ and therefore

$$\langle E \rangle = \frac{E_1 |\langle \psi | \psi_1 \rangle|^2 + \sum_{n>1} |\langle \psi | \psi_n \rangle|^2 E_n}{|\langle \psi | \psi_1 \rangle|^2 + \sum_{n>1} |\langle \psi | \psi_n \rangle|^2} \implies \langle E \rangle \ge E_1 . \tag{18.115}$$

Using the odd trial wave function $\psi_1(x, \alpha) = \sqrt{2} \left(\frac{\alpha^3}{\pi} \right)^{1/4} x\, e^{-\frac{\alpha}{2} x^2}$ we compute the expectation value of the energy $\mathcal{E}(\alpha)$ minimizing it,t we obtain

$$E_1 \le 3 \left(\frac{3g^4 \hbar^2}{4m\pi} \right)^{1/3} \approx 2.72 \left(\frac{g^4 \hbar^2}{m\pi} \right)^{1/3} . \tag{18.116}$$

This is correctly above the previously estimated ground state energy

$$E_0 = \frac{3}{2} \left(\frac{\hbar^2 g^4}{2\pi m} \right)^{1/3} \approx 1.19 \left(\frac{\hbar^2 g^4}{\pi m} \right)^{1/3} . \tag{18.117}$$

18.4.1 Application on the Helium Atom

The Hamilton operator for the Helium atom

$$\hat{H} = \frac{\mathbf{p}_1^2}{2m_e} + \frac{\mathbf{p}_2^2}{2m_e} - \frac{\mathcal{Z}e^2}{r_1} - \frac{\mathcal{Z}e^2}{r_2} + \frac{e^2}{|\mathbf{r}_1 - \mathbf{r}_2|} \qquad (18.118)$$

is made of a *solvable* part, consisting of two hydrogen-type Hamiltonians, and the electron repulsion term. This term is not a priori small and cannot be safely treated as a perturbation. The obvious alternative is to apply the variational method. A starting point would be to observe that the effective charge of the nuclei felt by each of the electrons is not exactly $\mathcal{Z} = 2$ but is lowered due to the presence of the second electron. This motivates us to treat $\mathcal{Z}$ as a variable parameter and introduce a trial wave function corresponding to two independent copies of a hydrogen-like atom with atomic number $\mathcal{Z}$. Thus, we embark in an application of the variational method, taking as our trial wave function

$$\Psi(\mathbf{r}_1, \mathbf{r}_2) = \psi_{100}(\mathbf{r}_1) \, \psi_{100}(\mathbf{r}_2) \qquad (18.119)$$

with

$$\psi_{100}(r) = \left(\frac{\mathcal{Z}'}{a_0}\right)^{3/2} \frac{e^{-\mathcal{Z}'r/a_0}}{\sqrt{\pi}} . \qquad (18.120)$$

We can rewrite the Hamiltonian as

$$\hat{H} = \frac{\mathbf{p}_1^2}{2m_e} + \frac{\mathbf{p}_2^2}{2m_e} - \frac{\mathcal{Z}'e^2}{r_1} - \frac{\mathcal{Z}'e^2}{r_2} - e^2(2 - \mathcal{Z}')\left(\frac{1}{r_1} + \frac{1}{r_2}\right) + \frac{e^2}{|\mathbf{r}_1 - \mathbf{r}_2|} . \qquad (18.121)$$

The expectation value of the energy in this state is

$$\mathcal{E}(\mathcal{Z}') = -\frac{\mathcal{Z}'^2 e^2}{a_0} - 2e^2(2 - \mathcal{Z}')\left\langle\frac{1}{r}\right\rangle + e^2 \int d^3r_1 \int d^3r_2 \frac{|\psi_{100}(r_1)|^2|\psi_{100}(r_2)|^2}{|\mathbf{r}_1 - \mathbf{r}_2|} . \qquad (18.122)$$

The quantity $\langle\frac{1}{r}\rangle$ is

$$\left\langle\frac{1}{r}\right\rangle = \frac{1}{\pi}\left(\frac{\mathcal{Z}'}{a_0}\right)^3 \int d^3r \frac{e^{-2\mathcal{Z}'r/a_0}}{r} = \frac{\mathcal{Z}'}{a_0} . \qquad (18.123)$$

The repulsion integral has been computed before in Sect. 15.5.2 for the case $\mathcal{Z} = 2$. We can obtain its value here by the replacement $a_0 \to a_0(2/\mathcal{Z}')$. We obtain

$$e^2 \int d^3r_1 \int d^3r_2 \frac{|\psi_{100}(r_1)|^2|\psi_{100}(r_2)|^2}{|\mathbf{r}_1 - \mathbf{r}_2|} = \frac{5}{8}\mathcal{Z}'\frac{e^2}{a_0} . \qquad (18.124)$$

Summing up we have

$$\mathcal{E}(\mathcal{Z}') = -\frac{e^2}{a_0}\left(\mathcal{Z}'^2 + 2\mathcal{Z}'(2 - \mathcal{Z}') - \frac{5}{8}\mathcal{Z}' \right). \tag{18.125}$$

Minimizing we obtain

$$\bar{\mathcal{Z}} = 2 - \frac{5}{16} \implies E_0 = \mathcal{E}(\bar{\mathcal{Z}}) = -\frac{e^2}{a_0}\left(2 - \frac{5}{16} \right)^2. \tag{18.126}$$

The corresponding numerical value is $-77.5\,\text{eV}$, not far from the experimental value of $-79.0\,\text{eV}$.

18.4.2 Variational Arguments on the Existence of Bound States

The variational method is based on the fact that the ground state energy is always smaller than the expectation value of the energy in any (trial) state. In the case of a particle that moves in a localized potential with its value at infinity taken to be zero, a bound state is a state with negative energy. Since the ground state energy is smaller than the expectation value of the energy of any state, finding one trial wave function with negative energy would be sufficient proof for the existence of a bound state. As an example we may consider a particle moving under the influence of the *Yukawa potential* $V(r) = -\frac{g^2}{r}e^{-r/r_0}$. Using a hydrogen-like trial wave function

$$\psi(r, \alpha) = \frac{\alpha^{3/2}}{\sqrt{\pi}}e^{-\alpha r} \tag{18.127}$$

we calculate the expectation value of the energy as

$$\langle E \rangle_\alpha = \frac{\hbar^2 \alpha^3}{2m} - \frac{4g^2\alpha^3}{\left(2\alpha + \frac{1}{r_0} \right)^2}. \tag{18.128}$$

It is clear that for

$$g^2 > \frac{\hbar^2}{8mr_0^2} \implies \langle E \rangle_\alpha < 0 \tag{18.129}$$

and, therefore,

$$E_0 < \langle E \rangle_\alpha < 0 \implies E_0 < 0. \tag{18.130}$$

It is useful to note that the *Virial Theorem*, combined with the above argument, can also be used to answer the question of the existence of bound states. For monomial nonnegative central potentials of the form $V(r) = \lambda r^n$ the Virial theorem takes the form $\langle T \rangle = \frac{n}{2}\langle V \rangle$, which implies that $\langle E \rangle = \left(\frac{2}{n} + 1\right)\langle T \rangle$ and negative energy arises for $n \geq -2$.

18.5 Time-Independent Perturbation Theory

Consider a physical system with a Hamiltonian

$$\hat{H} = \hat{H}_0 + \lambda \hat{V} \tag{18.131}$$

that is composed of two parts, a *solvable* part $\hat{H}_0$, with a known set of eigenstates and eigenvalues, and a part $\lambda \hat{V}$ that can be considered as a perturbation,[5] characterized by a *"small"* parameter λ. Let $|\psi_n^{(0)}\rangle$ and $E_n^{(0)}$ be the eigenstates and eigenvalues of the soluble part[6]

$$\hat{H}_0|\psi_n^{(0)}\rangle = E_n^{(0)}|\psi_n^{(0)}\rangle \tag{18.132}$$

and let $|\psi_n\rangle$ and E_n be the exact eigenstates and eigenvalues of the full Hamiltonian

$$\hat{H}|\psi_n\rangle = E_n|\psi_n\rangle. \tag{18.133}$$

The basic idea of *Perturbation Theory* is to obtain approximate solutions to (18.133) in the form of power series in λ, namely,

$$E_n = E_n^{(0)} + \lambda E_n^{(1)} + \lambda^2 E_n^{(2)} + \dots$$
$$|\psi_n\rangle = |\psi_n^{(0)}\rangle + \lambda|\psi_n^{(1)}\rangle + \lambda^2|\psi_n^{(2)}\rangle + \dots \tag{18.134}$$

The success of the perturbation theory procedure depends on the form of each particular perturbation V and from a mathematical point of view relies on the convergence or, at least, on the asymptotic character of the perturbation series (18.134).

We assume that the unperturbed eigenstates are normalized in the standard way, namely,

$$\langle \psi_n^{(0)}|\psi_{n'}^{(0)}\rangle = \delta_{nn'}. \tag{18.135}$$

Although we could normalize the exact eigenstates $|\psi_n\rangle$ in the same way, we are free to adopt an unconventional normalization, the merit of which will become clear in what is to follow. Therefore, we normalize the exact eigenstates according to

$$\langle \psi_n^{(0)}|\psi_n\rangle = 1. \tag{18.136}$$

Note that this implies

$$\langle \psi_n^{(0)}|\left(|\psi_n^{(0)}\rangle + \lambda|\psi_n^{(1)}\rangle + \lambda^2|\psi_n^{(2)}\rangle + \dots \right) = 1$$

---

[5]For the treatment of perturbation theory see also [1, 3–5].

[6]We assume that the spectrum is discrete. The formulation of perturbation theory in the case of continuous spectrum will be presented in the chapter on Scattering. In addition, here, we assume that the eigenvalue spectrum is *nondegenerate*. We shall deal with the case of degeneracies in a particular subsection.

or

$$\langle \psi_n^{(0)} | \psi_n^{(j)} \rangle = 0 \ for \quad j \geq 1 . \tag{18.137}$$

Substituting the expansions (18.134) into (18.133) we obtain

$$\left(\hat{H}_0 + \lambda \hat{V} \right) \left(|\psi_n^{(0)}\rangle + \lambda |\psi_n^{(1)}\rangle + \lambda^2 |\psi_n^{(2)}\rangle + \dots \right) =$$

$$\left(E_n^{(0)} + \lambda E_n^{(1)} + \lambda^2 E_n^{(2)} + \dots \right) \left(|\psi_n^{(0)}\rangle + \lambda |\psi_n^{(1)}\rangle + \lambda^2 |\psi_n^{(2)}\rangle + \dots \right). \tag{18.138}$$

After some manipulations, equating the coefficients of each power of λ, we obtain

$$\hat{H}_0 |\psi_n^{(1)}\rangle + \hat{V} |\psi_n^{(0)}\rangle = E_n^{(0)} |\psi_n^{(1)}\rangle + E_n^{(1)} |\psi_n^{(0)}\rangle$$

$$\hat{H}_0 |\psi_n^{(2)}\rangle + \hat{V} |\psi_n^{(1)}\rangle = E_n^{(0)} |\psi_n^{(2)}\rangle + E_n^{(1)} |\psi_n^{(1)}\rangle + E_n^{(2)} |\psi_n^{(0)}\rangle \tag{18.139}$$

......

or, generally for $j \geq 1$

$$\hat{H}_0 |\psi_n^{(j)}\rangle + \hat{V} |\psi_n^{(j-1)}\rangle = E_n^{(0)} |\psi_n^{(j)}\rangle + E_n^{(1)} |\psi_n^{(j-1)}\rangle + \dots + E_n^{(j)} |\psi_n^{(0)}\rangle . \tag{18.140}$$

The first of the relations (18.139) allows us to determine the first-order correction to the energy eigenvalue. Taking the inner product with $|\psi_n^{(0)}\rangle$ we obtain

$$\langle \psi_n^{(0)} | \hat{H}_0 | \psi_n^{(1)} \rangle + \langle \psi_n^{(0)} | \hat{V} | \psi_n^{(0)} \rangle = E_n^{(0)} \langle \psi_n^{(0)} | \psi_n^{(1)} \rangle + E_n^{(1)} \langle \psi_n^{(0)} | \psi_n^{(0)} \rangle \tag{18.141}$$

or, after we take into account (18.137)

$$E_n^{(1)} = \langle \psi_n^{(0)} | \hat{V} | \psi_n^{(0)} \rangle . \tag{18.142}$$

We may also use the first-order relation of the set (18.139) to obtain the first-order correction to the corresponding eigenstate. Multiplying with a different unperturbed eigenstate $|\psi_m^{(0)}\rangle$, we obtain

$$\langle \psi_m^{(0)} | \hat{H}_0 | \psi_n^{(1)} \rangle + \langle \psi_m^{(0)} | \hat{V} | \psi_n^{(0)} \rangle = E_n^{(0)} \langle \psi_m^{(0)} | \psi_n^{(1)} \rangle + E_n^{(1)} \langle \psi_m^{(0)} | \psi_n^{(0)} \rangle \tag{18.143}$$

or

$$E_m^{(0)} \langle \psi_m^{(0)} | \psi_n^{(1)} \rangle + \langle \psi_m^{(0)} | \hat{V} | \psi_n^{(0)} \rangle = E_n^{(0)} \langle \psi_m^{(0)} | \psi_n^{(1)} \rangle \tag{18.144}$$

or

$$\langle \psi_m^{(0)} | \psi_n^{(0)} \rangle = \frac{\langle \psi_m^{(0)} | \hat{V} | \psi_n^{(0)} \rangle}{E_n^{(0)} - E_m^{(0)}} . \tag{18.145}$$

Using completeness, we have

$$|\psi_n^{(1)}\rangle = \sum_{m \neq n} |\psi_m^{(0)}\rangle \langle \psi_m^{(0)}|\psi_n^{(1)}\rangle \qquad (18.146)$$

or

$$|\psi_n^{(1)}\rangle = \sum_{m \neq n} \frac{\langle \psi_m^{(0)}|\hat{V}|\psi_n^{(0)}\rangle}{E_n^{(0)} - E_m^{(0)}} |\psi_m^{(0)}\rangle . \qquad (18.147)$$

Note that, if degeneracies occur and more than one state corresponds to the same energy level, the denominator vanishes and the above expression is not valid. Thus, here we explicitly assume that no degeneracies occur. The treatment of degeneracies will be treated in a subsequent subsection on *Degenerate Perturbation Theory*.

Having obtained these first-order corrections, we can proceed and compute the next order corrections. For example, the second-order correction to the energy eigenvalues is obtained by multiplying the corresponding relation in (18.139) with $|\psi_n^{(0)}\rangle$

$$\langle \psi_n^{(0)}|\hat{H}_0|\psi_n^{(2)}\rangle + \langle \psi_n^{(0)}|\hat{V}|\psi_n^{(1)}\rangle = E_n^{(0)}\langle \psi_n^{(0)}|\psi_n^{(2)}\rangle + E_n^{(1)}\langle \psi_n^{(0)}|\psi_n^{(1)}\rangle + E_n^{(2)}\langle \psi_n^{(0)}|\psi_n^{(0)}\rangle$$

or

$$E_n^{(2)} = \langle \psi_n^{(0)}|\hat{V}|\psi_n^{(1)}\rangle \qquad (18.148)$$

and, using (18.147),

$$E_n^{(2)} = \sum_{m=\neq n} \frac{|\langle \psi_n^{(0)}|\hat{V}|\psi_m^{(0)}\rangle|^2}{E_n^{(0)} - E_m^{(0)}} . \qquad (18.149)$$

Similarly, multiplying the second equation of the set (18.139) by an eigenstate $|\psi_m^{(0)}\rangle$ for $m \neq n$, we obtain

$$\langle \psi_m^{(0)}|\hat{H}_0|\psi_n^{(2)}\rangle + \langle \psi_m^{(0)}|\hat{V}|\psi_n^{(1)}\rangle = E_n^{(0)}\langle \psi_m^{(0)}|\psi_n^{(2)}\rangle + E_n^{(1)}\langle \psi_m^{(0)}|\psi_n^{(1)}\rangle + E_n^{(2)}\langle \psi_m^{(0)}|\psi_n^{(0)}\rangle$$

or

$$\langle \psi_m^{(0)}|\psi_n^{(2)}\rangle = \frac{\langle \psi_m^{(0)}|\hat{V}|\psi_n^{(1)}\rangle}{E_n^{(0)} - E_m^{(0)}} \qquad (18.150)$$

and using completeness

$$|\psi_n^{(2)}\rangle = \sum_{m \neq n} \frac{\langle \psi_m^{(0)}|\hat{V}|\psi_n^{(1)}\rangle}{E_n^{(0)} - E_m^{(0)}} |\psi_m^{(0)}\rangle . \qquad (18.151)$$

The complete second-order expressions are

$$E_n^{(2)} = \sum_{m \neq n} \frac{|\langle \psi_n^{(0)} | \hat{V} | \psi_m^{(0)} \rangle|^2}{E_n^{(0)} - E_m^{(0)}}$$

$$|\psi_n^{(2)}\rangle = \sum_{\substack{m' \neq n \\ m \neq n}} \frac{\langle \psi_m^{(0)} | \hat{V} | \psi_{m'}^{(0)} \rangle \langle \psi_{m'}^{(0)} | \hat{V} | \psi_n^{(0)} \rangle}{(E_n^{(0)} - E_{m'}^{(0)})(E_n^{(0)} - E_m^{(0)})} |\psi_m^{(0)}\rangle - \sum_{m \neq n} \frac{\langle \psi_m^{(0)} | \hat{V} | \psi_n^{(0)} \rangle \langle \psi_n^{(0)} | \hat{V} | \psi_n^{(0)} \rangle}{(E_n^{(0)} - E_m^{(0)})^2} |\psi_m^{(0)}\rangle.$$

(18.152)

Example 18.6 Consider a simple harmonic oscillator subject to an anharmonic perturbation λx^4 and calculate the first-order correction to the energy eigenvalue and eigenstate of the ground state.

We have

$$E_0^{(1)} = \langle 0 | x^4 | 0 \rangle = \left\| x^2 | 0 \rangle \right\|^2 = \frac{\hbar^2}{4(m\omega)^2} \left\| \left(a + a^\dagger \right)^2 | 0 \rangle \right\|^2$$

$$= \frac{\hbar^2}{4(m\omega)^2} \left\| \left(a^2 + a^{\dagger 2} + a^\dagger a + a a^\dagger \right) | 0 \rangle \right\|^2 = \frac{\hbar^2}{4(m\omega)^2} \left\| \left(a^2 + a^{\dagger 2} + 2 a^\dagger a + 1 \right) | 0 \rangle \right\|^2$$

$$= \frac{\hbar^2}{4(m\omega)^2} \left\| \sqrt{2} | 2 \rangle + | 0 \rangle \right\|^2 = \frac{3\hbar^2}{4(m\omega)^2}.$$

(18.153)

For the corresponding correction to the ground state, we have

$$|\psi_0^{(1)}\rangle = \sum_{n=1}^{\infty} \frac{\langle n | x^4 | 0 \rangle}{E_0^{(0)} - E_n^{(0)}} | n \rangle.$$

(18.154)

In particular, we have

$$x^2 | n \rangle = \frac{\hbar}{2m\omega} \left(a^2 + a^{\dagger 2} + 2 a^\dagger a + 1 \right) | n \rangle =$$

$$\frac{\hbar}{2m\omega} \left(\sqrt{n(n-1)} | n - 2 \rangle + \sqrt{(n+1)(n+2)} | n + 2 \rangle + (2n+1) | n \rangle \right)$$

(18.155)

and

$$x^2 | 0 \rangle = \frac{\hbar}{2m\omega} \left(\sqrt{2} | 2 \rangle + | 0 \rangle \right).$$

(18.156)

Therefore,

$$\langle n | x^4 | 0 \rangle = \frac{\hbar^2}{4(m\omega)^2} \left(\delta_{n,4} \sqrt{4} \sqrt{3} \sqrt{2} + 6 \sqrt{2} \delta_{n,2} \right).$$

(18.157)

Finally, we have

$$|\psi_0^{(1)}\rangle = \frac{-\hbar}{4m^2\omega^3} \sum_{n=1}^{\infty} \frac{1}{n} \left(\delta_{n,4} \sqrt{4} \sqrt{3} \sqrt{2} + 6 \sqrt{2} \delta_{n,2} \right) | n \rangle = \frac{-\hbar}{8m^2\omega^3} \left(6 \sqrt{2} | 2 \rangle + \sqrt{6} | 4 \rangle \right).$$

(18.158)

The complete first-order eigenvalue and state are

$$E_0 = \tfrac{1}{2} \hbar \omega + \frac{3\lambda\hbar^2}{4(m\omega)^2}$$

$$|\psi_0\rangle = | 0 \rangle - \frac{\hbar\lambda}{8m^2\omega^3} \left(6 \sqrt{2} | 2 \rangle + \sqrt{6} | 4 \rangle \right).$$

(18.159)

18.5.1 An Application: The Stark Effect

Consider a hydrogen atom subject to a homogeneous electric field $\mathcal{E}$. The corresponding potential energy term is $-e\mathcal{E} \cdot \mathbf{r}$. Assuming that the field is weak enough, it can be treated as a perturbation. This assumption is substantiated as

$$\frac{e^2}{a_0} \gg |e\mathcal{E}|a_0 \quad \Longrightarrow \quad |e\mathcal{E}| \ll e^2/a_0^3. \tag{18.160}$$

Although the unperturbed energy spectrum is degenerate, it is instructive to apply the previously analyzed machinery of perturbation theory to explore the basic features of the corrections. The lowest order correction to the energy eigenvalues has to be proportional to the matrix elements $\langle n, \ell, m|\mathbf{r}|n, \ell', m'\rangle$. Note, however, that $\mathbf{r}$ is odd under parity and $Pr P = -\mathbf{r}$. Thus, we have

$$\langle n, \ell, m|\mathbf{r}|n, \ell', m'\rangle = -\langle n, \ell, m|Pr P|n, \ell', m'\rangle = -(-1)^{\ell+\ell'}\langle n, \ell, m|\mathbf{r}|n, \ell', m'\rangle,$$

which implies that this matrix element vanishes unless $\ell + \ell'$ is a odd number. Without loss of generality we can choose the $\hat{z}$-axis along the direction of the electric field. Then, the relevant matrix elements are $\langle n, \ell, m|z|n., \ell', m'\rangle$. Since z is unaffected by rotations around the $\hat{z}$-axis (i.e., $e^{\frac{i}{\hbar}\alpha \hat{L}_z} z e^{-\frac{i}{\hbar}\alpha \hat{L}_z} = z$), we have

$$\langle n\ell m|z|n'\ell'm'\rangle = \langle n\ell m|e^{\frac{i}{\hbar}\alpha \hat{L}_z} z e^{-\frac{i}{\hbar}\alpha \hat{L}_z}|n'\ell'm'\rangle = e^{i(m-m')\alpha}\langle n\ell m|z|n'\ell'm'\rangle,$$

which implies that matrix elements with only $m' = m$ arise. Thus, the first-order correction to the energy levels (*Linear Stark Effect*) is

$$E_{n\ell\ell'}^{(1)} = -e\mathcal{E}\langle n, \ell, m|z|n, \ell', m\rangle \tag{18.161}$$

with $\ell' + \ell = odd$. Note that the operator involved in the linear Stark effect is the *electric dipole moment* operator $\mathbf{d}_e = e\,\mathbf{r}$. Specifically, for the ground state energy the first-order correction vanishes, since it is proportional to $\langle 100|z|100\rangle$. In contrast, for the first excited set of states $|2, \ell, m\rangle$ there is nonvanishing first-order correction

$$E_2^{(1)} = -e\mathcal{E}\begin{pmatrix} \langle 2, 0, 0|z|2, 0, 0\rangle & \langle 2, 0, 0|z|2, 1, 0\rangle \\ \langle 2, 1, 0|z|2, 0, 0\rangle & \langle 2, 1, 0|z|2, 1, 0\rangle \end{pmatrix} = 3e\mathcal{E}a_0 \begin{pmatrix} 0 & 1 \\ 1 & 0 \end{pmatrix}, \tag{18.162}$$

where, inserting the specific wave functions, we obtain $\langle 2, 0, 0|z|2, 1, 0\rangle = -3a_0$. Thus, to first order, the $n = 2$ energy eigenstates are

$$|2, \pm\rangle = \frac{1}{\sqrt{2}}\left(|2, 0, 0\rangle \pm |2, 1, 0\rangle\right) \quad \Longrightarrow \quad E_{n=2} = -\frac{e^2}{4a_0} \pm 3\mathcal{E}a_0. \tag{18.163}$$

Since there is no first-order correction to the ground state energy, we proceed to consider the second-order correction

$$E_{100}^{(2)} = e^2 \mathcal{E}^2 \sum_{n=2}^{\infty} \sum_{\ell=odd}^{n-1} \frac{|\langle 1, 0, 0|z|n, \ell, 0\rangle|^2}{E_1^{(0)} - E_n^{(0)}} . \qquad (18.164)$$

The matrix elements involved are

$$\langle 1, 0, 0|z|n, \ell, 0\rangle = \int d\Omega\, Y_{00}^* Y_{\ell 0} \cos\theta \int_0^\infty dr\, r^3\, R_{10}(r)\, R_{n\ell}(r)$$

$$= \frac{1}{\sqrt{3}} \int d\Omega\, Y_{10}^* Y_{\ell 0} \int_0^\infty dr\, r^3\, R_{10}(r)\, R_{n\ell}(r)$$

or

$$\langle 1, 0, 0|z|n, \ell, 0\rangle = \frac{\delta_{\ell 1}}{\sqrt{3}} \int_0^\infty dr\, r^3\, R_{10}(r)\, R_{n1}(r) . \qquad (18.165)$$

Thus, the above sum reduces to

$$E_{100}^{(2)} = \frac{e^2 \mathcal{E}^2}{3 E_1^{(0)}} \sum_{n=2}^{\infty} \frac{\left|\int_0^\infty dr\, r^3 R_{10} R_{n1}\right|^2}{1 - 1/n^2} . \qquad (18.166)$$

The integral involved on dimensional grounds will be proportional to a_0. Therefore, the final expression will be in the form

$$E_{100}^{(2)} = \mathcal{E}^2 a_0^3\, C^2 , \qquad (18.167)$$

where C^2 stands for

$$C^2 = \frac{1}{3} \sum_{n=2}^{\infty} \frac{|\int_0^\infty dx\, x^3\, R_{10}(x) R_{n1}(x)|^2}{1 - 1/n^2}$$

with $x = r/a_0$ in the argument of radial wave functions. The exact value of this sum is $C^2 = 9/4$. Therefore, the second-order correction of the ground state (*Quadratic Stark Effect*) is

$$E_{100}^{(2)} = \frac{9}{4}\mathcal{E}^2 a_0^3 . \qquad (18.168)$$

18.5.2   An Application: van der Waals Forces

Consider two hydrogen atoms. The Hamiltonian of the full system can be written as

$$\hat{H} = \hat{H}_0 + \hat{V}, \tag{18.169}$$

where $\hat{H}_0$ is the soluble part of two hydrogen atom Hamiltonians

$$\hat{H}_0 = \frac{p_1^2}{2me} + \frac{p_2^2}{2m_e} - \frac{e^2}{r_1} - \frac{e^2}{r_2} \tag{18.170}$$

and $\hat{V}$ is

$$\hat{V} = \frac{e^2}{R} - \frac{e^2}{|\mathbf{R} + \mathbf{r}_2|} - \frac{e^2}{|\mathbf{R} - \mathbf{r}_1|} + \frac{e^2}{|\mathbf{R} + \mathbf{r}_2 - \mathbf{r}_1|}. \tag{18.171}$$

We are treating the distance R between the atoms as a parameter and assume that it is quite large in comparison to the size of the atoms ($R \gg a_0$). Thus, the interaction term is proportional to inverse powers of R and can be treated as a perturbation. Expanding in $1/R$ we obtain

$$\hat{V} \approx \frac{e^2}{R^3} (x_1 x_2 + y_1 y_2 - 2z_1 z_2) + O(R^{-4}). \tag{18.172}$$

For the ground state of $\hat{H}_0$, i.e., $\psi_{100}(\mathbf{r}_1) \psi_{100}(\mathbf{r}_2)$, the first-order shift to the energy vanishes and we must consider the second-order correction

$$E_1^{(2)} = -\frac{a_0 e^2}{R^6} \sum_{n_1, n_2 \neq 1} \frac{|\langle 100, 100| (x_1 x_2 + y_1 y_2 - 2z_1 z_2) |n_1 \ell_1 m_1, n_2 \ell_2 m_2\rangle|^2}{2 - \frac{1}{n_1^2} - \frac{1}{n_2^2}}. \tag{18.173}$$

This sum is of the form

$$E_1^{(2)} = -\frac{e^2}{2a_0} \left(\frac{a_0}{R}\right)^6 C^2, \tag{18.174}$$

where C^2 is a dimensionless number to be computed by performing the integral.

18.5.3 Neighboring Energy Levels

It often occurs that two energy levels are very close. As a result, very large terms of the form $\langle \psi_n^{(0)}|\hat{V}|\psi_m^{(0)}\rangle / (E_n^{(0)} - E_m^{(0)})$ may arise, something that would make the convergence of the perturbation series problematic. A way to circumvent this problem is to consider separately this set of states. We shall illustrate this procedure in the simple case of two such states.

The potential operator can be written as

$$\hat{V} = \sum_{i,j} |\psi_i^{(0)}\rangle V_{ij} \langle \psi_j^{(0)}|, \tag{18.175}$$

where $V_{ij} = \langle \psi_i^{(0)} | \hat{V} | \psi_j^{(0)} \rangle$. Isolating the two problematic states $| \psi_n^{(0)} \rangle$, $| \psi_m^{(0)} \rangle$, we have

$$\hat{V} = \hat{V}_1 + \hat{V}_2, \tag{18.176}$$

where

$$\hat{V}_1 = | \psi_m^{(0)} \rangle \langle \psi_m^{(0)} | V_{mm} + | \psi_m^{(0)} \rangle \langle \psi_n^{(0)} | V_{mn} + | \psi_n^{(0)} \rangle \langle \psi_m^{(0)} | V_{nm} + | \psi_n^{(0)} \rangle \langle \psi_n^{(0)} | V_{nn} \tag{18.177}$$

and

$$\hat{V}_2 = \sum_{i,j \neq n,m} | \psi_i^{(0)} \rangle V_{ij} \langle \psi_j^{(0)} |. \tag{18.178}$$

The Hamiltonian $\hat{H}_1 = \hat{H}_0 + \hat{V}_1$ can be solved exactly. This goes as follows: Any state $| \psi_j^{(0)} \rangle$ with $j \neq n, m$ is an eigenstate of $\hat{H}_1$ because of orthogonality. Therefore,

$$\hat{H}_1 | \psi_j^{(0)} \rangle = \hat{H}_0 | \psi_j^{(0)} \rangle = E_j^{(0)} | \psi_j^{(0)} \rangle. \tag{18.179}$$

Then, acting on a linear combination of the neighboring states and demanding that it corresponds to an eigenstate

$$\hat{H}_1 \left(C | \psi_n^{(0)} \rangle + D | \psi_m^{(0)} \rangle \right) = E \left(C | \psi_n^{(0)} \rangle + D | \psi_m^{(0)} \rangle \right), \tag{18.180}$$

we obtain the system of equations

$$\begin{aligned} \left(E_n^{(0)} + V_{nm} - E \right) C + V_{nm} D &= 0 \\ V_{mn} C + \left(E_m^{(0)} + V_{mn} - E \right) D &= 0. \end{aligned} \tag{18.181}$$

The corresponding determinant of the system leads to the E-eigenvalue equation

$$\begin{vmatrix} E_n^{(0)} + V_{nn} - E & V_{nm} \\ V_{mn} & E_m^{(0)} + V_{mn} - E \end{vmatrix} = 0 \tag{18.182}$$

$$E_\pm = \frac{1}{2} \left(E_n^{(0)} + E_m^{(0)} + V_{nn} + V_{mm} \pm \sqrt{\Delta} \right), \tag{18.183}$$

where $\Delta = \left(E_n^{(0)} - E_m^{(0)} + V_{nm} - V_{mn} \right)^2 + | V_{nm} |^2$. The corresponding solution for the coefficients reads

$$C = -D \frac{V_{nm}}{E_n^{(0)} + V_{nn} - E_\pm} \tag{18.184}$$

and, after normalization, $D = 1/\sqrt{1 + |V_{nm}|^2/(E_n^{(0)} + V_{nn} - E_\pm)^2}$.

Next, we can replace the states $| \psi_n^{(0)} \rangle$, $| \psi_m^{(0)} \rangle$ with the states

$$|\psi_{\pm}^{(0)}\rangle = C|\psi_n^{(0)}\rangle + D|\psi_m^{(0)}\rangle \implies E_{\pm}. \tag{18.185}$$

Now, the set of unperturbed eigenstates $\left\{|\psi_{\pm}^{(0)}\rangle, \, |\psi_j^{(0)}\rangle\right\}$ with $j \neq n, m$ does not contain any states of neighboring eigenvalues. Note that

$$E_+ - E_- = \sqrt{(V_{mm} - V_{nn})^2 + 4|V_{nm}|^2}, \tag{18.186}$$

which is not necessarily small. Thus, we may proceed to formulate perturbation theory for $\hat{V}_2$ using as the *"unperturbed set"* the set $\left\{|\psi_{\pm}^{(0)}\rangle, \, |\psi_j^{(0)}\rangle\right\}$.

Example 18.7 Consider the case of a three-level system[7] with the Hamiltonian

$$H = H_0 + \lambda V = \begin{pmatrix} E + \delta & 0 & 0 \\ 0 & E - \delta & 0 \\ 0 & 0 & E' \end{pmatrix} + \lambda \begin{pmatrix} 0 & 1 & 1 \\ 1 & 0 & 1 \\ 1 & 1 & 1 \end{pmatrix}. \tag{18.187}$$

Calculate the second-order corrections to the energy eigenvalues.

If δ is small (i.e., of the order of λ), in order to avoid large numbers of $O(1/\delta)$ occurring to second order, we may apply the above procedure and redefine the problem as

$$H = \bar{H}_0 + \lambda \bar{V}$$

with

$$\bar{H}_0 = \begin{pmatrix} E + \delta & \lambda & 0 \\ \lambda & E - \delta & 0 \\ 0 & 0 & E' \end{pmatrix} \quad and \quad \bar{V} = \begin{pmatrix} 0 & 0 & 1 \\ 0 & 0 & 1 \\ 1 & 1 & 1 \end{pmatrix}$$

The eigenvalues of $\bar{H}_0$ are given by

$$\begin{vmatrix} E + \delta - \bar{E} & \lambda \\ \lambda & E - \delta - \bar{E} \end{vmatrix} = 0 \implies \bar{E} = E \pm \sqrt{\lambda^2 + \delta^2}, \tag{18.188}$$

while the corresponding eigenstates are

$$\begin{aligned} |\bar{1}\rangle &= N_1\left(\lambda|1\rangle + (\bar{E}_+ - E - \delta)|2\rangle\right) \\ |\bar{2}\rangle &= N_2\left(\lambda|1\rangle + (\bar{E}_- - E - \delta)|2\rangle\right) \end{aligned} \tag{18.189}$$

with $N_{1,2} = \left(\lambda^2 + (\bar{E}_{\pm} - E - \delta)^2\right)^{-1/2}$.

---

[7]The eigenvalue problem of any two-level system is trivially soluble. This is not the case, however, of a general three-level system, since it involves a cubic equation, the algebraic solution of which is rather hairy.

The first-order corrections are

$$\bar{E}_1^{(1)} = \langle \bar{1}| \, \bar{V} \, |\bar{1}\rangle = 0, \quad \bar{E}_2^{(1)} = \langle \bar{2}| \, \bar{V} \, |\bar{2}\rangle = 0 \tag{18.190}$$

and

$$\bar{E}_3^{(1)} = \langle 3|\hat{\bar{V}}|3\rangle = 1. \tag{18.191}$$

The second-order correction of E_1 is

$$E_1^{(2)} = \frac{|\langle \bar{1}| \, \bar{V} \, |\bar{2}\rangle|^2}{\bar{E}_1 - \bar{E}_2} + \frac{\left|\langle \bar{1}| \, \hat{\bar{V}} \, |3\rangle\right|^2}{\bar{E}_1 - E_3} = |N_+|^2 \frac{|\lambda + (\bar{E}_+ - E - \delta)|^2}{\bar{E}_+ - E'} \tag{18.192}$$

and similarly for

$$E_2^{(2)} = |N_-|^2 \frac{|\lambda + (\bar{E}_- - E - \delta)|^2}{\bar{E}_- - E'}, \quad E_3^{(2)} = \frac{|\langle 3|\bar{V} \, |\bar{1}\rangle|^2}{E_3 - \bar{E}_+} + \frac{|\langle 3|\bar{V} \, |\bar{2}\rangle|^2}{E_3 - \bar{E}_-}$$

$$= |N_+|^2 \frac{|\lambda + \bar{E}_+ - E - \delta|^2}{E' - \bar{E}_+} + |N_-|^2 \frac{|\lambda + \bar{E}_- - E - \delta|^2}{E' - \bar{E}_-}. \tag{18.193}$$

18.5.4 Degenerate Perturbation Theory

Degeneracies in the energy spectrum are very common among most soluble three-dimensional problems for the same reason that these problems are soluble, i.e., because of the occurrence of symmetries. For example, in the case of the hydrogen atom there are n^2 different eigenstates $|n, \ell, m\rangle$ of the same energy $E_n = -e^2/2n^2 a_0$. As a result, in the expressions of perturbative expansions as presented so far, there will be terms with vanishing denominators like

$$\frac{\langle n\ell m|\hat{V}|n\ell'm'\rangle}{E_n^{(0)} - E_n^{(0)}} = \infty.$$

A method to treat this problem, similar to the one employed for neighboring energy levels, is to isolate the group of degenerate states and introduce an equal number of linear combinations of them that diagonalize the perturbing potential, since only off-diagonal matrix elements arise in the perturbative expansion series.

Consider a set of eigenstates

$$|\psi_{n_a}^{(0)}\rangle, \; |\psi_{n_b}^{(0)}\rangle, \; \ldots, \; |\psi_{n_f}^{(0)}\rangle, \tag{18.194}$$

all corresponding to one and the same eigenvalue $E_n^{(0)}$, namely,

$$\hat{H}_0|\psi_{n_i}^{0}\rangle = E_n^{(0)}|\psi_{n_i}^{(0)}\rangle \quad (i = a, \, b, \, \ldots, \, f). \tag{18.195}$$

Note that any linear combination of these states will also be an eigenstate with the same eigenvalue. We therefore are free to introduce an equal number of eigenstates

$$|\psi_{n_\alpha}^{(0)}\rangle = \sum_{i=a,b,...f} D_{\alpha i}|\psi_{n_i}^{(0)}\rangle. \tag{18.196}$$

We demand that the matrix $D_{\alpha i}$ is such that the above states diagonalize the operator $\hat{V}$, namely,

$$\langle\psi_{n_\alpha}^{(0)}|\hat{V}|\psi_{n_\beta}^{(0)}\rangle = V_\alpha\delta_{\alpha\beta}. \tag{18.197}$$

This leads to the condition

$$\sum_{i,j} D_{\alpha i}^* V_{ij}^{(n)} D_{\beta j} = V_\alpha^{(n)}\delta_{\alpha\beta}. \tag{18.198}$$

We have introduced the notation

$$V_{ij}^{(n)} = \langle\psi_{n_i}^{(0)}|\hat{V}|\psi_{n_j}^{(0)}\rangle, \quad V_\alpha^{(n)} = \langle\psi_{n_\alpha}^{(0)}|\hat{V}|\psi_{n_\alpha}^{(0)}\rangle. \tag{18.199}$$

Orthonormality of the states $|\psi_{n_\alpha}^{(0)}\rangle$ implies

$$\sum_i D_{\alpha i}^* D_{\beta i} = \delta_{\alpha\beta}. \tag{18.200}$$

Thus, the relation (18.198) can be written as

$$\sum_i D_{\alpha i}^* \left(\sum_j V_{ij}^{(n)} D_{\beta j} - V_\alpha^{(n)} D_{\beta i} \right) = 0, \tag{18.201}$$

which in turn implies

$$\sum_j V_{ij}^{(n)} D_{\beta j} = V_\beta^{(n)} D_{\beta i}. \tag{18.202}$$

Thus, for each β the corresponding D-column is the eigenvector of the matrix $V_{ij}^{(n)}$ with eigenvalue $V_\beta^{(n)}$.

Example 18.8 Consider the three-level system of Example 18.7 in the degenerate case of $\delta = 0$. Apply the above procedure and compute the second-order corrections to the energy eigenvalues.

We have

$$\hat{H}_0 = \begin{pmatrix} E & 0 & 0 \\ 0 & E & 0 \\ 0 & 0 & E' \end{pmatrix} \quad \hat{V} = \lambda \begin{pmatrix} 0 & 1 & 1 \\ 1 & 0 & 1 \\ 1 & 1 & 1 \end{pmatrix}. \tag{18.203}$$

Instead of the unperturbed eigenstates

$$|1\rangle \implies \begin{pmatrix} 1 \\ 0 \\ 0 \end{pmatrix}, \ |2\rangle \implies \begin{pmatrix} 0 \\ 1 \\ 0 \end{pmatrix}, \ |3\rangle \implies \begin{pmatrix} 0 \\ 0 \\ 1 \end{pmatrix}, \tag{18.204}$$

we use as a basis the set

$$|+\rangle \implies \frac{1}{\sqrt{2}} \begin{pmatrix} 1 \\ 1 \\ 0 \end{pmatrix}, \ |-\rangle \implies \frac{1}{\sqrt{2}} \begin{pmatrix} 1 \\ -1 \\ 0 \end{pmatrix}, \ |3\rangle \implies \begin{pmatrix} 0 \\ 0 \\ 1 \end{pmatrix}, \tag{18.205}$$

which correspond to the same set of eigenvalues of $\hat{H}_0$, namely, E, E, E' and have the property $\langle +|\hat{V}|-\rangle = 0$. Thus, we have

$$E_1^{(1)} = \langle +|\hat{V}|+\rangle = \frac{1}{2}(V_{11} + V_{22} + V_{12} + V_{21}) = 1 \tag{18.206}$$

$$E_2^{(1)} = \langle -|\hat{V}|-\rangle = \frac{1}{2}(V_{11} + V_{22} - V_{12} - V_{21}) = -1 \tag{18.207}$$

$$E_3^{(1)} = \langle 3|\hat{V}|3\rangle = V_{33} = 1 \tag{18.208}$$

and

$$E_1^{(2)} = \frac{|V_{+3}|^2}{E - E'} = \frac{1}{2}\frac{(V_{13} + V_{23})^2}{E - E'} = \frac{2}{E - E'} \tag{18.209}$$

$$E_2^{(2)} = \frac{|V_{-3}|^2}{E - E'} = \frac{1}{2}\frac{(V_{13} - V_{23})^2}{E - E'} = 0 \tag{18.210}$$

$$E_3^{(2)} = \frac{|V_{3+}|^2}{E' - E} + \frac{|V_{3-}|^2}{E' - E} = \frac{1}{2(E' - E)}\left((V_{31} + V_{32})^2 + (V_{31} - V_{32})^2\right) = \frac{2}{E' - E}. \tag{18.211}$$

18.6   Time-Dependent Perturbation Theory

When atoms are subject to electromagnetic radiation, we have a case of charged particles (electrons) under the influence of time-dependent electric and magnetic fields. In view of the smallness of the electron charge ($e^2/\hbar c \sim 1/137$), a problem like this can be treated in the framework of *Time-Dependent Perturbation Theory* [1, 4, 5].

A problem of time-dependent perturbations corresponds to a Hamiltonian of the form

$$\hat{H} = \hat{H}_0 + \hat{H}_{int}(t), \tag{18.212}$$

consisting of a known part and a time-dependent perturbation $\hat{H}_{int}(t)$, which is somehow "small". An ideal starting point to formulate time-dependent perturbation theory is the interaction picture in which the Schroedinger equation takes the form

$$\frac{d}{dt}|\psi_I(t)\rangle = \hat{H}_I(t)|\psi_I(t)\rangle, \tag{18.213}$$

where

$$\hat{H}_I(t) = \hat{U}_0^\dagger(t, t_0)\, \hat{H}_{int}(t)\, \hat{U}_0(t, t_0) \quad \left(\hat{U}_0(t, t_0) = e^{-\frac{i}{\hbar}\hat{H}_0(t-t_0)}\right). \tag{18.214}$$

The relation of interaction picture state vectors to the Schroedinger ones is quite simple, namely,

$$|\psi_I(t)\rangle = \hat{U}_0^\dagger(t, t_0)|\psi_S(t)\rangle. \tag{18.215}$$

We may transform (18.213) into an integral equation by integrating with respect to time. We have

$$|\psi_I(t)\rangle = |\psi_I(t_0)\rangle - \frac{i}{\hbar}\int_{t_0}^t dt'\, \hat{H}_I(t')|\psi_I(t')\rangle. \tag{18.216}$$

A series of successive approximations to this equation is

$$|\psi_I(t)\rangle^{(0)} = |\psi_I(t_0)\rangle$$

$$|\psi_I(t)\rangle^{(1)} = |\psi_I(t_0)\rangle - \frac{i}{\hbar}\int_{t_0}^t dt'\hat{H}_I(t')\,|\psi_I(t_0)\rangle$$

$$|\psi_I(t)\rangle^{(2)} = |\psi_I(t_0)\rangle - \frac{i}{\hbar}\int_{t_0}^t dt'\hat{H}_I(t')\,|\psi_I(t_0)\rangle - \frac{1}{\hbar^2}\int_{t_0}^t dt'\int_{t_0}^{t'} dt''\,\hat{H}_I(t')\hat{H}_I(t'')|\psi_I(t_0)\rangle$$

$$\cdots$$

$$\tag{18.217}$$

Assume now that initially, at t_0, the system occupies an eigenstate[8] $|0\rangle$ of $\hat{H}_0$, corresponding to the unperturbed energy eigenvalue E_0

$$|\psi_I(t_0)\rangle = |\psi_S(t_0)\rangle = |0\rangle. \tag{18.218}$$

A quantity of interest is the probability amplitude to make a transition to another eigenstate $|n\rangle \neq |0\rangle$ of $\hat{H}_0$ at time $t > t_0$

$$\langle n|\psi_S(t)\rangle = \langle n|\hat{U}_0(t, t_0)|\psi_I(t)\rangle = e^{-\frac{i}{\hbar}E_n(t-t_0)}\langle n|\psi_I(t)\rangle. \tag{18.219}$$

To a first approximation the corresponding probability of such a transition is

$$\mathcal{P}_{0\to n}(t) = |\langle n|\psi_S(t)\rangle|^2 = \frac{1}{\hbar^2}\left|\int_{t_0}^t dt'\langle n|\hat{H}_I(t')|0\rangle\right|^2 \tag{18.220}$$

or

$$\mathcal{P}_{0\to n}(t) = |\langle n|\psi_S(t)\rangle|^2 = \frac{1}{\hbar^2}\left|\int_{t_0}^t dt' e^{\frac{i}{\hbar}(E_n-E_0)(t'-t_0)}\langle n|\hat{H}_{int}(t')|0\rangle\right|^2. \tag{18.221}$$

---

[8] $|0\rangle$ should not be confused with the harmonic oscillator ground state.

18.6.1 Temporal Step-Function Potential

Let's consider the case that a potential is turned on at some time t_0 and stays constant for all subsequent times

$$\hat{H}_{int}(t) = \hat{V}_0 \, \Theta(t - t_0). \tag{18.222}$$

Taking t_0 as our initial time we have

$$\mathcal{P}_{0 \to n}(t) = |\langle n | \psi_S(t) \rangle|^2 = \frac{1}{\hbar^2} \left| \langle n | \hat{V}_0 | 0 \rangle \right|^2 \left| \int_{t_0}^{t} dt' e^{\frac{i}{\hbar}(E_n - E_0)(t' - t_0)} \right|^2. \tag{18.223}$$

The time integral is

$$\int_{t_0}^{t} dt' e^{\frac{i}{\hbar}(E_n - E_0)(t' - t_0)} = -\frac{i\hbar}{(E_n - E_0)} \left(e^{i(t - t_0)(E_n - E_0)/\hbar} - 1 \right)$$

$$= \frac{2\hbar e^{\frac{i}{2\hbar}(E_n - E_0)(t - t_0)}}{(E_n - E_0)} \sin \left(\frac{(E_n - E_0)(t - t_0)}{2\hbar} \right) \tag{18.224}$$

and the probability is given by

$$\mathcal{P}_{0 \to n}(t) = \frac{1}{\hbar^2} \left| \langle n | \hat{V}_0 | 0 \rangle \right|^2 \left(\frac{\sin \left(\frac{(E_n - E_0)(t - t_0)}{2\hbar} \right)}{\frac{(E_n - E_0)}{2\hbar}} \right)^2. \tag{18.225}$$

In Fig. 18.5, we have plotted the transition probability as a function of time. For small time intervals $t - t_0 \sim 0$ the transition probability $\mathcal{P}_{0 \to n}$ behaves as t^2 for all E_n's. At later times the probability develops maxima at $E_0 \pm \frac{(2n+1)\pi\hbar}{(t - t_0)}$, proportional to t^2 and with a spread $\sim 1/t$. The eigenvalues E_n of maximal probability are contained in the region

$$|E_n - E_0| < \frac{2\pi\hbar}{(t - t_0)}. \tag{18.226}$$

Suppose now that the state $|n\rangle$ is a member of continuum of states $\{n\}$. Then, the total transition rate $0 \to \{n\}$ will be

$$\mathcal{P}_{0 \to \{n\}}(t) = \frac{1}{\hbar^2} \sum_n \left| \langle n | \hat{V}_0 | 0 \rangle \right|^2 \left(\frac{\sin \left(\frac{(E_n - E_0)(t - t_0)}{2\hbar} \right)}{\frac{(E_n - E_0)}{2\hbar}} \right)^2. \tag{18.227}$$

For a continuum of final states the sum is symbolic and should be replaced by an integral $\int dn \ldots$ or an integral over the energy $\int dE \, \rho(E) \ldots$, where $\rho(E)$ is the *density of states*, defined as the "number of states" per unit of energy, i.e., dn/dE. Thus, we may write

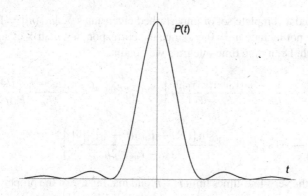

Fig. 18.5 Transition probability

$$\mathcal{P}_{0\to\{n\}}(t) = \frac{1}{\hbar^2} \int dE\, \rho(E)\, \left|\langle E|\hat{V}_0|0\rangle\right|^2 \left(\frac{\sin\left(\frac{(E-E_0)(t-t_0)}{2\hbar}\right)}{\frac{(E-E_0)}{2\hbar}}\right)^2. \tag{18.228}$$

At very late times $t - t_0 \to +\infty$ we may use the fact that

$$\lim_{t\to\infty}\left(\frac{\sin^2(xt)}{x^2 t}\right) = \pi\,\delta(x). \tag{18.229}$$

Then, we have

$$\mathcal{P}_{0\to\{n\}}(t) = \frac{2\pi}{\hbar}\int dE\rho(E)|\langle E|\hat{V}_0|0\rangle|^2\delta(E-E_0)(t-t_0) = \frac{2\pi}{\hbar}(t-t_0)|\langle 0|\hat{V}_0|0\rangle|^2\rho(E_0). \tag{18.230}$$

Dividing this probability by the time, we obtain the *total rate of transitions*

$$\Gamma_{0\to\{n\}} = \frac{2\pi}{\hbar}\rho(E_0)\left|\langle 0|\hat{V}_0|0\rangle\right|^2. \tag{18.231}$$

This is the so-called *Fermi's Golden Rule*.

It is possible that the above first-order contribution to the transition probability may vanish. Then, assuming that $\langle 0|\hat{V}_0|n\rangle = 0$, the second-order contribution becomes important. Going back to (18.217) we have

$$\langle n|\psi_I(t)\rangle^{(2)} = -\frac{1}{\hbar^2}\int_{t_0}^{t}dt'\int_{t_0}^{t'}dt''\, e^{\frac{i}{\hbar}E_n(t'-t_0)}e^{-\frac{i}{\hbar}E_0(t''-t_0)}\langle n|\hat{V}_0 e^{-\frac{i}{\hbar}\hat{H}_0(t'-t'')}\hat{V}_0|0\rangle$$

$$= -\frac{1}{\hbar^2}\sum_m\int_{t_0}^{t}dt'\int_{t_0}^{t'}dt''\, e^{\frac{i}{\hbar}E_n(t'-t_0)}e^{-\frac{i}{\hbar}E_0(t''-t_0)}\langle n|\hat{V}_0|m\rangle\langle m|\hat{V}_0|0\rangle\, e^{-\frac{i}{\hbar}E_m(t'-t'')},$$

having inserted a complete set of unperturbed eigenstates $\sum |m\rangle\langle m| = \mathbf{I}$. Note that the sum does not include $m = 0, n$, since the corresponding matrix element contributions vanish. Doing the time integrals we obtain

$$\langle n|\psi_I(t)\rangle^{(2)} = i\frac{1}{\hbar}\sum_m \frac{\langle n|\hat{V}_0|m\rangle\langle m|\hat{V}_0|0\rangle}{(E_m - E_0)}\left\{ e^{\frac{i}{2\hbar}(E_n - E_0)(t - t_0)} \frac{\sin\left[(E_n - E_0)(t - t_0)/2\hbar\right]}{(E_n - E_0)/2\hbar} \right.$$

$$\left. - e^{\frac{i}{2\hbar}(E_n - E_m)(t - t_0)} \frac{\sin\left[(E_n - E_m)(t - t_0)/2\hbar\right]}{(E_n - E_m)/2\hbar} \right\}. \tag{18.232}$$

Considering the very-late-times limit $t \gg t_0$ and making use of the property (18.229), we see that the second term is proportional to $\delta(E_n - E_m)$. However, since $E_m \neq E_n$, this term will necessarily vanish. Thus, our result is

$$\langle n|\psi_I(t)\rangle^{(2)} = \frac{i}{\hbar}e^{\frac{i}{2\hbar}(E_n - E_0)(t - t_0)} \frac{\sin\left[(E_n - E_0)(t - t_0)/2\hbar\right]}{(E_n - E_0)/2\hbar}\sum_m \frac{\langle n|\hat{V}_0|m\rangle\langle m|\hat{V}_0|0\rangle}{(E_m - E_0)} \tag{18.233}$$

and the corresponding probability for $t \gg t_0$ will be

$$\mathcal{P}_{0\to n}(t) = \frac{1}{\hbar^2}\left(\frac{\sin\left[(E_n - E_0)(t - t_0)/2\hbar\right]}{(E_n - E_0)/2\hbar}\right)^2 \left|\sum_m \frac{\langle n|\hat{V}_0|m\rangle\langle m|\hat{V}_0|0\rangle}{(E_m - E_0)}\right|^2 \tag{18.234}$$

or

$$\mathcal{P}_{0\to n}(t) = \frac{2\pi}{\hbar}\delta(E_n - E_0) \left.\frac{\sin\left[(E_n - E_0)(t - t_0)/2\hbar\right]}{(E_n - E_0)/2\hbar}\right|_{E_n = E_0} \left|\sum_m \frac{\langle n|\hat{V}_0|m\rangle\langle m|\hat{V}_0|0\rangle}{(E_m - E_0)}\right|^2$$

$$= \frac{2\pi}{\hbar}\delta(E_n - E_0)\,(t - t_0) \left|\sum_m \frac{\langle n|\hat{V}_0|m\rangle\langle m|\hat{V}_0|0\rangle}{(E_m - E_0)}\right|^2. \tag{18.235}$$

The total probability of transitions to a continuum $\{n\}$ will be

$$\mathcal{P}_{0\to\{n\}}(t) = \int dE\,\rho(E)\frac{2\pi}{\hbar}\delta(E - E_0)\,(t - t_0) \left|\sum_m \frac{\langle n|\hat{V}_0|m\rangle\langle m|\hat{V}_0|0\rangle}{(E_m - E_0)}\right|^2 \tag{18.236}$$

or

$$\mathcal{P}_{0\to\{n\}}(t) = \frac{2\pi}{\hbar}\rho(E_0)\,(t - t_0) \left|\sum_m \frac{\langle n|\hat{V}_0|m\rangle\langle m|\hat{V}_0|0\rangle}{(E_m - E_0)}\right|^2. \tag{18.237}$$

Thus, the corresponding rate will be

$$\Gamma_{0 \to \{n\}} = \frac{2\pi}{\hbar} \rho(E_0) \left| \sum_m \frac{\langle n|\hat{V}_0|m\rangle\langle m|\hat{V}_0|0\rangle}{(E_m - E_0)} \right|^2 . \tag{18.238}$$

This second-order transition amplitude can be interpreted as a two-stage process, consisting of a transition $|0\rangle \to |m\rangle$ and a second transition $|m\rangle \to |n\rangle$. Note also that although the two intermediate transitions (virtual transitions) do not conserve the energy ($E_0 \neq E_m$ and $E_m \neq E_n$), the overall process conserves the energy, since $E_n = E_0$.

Example 18.9 A simple harmonic oscillator is subject to an external time-dependent force

$$\hat{H}_{int}(t) = J(t)\,x(t) \quad \text{with} \quad J(t) = J_0 \,\Theta(t - t_0) . \tag{18.239}$$

At times $t < t_0$ the system occupies the harmonic oscillator ground state $|0\rangle$.

(a) Use the interaction picture time-ordered exponential formula to calculate the probability that the system stays in this state at times $t > t_0$.

(b) Calculate the above probability to lowest nontrivial order in time-dependent perturbation theory.

For the part (a) we have $\mathcal{P}(t) = |\langle 0|\psi_I(t)\rangle|^2 = |\langle 0|\hat{U}_I(t, t_0)|0\rangle|^2$. The evolution operator is

$$\hat{U}_I(t, t_0) = \mathcal{T}\left(e^{-\frac{i}{\hbar} \int_{t_0}^{t} dt' \, J_0 \, \hat{x}_I(t')} \right) = \mathcal{T}\left(e^{\hat{A} + \hat{B}} \right), \tag{18.240}$$

where

$$\hat{A}(t) = -\frac{i J_0}{\sqrt{2\hbar m\omega}} \int_{t_0}^{t} dt' \hat{a}_I^\dagger(t'), \quad \hat{B}(t) = -\frac{i J_0}{\sqrt{2\hbar m\omega}} \int_{t_0}^{t} dt' \hat{a}_I(t') . \tag{18.241}$$

From the interaction picture equation of motion for operators we have

$$\frac{d\hat{a}_I}{dt} = \frac{i}{\hbar}\left[\hat{H}_0, \, \hat{a}_I \right] = -i\omega\hat{a}_I \implies \hat{a}_I(t) = e^{-i\omega(t-t_0)}\hat{a}_I(t_0) . \tag{18.242}$$

Then, we have

$$\hat{A}(t) = -\frac{J_0}{\sqrt{2m\hbar\omega}} \hat{a}_I^\dagger(t_0) \, e^{i\omega(t-t_0)/2} \frac{\sin(\omega(t-t_0)/2)}{\omega}$$

$$\hat{B}(t) = \frac{J_0}{\sqrt{2m\hbar\omega}} \hat{a}_I(t_0) \, e^{-i\omega(t-t_0)/2} \frac{\sin(\omega(t-t_0)/2)}{\omega} \tag{18.243}$$

$$\left[\hat{A}, \, \hat{B} \right] = -\frac{J_0^2}{2\hbar m\omega} \int_{t_0}^{t} dt' \int_{t_0}^{t} dt'' \left[\hat{a}_I^\dagger(t_0), \, \hat{a}_I(t_0) \right] e^{i\omega(t'-t_0)} e^{-i\omega(t''-t_0)} .$$

The commutator, since $\hat{a}_I(t_0) = \hat{a}_S$, reduces to

$$\left[\hat{A}, \, \hat{B} \right] = \frac{J_0^2}{2\hbar m\omega} \left| \int_{t_0}^{t} dt' \, e^{i\omega(t'-t_0)} \right|^2 = \frac{2J_0^2}{\hbar\omega m} \left(\frac{\sin(\omega(t - t_0)/2)}{\omega} \right)^2 . \tag{18.244}$$

Since the commutator of $\hat{A}$ and $\hat{B}$ is a c-number, the following formula is true

$$e^{\hat{A} + \hat{B}} = e^{\hat{A}} \, e^{\hat{B}} \, e^{-\frac{1}{2}\left[\hat{A}, \hat{B}\right]} \tag{18.245}$$

and we obtain

$$
\hat{U}_I(t, t_0) = \mathcal{T}\left(e^{\hat{A}} e^{\hat{B}}\right) e^{-\frac{J_0^2}{m\hbar\omega}\left(\frac{\sin(\omega(t-t_0)/2)}{\omega}\right)^2} = e^{\hat{A}} e^{\hat{B}} e^{-\frac{J_0^2}{m\hbar\omega}\left(\frac{\sin(\omega(t-t_0)/2)}{\omega}\right)^2}. \tag{18.246}
$$

We have dropped the time-ordering since the operators in $\hat{A}$ and $\hat{B}$ are at the same time t_0. Furthermore, we have

$$
\langle 0|e^{\hat{A}} = \langle 0| \quad and \quad e^{\hat{B}}|0\rangle = |0\rangle. \tag{18.247}
$$

Thus, finally

$$
\mathcal{P}(t) = e^{-\frac{J_0^2}{2m\hbar\omega}\left(\frac{\sin(\omega(t-t_0)/2)}{\omega/2}\right)^2}. \tag{18.248}
$$

For the (b) part, we may start from the lowest order formula

$$
|\langle 0|\psi_S(t)\rangle|^2 = |\langle 0|\hat{U}_0(t, t_0)|\psi_I(t)\rangle|^2 = \left|\langle 0|\left(1 - \frac{i}{\hbar}\int_{t_0}^{t} dt'\hat{H}_I(t') + \dots\right)|0\rangle\right|^2. \tag{18.249}
$$

The first-order vanishes, since

$$
\langle 0|\hat{H}_I(t')|0\rangle = J(t')\langle 0|\hat{U}_0^\dagger(t', t_0)\hat{x}_I(t_0)\hat{U}_0(t', t_0)|0\rangle = J(t')\langle 0|\hat{x}_S|0\rangle = 0
$$

because of parity. Thus, we have the second-order

$$
\mathcal{P}(t) = 1 - \frac{2}{\hbar^2}\mathrm{Re}\left\{\int_{t_0}^{t} dt' J(t')\int_{t_0}^{t'} dt'' J(t'')\langle 0|\hat{x}_I(t')\hat{x}_I(t'')|0\rangle\right\}
$$

$$
= 1 - \frac{2J_0^2}{\hbar^2}\mathrm{Re}\left\{\int_{t_0}^{t} dt'\int_{t_0}^{t'} dt'' \langle 0|e^{i\hat{H}_0(t'-t_0)/\hbar}\,\hat{x}_I(t_0)e^{-i\hat{H}_0(t'-t'')/\hbar}\hat{x}_I(t_0)e^{-i\hat{H}_0(t''-t_0)/\hbar}|0\rangle\right\}
$$

$$
= 1 - \frac{2J_0^2}{2\hbar m\omega}\mathrm{Re}\left\{\int_{t_0}^{t} dt'\int_{t_0}^{t'} dt'' e^{i\omega(t'-t_0)/2}e^{-i\omega(t''-t_0)/2} \langle 0|\hat{a}_S\,e^{-i\hat{H}_0(t'-t'')/\hbar}\,\hat{a}_S^\dagger|0\rangle\right\}
$$

$$
= 1 - \frac{J_0^2}{\hbar m\omega}\mathrm{Re}\left\{\int_{t_0}^{t} dt'\int_{t_0}^{t'} dt'' e^{i\omega(t'-t_0)/2}e^{-i\omega(t''-t_0)/2} \langle 1|e^{-i\hat{H}_0(t'-t'')/\hbar}|1\rangle\right\}
$$

$$
= 1 - \frac{J_0^2}{\hbar m\omega}\mathrm{Re}\left\{\int_{t_0}^{t} dt'\int_{t_0}^{t'} dt'' e^{i\omega(t'-t_0)/2}e^{-i\omega(t''-t_0)/2} e^{-3i\omega(t'-t'')/2}\right\}
$$

or

$$
\mathcal{P}(t) = 1 - \frac{J_0^2}{\hbar m\omega^2}\int_{t_0}^{t} dt' \sin(\omega(t'-t_0)) = 1 - \frac{J_0^2}{2\hbar m\omega}\left(\frac{\sin(\omega(t-t_0)/2)}{\omega/2}\right)^2, \tag{18.250}
$$

which is just the lowest nontrivial approximation of the exact formula obtained in (a).

18.6.2 Sinusoidal Perturbations

The electrons in an atom, subject to an external electromagnetic field

$$\mathcal{E} = \mathcal{E}_0 \cos(\mathbf{k} \cdot \mathbf{r} - \omega t), \quad \mathcal{B} = \mathcal{B}_0 \cos(\mathbf{k} \cdot \mathbf{r} - \omega t), \tag{18.251}$$

will acquire a time-dependent electric dipole moment—as well as a magnetic dipole moment—interaction terms in the Hamiltonian that can be put in the general form

$$\hat{H}_{int}(t) = \hat{V}_0 e^{i\omega t} + \hat{V}_0^\dagger e^{-i\omega t}, \tag{18.252}$$

where $\hat{V}_0$ will depend on the atomic variables. The first-order probability of transition $|0\rangle \to |n\rangle$ is given by (18.221). In our case it is

$$\mathcal{P}_{0 \to n}(t) = \frac{1}{\hbar^2} \left| \int_{t_0}^{t} dt'\, e^{i(E_n - E_0)(t' - t_0)/\hbar} \langle n| \left(\hat{V}_0 e^{i\omega t'} + \hat{V}_0^\dagger e^{-i\omega t'} \right) |0\rangle \right|^2. \tag{18.253}$$

Integrating with respect to time, we obtain

$$\mathcal{P}_{0 \to n}(t) = \frac{1}{\hbar^2} \left| \frac{\langle n|\hat{V}_0|0\rangle}{(E_n - E_0 + \hbar\omega)} \left(e^{i(E_n - E_0 + \hbar\omega)t/\hbar} - e^{i(E_n - E_0 + \hbar\omega)t_0/\hbar} \right) + \right.$$

$$\left. \frac{\langle n|\hat{V}_0^\dagger|0\rangle}{(E_n - E_0 - \hbar\omega)} \left(e^{i(E_n - E_0 - \hbar\omega)t/\hbar} - e^{i(E_n - E_0 - \hbar\omega)t_0/\hbar} \right) \right|^2$$

$$= \left| \langle n|\hat{V}_0|0\rangle\, e^{i(E_n - E_0 + \hbar\omega)(t + t_0)/2\hbar} \frac{\sin\left[(E_n - E_0 + \hbar\omega)(t - t_0)/2\hbar \right]}{(E_n - E_0 + \hbar\omega)/2\hbar} + \right.$$

$$\left. \langle 0|\hat{V}_0|n\rangle^*\, e^{i(E_n - E_0 - \hbar\omega)(t + t_0)/2\hbar} \frac{\sin\left[(E_n - E_0 - \hbar\omega)(t - t_0)/2\hbar \right]}{(E_n - E_0 - \hbar\omega)/2\hbar} \right|^2. \tag{18.254}$$

For very late times $t \gg t_0$ we have

$$\frac{\sin\left[(E_n - E_0 \pm \hbar\omega)(t - t_0)/2\hbar \right]}{(E_n - E_0 \pm \hbar\omega)/2\hbar} \to 2\pi\hbar\delta(E_n - E_0 \pm \hbar\omega) \tag{18.255}$$

and

$$\left(\frac{\sin\left[(E_n - E_0 \pm \hbar\omega)(t - t_0)/2\hbar \right]}{(E_n - E_0 \pm \hbar\omega)/2\hbar} \right)^2 \to (t - t_0)\, 2\pi\hbar\delta(E_n - E_0 \pm \hbar\omega). \tag{18.256}$$

Thus, we have

$$\Gamma_{0 \to n}(t) = \frac{2\pi}{\hbar} \delta(E_n - E_0 + \hbar\omega) |\langle n|\hat{V}_0|0\rangle|^2_{E_n = E_0 - \hbar\omega}$$

$$+ \frac{2\pi}{\hbar} \delta(E_n - E_0 - \hbar\omega) |\langle 0|\hat{V}_0|n\rangle|^2_{E_n = E_0 + \hbar\omega} \tag{18.257}$$

or

$$\Gamma_{0\to\{n\}} = \frac{2\pi}{\hbar}\rho(E_n)|\langle n|\hat{V}_0|0\rangle|^2_{E_n=E_0-\hbar\omega} + \frac{2\pi}{\hbar}\rho(E_n)|\langle 0|\hat{V}_0|n\rangle|^2_{E_n=E_0+\hbar\omega}.$$

(18.258)

The two terms have a different interpretation, the first may be interpreted as *induced emmission*, where from the initial state of energy E_0 the system goes to a final state of reduced energy by emitting a quantum of energy $\hbar\omega$, and the second as *induced absorption*, where from the initial state of energy E_0 the system makes a transition to a state of increased energy by absorbing a quantum of energy $\hbar\omega$

$$\Gamma^{(em)}_{0\to\{n\}} = \frac{2\pi}{\hbar}\rho(E_n)|\langle n|\hat{V}_0|0\rangle|^2_{E_n=E_0-\hbar\omega}$$

$$\Gamma^{(abs)}_{0\to\{n\}} = \frac{2\pi}{\hbar}\rho(E_n)|\langle 0|\hat{V}_0|n\rangle|^2_{E_n=E_0+\hbar\omega}$$

(18.259)

Note that both the electric and magnetic dipole interactions involve Hermitian operators and, therefore, $\langle 0|\hat{V}_0|n\rangle = \langle n|\hat{V}_0|0\rangle^*$. Then, the following relation is true:

$$\Gamma^{(em)}_{0\to\{n\}}/\Gamma^{(abs)}_{0\to\{n\}} = \rho(E_0+\hbar\omega)/\rho(E_0-\hbar\omega),$$

(18.260)

expressing the *equilibrium* between the processes of emission and absorption.

Example 18.10 A hydrogen atom is initially in its ground state $|1,0,0\rangle$, while it interacts with a monochromatic electromagnetic wave of wavelength much larger than the Bohr radius $\lambda \gg a_0$ through the electric dipole interaction term

$$\hat{H}_{int}(t) = -e\mathbf{r}\cdot\mathcal{E}_0\cos(\mathbf{k}\cdot\mathbf{r}-\omega t).$$

(18.261)

Calculate the rate of transitions to the first excited state $|2,\ell,m\rangle$ in the first order of perturbation theory.

By hypothesis we have $\mathbf{k}\cdot\mathbf{r} \sim a_0/\lambda \ll 1$ and, therefore, $e^{i\mathbf{k}\cdot\mathbf{r}} \approx 1$. The corresponding probability for $t \gg t_0$ will be

$$\mathcal{P}_{(1,0,0)\to(2,\ell,m)} = \frac{e^2\mathcal{E}_0^2}{\hbar^2}\left|\int_{t_0}^t dt' e^{i(E_2-E_1)t'/\hbar}\langle 2,\ell,m|z|1,0,0\rangle\cos(\omega t')\right|^2$$

$$= \frac{e^2\mathcal{E}_0^2}{4\hbar^2}\left|\int_{t_0}^t dt'\left(e^{i(3|E_1|/4+\hbar\omega)t'/\hbar} + e^{i(3|E_1|/4-\hbar\omega)t'/\hbar}\right)\right|^2$$

(18.262)

or

$$\mathcal{P}_{(1,0,0)\to(2,\ell,m)} = \frac{e^2\pi\mathcal{E}_0^2}{2\hbar}\delta(3|E_1|/4-\hbar\omega)(t-t_0)|\langle 2,\ell,m|z|1,0,0\rangle|^2.$$

(18.263)

The corresponding rate will be

$$\Gamma_{(1,0,0)\to(2,\ell,m)} = \frac{e^2\pi\mathcal{E}_0^2}{2\hbar}\delta(3|E_1|/4-\hbar\omega)\begin{cases} 0 & \ell=0 \\ 0 & \ell=1,\, m=\pm 1 \\ |\langle 2,1,0|z|1,0,0\rangle|^2 & \ell=1,\, m=0 \end{cases}$$

(18.264)

Note that

$$\begin{cases} \langle 100|z|200\rangle = -\langle 100|z|200\rangle = 0 \\ 0 = \langle 100|\left[L_z, z\right]|2, 1, \pm 1\rangle = -\langle 100|z|2, 1, \pm 1\rangle (\pm\hbar) \implies \langle 100|z|2, 1, \pm 1\rangle = 0 \\ \langle 100|z|2, 1, 0\rangle = \int d^3r\, \psi_{100}(r)z\psi_{210}(\mathbf{r}) = \ldots = 4\sqrt{2}\left(\tfrac{2}{3}\right)^5 a_0 \end{cases}$$

$$(18.265)$$

Problems and Exercises

18.1 Consider a particle of mass m bound in the potential

$$V(x) = \begin{cases} +\infty & (x \le 0) \\ \tfrac{1}{2}m\omega^2 x^2 & (x > 0) \end{cases}$$

Apply the WKB method to derive the allowed energy levels.

18.2 A particle of mass m is subject to the potential

$$V(x) = \begin{cases} +\infty & (x \le 0) \\ mgx & (x > 0) \end{cases}$$

where $g > 0$ is a coupling parameter. Apply the WKB method to derive the allowed energies of the system. This is a quantum analogue of a bouncing ball subject to gravity.

18.3 Use the WKB approximation to calculate the transmission coefficient of a particle of mass m and energy $E < V_0$ in a square barrier of height V_0 and width $[-a, a]$. Compare to the exact result.

18.4 Consider electrons subject to a spatially constant magnetic field. Ignoring spatial degrees of freedom the Hamiltonian is just $H = -\frac{e}{mc}\mathbf{B} \cdot \mathbf{S}$. The direction of the magnetic field is parametrized as $\hat{\mathbf{B}} = \hat{x}\sin\theta\cos\phi + \hat{y}\sin\theta\sin\phi + \hat{z}\cos\theta$. Assume that the beam of electrons, which are in an eigenstate of the energy, is split into two parts, the second of which is subject to an adiabatic rotation of the magnetic field in a complete circle in the x, y plane. Calculate the acquired geometric phase for the second beam. Consider now the interference of the two beams and relate the distance between maxima to the rest of parameters.

18.5 Consider a particle in the ground state of an infinite square well. The width of the well increases adiabatically at a constant rate $\dot{L}/L = \frac{1}{\tau} \ll \hbar\pi^2/8mL_i^2$ from its initial value $2L_i$ at $t = 0$ to a final value $2L_f$ at $t > 0$. Calculate the *dynamical phase* acquired by the wave function. Is there a corresponding *geometrical phase*?

18.6 Estimate the ground state energy of a hydrogen atom using a Gaussian trial wave function $\psi(r, \beta) = N(\beta)\, e^{-\beta r^2}$.

18.7 Consider a system with a Hamiltonian $\hat{H}_0$ that has a ground state $|\psi_0\rangle$ of energy ϵ_0 and a first excited state $|\psi_1\rangle$ of energy $\epsilon_1 > \epsilon_0$. Assume now that a perturbation $\hat{V}$ acts on the system with the property $\langle\psi_0|\hat{V}|\psi_0\rangle = \langle\psi_1|\hat{V}|\psi_1\rangle = 0$. The matrix element $\langle\psi_0|\hat{V}|\psi_1\rangle = \eta < 0$ is known. Apply the variational method to estimate the ground state energy using the trial wave function $\psi_\beta = \cos\beta\,\psi_0 + \sin\beta\,\psi_1$.

18.8 A particle of mass m is bound in a central potential $V(r) = \lambda^2 r^3$, where λ is a known coupling parameter. Estimate the energy of the ground state using a trial wave function $\psi(r) = N(\alpha)\, e^{-\alpha r^2}$, where $\alpha > 0$ is a variable parameter.

18.9 Consider an electron of a linear polymer subject to forces that can be approximated by a potential $V(x) = \frac{1}{2}m\omega_0^2 x^2 + \lambda(1 - \cos(kx))$, where the appearing parameters satisfy $\lambda \ll \hbar\omega$. Treat the term $\lambda(1 - \cos kx)$ as a perturbation and calculate the first-order corrections to the energy eigenvalues and to the eigenstates $|n\rangle$.

18.10 A harmonic oscillator is at times $t < 0$ in its ground state. At $t = 0$, a perturbation $\Delta\hat{H} = \lambda x^2 e^{-\frac{t}{\tau}}$ acts on the system. Calculate the probability for the system to make a transition to any excited state at times $t \gg \tau$.

18.11 A particle trapped in a two-dimensional infinite square well is subject to a perturbing potential $V(x, y) = \lambda x y$ with $\lambda \ll \hbar^2/mL^4$, where m is the mass of the particle and $2L \times 2L$ is the size of the well. Calculate the first nontrivial correction to the energy levels.

18.12 A two-dimensional isotropic harmonic oscillator is subject to a perturbation $\Delta V = \frac{1}{2}m\omega^2 \delta x y \left(x^2 + y^2\right)$, where $\delta \ll \sqrt{\hbar/m\omega}$. Calculate the first-order corrections to the first three energy levels.

18.13 A hydrogen atom is subject to an external time-dependent electric field $\mathbf{E}(t) = \mathbf{E}_0\, e^{-t/\tau}\, \Theta(t)$. The atom at $t < 0$ is in its ground state. Calculate, to first order in the electric field, the probability to make a transition to the state $n = 2$, $\ell = 1$, $m = -1, 0, 1$ at times $t \gg \tau$. What about transitions to the state $n = 2$, $\ell = m = 0$?

18.14 The relativistic correction to the kinetic energy of a particle of mass m is $\Delta H = p^4/8m^3 c^2$. For a simple harmonic oscillator, calculate the first-order correction to the energy levels.

18.15 Calculate the first-order correction to the energy levels of the infinite square well $[-L, L]$ in the presence of the perturbation $V(x) = \lambda |x|$.

References

1. E. Merzbacher, *Quantum Mechanics*, 3rd edn. (Wiley, Hoboken, 1998)
2. A. Messiah, *Quantum Mechanics* (Dover publications, Mineola, 1958). Single-volume reprint of the Wiley, New York, two-volume 1958 edn
3. D. Griffiths, *Introduction to Quantum Mechanics*, 2nd edn. (Cambridge University Press, Cambridge, 1992)
4. F. Levin, *An Introduction to Quantum Theory*, 3rd edn. (Cambridge University Press, Cambridge, 2002)
5. G. Baym, *Lectures in Quantum Mechanics*, Lecture Notes and Supplements in Physics (ABP, 1969)

Chapter 19
Symmetries

19.1 General Considerations

The conservation of a particular physical observable is intimately connected with the existence of a *symmetry* and a set of transformations that leave measurable quantities invariant. The importance of symmetries, already greatly appreciated in Classical Physics, is even more important in Quantum Mechanics. The absence of a classical analogue in many cases makes it difficult not only to determine the correct Hamiltonian of the system but more importantly to proceed with the solution of the motion of the system, something that makes the use of symmetries a powerful tool enabling us to circumvent the difficult problem of the detailed solution of a system and arrive at many useful conclusions about its measurable properties.

A set of transformations corresponding to a symmetry of a quantum system is by definition a mapping of states and operators that leaves the measurable quantities (such as inner products) invariant. The following theorem is true (*Wigner's Theorem*): *Any symmetry transformation is represented by either a linear and unitary operator or antilinear and antiunitary operator*[1] *acting in the Hilbert space of states.* One of the most interesting cases of this mapping of transformations to operators is the case of rotations. In this case, the corresponding unitary operators constitute a *group* mapped isomorphically to the group of transformations. These operators are differentiable functions of the parameters of rotation (angles) and belong to a special type of groups named *Lie groups*. Any element of a Lie group of operators can be put in the form $\hat{U}(\alpha_1, \alpha_2, \dots) = e^{i \sum_i \alpha_i \Theta_i}$, where Θ_i are Hermitian operators, the so-called *group generators*. Note, however, that there exist groups of interest which are not characterized by any continuous parameters. These are called *discrete groups*.

---

[1] An *antilinear operator* is defined by the property $\mathcal{Q}c = c^*\mathcal{Q}$ for any complex number c. This is in contrast to linear operators which commute with complex numbers.

© Springer Nature Switzerland AG 2019
K. Tamvakis, *Basic Quantum Mechanics*, Undergraduate Texts in Physics,
https://doi.org/10.1007/978-3-030-22777-7_19

19.2   Rotations and Angular Momentum

Consider the system of a particle with a wave function $\psi(\mathbf{r})$. A *rotation* of the system consists of a transformation of the spatial coordinates of the system as well as a transformation on its wave function

$$x'_i = \mathcal{R}_{ij}\, x_j \ \ \text{and} \ \ \psi \to \psi', \tag{19.1}$$

where $\mathcal{R}$ is an *orthogonal matrix* $(\mathcal{R}^{\perp}\mathcal{R} = \mathcal{R}\mathcal{R}^{\perp} = \mathbf{I})$. Since all measurable physical quantities should be preserved by the rotation, we must have

$$\psi'(\mathbf{r}') = \psi(\mathbf{r}). \tag{19.2}$$

The relation between the corresponding state vectors will be necessarily expressed in terms of a *unitary operator* [5]

$$|\psi'\rangle = \hat{R}|\psi\rangle. \tag{19.3}$$

Indeed, we have[2]

$$\langle\psi'|\psi'\rangle = \int d^3x' \, |\psi'(\mathbf{r}')|^2 = \int d^3r \, |\psi(\mathbf{r})|^2 = \langle\psi|\psi\rangle = \langle\psi'|\hat{R}\hat{R}^{\dagger}|\psi'\rangle$$

and, therefore

$$\hat{R}\hat{R}^{\dagger} = \hat{R}^{\dagger}\hat{R} = \mathbf{I}. \tag{19.4}$$

The various observables of the physical system will also be transformed under a rotation

$$\mathcal{R} : \hat{A} \to \hat{A}'. \tag{19.5}$$

Nevertheless, measurable quantities should not change. Therefore, we must have

$$\langle\psi'|\hat{A}'|\psi'\rangle = \langle\psi|\hat{A}|\psi\rangle. \tag{19.6}$$

This is equivalent to

$$\hat{A}' = \hat{R}\,\hat{A}\,\hat{R}^{\dagger}. \tag{19.7}$$

Different observables behave differently under rotations. A *scalar* $\hat{\Sigma}$ is defined as simply an operator that does not change at all under rotations, namely,

$$\hat{\Sigma}' = \hat{\Sigma} \ \ or \ \ \hat{\Sigma} = \hat{R}\,\hat{\Sigma}\,\hat{R}^{\dagger} \tag{19.8}$$

---

[2]Since $\mathcal{R}$ is orthogonal, the Jacobian of the transformation is unity (det $\mathcal{R} = 1$) and $d^3r' = d^3r$.

or

$$\left[\hat{\Sigma}, \hat{R} \right] = 0. \tag{19.9}$$

A *vector operator* is one that transforms as the position operator transforms. It is not difficult to see that the correct transformation for the position operator corresponding to a rotation $x_i' = \mathcal{R}_{ij} x_j$ is

$$\hat{R} \hat{x}_i \hat{R}^\dagger = \mathcal{R}_{ij}^{-1} \hat{x}_j = \hat{x}_i'. \tag{19.10}$$

Indeed, from the condition $\psi(\mathbf{r}) = \psi'(\mathbf{r}')$ we have $|\mathbf{r}'\rangle = \hat{R}|\mathbf{r}\rangle$ and

$$\hat{R} \hat{x}_i \hat{R}^\dagger |\mathbf{r}'\rangle = \mathcal{R}_{ij}^{-1} \hat{x}_j |\mathbf{r}'\rangle \implies \hat{R} \hat{x}_i |\mathbf{r}\rangle = \mathcal{R}_{ij}^{-1} x_j' |\mathbf{r}'\rangle$$

or

$$x_i |\mathbf{r}'\rangle = \mathcal{R}_{ij}^{-1} x_j' |\mathbf{r}'\rangle \implies x_i = \mathcal{R}_{ij}^{-1} x_j',$$

which is the correct coordinate rotation. Therefore, a *vector operator* is generally defined by the law

$$\hat{R} \hat{V}_i \hat{R}^\dagger = \mathcal{R}_{ij}^{-1} \hat{V}_j = \hat{V}_i'. \tag{19.11}$$

This definition preserves the expectation values

$$\langle \psi' | \hat{V}' | \psi' \rangle = \langle \psi | \hat{V} | \psi \rangle. \tag{19.12}$$

More general *tensor operators* are defined accordingly. A *tensor operator of rank k* $Q_{i_1 i_2 \ldots i_k}$ is defined by the rotation transformation rule

$$\hat{R} \hat{Q}_{i_1 i_2 \ldots i_k} \hat{R}^\dagger = \mathcal{R}_{i_1 j_1}^{-1} \mathcal{R}_{i_2 j_2}^{-1} \ldots \mathcal{R}_{i_k j_k}^{-1} \hat{Q}_{j_1 j_2 \ldots j_k}. \tag{19.13}$$

In Chap. 14, we discussed rotations around a given axis and their relation to the orbital angular momentum. There, we saw that a rotation around the $\hat{z}$-axis by an infinitesimal angle α is described by the matrix

$$\mathcal{R} = \begin{pmatrix} 1 & -\alpha & 0 \\ \alpha & 1 & 0 \\ 0 & 0 & 1 \end{pmatrix} \tag{19.14}$$

and leads to a rotated wave function

$$\psi'(\mathbf{r}) \sim \psi(\mathbf{r}) - \frac{i}{\hbar} \alpha \, (\mathbf{r} \times (-i\hbar\nabla))_z \, \psi(\mathbf{r}) + O(\alpha^2) \tag{19.15}$$

or

$$|\psi'\rangle = \hat{R}_{\hat{z}}(\alpha)|\psi\rangle \quad \text{with} \quad \hat{R}_{\hat{z}}(\alpha) \sim 1 - \frac{i}{\hbar}\alpha \hat{L}_z. \tag{19.16}$$

If the angle α is finite, we may consider a superposition of $N \to \infty$ infinitesimal rotations by an infinitesimal angle $\frac{\alpha}{N}$ around the $\hat{z}$-axis and obtain

$$\hat{R}_{\hat{z}}(\alpha) = \Pi_{N \to \infty} \left(1 - \frac{i\alpha}{\hbar N}\hat{L}_z\right)^N = e^{-\frac{i}{\hbar}\alpha \hat{L}_z}. \tag{19.17}$$

Furthermore, for a rotation around an arbitrary axis we may rewrite (19.17) as

$$\hat{R}(\mathbf{a}) = e^{-\frac{i}{\hbar}\mathbf{a}\cdot\mathbf{L}}, \tag{19.18}$$

where $\mathbf{a}$ is a vector with direction along the axis of rotation and magnitude equal to the rotation angle.

For a system with nonzero spin, the above formula of the rotation operator is generalized to

$$\hat{R}(\mathbf{a}) = e^{-\frac{i}{\hbar}\mathbf{a}\cdot\mathbf{J}}, \tag{19.19}$$

where $\mathbf{J} = \mathbf{L} + \mathbf{S}$ stands for the total angular momentum. Since any rotation around an axis can be expressed in terms of $\mathbf{J}$ as in (19.19), the total angular momentum is called the *generator of rotations*. Note that the unitarity of $\hat{R}$ is a direct consequence of the Hermiticity of $\mathbf{J}$.

The various observables, classified under their distinct behavior under rotations, in view of (19.19) and its infinitesimal version

$$\hat{R}(\mathbf{a}) = 1 - \frac{i}{\hbar}\mathbf{a}\cdot\mathbf{J} \tag{19.20}$$

lead to the following commutation relations[3]:

$$\hat{R}\hat{\Sigma}\hat{R}^\dagger = \hat{\Sigma} \implies \left[\hat{\Sigma}, \hat{J}_i\right] = 0$$

$$\hat{R}\hat{V}_i\hat{R}^\dagger = \mathcal{R}_{ij}^{-1}\hat{V}_j \implies \left[\hat{V}_i, \hat{J}_j\right] = i\hbar\epsilon_{ijk}\hat{V}_k \tag{19.21}$$

$$\hat{R}\hat{Q}_{ij}\hat{R}^\dagger = \mathcal{R}_{ii'}^{-1}\mathcal{R}_{jj'}^{-1}\hat{Q}_{i'j'} \implies \left[\hat{Q}_{ij}, \hat{J}_k\right] = i\hbar\left(\epsilon_{ik\ell}\hat{Q}_{k\ell} + \epsilon_{jk\ell}\hat{Q}_{i\ell}\right).$$

The rotation by 2π around an axis $\hat{n}$, represented by the operator

---

[3]Note that

$$\mathcal{R} \approx 1 + \begin{pmatrix} 0 & -\alpha & 0 \\ \alpha & 0 & 0 \\ 0 & 0 & 0 \end{pmatrix} \implies \mathcal{R}_{ij} = \delta_{ij} - \alpha_k \epsilon_{ijk} + O(\alpha^2).$$

$$\hat{R}_{\hat{n}}(2\pi) = e^{-\frac{2\pi i}{\hbar}\hat{n}\cdot\mathbf{J}},\qquad(19.22)$$

should leave all probability densities unaffected. Note, however, that (19.22) is not necessarily the unit operator.[4] Taking $\hat{n} = \hat{z}$ and acting on a J_z eigenstate, we obtain

$$\hat{R}_{\hat{z}}(2\pi) \to e^{-2i\pi m} = \pm 1,\qquad(19.23)$$

the sign depending on whether the azimuthal quantum number m obtains an integer or a half-odd value. The fact that a rotation by 2π may change the sign of the wave function does not necessarily produce any observable result, since, for instance, the probability density stays invariant. A way to express the invariance under rotations by 2π is to demand that for any observable $\hat{A}$ we must have

$$\left[\hat{A},\ \hat{R}_{\hat{n}}(2\pi)\right] = 0.\qquad(19.24)$$

19.2.1 The Rotation Group

The matrices $\mathcal{R}$ that define rotations satisfy the requirements of the mathematical structure called *a group*. The basic requirement of a *group*[5] is the existence of an operation—in this case, the matrix multiplication—in terms of which the product of two rotations is also a rotation. In mathematical language, symbolizing the group as

$$\mathcal{G} = \{\mathcal{R}_1,\ \mathcal{R}_2,\ \ldots\},\qquad(19.25)$$

its basic property is

$$\mathcal{R}_1,\ \mathcal{R}_2 \in \mathcal{G} \implies (\mathcal{R}_1)_{ij}\,(\mathcal{R}_2)_{jk} = (\mathcal{R}_3)_{ik} \in \mathcal{G}.\qquad(19.26)$$

The rest of the *group "axioms"* are

$$(\mathcal{R}_1\mathcal{R}_2)\,\mathcal{R}_3 = \mathcal{R}_1\,(\mathcal{R}_2\mathcal{R}_3)$$

$$\exists\,\mathbf{I} \implies \mathcal{R}\mathbf{I} = \mathbf{I}\mathcal{R} = \mathcal{R}\qquad(19.27)$$

$$\forall \mathcal{R},\ \exists\,\mathcal{R}^{-1} \implies \mathcal{R}\mathcal{R}^{-1} = \mathcal{R}^{-1}\mathcal{R} = \mathbf{I}.$$

Rotations are not an *Abelian Group*, meaning that general rotations do not necessarily commute[6]

---

[4]Nevertheless $\left(\hat{R}(2\pi)\right)^2 = \hat{R}(4\pi) = \mathbf{I}$, meaning that the rotation by 4π must be the unit operator.

[5]See any of the standard Mathematical Methods textbooks, e.g., Mathews and Walker [3].

[6]Nevertheless, rotations around the same axis are commutative.

$$\mathcal{R}_1\mathcal{R}_2 \neq \mathcal{R}_2\mathcal{R}_1 .$$

The matrices representing rotations in the three-dimensional space are the orthogonal 3×3 matrices with determinant equal to $+1$. This is a so-called *representation of the $O(3)$ group*.

For a quantum system, for every rotation described by $\mathcal{R}$, there is a corresponding unitary operator $\hat{R}$ in Hilbert space acting on the state vectors. The set of these operators shares the group property of the rotation matrices, the corresponding operation being that of operator multiplication in Hilbert space.[7] All the group axioms satisfied by the rotation matrices correspond to analogous relations satisfied by the rotation operators which supply a *unitary representation of the rotation group*

$$\forall\, \hat{R}(\mathbf{a}),\ \hat{R}(\mathbf{b}) \in \mathcal{G} \implies \hat{R}(\mathbf{a})\,\hat{R}(\mathbf{b}) = \hat{R}(\mathbf{c}) \in \mathcal{G}. \tag{19.28}$$

In the case of general rotations, with the axes of rotation not being the same, the actual law of rotation composition is quite complicated.

The rotation operators are differentiable functions of the rotation angles. In that sense they belong to a special class of groups, the so-called *Lie groups*. It can be shown that every element of a Lie group can be written as the exponential of an anti-Hermitian operator or, equivalently, as

$$\hat{U}(\lambda) = e^{i \sum_a \lambda_a \hat{H}_a} \ \text{ with } \ \hat{H}_a^\dagger = \hat{H}_a , \tag{19.29}$$

where a sums over the possible parameters defining the group elements (in the case of rotations, three angles). The Hermitian operators $\hat{H}_a$ in the exponent are called the *generators* of the group and they satisfy the so-called *Lie Algebra*. In the case of rotations, each rotation operator is cast in the form

$$\hat{R}(\mathbf{a}) = e^{-\frac{i}{\hbar}\mathbf{a}\cdot\mathbf{J}}$$

and the three generators of the rotation group are the three components of the angular momentum $\mathbf{J}$, satisfying the corresponding Lie Algebra

$$\left[\hat{J}_i,\ \hat{J}_j\right] = i\hbar\epsilon_{ijk}\hat{J}_k ,$$

which is just the familiar angular momentum commutation relation.

---

[7]The relation of the rotation operator to the rotation matrix was introduced through $\langle \mathcal{R}_1\{\mathbf{r}\}|\hat{R}_1|\psi\rangle = \langle \mathbf{r}_1|\psi_1\rangle = \langle \mathbf{r}|\psi\rangle$. Obviously, we have $\langle \mathcal{R}_2\mathcal{R}_1\{\mathbf{r}\}|\hat{R}_2\hat{R}_1|\psi\rangle = \langle \mathbf{r}|\psi\rangle$. Thus, the product $\hat{R}_2\hat{R}_1$ will be the rotation operator corresponding to the rotation matrix $\mathcal{R}_2\mathcal{R}_1$. The product operator is unitary, since $\hat{R}_2\hat{R}_1\left(\hat{R}_2\hat{R}_1\right)^\dagger = \hat{R}_2\hat{R}_1\hat{R}_1^\dagger\hat{R}_2^\dagger = \mathbf{I}$.

19.2.2 The Rotation Matrices $d^{(j)}$

A rotation cannot change the magnitude of the angular momentum but only its direction. This corresponds to the mathematical fact that for $\hat{R}(\mathbf{a}) = e^{-\frac{i}{\hbar}\mathbf{a}\cdot\mathbf{J}}$, we have

$$\left[\hat{R}(\mathbf{a}), \mathbf{J}^2\right] = 0. \tag{19.30}$$

Acting with $\hat{R}(\mathbf{a})$ on an angular momentum eigenstate $|jm\rangle$, we shall obtain a superposition of eigenstates of the same j but of different m, namely,

$$\hat{R}(\mathbf{a})|jm\rangle = \sum_{m=-j}^{j} |jm'\rangle\, d_{m'm}^{(j)}(\mathbf{a}). \tag{19.31}$$

The coefficients $d_{m'm}^{(j)}$ of this superposition are $(2j + 1) \times (2j + 1)$ matrices and will be referred to as *rotation matrices* [2, 4, 5]. They are obviously the matrix elements

$$d_{mm'}^{(j)}(\mathbf{a}) = \langle jm|\hat{R}(\mathbf{a})|jm'\rangle = \langle jm|e^{-\frac{i}{\hbar}\mathbf{a}\cdot\mathbf{J}}|jm'\rangle \tag{19.32}$$

and are fully determined by the parameters defining the rotation at hand, independent of the physical system.

It is expected that the group property of rotations expressed by (19.28) will be reflected on an analogous expression among the $d^{(j)}$'s. Indeed, we have

$$\langle jm|\hat{R}(\mathbf{c})|jm'\rangle = \langle jm|\hat{R}(\mathbf{a})\left(\sum_{j''}\sum_{m''=-j''}^{\overline{m''=j''}} |j''m''\rangle\langle j''m''|\right)\hat{R}(\mathbf{b})|jm'\rangle$$

$$= \langle jm|\hat{R}(\mathbf{a})\left(\sum_{m''=-j}^{m''=j} |jm''\rangle\langle jm''|\right)\hat{R}(\mathbf{b})|jm'\rangle = \sum_{m''=-j}^{m''=j} d_{mm''}^{(j)}(\mathbf{a})\, d_{m''m'}^{(j)}(\mathbf{b}).$$

Thus, we have arrived at the relation

$$d_{mm'}^{(j)}(\mathbf{c}) = \sum_{m''=-j}^{m''=j} d_{mm''}^{(j)}(\mathbf{a})\, d_{m''m'}^{(j)}(\mathbf{b}) \tag{19.33}$$

or in matrix shorthand

$$d^{(j)}(\mathbf{c}) = d^{(j)}(\mathbf{a})\, d^{(j)}(\mathbf{b}). \tag{19.34}$$

These matrices are obviously unitary.[8] They constitute a $(2j + 1) \times (2j + 1)$-dimensional representation of the rotation group.

It should be stressed that a rotation acting on any $|jm\rangle$ reduces it into a linear combination of $2j + 1$ states corresponding to the different values of m. There is no subset of these states that transforms to itself under rotation. This can also be seen by the fact that we can construct all the $|jm\rangle$ states starting from $|jj\rangle$ and acting with $\hat{J}_-$. We say that the rotation operators $\hat{R}(\mathbf{a})$ act on the states $|jm\rangle$ for fixed j in an *irreducible fashion*. This reflects on the rotation matrices $d^{(j)}(\mathbf{a})$ which constitute an *irreducible representation* of rotation group. In contrast, a reducible representation would have the property that a basis exists in which they would all have a "*block-diagonal*" form. This is not the case and the $d^{(j)}$'s are irreducible.

Consider now a system of two independent angular momenta $\mathbf{J}_1$ and $\mathbf{J}_2$. A rotation operator for the system would be $e^{-\frac{i}{\hbar}\mathbf{a}\cdot\mathbf{J}}$, where $\mathbf{J} = \mathbf{J}_1 + \mathbf{J}_2$ is the total angular momentum of the system. Rotations will not act irreducibly on states $|j_1 j_2 m_1 m_2\rangle$, since these states are linear combinations of states $|j_1 j_2 jm\rangle$ and we know that states with different j's do not mix. The set of the $(2j_1 + 1)(2j_2 + 1)$ states $|j_1 j_2 m_1 m_2\rangle$ will break up into groups of states, each corresponding to the different values of j. These subgroups will transform among themselves under rotations. A rotation $e^{-\frac{i}{\hbar}\mathbf{a}\cdot\mathbf{J}} = e^{-\frac{i}{\hbar}\mathbf{a}\cdot\mathbf{J}_1} e^{-\frac{i}{\hbar}\mathbf{a}\cdot\mathbf{J}_2}$ on $|j_1 j_2 m_1 m_2\rangle$ will give

$$e^{-\frac{i}{\hbar}\mathbf{a}\cdot\mathbf{J}}|j_1 j_2 m_1 m_2\rangle = \sum_{m_1'}|j_1 m_1'\rangle d^{(j_1)}_{m_1' m_1} \sum_{m_2'}|j_2 m_2'\rangle d^{(j_2)}_{m_2' m_2}$$

or

$$e^{-\frac{i}{\hbar}\mathbf{a}\cdot\mathbf{J}}|j_1 j_2 m_1 m_2\rangle = \sum_{m_1' m_2'}|j_1 j_2 m_1' m_2'\rangle d^{(j_1)}_{m_1' m_1} d^{(j_2)}_{m_2' m_2}. \qquad (19.35)$$

It is clear that the rotation matrices of the system

$$\langle j_1 j_2 m_1' m_2'|e^{-\frac{i}{\hbar}\mathbf{a}\cdot\mathbf{J}}|j_1 j_2 m_1 m_2\rangle = d^{(j_1)}_{m_1' m_1} d^{(j_2)}_{m_2' m_2} \qquad (19.36)$$

are the *tensor product*

$$\left(d^{(j_1)}(\mathbf{a}) \otimes d^{(j_2)}(\mathbf{a})\right)_{m_1' m_2', m_1 m_2} = d^{(j_1)}_{m_1' m_1} d^{(j_2)}_{m_2' m_2}. \qquad (19.37)$$

Such tensor product matrices are automatically block-diagonal and, therefore, a reducible representation, i.e.,

$$d^{(j_1)} \otimes d^{(j_2)} = \begin{pmatrix} \begin{pmatrix} \cdot\cdot \\ \cdot\cdot \end{pmatrix} & 0 \\ \\ 0 & \begin{pmatrix} \cdot\cdot \\ \cdot\cdot \end{pmatrix} \end{pmatrix}, \qquad (19.38)$$

where the submatrices refer to different j values.

---

[8]Note that $d^{(j)}(-\mathbf{a}) = \left(d^{(j)}(\mathbf{a})\right)^{\dagger}$.

19.2.3 General Rotations and Euler Angles

An arbitrary rotation in three-dimensional space can be performed in three succes-sive steps. These are conveniently parametrized in terms of the three *Euler angles*, corresponding to each of these steps [1, 2, 4, 5]. Thus, a general rotation can be performed by

(1) rotating around the $\hat{z}$-axis by an angle φ,
(2) rotating around $\hat{y}'$-axis (the $\hat{y}$-axis after the first rotation) by an angle θ, and
(3) rotating around $\hat{z}'$-axis (the $\hat{z}$-axis after the second rotation) by an angle ψ.

This is depicted in Fig.19.1.
The corresponding rotation operator is

$$\hat{R}(\varphi, \theta, \psi) \;=\; e^{-\frac{i}{\hbar}\psi \hat{J}_{z'}}\, e^{-\frac{i}{\hbar}\theta \hat{J}_{y'}}\, e^{-\frac{i}{\hbar}\varphi \hat{J}_z} \tag{19.39}$$

At this point we can make use of the relation $\hat{R}\hat{V}_i\hat{R}^\dagger = \hat{V}_i'$ for $\hat{J}_y$ and have

$$e^{-\frac{i}{\hbar}\varphi \hat{J}_z}\, \hat{J}_y e^{\frac{i}{\hbar}\varphi \hat{J}_z} \;=\; \hat{J}_{y'} . \tag{19.40}$$

Note also that

$$e^{-\frac{i}{\hbar}\varphi \hat{J}_z}\, \hat{J}_y^2 e^{\frac{i}{\hbar}\varphi \hat{J}_z} \;=\; e^{-\frac{i}{\hbar}\varphi \hat{J}_z}\, \hat{J}_y e^{\frac{i}{\hbar}\varphi \hat{J}_z} e^{-\frac{i}{\hbar}\varphi \hat{J}_z}\, \hat{J}_y e^{\frac{i}{\hbar}\varphi \hat{J}_z} \;=\; \hat{J}_{y'}^2 \tag{19.41}$$

and, therefore, we shall have

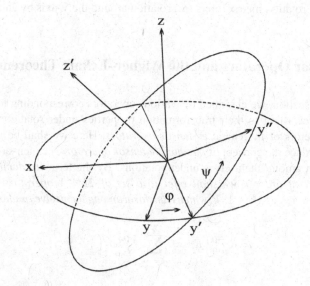

Fig. 19.1 The Euler angles

$$e^{-\frac{i}{\hbar}\theta\hat{J}_{y'}} = e^{-\frac{i}{\hbar}\varphi\hat{J}_z} e^{-\frac{i}{\hbar}\theta\hat{J}_y} e^{\frac{i}{\hbar}\varphi\hat{J}_z} . \tag{19.42}$$

Similarly, we also have

$$e^{-\frac{i}{\hbar}\psi\hat{J}_{z'}} = e^{-\frac{i}{\hbar}\theta\hat{J}_{y'}} e^{-\frac{i}{\hbar}\psi\hat{J}_z} e^{\frac{i}{\hbar}\theta\hat{J}_{y'}} \tag{19.43}$$

or

$$e^{-\frac{i}{\hbar}\psi\hat{J}_{z'}} = e^{-\frac{i}{\hbar}\varphi\hat{J}_z} e^{-\frac{i}{\hbar}\theta\hat{J}_y} e^{\frac{i}{\hbar}\varphi\hat{J}_z} e^{-\frac{i}{\hbar}\psi\hat{J}_z} e^{-\frac{i}{\hbar}\varphi\hat{J}_z} e^{\frac{i}{\hbar}\theta\hat{J}_y} e^{\frac{i}{\hbar}\varphi\hat{J}_z} . \tag{19.44}$$

Substituting in $\hat{R}$ we obtain

$$\hat{R} = e^{-\frac{i}{\hbar}\varphi\hat{J}_z} e^{-\frac{i}{\hbar}\theta\hat{J}_y} e^{\frac{i}{\hbar}\varphi\hat{J}_z} e^{-\frac{i}{\hbar}\psi\hat{J}_z} e^{-\frac{i}{\hbar}\varphi\hat{J}_z} \tag{19.45}$$

and, since $e^{-\frac{i}{\hbar}\psi\hat{J}_z} e^{-\frac{i}{\hbar}\varphi\hat{J}_z} = e^{-\frac{i}{\hbar}\varphi\hat{J}_z} e^{-\frac{i}{\hbar}\psi\hat{J}_z}$,

$$\hat{R}(\varphi, \theta, \psi) = e^{-\frac{i}{\hbar}\varphi\hat{J}_z} e^{-\frac{i}{\hbar}\theta\hat{J}_y} e^{-\frac{i}{\hbar}\psi\hat{J}_z} . \tag{19.46}$$

From this expression it is not difficult to deduce the corresponding rotation matrix. We have

$$d^{(j)}_{mm'}(\varphi, \theta, \psi) = \langle jm|e^{-\frac{i}{\hbar}\varphi\hat{J}_z} e^{-\frac{i}{\hbar}\theta\hat{J}_y} e^{-\frac{i}{\hbar}\psi\hat{J}_z}|jm'\rangle = e^{-im\varphi - im'\psi}\langle jm|e^{-\frac{i}{\hbar}\theta\hat{J}_y}|jm'\rangle \tag{19.47}$$

or

$$d^{(j)}_{mm'}(\varphi, \theta, \psi) = e^{-im\varphi - im'\psi} d^{(j)}_{mm'}(\theta) , \tag{19.48}$$

where the last rotation matrix refers to a rotation around the $\hat{y}$-axis by an angle θ.

19.3 Tensor Operators and the Wigner–Eckart Theorem

In a previous section, we classified the various operators corresponding to physical observables according to their transformation properties under rotations, namely, as scalars, vectors or in general as *tensor operators*. Here we shall be more specific and introduce the concept of *irreducible tensor operators*, i.e., tensor operators that transform among themselves under rotations. By the term *irreducible tensor operators $\hat{T}^{(k)}$ of order k we shall refer to a set of $2k + 1$ operators $\hat{T}^{(k)}_q$ with $q = -k, -k + 1, \ldots, k - 1, k$, which transform among themselves under rotations according to*

$$\hat{R}\,\hat{T}^{(k)}_q\,\hat{R}^\dagger = \sum_{q'=-k}^{k} \hat{T}^{(k)}_{q'}\, d^{(k)}_{q'q} . \tag{19.49}$$

The irreducibility of the set of $\hat{T}^{(k)}_q$ follows directly from the irreducibility of the rotation matrices $d^{(k)}_{q'q}$.

Let's apply the transformation law (19.49) in the case of an infinitesimal rotation

$$\hat{R}(\mathbf{a}) \approx 1 - \frac{i}{\hbar}\mathbf{a} \cdot \mathbf{J}. \tag{19.50}$$

We have

$$\hat{T}_q^{(k)} - \frac{i}{\hbar}\left[\mathbf{a} \cdot \mathbf{J}, \hat{T}_q^{(k)}\right] = \sum_{q'} \hat{T}_{q'}^{(k)} \langle kq' | \left(1 - \frac{i}{\hbar}\mathbf{a} \cdot \mathbf{J}\right) | kq \rangle \tag{19.51}$$

or

$$\left[\mathbf{J}, \hat{T}_q^{(k)}\right] = \sum_{q'} \hat{T}_{q'}^{(k)} \langle kq' | \mathbf{J} | kq \rangle, \tag{19.52}$$

leading to

$$\left[\hat{J}_z, \hat{T}_q^{(k)}\right] = \hbar q \hat{T}_q^{(k)}$$

$$\left[\hat{J}_\pm, \hat{T}_q^{(k)}\right] = \hbar \hat{T}_{q\pm1}^{(k)} \sqrt{k(k+1) - q(q\pm1)} \tag{19.53}$$

Any vector operator, defined through the commutation relation of its Cartesian components

$$\left[\hat{J}_i, \hat{V}_j\right] = i\hbar\epsilon_{ijk}\hat{V}_k,$$

can be expressed as an irreducible *order*-1 *tensor* in terms of the components

$$\hat{V}_{q=1} = -\frac{1}{\sqrt{2}}\left(\hat{V}_x + i\hat{V}_y\right)$$

$$\hat{V}_{q=0} = \hat{V}_z \tag{19.54}$$

$$\hat{V}_{q=-1} = \frac{1}{\sqrt{2}}\left(\hat{V}_x - i\hat{V}_y\right).$$

It is straightforward to verify for $\hat{V}_q$ the relations (19.53). Along these lines we may introduce the so-called *spherical components* of the position operator as

$$r_1 = -\frac{1}{\sqrt{2}}(x + iy) = -\frac{r}{2}\sin\theta\, e^{i\phi}$$

$$r_0 = z = r\cos\theta \tag{19.55}$$

$$r_{-1} = \frac{1}{\sqrt{2}}(x - iy) = \frac{r}{2}\sin\theta e^{-i\phi}.$$

Consider now an eigenstate of angular momentum $|\lambda, j_1, m_1\rangle$, where λ denotes collectively all other quantum numbers. When we act with a tensor operator $T_q^{(k)}$ on this state, we obtain an eigenstate of $\hat{J}_z$ with eigenvalue $\hbar(q + m_1)$. This follows immediately from the first of the relations (19.53). Indeed, we have

$$\left[\hat{J}_z, \hat{T}_q^{(k)}\right]|\lambda, j_1, m_1\rangle = \hbar q \hat{T}_q^{(k)}|\lambda, j_1, m_1\rangle$$

or

$$\hat{J}_z \hat{T}_q^{(k)}|\lambda, j_1, m_1\rangle - \hat{T}_q^{(k)}(\hbar m_1)|\lambda, j_1, m_1\rangle = \hbar q \hat{T}_q^{(k)}|\lambda, j_1, m_1\rangle$$

or

$$\hat{J}_z \hat{T}_q^{(k)}|\lambda, j_1, m_1\rangle = \hbar(q + m_1)\hat{T}_q^{(k)}|\lambda, j_1, m_1\rangle.$$

It follows then that the matrix elements

$$\langle\lambda', j', m'|\hat{T}_q^{(k)}|\lambda, j, m\rangle = 0 \quad unless \quad m' = q + m. \tag{19.56}$$

So, the action of the tensor operator on the eigenstate $|\lambda, j_1, m_1\rangle$ is analogous to the addition of angular momenta J_z. Note, however, that the states $\hat{T}_q^{(k)}|\lambda, j, m\rangle$ are not eigenstates of J^2.

Consider now the second of the relations (19.53) applied on a state $|\lambda, j, m\rangle$ and take the inner product with a state $\langle\lambda', j', m'|$. We have

$$\langle\lambda', j', m'|\left[\hat{J}_\pm, \hat{T}_q^{(k)}\right]|\lambda, j, m\rangle = \hbar \langle\lambda', j', m'|\hat{T}_{q\pm1}^{(k)}|\lambda, jm\rangle \sqrt{k(k + 1) - q(q \pm 1)} \tag{19.57}$$

or

$$\sqrt{j'(j' + 1) - m'(m' \mp 1)}\langle\lambda'j'm' \mp 1|\hat{T}_q^{(k)}|\lambda jm\rangle - \sqrt{j(j + 1) - m(m \pm 1)}\langle\lambda'j'm'|\hat{T}_q^{(k)}|\lambda jm \pm 1\rangle$$

$$= \langle\lambda', j', m'|\hat{T}_{q\pm1}^{(k)}|\lambda, jm\rangle \sqrt{k(k + 1) - q(q \pm 1)}.$$

Notice that if we make the substitution $j' \to j$, $m' \to m$, $j \to j_1, m \to m_1, k \to j_2$ and $q \to m_2$, the irreducible tensor operator relation becomes

$$\sqrt{j(j + 1) - m(m \mp 1)}\langle\lambda'jm \mp 1|\hat{T}_{m_2}^{(j_2)}|\lambda j_1 m_1\rangle - \sqrt{j_1(j_1 + 1) - m_1(m_1 \pm 1)}\langle\lambda'j'm'|\hat{T}_{m_2}^{(j_2)}|\lambda jm \pm 1\rangle$$

$$= \langle\lambda', j, m|\hat{T}_{m_2\pm1}^{(j_2)}|\lambda, j_1 m_1\rangle \sqrt{j_2(j_2 + 1) - m_2(m_2 \pm 1)}$$

and these two relations are similar. They are both of the form

$$\sum_j C_{ij} x_j = 0 \quad and \quad \sum_j C_{ij} y_j = 0 \tag{19.58}$$

with the same coefficients C_{ij}. Therefore, their solutions must be proportional or $x_j = C y_j$ with C a universal coefficient. Thus, we may conclude that

$$\langle\lambda', j', m'|\hat{T}_{q\pm1}^{(k)}|\lambda, j, m\rangle = C \langle j, k, m, q \pm 1|j, k, j', m'\rangle, \tag{19.59}$$

where the coefficient C will not depend on the azimuthal quantum numbers m, q, m'. This is the *Wigner–Eckart Theorem* [2, 4, 5], usually stated as

$$\langle \lambda', j', m' | \hat{T}_q^{(k)} | \lambda, j, m \rangle = \langle j, k, m, q | j, k, j', m' \rangle \frac{\langle \lambda' j' || \hat{T}^{(k)} || \lambda j \rangle}{\sqrt{2j+1}}, \qquad (19.60)$$

the last factor being just a way to symbolize the universal coefficient.

19.4 Translations in Time and Space

We are already quite familiar with the evolution in time of quantum systems in terms of the unitary time-evolution operators. Time evolution can be seen as a *time translation* transformation

$$t \rightarrow t + T. \qquad (19.61)$$

In the Schroedinger picture, this amounts to

$$|\psi(t + T)\rangle = \hat{U}(t + T, t) |\psi(t)\rangle. \qquad (19.62)$$

For a conservative system, the evolution operator is simply

$$\hat{U}(t + T, t) = e^{-\frac{i}{\hbar} \hat{H} T}. \qquad (19.63)$$

Invariance under the transformations (19.61), often referred to as *homogeneity of time*, implies the conservation of a physical quantity. This is a general rule of symmetries and here it simply corresponds to the fact that the Hamiltonian is a constant of the motion or, equivalently, to the conservation of energy.

The set of operators $\hat{U}$ is a *one-parameter* Lie group with a single generator, namely, the Hamiltonian $\hat{H}$. The group property can be expressed as

$$\hat{U}(t, t') \hat{U}(t', t'') = \hat{U}(t, t''). \qquad (19.64)$$

The group of time translations is commutative (Abelian). This, together with the rest of group properties, goes as follows:

$$\hat{U}(t_1, t_2) \hat{U}(t_2, t_3) = \hat{U}(t_2, t_3) \hat{U}(t_1, t_2)$$

$$\left(\hat{U}(t_1, t_2) \hat{U}(t_2, t_3) \right) \hat{U}(t_3, t_4) = \hat{U}(t_1, t_2) \left(\hat{U}(t_2, t_3) \hat{U}(t_3, t_4) \right)$$

$$\hat{U}(t_1, t_2) \hat{U}(t_2, t_1) = \mathbf{I} \implies \hat{U}^{-1}(t_1, t_2) = \hat{U}(t_2, t_1) \qquad (19.65)$$

$$\hat{U}(t, t) = \mathbf{I}.$$

In addition to these, we also have unitarity

$$\hat{U}^\dagger(t, t') = \hat{U}^{-1}(t, t') = \hat{U}(t', t). \tag{19.66}$$

The effect of time translations on physical observables can be seen in the Heisenberg picture as

$$\hat{A}(t) = \hat{U}^\dagger(t, t_0)\, \hat{A}(t_0)\, \hat{U}(t, t_0). \tag{19.67}$$

An observable that is invariant in time (*constant of the motion*) will necessarily have to commute with all the $\hat{U}(t, t_0)$, namely,

$$\hat{C}(t) = \hat{U}^\dagger(t, t_0)\hat{C}(t_0)\hat{U}(t, t_0) = \hat{C}(t_0) \implies \left[\hat{U}(t, t_0),\, \hat{C}(t_0)\right] = 0. \tag{19.68}$$

Next, we may consider the case of *spatial translations*

$$\mathbf{r} \to \mathbf{r} + \mathbf{a}. \tag{19.69}$$

We are already familiar with these transformations and especially in the case of periodic systems. For the system of a quantum particle, the corresponding set of unitary operators is

$$\hat{T}(\mathbf{a}) = e^{\frac{i}{\hbar}\mathbf{a}\cdot\mathbf{p}}. \tag{19.70}$$

This is a three-parameter commutative (Abelian) Lie group. The basic group property is

$$\hat{T}(\mathbf{a}_1)\, \hat{T}(\mathbf{a}_2) = \hat{T}(\mathbf{a}_1 + \mathbf{a}_2). \tag{19.71}$$

The group axioms read

$$\hat{T}(\mathbf{a}_1)\hat{T}(\mathbf{a}_2) = \hat{T}(\mathbf{a}_2)\hat{T}(\mathbf{a}_1)$$

$$\left(\hat{T}(\mathbf{a}_1)\hat{T}(\mathbf{a}_2)\right)\hat{T}(\mathbf{a}_3) = \hat{T}(\mathbf{a}_1)\left(\hat{T}(\mathbf{a}_2)\hat{T}(\mathbf{a}_3)\right)$$

$$\hat{T}(-\mathbf{a})\hat{T}(\mathbf{a}) = \mathbf{I} \implies \hat{T}^{-1}(\mathbf{a}) = \hat{T}(-\mathbf{a}) \tag{19.72}$$

$$\hat{T}(0) = \mathbf{I}.$$

In addition to these, we also have unitarity

$$\hat{T}^\dagger(\mathbf{a}) = \hat{T}^{-1}(\mathbf{a}) = \hat{T}(-\mathbf{a}). \tag{19.73}$$

It is straightforward to see that the action of $\hat{T}(\mathbf{a})$ on the position operator amounts to a translation by $\mathbf{a}$, namely,

$$\hat{T}(\mathbf{a}) \, \mathbf{r} \, \hat{T}^\dagger(\mathbf{a}) = \mathbf{r} + \mathbf{a}. \tag{19.74}$$

The general action of $\hat{T}(\mathbf{a})$ on an arbitrary observable will be

$$\hat{T}(\mathbf{a}) \, \hat{O} \, \hat{T}^\dagger(\mathbf{a}) = \hat{O}'. \tag{19.75}$$

In the case that the observable $\hat{O}$ is *translationally invariant*, we shall have

$$\hat{T}(\mathbf{a}) \, \hat{O} \, \hat{T}^\dagger(\mathbf{a}) = \hat{O} \implies \left[\hat{O}, \, \hat{T}(\mathbf{a}) \right] = 0. \tag{19.76}$$

Invariance in space translations is not a property as general as time translational invariance. In the absence of forces, spatial translational invariance is true and it is often referred to as *homogeneity of space*. The conserved quantity in the case of translational invariance is the momentum. Thus, translational invariance is linked to momentum conservation.

A set of transformations similar to spatial translations, which, however, do not by themselves correspond to any common symmetry, is the *translations in momentum space*. For the system of a quantum particle, the corresponding unitary operators are $\hat{\Gamma}(\mathbf{k}) = e^{-i\mathbf{k}\cdot\mathbf{r}}$. They act on the momentum operator as

$$\hat{\Gamma}(\mathbf{k}) \, \mathbf{p} \, \hat{\Gamma}^\dagger(\mathbf{k}) = \mathbf{p} + \hbar\mathbf{k}. \tag{19.77}$$

Example 19.1 Consider the group of *Galilean Transformations* to a (primed) reference frame moving with velocity $\mathbf{V}$

$$\mathbf{r} \to \mathbf{r}' = \mathbf{r} - \mathbf{V}t, \quad \mathbf{p} \to \mathbf{p}' = \mathbf{p} - m\mathbf{V}. \tag{19.78}$$

(a) Show that the unitary operator that can realize the above Galilean Transformation on the position and momentum operators is

$$\hat{G} = e^{\frac{i}{\hbar}\mathbf{V}\cdot(m\mathbf{r} - t\mathbf{p})}. \tag{19.79}$$

(b) Find how $\hat{G}$ acts on the wave function of a particle.
(c) Show that the operators $\hat{G}(\mathbf{V})$ are an Abelian group.

(a) The proposed Galilean operator is of the form $\hat{G} = e^{\hat{A}+\hat{B}}$ with $[\hat{A}, \hat{B}]$ being a c-number. Therefore, the special case of the Baker–Hausdorff formula $e^{\hat{A}+\hat{B}} = e^{\hat{A}} e^{\hat{B}} e^{-\frac{1}{2}[\hat{A},\hat{B}]}$ applies and we have

$$\hat{G}(\mathbf{V}) = e^{-\frac{i}{\hbar}\left(\frac{mV^2}{2}\right)t} \, e^{\frac{i}{\hbar}m\mathbf{V}\cdot\mathbf{r}} \, e^{-\frac{i}{\hbar}t\mathbf{V}\cdot\mathbf{p}}. \tag{19.80}$$

Thus, we have

$$\hat{G}\mathbf{r}\hat{G}^\dagger = e^{\frac{i}{\hbar}m\mathbf{V}\cdot\mathbf{r}}e^{-\frac{i}{\hbar}t\mathbf{V}\cdot\mathbf{p}}\mathbf{r}\,e^{\frac{i}{\hbar}t\mathbf{V}\cdot\mathbf{p}}e^{-\frac{i}{\hbar}m\mathbf{V}\cdot\mathbf{r}} = e^{\frac{i}{\hbar}m\mathbf{V}\cdot\mathbf{r}}\,(\mathbf{r}-\mathbf{V}t)\,e^{-\frac{i}{\hbar}m\mathbf{V}\cdot\mathbf{r}} = \mathbf{r}-\mathbf{V}t$$

$$\hat{G}\mathbf{p}\hat{G}^\dagger = e^{\frac{i}{\hbar}m\mathbf{V}\cdot\mathbf{r}}e^{-\frac{i}{\hbar}t\mathbf{V}\cdot\mathbf{p}}\mathbf{p}\,e^{\frac{i}{\hbar}t\mathbf{V}\cdot\mathbf{p}}e^{-\frac{i}{\hbar}m\mathbf{V}\cdot\mathbf{r}} = e^{\frac{i}{\hbar}m\mathbf{V}\cdot\mathbf{r}}\,\mathbf{p}\,e^{-\frac{i}{\hbar}m\mathbf{V}\cdot\mathbf{r}} = \mathbf{p}-m\mathbf{V}.$$

(19.81)

(b) We have

$$\hat{G}|\psi\rangle = |\psi'\rangle \implies e^{-\frac{i}{\hbar}\left(\frac{mV^2}{2}\right)t}\,e^{\frac{i}{\hbar}m\mathbf{V}\cdot\mathbf{r}}\,e^{-t\mathbf{V}\cdot\nabla}\psi(\mathbf{r},t) = \psi'(\mathbf{r},t)\,. \tag{19.82}$$

or

$$\psi'(\mathbf{r},t) = e^{-\frac{i}{\hbar}\left(\frac{mV^2}{2}\right)t}\,e^{\frac{i}{\hbar}m\mathbf{V}\cdot\mathbf{r}}\,\psi(\mathbf{r}-\mathbf{V}t,t)\,. \tag{19.83}$$

(c) We have

$$\hat{G}(\mathbf{V}_1)\hat{G}(\mathbf{V}_2) = e^{-\frac{im}{2\hbar}(V_1^2+V_2^2)t}\,e^{\frac{i}{\hbar}m\mathbf{V}_1\mathbf{r}}\,e^{-\frac{i}{\hbar}V_1\mathbf{p}t}\,e^{\frac{i}{\hbar}m\mathbf{V}_2\mathbf{r}}\,e^{-\frac{i}{\hbar}V_2\mathbf{p}t}$$

$$= e^{-\frac{im}{2\hbar}(V_1^2+V_2^2)t}\,e^{\frac{i}{\hbar}m\mathbf{V}_1\mathbf{r}}\,e^{\frac{i}{\hbar}m\mathbf{V}_2\mathbf{r}}\,\underbrace{e^{-\frac{i}{\hbar}m\mathbf{V}_2\mathbf{r}}e^{-\frac{i}{\hbar}V_1\mathbf{p}t}\,e^{\frac{i}{\hbar}m\mathbf{V}_2\mathbf{r}}}\,e^{-\frac{i}{\hbar}V_2\mathbf{p}t}$$

$$= e^{-\frac{im}{2\hbar}(V_1^2+V_2^2)t}\,e^{\frac{i}{\hbar}m(\mathbf{V}_1+\mathbf{V}_2)\cdot\mathbf{r}}\,e^{-\frac{i}{\hbar}V_1\cdot(\mathbf{p}-m\mathbf{V}_2)t}\,e^{-\frac{i}{\hbar}V_2\cdot\mathbf{p}t}$$

$$= e^{-\frac{im}{2\hbar}(\mathbf{V}_1+\mathbf{V}_2)^2t}\,e^{\frac{i}{\hbar}m(\mathbf{V}_1+\mathbf{V}_2)\cdot\mathbf{r}}\,e^{-\frac{i}{\hbar}(\mathbf{V}_1+\mathbf{V}_2)\cdot\mathbf{p}t} = \hat{G}(\mathbf{V}_1+\mathbf{V}_2)\,. \tag{19.84}$$

19.5   Discrete Transformations

We have already met a transformation that does not depend on any continuous parameter. This is the familiar transformation of *spatial reflection or parity*

$$\mathcal{P}: \quad \mathbf{r} \to -\mathbf{r}\,. \tag{19.85}$$

The corresponding unitary operator, defined by

$$\langle\mathbf{r}|\hat{P}|\psi\rangle = \langle-\mathbf{r}|\psi\rangle \cdot \forall|\psi\rangle\,, \tag{19.86}$$

is both unitary and Hermitian

$$\hat{P}^{-1} = \hat{P}^\dagger = \hat{P}\,. \tag{19.87}$$

From its definition, we can immediately deduce its action on the position operator

$$\hat{P}\,\mathbf{r}\,\hat{P} = -\mathbf{r} \tag{19.88}$$

as well as on other observables

$$\hat{P}\mathbf{p}\hat{P} = -\mathbf{p}, \quad \left[\mathbf{L}, \hat{P}\right] = \left[\mathbf{S}, \hat{P}\right] = \left[\mathbf{J}, \hat{P}\right] = 0\,. \tag{19.89}$$

All these observables, characterized as vector operators due to their behavior under rotations, are divided as *polar vectors*, if they anticommute with parity, or *axial vectors*, if they commute.

It is possible to define transformations that correspond to a *reflection with respect to a plane*, for example, a reflection on the (x, z)-plane

$$\mathcal{P}_{\hat{y}}: \quad x, y, z \to x, -y, z. \tag{19.90}$$

We may symbolize such a transformation with $\mathcal{P}_{\hat{n}}$ where $\hat{n}$ is the unit vector perpendicular on the plane of reflection. The corresponding unitary operator has all the properties of the usual parity (unitarity, Hermiticity) and is symbolized as $\hat{P}_{\hat{n}}$. Note, however, that it can always be expressed as the product of the standard parity and a rotation

$$\hat{P}_{\hat{n}} = \hat{P}\,\hat{R}(\pi\hat{n}) = \hat{P}\,e^{-\frac{i}{\hbar}\pi\hat{n}\cdot\mathbf{J}}. \tag{19.91}$$

As an example, consider $\hat{n} = \hat{x}$. Acting with $\hat{P}_{\hat{x}}$ on the wave function, we obtain

$$\hat{P}_{\hat{x}}\psi(x, y, z) = \psi(-x, y, z).$$

However, acting with $\hat{P}\,\hat{R}(\pi\hat{x})$ we obtain exactly the same result

$$\hat{P}\,\hat{R}(\pi\hat{x})\psi(x, y, z) = \hat{P}\psi(x, -y, -z) = \psi(-x, y, z).$$

An interesting result appears when we consider the square of these operators. We get

$$\hat{P}_{\hat{n}}^2 = \hat{P}e^{\frac{i}{\hbar}\pi\hat{n}\cdot\mathbf{J}}\hat{P}e^{\frac{i}{\hbar}\pi\hat{n}\cdot\mathbf{J}} = \hat{P}^2e^{\frac{i}{\hbar}2\pi\hat{n}\cdot\mathbf{J}}$$

or

$$\hat{P}_{\hat{n}}^2 = e^{\frac{2\pi i}{\hbar}\hat{n}\cdot\mathbf{J}}. \tag{19.92}$$

If j obtains half-odd values, this rotation by 2π gives $\hat{P}_{\hat{n}}^2 = -1$. This is a property we have encountered earlier. Spinors can come out with a reversed sign after a rotation by 2π, nevertheless, without any unwanted observable consequences.

Consider now the Schroedinger equation for a particle

$$\left\{-\frac{\hbar^2}{2m}\nabla^2 + V(\mathbf{r})\right\}\psi(\mathbf{r}, t) = i\hbar\frac{\partial}{\partial t}\psi(\mathbf{r}, t). \tag{19.93}$$

Its complex conjugate

$$\left\{-\frac{\hbar^2}{2m}\nabla^2 + V(\mathbf{r})\right\}\psi^*(\mathbf{r}, t) = -i\hbar\frac{\partial}{\partial t}\psi^*(\mathbf{r}, t) = i\hbar\frac{\partial}{\partial(-t)}\psi^*(\mathbf{r}, t) \tag{19.94}$$

takes the same form if we make the change $t \to -t$. Thus, apart from (19.93) we also have the equation

$$\left\{ -\frac{\hbar^2}{2m} \nabla^2 + V(\mathbf{r}) \right\} \psi^*(\mathbf{r}, -t) = i\hbar \frac{\partial}{\partial t} \psi^*(\mathbf{r}, -t), \qquad (19.95)$$

meaning that the wave function $\psi^*(\mathbf{r}, -t)$ is also a solution of the same Schroedinger equation. This is not very different from what happens for a classical particle for which the variables $\mathbf{r}(-t)$, $-\mathbf{v}(-t)$ (describing a *time-inverted* course of motion) satisfy the same equations of motion with the variables $\mathbf{r}(t)$, $\mathbf{v}(t)$. This reflects the property of both Newtonian physics and the Schroedinger equation that they are invariant in the transformation

$$\mathcal{C}: \quad t \to -t. \qquad (19.96)$$

This is the *Time Reflection* transformation [2, 4, 5] and the corresponding symmetry is the *Time Reflection Invariance*. Since $\psi(\mathbf{r}, t)$ and $\psi^*(\mathbf{r}, -t)$ are both solutions of the same equation, there should be a relation between them in terms of an operator, representing the above transformation in Hilbert space. From (19.93) and (19.95), we have for the *time-reversed ket* $|\psi'\rangle$

$$|\psi'(t)\rangle = \hat{C}_0 |\psi(-t)\rangle. \qquad (19.97)$$

Nevertheless, these two solutions differ by a complex conjugation and a relation in terms of a linear operator is not possible. Therefore, they must be related through the action of an antilinear operator[9] (and according to Wigner's theorem, antiunitary) that involves a complex conjugation operation

$$\hat{C}_0 \{ \psi(\mathbf{r}) \} = \psi^*(\mathbf{r}) \quad \Longrightarrow \quad \langle \mathbf{r} | \hat{C}_0 | \psi \rangle = \langle \psi | \mathbf{r} \rangle. \qquad (19.98)$$

A repeated action of $\hat{C}_0$ should take us back to the original state, namely,

$$\langle \mathbf{r} | \hat{C}_0^2 | \psi \rangle = \langle \mathbf{r} | \hat{C}_0 | \psi \rangle^* = \langle \mathbf{r} | \psi \rangle \quad \Longrightarrow \quad \hat{C}_0^2 = \mathbf{I}. \qquad (19.99)$$

For an antilinear operator, it is important to indicate whether the operator acts on the left or on the right of a matrix element. The following identity concerning the inner products involving antilinear operators is true

$$\left\{ \langle \psi_1 | \hat{A} \right\} | \psi_2 \rangle = \langle \psi_1 | \left\{ \hat{A} | \psi_2 \rangle \right\}^*. \qquad (19.100)$$

---

[9]For an antilinear operator, we have $\hat{C} i = -i \hat{C}$.

Thus, the definition of Hermitian conjugation for an antilinear operator $\hat{A}$ is modified to[10]

$$\langle \psi_1 | \left\{ \hat{A}^\dagger | \psi_2 \rangle \right\} = \langle \psi_2 | \left\{ \hat{A} | \psi_1 \rangle \right\} . \tag{19.101}$$

Applying this for the operator $\hat{C}_0$, we get

$$\langle \psi_1 | \left\{ \hat{C}_0^\dagger | \psi_2 \rangle \right\} = \left\{ \langle \psi_2 | \hat{C}_0 \right\} | \psi_1 \rangle = \langle \psi_1 | \psi_2 \rangle^* = \langle \psi_2 | \psi_1 \rangle = \langle \psi_1 | \left\{ \hat{C}_0 | \psi_2 \rangle \right\} , \tag{19.102}$$

implying that $\hat{C}_0$ is Hermitian

$$\hat{C}_0^\dagger = \hat{C}_0 . \tag{19.103}$$

Summarizing we have

$$\hat{C}_0^{-1} = \hat{C}_0^\dagger = \hat{C}_0 . \tag{19.104}$$

Since $\hat{C}_0$ only amounts to a complex conjugation, we may define its relation to the position and momentum operators of a particle according to

$$\hat{C}_0 \, \mathbf{r} \, \hat{C}_0 = \mathbf{r}$$
$$\hat{C}_0 \, \mathbf{p} \, \hat{C}_0 = -\mathbf{p}. \tag{19.105}$$

As a result, the orbital angular momentum will obey

$$\hat{C}_0 \, \mathbf{L} \, \hat{C}_0 = -\mathbf{L} . \tag{19.106}$$

Having not included spin yet in our analysis, we move to generalize the time reversal operator in the case of spin. For spin operators we note that under complex conjugation the Pauli matrices change according to

$$\sigma_{1,3}^* = \sigma_{1,3} \quad and \quad \sigma_2^* = -\sigma_2 .$$

Thus, we may impose

$$\hat{C}_0 \hat{S}_x \hat{C}_0 = \hat{S}_x, \quad \hat{C}_0 \hat{S}_y \hat{C}_0 = -\hat{S}_y, \quad \hat{C}_0 \hat{S}_z \hat{C}_0 = \hat{S}_z .$$

Note, however, the identity

$$\sigma_2 \sigma_i \sigma_2 = -\sigma_i^* .$$

Thus, we are led to define the complete time reversal operator as

$$\hat{C} = -i \sigma_2 \, \hat{C}_0 . \tag{19.107}$$

---

[10]Compare this to Hermitian conjugation for a linear operator $\langle \psi_1 | \hat{L}^\dagger | \psi_2 \rangle = \langle \psi_2 | \hat{L} | \psi_1 \rangle^*$.

Under this definition

$$\hat{C}\,\mathbf{r}\,\hat{C} = \mathbf{r} \quad \hat{C}\,\mathbf{p}\,\hat{C} = -\mathbf{p}$$
$$\hat{C}\,\mathbf{L}\,\hat{C} = -\mathbf{L}, \ \hat{C}\,\mathbf{S}\,\hat{C} = -\mathbf{S}. \tag{19.108}$$

An equivalent expression for $\hat{C}$ is[11]

$$\hat{C} = e^{-\frac{i}{\hbar}\pi \hat{S}_y}\,\hat{C}_0. \tag{19.109}$$

Note that now

$$\hat{C}^2 = -i\sigma_2\hat{C}_0(-i\sigma_2)\hat{C}_0 = (-i\sigma_2)^2\hat{C}_0^2 = -1.$$

This happens generally if the system contains an odd number of fermions. The general result is

$$\hat{C}^2 = (-1)^{N_F}. \tag{19.110}$$

For a system with $\hat{C}^2 = -1$, the time reversal operator has to be *anti-Hermitian*, namely,

$$\hat{C}^\dagger = \left(-i\sigma_2\hat{C}_0\right)^\dagger = \hat{C}_0 i\sigma_2 = i\sigma_2\hat{C}_0 = -\hat{C}. \tag{19.111}$$

In summary, for systems with odd number of fermions

$$\hat{C}^{-1} = \hat{C}^\dagger = -\hat{C}. \tag{19.112}$$

An observable $\hat{Q}$ is designated as *real* if it commutes with $\hat{C}$

$$\left[\hat{Q}, \hat{C}\right] = 0 \implies \hat{Q} = real. \tag{19.113}$$

Also a state $|R\rangle$ is characterized as *real* if

$$\hat{C}|R\rangle = |R\rangle. \tag{19.114}$$

Note, however, that if the system is such that $\hat{C}^2 = -1$, there can be no real vectors. Applying the definition of Hermitian conjugation, we have

$$\langle R|\left\{\hat{C}^\dagger|R\rangle\right\} = \langle R|\left\{\hat{C}|R\rangle\right\} = \langle R|R\rangle$$

but since $\hat{C}^\dagger = -\hat{C}$ the left-hand side is just $-\langle R|R\rangle$ implying that $|R\rangle$ is the null vector. Note, however, that for a general vector $|a\rangle$, its complex conjugate $\hat{C}|a\rangle$ will be orthogonal to it, namely,

---

[11] $e^{-\frac{i}{\hbar}\pi\hat{S}_y} = e^{-i\pi\sigma_2/2} = \cos(\pi/2) - i\sigma_2\sin(\pi/2) = -i\sigma_2.$

$$-\langle a| \left\{ \hat{C}|a\rangle \right\} = \langle a| \left\{ \hat{C}^\dagger|a\rangle \right\} = \langle a| \left\{ \hat{C}|a\rangle \right\} = 0.$$

Thus, in such a system, we may construct a basis of mutually complex conjugate vectors.

Consider now a conservative system with a real Hamiltonian

$$\left[\hat{H}, \hat{C} \right] = 0. \tag{19.115}$$

The following relation is true

$$\hat{U}(t, t_0)\hat{C} = \hat{C}\, \hat{U}(t_0, t) = \hat{C}\, \hat{U}^\dagger(t, t_0). \tag{19.116}$$

Multiplying with $\hat{C}^\dagger$ and using unitarity, we have

$$\hat{C}^\dagger\, \hat{U}(t, t_0)\hat{C} = \hat{U}^\dagger(t, t_0). \tag{19.117}$$

This relation is referred to as the *principle of microreversibility* and embodies the time reversal invariance of conservative systems with real Hamiltonians. Consider next the energy eigenvalue problem of such a system $\hat{H}|\psi_E\rangle = E|\psi_E\rangle$. Since $\hat{C}$ commutes with $\hat{H}$, the vector $\hat{C}|\psi_E\rangle$ will also be an eigenstate of the same energy

$$\hat{H}\left(\hat{C}|\psi_E\rangle \right) = E\left(\hat{C}|\psi_E\rangle \right). \tag{19.118}$$

If the system contains an odd number of fermions and $\hat{C}^2 = -1$, we know that there can be no real eigenvectors. Therefore, in the subspace spanned by the degenerate eigenvectors of each eigenvalue, we may choose a basis of complex conjugate eigenvectors, meaning that this subspace will be of even dimension. This type of degeneracy is the so-called *Kramers Degeneracy*.

19.6 Dynamical Symmetry

As we have seen, the symmetries of physical systems correspond to the existence of transformations that leave the Hamiltonian invariant. This is equivalent to the existence of physical observables that commute with the Hamilton operator and, therefore, are constants of the motion. Thus, symmetry under rotations (*spatial isotropy*) leads to angular momentum conservation, symmetry under spatial translations (*homogeneity of space*) leads to momentum conservation, and symmetry under time translations (*homogeneity of time*) leads to energy conservation. Nevertheless, these are not the only physical quantities that could be conserved and it is possible that there are additional physical observables that are conserved depending on the details of the physical system at hand. In contrast to temporal and spatial homogeneity or spatial isotropy which are characterized as *kinematical symmetries*, the additional symmetries arising in particular systems are often called *dynamical symmetries*. Note, however, that this distinction is purely a convention, since, e.g., angular momentum conservation is intimately connected with the dynamics, i.e., the potential of the model.

A very important consequence of the existence of a conserved quantity is that it will have common eigenstates with the Hamiltonian, since it commutes with it. As a result the energy spectrum will be *degenerate*. Therefore, a symmetry implies degeneracy. Note, however, that the opposite is also true. The existence of degeneracies revealed in the calculation of the energy spectrum points out to the existence of symmetries, even if these symmetries were not manifest to begin with. In this section, we will discuss one such example of dynamical symmetry, namely, the case of the hydrogen atom [1].

The large degeneracy of the hydrogen atom is directly related to an additional symmetry that is already present in the corresponding classical problem of the attractive $1/r$ potential. There, the motion is planar because of angular momentum conservation, the plane of motion being perpendicular to the constant angular momentum vector. Nevertheless, the closed elliptical trajectories do not exhibit any *precession*, i.e., the direction of their principal axis is a constant of the motion as well. This would not be true if the attractive potential were not exactly $1/r$. The classical conserved quantity (*Runge–Lenz vector*) is

$$\mathbf{R} = \frac{1}{m}(\mathbf{p} \times \mathbf{L}) - e^2 \frac{\mathbf{r}}{r} \tag{19.119}$$

but, as a quantum operator, because of the non-commutativity of momentum and angular momentum, it has to be modified to

$$\mathbf{R} = \frac{1}{2m}(\mathbf{p} \times \mathbf{L} - \mathbf{L} \times \mathbf{p}) - e^2 \frac{\mathbf{r}}{r}. \tag{19.120}$$

It is straightforward, although tedious, to see that indeed this is a constant of the motion and

$$\left[\hat{H}, \mathbf{R}\right] = 0. \tag{19.121}$$

The existence of this additional constant explains why the degeneracy, although expected to be only $2\ell + 1$ due to rotational symmetry, is as large as

$$\sum_{\ell=1}^{n-1}(2\ell + 1) = n^2.$$

Note also that

$$\mathbf{L} \cdot \mathbf{R} = \mathbf{R} \cdot \mathbf{L} = 0, \tag{19.122}$$

this being the quantum analogue of the fact that at the classical level the Runge–Lenz vector, signifying the direction of the principal elliptical axis, is perpendicular to the angular momentum. Next, we may compute the square of **R**. We find

$$\mathbf{R}^2 = e^4 + \frac{2}{m} \hat{H} \left(\mathbf{L}^2 + \hbar^2 \right) . \tag{19.123}$$

Let us now try to analyze the operator **R** further by deriving its commutators with the angular momentum. We obtain

$$\left[\hat{L}_i, \ \hat{R}_j \right] = i\hbar \epsilon_{ijk} \hat{R}_k , \tag{19.124}$$

verifying the **R** is indeed a vector operator, satisfying the same type of commutation relation with angular momentum as the position and momentum operators.

Next, we may compute the commutator of the different components of **R**. After some effort, we obtain

$$\left[\hat{R}_i, \ \hat{R}_j \right] = -\frac{2i}{m} \hat{H} \epsilon_{ijk} \hat{L}_k . \tag{19.125}$$

The six commutation relations (19.124), (19.125) between the six operators **L**, **R** is a so-called *closed algebra*, since these operators through these relations transform only among themselves. This does not change by the presence of the Hamilton operator, since this is a constant of the motion, commuting with all six of them and acting just as a number.

Let us restrict ourselves to the Hilbert subspace spanned by the discrete energy eigenstates (*bound states*), where H is just a negative number. Then, instead of the **R**'s let's introduce the three operators

$$\mathbf{K} = \sqrt{\frac{-m}{2H}} \, \mathbf{R} . \tag{19.126}$$

The set of coupled commutation relations satisfied by all six **K**, **L** is

$$\left[\hat{K}_i, \ \hat{K}_j \right] = i\hbar \epsilon_{ijk} \hat{L}_k$$

$$\left[\hat{L}_i, \ \hat{K}_j \right] = i\hbar \epsilon_{ijk} \hat{L}_k \tag{19.127}$$

$$\left[\hat{L}_i, \ \hat{L}_j \right] = i\hbar \epsilon_{ijk} \hat{L}_k .$$

These relations can be decoupled if we introduce the operators

$$\mathbf{J}^{(\pm)} = \frac{1}{2} \left(\mathbf{L} \pm \mathbf{K} \right) . \tag{19.128}$$

It turns out that $\mathbf{J}^{(+)}$ and $\mathbf{J}^{(-)}$ are independent and that they both satisfy standard angular momentum-type commutation relations. We have

$$\left[\mathbf{J}^{(+)}, \mathbf{J}^{(-)}\right] = 0$$

$$\left[\hat{J}_i^{(+)}, \hat{J}_j^{(+)}\right] = i\hbar\epsilon_{ijk}\hat{J}_k^{(+)} \tag{19.129}$$

$$\left[\hat{J}_i^{(-)}, \hat{J}_j^{(-)}\right] = i\hbar\epsilon_{ijk}\hat{J}_k^{(-)}.$$

These two angular momenta will have common standard eigenstates and eigenvalues, which may label accordingly

$$\left(\hat{J}^{(+)}\right)^2 |j_+, m_+; j_-, m_-\rangle = \hbar^2 j_+(j_+ + 1)|j_+, m_+; j_-, m_-\rangle$$

$$\hat{J}_z^{(+)}|j_+, m_+; j_-, m_-\rangle = \hbar m_+|j_+, m_+; j_-, m_-\rangle \tag{19.130}$$

$$\left(\hat{J}^{(-)}\right)^2 |j_+, m_+; j_-, m_-\rangle = \hbar^2 j_-(j_- + 1)|j_+, m_+; j_-, m_-\rangle$$

$$\hat{J}_z^{(-)}|j_+, m_+; j_-, m_-\rangle = \hbar m_-|j_+, m_+; j_-, m_-\rangle.$$

Note, however, that there is a constraint between them since $\mathbf{K} \cdot \mathbf{L} = 0$ and

$$\left(\mathbf{J}^{(+)}\right)^2 = \frac{1}{4}\left(\mathbf{L}^2 + \mathbf{K}^2\right) = \left(\mathbf{J}^{(-)}\right)^2. \tag{19.131}$$

As a result of this constraint the angular momentum quantum numbers will be equal, i.e., $j_+ = j_-$. Returning to the expression (19.123), we rewrite it as

$$\mathbf{R}^2 - \frac{2}{m}\hat{H}\mathbf{L}^2 = -\frac{2}{m}\hat{H}(\mathbf{K}^2 + \mathbf{L}^2) = e^4 + \frac{2\hbar^2}{m}\hat{H}$$

or

$$-\frac{2}{m}\hat{H}\left(\left(\mathbf{J}^{(+)}\right)^2 + \left(\mathbf{J}^{(-)}\right)^2\right) = e^4 + \frac{2\hbar^2}{m}\hat{H}. \tag{19.132}$$

Solving for $\hat{H}$, we obtain

$$\hat{H} = -\frac{me^4}{2\hbar^2 + 4\left(\left(\mathbf{J}^{(+)}\right)^2 + \left(\mathbf{J}^{(-)}\right)^2\right)}. \tag{19.133}$$

The corresponding relation for the energy eigenvalues is

$$E = -\frac{me^4}{2\hbar^2(2j_+ + 1)^2}.$$ (19.134)

Since j_+ is the principal quantum number of a generalized angular momentum operator $2j_+ + 1$ must be an integer $n = 1, 2, \ldots$. Thus, we recover the hydrogen atom spectrum

$$E_n = -\frac{me^4}{2\hbar^2 n^2}.$$ (19.135)

The degeneracy is immediately explained by the existence of the two angular momenta of equal principal quantum numbers. I will be

$$(2j_+ + 1) \times (2j_+ + 1) = (2j_+ + 1)^2 = n^2.$$ (19.136)

Example 19.2 Consider the three-dimensional isotropic harmonic oscillator

$$\hat{H} = \frac{\hat{p}^2}{2m} + \frac{1}{2}m\omega^2 r^2.$$

The corresponding classical system is known to have elliptical orbits. In contrast to the case of the attractive $1/r$ potential, the attraction center is not one of the focal points but coincides with the intersection of the principal and secondary axes of the ellipse. The corresponding conserved quantity is not a vector but a tensor. Show that this quantity is the traceless symmetric tensor

$$\hat{Q}_{ij} = m\omega \left(x_i x_j - \frac{1}{3}\delta_{ij}r^2 \right) + \frac{1}{m\omega}\left(\hat{p}_i \hat{p}_j - \frac{1}{3}\delta_{ij}\hat{p}^2 \right).$$

Derive the commutation algebra of $\hat{Q}_{ij}$ with the angular momentum.

It is straightforward but tedious to prove that

$$\left[\hat{H}, \hat{Q}_{ij}\right] = 0.$$ (19.137)

Writing the angular momentum operators as

$$\hat{L}_{ij} = x_i \hat{p}_j - x_j \hat{p}_i = \epsilon_{ijk}\hat{L}_k,$$ (19.138)

we may also derive the following commutator algebra:

$$\left[\hat{L}_{ij}, \hat{L}_{k\ell}\right] = i\hbar\left(\delta_{ik}\hat{L}_{j\ell} + \delta_{jk}\hat{L}_{\ell i} + \delta_{i\ell}\hat{L}_{kj} + \delta_{j\ell}\hat{L}_{ik}\right)$$

$$\left[\hat{L}_{ij}, \hat{Q}_{k\ell}\right] = i\hbar\left(\delta_{ik}\hat{Q}_{j\ell} - \delta_{jk}\hat{Q}_{\ell i} + \delta_{i\ell}\hat{L}_{kj} - \delta_{j\ell}\hat{Q}_{ik}\right)$$ (19.139)

$$\left[\hat{L}_{ij}, \hat{L}_{k\ell}\right] = -i\hbar\left(\delta_{jk}\hat{L}_{\ell i} + \delta_{\ell j}\hat{L}_{ki} + \delta_{ki}\hat{L}_{\ell j} + \delta_{i\ell}\hat{L}_{kj}\right).$$

This is a closed Lie Algebra among the $\hat{L}_{ij}$ and $\hat{Q}_{ij}$ of which eight components are independent. Out of these eight components we may define as linear combinations the eight generators of the $SU(3)$ group, i.e., the group of 3×3 unitary matrices with determinant equal to $+1$. The group elements are expressed as $e^{\frac{i}{2} \sum_{a=1}^{8} \theta_a \lambda_a}$ in terms of the Hermitian generators λ_a that satisfy the algebra $[\lambda_a, \lambda_b] = 2i\hbar f_{abc}\lambda_c$.

Problems and Exercises

19.1 If $\psi(\mathbf{r}, t)$ is the wave function of a particle, show that the wave function $\psi^*(\mathbf{r}, -t)$ describes a particle having the opposite momentum.

19.2 A particle of spin $s = 1$ is described by the Hamiltonian

$$\hat{H} = a\hat{S}_z^2 + b\left(\hat{S}_x^2 - \hat{S}_y^2\right),$$

with a, b real parameters. Is the energy invariant under time reversal? How do the energy eigenstates change under a time reversal?

19.3 Consider the three-dimensional isotropic harmonic oscillator. Prove that the angular momentum $\mathbf{L}$ and the tensor operators $\hat{Q}_{ij} = m\omega \left(x_i x_j - \frac{1}{3}\delta_{ij}r^2\right) + \frac{1}{m\omega}\left(\hat{p}_i \hat{p}_j - \frac{1}{3}\delta_{ij}\hat{p}^2\right)$ satisfy the relation

$$\hat{L}^2 + \frac{1}{2}Tr\left(\hat{Q}^2\right) = -3\hbar^2 + \frac{4\hat{H}^2}{3\omega^2}.$$

19.4 Show that, if the Hamiltonian of a system is time reversal invariant, we can always choose the wave function to be real. How is this compatible with the fact that at some particular time the wave function of a free particle can be a plane wave $e^{i\mathbf{k}\cdot\mathbf{r}}$?

19.5 A system consists of two particles of the same mass interacting through a potential $V(|\mathbf{r}_1 - \mathbf{r}_2|)$. Discuss the rotational properties of the system and the associated conserved quantities.

19.6 Let α, β, γ be the Euler angles. If $U = e^{-i\alpha X_1} e^{-i\beta X_2} e^{-i\gamma X_3}$ is to represent a rotation, find the commutation relations that have to be satisfied by the X_1, X_2, X_3. Relate them to angular momentum.

19.7 Consider the *Galilean Transformations* $\hat{G} = e^{\frac{i}{\hbar}\mathbf{V}\cdot\mathbf{K}}$, where their generator is $\mathbf{K} = m\mathbf{r} - t\mathbf{p}$. Show that $\mathbf{K}$ is a vector operator.

19.8 Consider a sequence of Euler rotations represented by

$$D^{(1/2)}(\alpha, \beta, \gamma) = e^{-i\sigma_3\alpha/2} e^{-i\sigma_2\beta/2} e^{-i\sigma_3\gamma/2}.$$

Show that this is equivalent to a *single* rotation around an axis by an angle θ. Find θ.

19.9 Consider the simple harmonic oscillator. Show that spatial reflection (parity) can be represented by the operator $-i e^{i\frac{\pi \hat{H}}{\hbar \omega}}$, where $\hat{H}$ is the Hamiltonian.

19.10 For the parity operator $\hat{P}_{\hat{n}} = \hat{P} e^{-\frac{i}{\hbar}\pi \hat{n}\cdot \mathbf{J}}$ prove the following:

$$\hat{P}_{\hat{n}} \mathbf{r} \hat{P}_{\hat{n}} = \mathbf{r} - 2\hat{n}\left(\hat{n}\cdot\mathbf{r}\right)$$

$$\hat{P}_{\hat{n}} \mathbf{p} \hat{P}_{\hat{n}} = \mathbf{p} - 2\hat{n}\left(\hat{n}\cdot\mathbf{p}\right)$$

$$\hat{P}_{\hat{n}} \mathbf{J} \hat{P}_{\hat{n}} = -\mathbf{J} + 2\hat{n}\left(\hat{n}\cdot\mathbf{J}\right)$$

References

1. G. Baym, *Lectures in Quantum Mechanics*. Lecture Notes and Supplements in Physics (ABP, 1969)
2. W. Greiner, B. Müller, *Quantum Mechanics Symmetries*, 2nd edn. (Springer, Berlin, 1992)
3. J. Mathews, R. Walker, *Mathematical Methods of Physics*, 2nd edn. (Pearson, London, 1970)
4. E. Merzbacher, *Quantum Mechanics*, 3rd edn. (Wiley, Hoboken, 1998)
5. A. Messiah, *Quantum Mechanics* (Dover publications, Mineola, 1958). Single-volume reprint of the Wiley, New York, two-volume 1958 edn

Chapter 20
Scattering

20.1 The Scattering Cross Section

Most of the experiments on short-distance phenomena are scattering experiments, where a target is struck by a beam of particles of well-defined energy. The density of incident particles almost always is sufficiently small so that mutual interactions can be neglected. In such an experiment, appropriately placed counters can count the number of scattered particles per unit time in various directions. In the majority of cases, the scattering centers composing the target are distant enough so that interference effects between waves scattered by each of them can be neglected. Let $\mathcal{J}$ be the *incident flux*, i.e., the number of particles per unit time per unit surface placed perpendicular to their direction of propagation. The number of particles $d\mathcal{N}$ scattered per unit time into a solid angle $d\Omega$ should be proportional to $\mathcal{J}$, i.e.,

$$d\mathcal{N} = \sigma \, \mathcal{J} \, d\Omega. \tag{20.1}$$

The proportionality constant σ is called *differential cross section* and depends on the direction Ω. The expression (20.1) can be considered as the definition of the differential scattering cross section, symbolized also as $\sigma(\Omega) = \frac{d\sigma}{d\Omega}$. Therefore, the differential scattering cross section is defined as *the number of scattered particles per unit time, per unit of incident flux, per unit of solid angle*. The *total scattering cross section* is defined as [1, 2]

$$\sigma = \int d\Omega \, \sigma(\Omega), \tag{20.2}$$

being the number of particles scattered *in all directions* per unit time, per unit of incident flux. Both the differential and the total scattering cross sections have dimensions of surface. Scattering experiments in Nuclear Physics involve distances of the order of $10^{-12} - 10^{-13}$ cm and, therefore, since cross sections are expected to be

© Springer Nature Switzerland AG 2019

K. Tamvakis, *Basic Quantum Mechanics*, Undergraduate Texts in Physics,
https://doi.org/10.1007/978-3-030-22777-7_20

$O(10^{-24}\,\mathrm{cm}^2)$, the established cross-sectional measurement unit is the *Barn* equal to $10^{-24}\,\mathrm{cm}^2$.

20.2   Two-Particle Collisions

Consider a particle of mass m_1 and velocity v colliding with a target particle of mass m_2, which is at rest [1–3]. This is the *Laboratory Frame* description of the collision. The *Center of Mass* of the two particles in this frame of reference moves with a velocity $\mathbf{V}$, which from momentum conservation is

$$(m_1 + m_2)\mathbf{V} = m_1\mathbf{v} \implies \mathbf{V} = \frac{m_1}{m_1 + m_2}\mathbf{v}. \tag{20.3}$$

In the *Center of Mass Frame*, the velocities of the incident and the target particles will be

$$v_1^{(CM)} = v - V = \frac{m_2}{m_1 + m_2}v, \quad v_2^{(CM)} = V = \frac{m_1}{m_1 + m_2}v. \tag{20.4}$$

The velocity of the incident particle after the collision in the Laboratory Frame will be

$$\mathbf{v}_1^{(L)} = \mathbf{v}_1^{(CM)} + \mathbf{V}. \tag{20.5}$$

This relation, written in terms of the angles shown in Fig. 20.1, gives us a relation between the scattering angles in the two reference frames

$$\begin{cases} \sin\theta^{(L)}\, v_1^{(L)} = \sin\theta^{(CM)}\, v_1^{(CM)} \\[2mm] \cos\theta^{(L)}\, v_1^{(L)} = V + \cos\theta^{(CM)}\, v_1^{(CM)} \end{cases} \implies \tan\theta^{(L)} = \frac{\sin\theta^{(CM)}}{\frac{m_1}{m_2} + \cos\theta^{(CM)}}. \tag{20.6}$$

It should be noted that the definition of the cross section introduced above is also valid in the case of a moving target provided that it refers to the relative incident flux. Since the same number of particles is scattered in the angle $d\Omega_L$ in the Laboratory Frame as scattered in the angle $d\Omega_{CM}$ in the center of mass frame and we can write

$$\sigma_L(\Omega_L)\, d\Omega_L = \sigma_{CM}(\Omega_{CM})\, d\Omega_{CM}. \tag{20.7}$$

Of course, the total cross section is *frame-independent*

$$\sigma_{tot} = \int d\Omega_L\, \sigma_L(\Omega_L) = \int d\Omega_{CM}\, \sigma_{CM}(\Omega_{CM}). \tag{20.8}$$

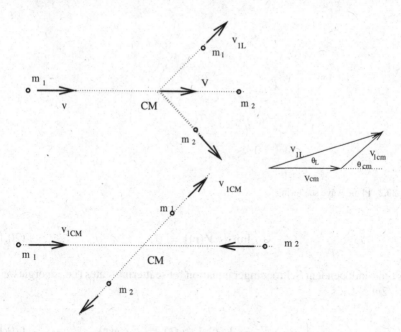

Fig. 20.1 Laboratory and center of mass frames kinematics

Using the relation (20.6) we have

$$\sigma_L(\Omega_L) = \sigma_{CM}(\Omega_{CM}) \left(\frac{d\cos\theta_{CM}}{d\cos\theta_L} \right) \qquad (20.9)$$

or

$$\sigma_L(\Omega_L) = \sigma_{CM}(\Omega_{CM}) \left(\frac{\left(1 + (m_1/m_2)^2 + 2(m_1/m_2)\cos\theta_{CM}\right)^{3/2}}{|1 + (m_1/m_2)\cos\theta_{CM}|} \right). \qquad (20.10)$$

20.3 The Scattering Amplitude

In the majority of collision problems, the forces between particles depend on the relative distance. In these cases, the problem of the scattering of two particles of masses m_1 and m_2 reduces to the scattering problem of one particle of mass $m_1 m_2/(m_1 + m_2)$ moving in a potential depending on the relative distance. We may also restrict ourselves in the case of *short-range forces*, i.e., forces corresponding to potentials that fall off faster than the Coulomb potential at great distances, namely, such that[1]

---

[1] This includes all forces relevant at the subatomic level. The *"long range"* Coulomb potential will be dealt with separately.

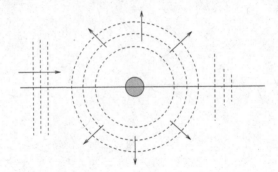

Fig. 20.2 Plane wave scattering

$$\lim_{r \to \infty} \{r\, V(\mathbf{r})\} = 0.$$ (20.11)

The time-independent Schroedinger equation for scattering states (i.e., energies $E = \hbar^2 k^2 / 2m > 0$) is

$$\left(-\frac{\hbar^2}{2m} \nabla^2 + V(\mathbf{r}) \right) \psi_E(\mathbf{r}) = E\, \psi_E(\mathbf{r}).$$ (20.12)

We may guess that in the asymptotic region ($r \to \infty$) the solution will consist of two terms, namely, a plane wave term, representing the incident particle, and a scattered wave term that has the form of a spherical wave multiplied by an amplitude that depends on the direction. This is depicted in Fig. 20.2. Thus, we may write down a trial asymptotic solution

$$\psi_E(\mathbf{r}) \sim \frac{e^{i\mathbf{k}\cdot\mathbf{r}}}{(2\pi)^{3/2}} + \frac{f(\Omega)}{(2\pi)^{3/2}} \frac{e^{ikr}}{r}.$$ (20.13)

Substituting (20.13) into the Schroedinger equation we see that it is indeed a solution in the asymptotic region, provided that the potential is short range as assumed by (20.11). The function $f(\Omega)$ is called *the Scattering Amplitude* [1, 3]. It is an important quantity, since it is directly related to the scattering cross section. If we calculate the probability current density $\mathcal{J} = \frac{\hbar}{2mi}(\psi^*\nabla\psi - \psi\nabla\psi^*)$, we arrive at an asymptotic (i.e., $r \to \infty$) expression with leading terms

$$\mathcal{J} = \frac{\hbar\mathbf{k}}{m(2\pi)^3} + \frac{\hbar k \hat{r}}{m(2\pi)^3} \frac{|f(\Omega)|^2}{r^2} + \cdots$$ (20.14)

The dots signify interference or subleading terms. Remembering that the differential cross section is defined as

$$\sigma(\Omega) = \frac{1}{\mathcal{J}_{in}} \frac{d\mathcal{N}_{sc}}{d\Omega},$$ (20.15)

we note that

$$\mathcal{J}_{in} = \frac{\hbar k}{m} \left| \frac{e^{i\mathbf{k}\cdot\mathbf{r}}}{(2\pi)^{3/2}} \right|^2 = \frac{\hbar k}{m(2\pi)^3}.$$

Also, note that the number of the scattered particles per unit time in a solid angle $d\Omega$ is just the *radial flux*[2] multiplied by the spherical surface area

$$d\mathcal{N}_{sc} = \left\{ \frac{\hbar k}{m} \left| \frac{f(\Omega)}{(2\pi)^{3/2}r} \right|^2 \hat{r} \right\} \cdot d\mathbf{S} = \frac{\hbar k}{m(2\pi)^3 r^2} |f(\Omega)|^2 r^2 d\Omega = \frac{\hbar k}{m(2\pi)^3} |f(\Omega)|^2 d\Omega$$

(20.16)

Therefore, we have

$$\sigma(\Omega) = \frac{1}{\mathcal{J}_{in}} \frac{d\mathcal{N}_{sc}}{d\Omega} = \left(\frac{\hbar k}{m(2\pi)^3} \right)^{-1} \left(\frac{\hbar k}{m(2\pi)^3} |f(\Omega)|^2 \right)$$

(20.17)

or just

$$\sigma(\Omega) = |f(\Omega)|^2.$$

(20.18)

Thus, the scattering amplitude $f(\Omega)$ is a quantity directly related to the scattering cross section, the latter being the absolute square of the former.

20.4 Wave Packet Scattering

Consider a wave packet with an initial ($t = 0$) wave function[3]

$$\psi(\mathbf{r}, 0) = \int \frac{d^3k}{(2\pi)^{3/2}} \tilde{\psi}(\mathbf{k}) e^{i\mathbf{k}\cdot(\mathbf{r}-\mathbf{r}_0)}.$$

(20.19)

The momentum wave function $\tilde{\psi}(\mathbf{k})$ is a smooth function centered around some wave number $\mathbf{k}_0$, fairly localized around it with a spread $(\Delta \mathbf{k})$. Similarly, the spatial wave function $\psi(\mathbf{r}, 0)$ is also fairly localized around a point $\mathbf{r}_0$, which, without loss of generality, can be taken to be parallel to $\mathbf{k}_0$. This wave packet represents the incident particle moving toward the scattering center of a short-range potential $V(\mathbf{r})$. As in many problems in physics, we are basically interested in determining the evolved wave function $\psi(\mathbf{r}, t)$ at times $t >> 0$. To do this it is necessary to solve the energy eigenvalue problem[4]

$$\hat{H} \psi_{\mathbf{k}} = E \psi_{\mathbf{k}}.$$

(20.20)

---

[2] Or, equivalently *the radial probability current density*.

[3] See also [1, 3].

[4] The eigenvalues of the energy corresponding to scattering states $E = \hbar^2 k^2 / 2m$ (continuous spectrum) cover the positive real axis $0 < E < \infty$, being the same as in the case where the potential is absent and the eigenstates are just the plane waves $e^{i\mathbf{k}\cdot\mathbf{r}}$. In other words, the spectrum of scattering states coincides with that of the free particle.

We already know the form of the eigenfunctions in the asymptotic region, namely,

$$\psi_{\mathbf{k}}(\mathbf{r}) \sim \frac{1}{(2\pi)^{3/2}} \left(e^{i\mathbf{k}\cdot\mathbf{r}} + f(\Omega)\frac{e^{ikr}}{r} \right). \tag{20.21}$$

Note that the initial wave function $\psi(\mathbf{r}, 0)$, although it is written as a superposition of plane waves $e^{i\mathbf{k}\cdot\mathbf{r}}/(2\pi)^{3/2}$, it can also be written as a superposition of the above scattering eigenstates as

$$\psi(\mathbf{r}, 0) = \int d^3k \, \tilde{\psi}(\mathbf{k}) \, e^{-i\mathbf{k}\cdot\mathbf{r}_0} \, \psi_{\mathbf{k}}(\mathbf{r}). \tag{20.22}$$

This is a very good approximation in the asymptotic region. It amounts to

$$\int d^3k \, \tilde{\psi}(\mathbf{k}) \, f(\Omega) \, e^{ikr} \approx 0. \tag{20.23}$$

This will be proven rigorously shortly.

Next, we write down a general expression for the evolved wave function

$$\psi(\mathbf{r}, t) = \int d^3k \, C(\mathbf{k}) \, e^{-i\frac{\hbar k^2}{2m}t} \, \psi_{\mathbf{k}}(\mathbf{r}) \tag{20.24}$$

with the coefficients

$$C(\mathbf{k}) = \int d^3r \, \psi(\mathbf{r}, 0) \, \psi_{\mathbf{k}}^*(\mathbf{r}). \tag{20.25}$$

We may use our approximation (20.22) and obtain

$$C(\mathbf{k}) = \int d^3r \int d^3k' \, \tilde{\psi}(\mathbf{k}') e^{-i\mathbf{k}'\cdot\mathbf{r}_0} \, \psi_{\mathbf{k}'}(\mathbf{r})\psi_{\mathbf{k}}^*(\mathbf{r}) \tag{20.26}$$

$$= \int d^3k' \, \tilde{\psi}(\mathbf{k}') e^{-i\mathbf{k}'\cdot\mathbf{r}_0} \int d^3r \, \psi_{\mathbf{k}'}(\mathbf{r})\psi_{\mathbf{k}}^*(\mathbf{r}) = \int d^3k' \, \tilde{\psi}(\mathbf{k}') e^{-i\mathbf{k}'\cdot\mathbf{r}_0} \delta(\mathbf{k} - \mathbf{k}')$$

or

$$C(\mathbf{k}) = \tilde{\psi}(\mathbf{k}) e^{-i\mathbf{k}\cdot\mathbf{r}_0}. \tag{20.27}$$

Thus, finally, we may have

$$\psi(\mathbf{r}, t) = \int d^3k \, \tilde{\psi}(\mathbf{k}) \, e^{-i\mathbf{k}\cdot\mathbf{r}_0} \, e^{-i\frac{\hbar k^2}{2m}t} \, \psi_{\mathbf{k}}(\mathbf{r}). \tag{20.28}$$

We may proceed now by doing some further approximations on the last expression. The energy in the exponent can be written as

$$\frac{\hbar k^2}{2m} = \frac{\hbar}{2m}\left((\mathbf{k} - \mathbf{k}_0)^2 - 2\mathbf{k}_0 \cdot (\mathbf{k} - \mathbf{k}_0) - k_0^2\right) \approx \frac{\hbar}{2m}\left(2\mathbf{k} \cdot \mathbf{k}_0 - k_0^2\right),$$

where we have neglected the $(k - k_0)^2$, being much smaller than the linear term, since by assumption the momentum distribution is localized around k_0. We may also introduce the symbols

$$\omega_0 = \frac{\hbar k_0^2}{2m} \quad and \quad \mathbf{v}_0 = \frac{\hbar \mathbf{k}_0}{m}. \tag{20.29}$$

Note that the approximation made in the exponent amounts to neglecting $\frac{\hbar(\Delta k)^2}{2m}t$ for the characteristic times t involved in the scattering process. These times correspond to traveling the distance to the scattering center r_0/v_0 and a distance of analogous magnitude from the scatterer to the detector, meaning that $t \sim 2r_0/v_0$. Thus, our approximation amounts to $2r_0\hbar(\Delta k)^2/2mv_0 \sim r_0(\Delta k)^2/k_0 << 1$.

Returning to the wave function, we have

$$\psi(\mathbf{r}, t) \sim e^{i\omega_0 t} \int d^3k \, \tilde{\psi}(\mathbf{k}) \, \psi_{\mathbf{k}}(\mathbf{r}) \, e^{-i\mathbf{k}\cdot(\mathbf{r}_0 + \mathbf{v}_0 t)} \tag{20.30}$$

or

$$\psi(\mathbf{r}, t) \sim e^{i\omega_0 t} \int \frac{d^3k}{(2\pi)^{3/2}} \, \tilde{\psi}(\mathbf{k}) \, e^{i\mathbf{k}\cdot(\mathbf{r} - (\mathbf{r}_0 + \mathbf{v}_0 t))}$$

$$+ \frac{e^{i\omega_0 t}}{r} \int \frac{d^3k}{(2\pi)^{3/2}} \, \tilde{\psi}(\mathbf{k}) \, f_{\mathbf{k}}(\hat{r}) e^{ikr - i\mathbf{k}\cdot(\mathbf{r}_0 + \mathbf{v}_0 t)}. \tag{20.31}$$

We have symbolized appropriately the scattering amplitude as $f_{\mathbf{k}}(\hat{r})$, since it depends on the wave number $\mathbf{k}$ and the direction $\hat{r}$, defined by the solid angle Ω. Since $\tilde{\psi}(\mathbf{k})$ has been assumed to be localized around $\mathbf{k}_0$, we can make an additional approximation and replace $f_{\mathbf{k}}(\hat{r})$ inside the integral of the second term with $f_{\mathbf{k}_0}(\hat{r})$ and move it outside the integral. We end up with

$$\psi(\mathbf{r}, t) \sim e^{i\omega_0 t} \int \frac{d^3k}{(2\pi)^{3/2}} \, \tilde{\psi}(\mathbf{k}) \, e^{i\mathbf{k}\cdot(\mathbf{r} - (\mathbf{r}_0 + \mathbf{v}_0 t))}$$

$$+ \frac{e^{i\omega_0 t}}{r} f_{\mathbf{k}_0}(\hat{r}) \int \frac{d^3k}{(2\pi)^{3/2}} \, \tilde{\psi}(\mathbf{k}) \, e^{ikr - i\mathbf{k}\cdot(\mathbf{r}_0 + \mathbf{v}_0 t)}. \tag{20.32}$$

The exponential in the second term can be written as $e^{i\mathbf{k}\cdot\left(\hat{k}r - \mathbf{r}_0 - \mathbf{v}_0 t\right)}$ and this can be approximated by $e^{i\mathbf{k}\cdot\left(\hat{k}_0 r - \mathbf{r}_0 - \mathbf{v}_0 t\right)}$ due to the localization around $\mathbf{k}_0$. Finally, the evolved wave function can be written in terms of the initial wave packet as

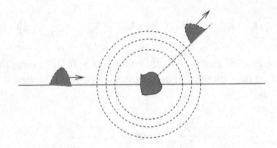

Fig. 20.3 Wave packet scattering

$$\psi(\mathbf{r}, t) \sim e^{i\omega_0 t}\, \psi(\mathbf{r} - \mathbf{v}_0 t,\, 0) + \frac{e^{i\omega_0 t}}{r}\, f_{\mathbf{k}_0}(\hat{r})\, \psi(\hat{k}_0 r - \mathbf{v}_0 t,\, 0)\,. \qquad (20.33)$$

This expression contains compactly all the physics of the scattering process as depicted in Fig. 20.3. The first term, apart from a trivial phase factor, is just the initial wave packet translated by $\mathbf{v}_0 t$ and represents the kinematical evolution of the initial wave if there was no scattering at all. The second term is again the initial wave packet

(1) translated by $\mathbf{v}_0 t$,
(2) rotated to the direction $\mathbf{r} \to \hat{k}_0 r$,
(3) multiplied by the scattering amplitude $f_{\hat{k}_0}(\hat{r})$, and
(4) decreased in proportion to $1/r$.

This term represents the scattered part of the wave packet.
The probability to observe a scattered particle in the time interval $[t,\, t + dt]$ and in a surface element $d\mathbf{S}$ will be the product

$$\mathcal{J} \cdot d\mathbf{S}\, dt = \mathcal{J}_r\, r^2\, d\Omega\, dt = v_0 \left| \frac{e^{i\omega_0 t}}{r}\, f_{\mathbf{k}_0}(\hat{r})\, \psi(\hat{k}_0 r - \mathbf{v}_0 t,\, 0) \right|^2 r^2\, d\Omega\, dt\,. \qquad (20.34)$$

For the total probability, we integrate over all time and get

$$v_0\, d\Omega\, |f_{\hat{k}_0}(\hat{r})|^2 \int_{-\infty}^{+\infty} dt\, \left| \psi\big(\hat{k}_0(r - v_0 t),\, 0\big) \right|^2 = d\Omega\, |f_{\mathbf{k}_0}(\hat{r})|^2 \int_{-\infty}^{+\infty} d\xi\, \left| \psi\big(\hat{k}_0 \xi,\, 0\big) \right|^2\,. \qquad (20.35)$$

Note, however, that this integral is just the probability per unit area for an incident particle to cross a surface perpendicular to the direction of incidence $\hat{k}_0$ or, equivalently, the number of incident particles per unit area

$$I_0 = \int_{-\infty}^{+\infty} d\xi\, \left| \psi\big(\hat{k}_0 \xi,\, 0\big) \right|^2\,. \qquad (20.36)$$

Then, the differential cross section will be the above scattering probability per unit area $|f_{\mathbf{k}_0}(\hat{r})|^2 I_0$ divided by the number of incident particles per unit area I_0, i.e.,

$$\sigma(\Omega) = |f_{\mathbf{k}_0}(\hat{r})|^2 . \tag{20.37}$$

Thus, for a rather sharply localized wave packet at $\mathbf{k}_0$, we have recovered the relation of the differential cross section and the scattering amplitude at the central wave number $\mathbf{k}_0$.

20.5 Integral Scattering Equation

The time-independent Schroedinger equation for scattering ($E = \hbar^2 k^2/2m$) can be written in the form

$$\left(\nabla^2 + k^2\right)\psi_{\mathbf{k}}(\mathbf{r}) = U(\mathbf{r})\,\psi_{\mathbf{k}}(\mathbf{r}) , \tag{20.38}$$

where we have introduced

$$U(\mathbf{r}) = \frac{2m}{\hbar^2} V(\mathbf{r}) . \tag{20.39}$$

In the absence of the potential, the corresponding free equation is

$$\left(\nabla^2 + k^2\right)\psi_{\mathbf{k}}^{(0)}(\mathbf{r}) = 0 . \tag{20.40}$$

It is possible to view (20.38) as an inhomogeneous version of (20.40), although this is not correct, since the *"inhomogeneous"* right-hand side contains the unknown wave function $\psi_{\mathbf{k}}(\mathbf{r})$. Nevertheless, this might be useful in order to set it into a form more susceptible to approximations. Thus, considering the right-hand side as an inhomogeneity, we apply the *Green's functions method*, consisting of the following steps. First, we determine the Green's function of the operator $\nabla^2 + k^2$ solving the Green's function equation with the appropriate boundary conditions

$$\left(\nabla^2 + k^2\right) G_{\mathbf{k}}(\mathbf{r}, \mathbf{r}') = -4\pi\,\delta(\mathbf{r} - \mathbf{r}') . \tag{20.41}$$

Next we have a *"solution"* of (20.38) as a sum of a solution of the homogeneous equation (20.40) and a convoluted integral of the Green's function and the *"inhomogeneity"* $U(\mathbf{r})\,\psi_{\mathbf{k}}(\mathbf{r})$

$$\psi_{\mathbf{k}}(\mathbf{r}) = \psi_{\mathbf{k}}^{(0)}(\mathbf{r}) - \frac{1}{4\pi}\int d^3r'\, G_{\mathbf{k}}(\mathbf{r}, \mathbf{r}')\, U(\mathbf{r}')\,\psi_{\mathbf{k}}(\mathbf{r}') . \tag{20.42}$$

Substituting (20.42) into the original Eq. (20.38), we verify that it is satisfied. Nevertheless, (20.42) is not a solution but simply an integral equation equivalent to the original Schroedinger differential equation.[5] Its merit will become clear shortly after we have determined the Green's function by solving (20.41).

---

[5]See also [1, 3].

Determination of the Green's function. Consider a Fourier transformation of the Green's function $G_{\mathbf{k}}(\mathbf{r}, \mathbf{r}')$

$$G_{\mathbf{k}}(\mathbf{r}, \mathbf{r}') = \int \frac{d^3k'}{(2\pi)^{3/2}} \, g(\mathbf{k}') \, e^{i\mathbf{k}' \cdot (\mathbf{r}-\mathbf{r}')} \,. \tag{20.43}$$

It is clear from the defining Eq. (20.41) that $G_{\mathbf{k}}(\mathbf{r}, \mathbf{r}')$ will be a function of the difference $\mathbf{r} - \mathbf{r}'$ only, since the operator ∇^2 is invariant in spatial translations. Introducing the Fourier transform in the Eq. (20.41), we obtain

$$g(k') = \left(\frac{2}{\pi}\right)^{1/2} \frac{1}{k'^2 - k^2} \tag{20.44}$$

and introducing this back to (20.43)

$$G_{\mathbf{k}}(\mathbf{r} - \mathbf{r}') = \frac{1}{2\pi^2} \int d^3k' \frac{e^{i\mathbf{k}' \cdot (\mathbf{r}-\mathbf{r}')}}{k'^2 - k^2} \,. \tag{20.45}$$

Taking the $\mathbf{r} - \mathbf{r}'$ direction to be the $\hat{z}_k$-axis, we have

$$d^3k' \, \exp[i\mathbf{k}' \cdot (\mathbf{r} - \mathbf{r}')] = d\phi_k d(\cos\theta_k) dk' k'^2 \, \exp[ik'R \cos\theta'_k] \,,$$

where

$$R = |\mathbf{r} - \mathbf{r}'| \,. \tag{20.46}$$

Then, we have

$$G_k(R) = \frac{1}{\pi} \int_0^\infty \frac{dk' k'^2}{k'^2 - k^2} \int_{-1}^1 d(\cos\theta'_k) e^{ik'R\cos\theta'_k} = \frac{1}{i\pi R} \int_0^\infty \frac{dk' k'}{k'^2 - k^2} \left(e^{ik'R} - e^{-ik'R}\right)$$

$$= \frac{1}{i\pi R} \int_{-\infty}^{+\infty} dk' \frac{k' e^{ik'R}}{k'^2 - k^2} = -\frac{1}{\pi R} \frac{d}{dR} \int_{-\infty}^{+\infty} dk' \frac{e^{ik'R}}{k'^2 - k^2} \,. \tag{20.47}$$

The integration at hand is ill-defined since the integrand has poles on the real axis at $k' = \pm k$. One way to define the integral is giving the poles a small imaginary part shifting them away from the real line. There are different choices leading to different Green's functions. Each of them corresponds to different boundary conditions at infinity for the original Green's function differential equation. The correct choice in our case is the one leading to outgoing spherical waves. Thus, we shift the poles according to[6]

$$k^2 \to k^2 + i\eta \,,$$

---

[6] k is taken to be positive.

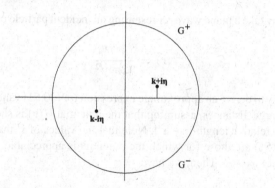

Fig. 20.4 Complex k-plane

corresponding to shifting the pole at k in the upper complex k'-plane and the pole at $-k$ in the lower complex k'-plane as shown in Fig. 20.4. Closing the contour in the upper complex k'-pane and applying the residue theorem, we obtain

$$\oint dz \, \frac{e^{izR}}{z^2 - k^2 - i\eta} = \int_{-\infty}^{+\infty} dk' \, \frac{e^{ik'R}}{k'^2 - k^2 - i\eta} + \lim_{\rho \to \infty} \int_0^\pi d\theta \, \frac{\rho e^{i\theta} e^{ik'\rho R \cos\theta}}{\rho^2 e^{2i\theta} - k^2} \, e^{-k'R\rho \sin\theta} .$$

Therefore

$$\int_{-\infty}^{+\infty} dk' \frac{e^{ik'R}}{k'^2 - k^2 - i\eta} = 2\pi i \, \frac{e^{ikR}}{2k} . \tag{20.48}$$

This leads us to the Green's function

$$G_k(R) = \frac{e^{ikR}}{R} . \tag{20.49}$$

If we had made the alternative choice of pole-shifting, namely, $k^2 \to k^2 - i\eta$, we would have ended with the Green's function $G_k(R) = e^{-ikR}/R$, exhibiting behavior of incoming spherical waves. We may then symbolize both choices as

$$G_k^{(\pm)}(R) = \frac{e^{\pm ikR}}{R} \tag{20.50}$$

the "+" sign corresponding to the presently physical behavior of outgoing waves.

Returning to the integral equation (20.42), we substitute the Green's function $G_k^{(+)}(R)$ and we obtain

$$\psi_{\mathbf{k}}^{(+)}(\mathbf{r}) = \psi_{\mathbf{k}}^{(0)}(\mathbf{r}) - \frac{1}{4\pi} \int d^3r' \, \frac{e^{ik|\mathbf{r}-\mathbf{r}'|}}{|\mathbf{r} - \mathbf{r}'|} \, U(\mathbf{r}') \, \psi_{\mathbf{k}}^{(+)}(\mathbf{r}') . \tag{20.51}$$

We have also designated the scattering wave function with the "+" sign underlining its association with $G_k^{(+)}$ and outgoing waves. For the solution of the free Schroedinger equation, we may take a plane wave representing an incident particle of wave number **k**, namely,

$$\psi_{\mathbf{k}}^{(0)}(\mathbf{r}) = \frac{e^{i\mathbf{k}\cdot\mathbf{r}}}{(2\pi)^{3/2}}. \qquad (20.52)$$

Next, we shall try to derive an approximate expression for (20.51) valid in the asymptotic region of large distances, assuming that the potential $U(\mathbf{r})$ is short range.

Consider the relative length $|\mathbf{r} - \mathbf{r}'|$. Note that the values of $\mathbf{r}'$ that contribute in the integral (20.51) are those for which the potential is appreciable. Therefore, we may assume that $r \gg r'$. Then, we have

$$|\mathbf{r} - \mathbf{r}'| = \sqrt{r^2 + r'^2 - 2\mathbf{r}\cdot\mathbf{r}'} \approx r\left(1 - \frac{\mathbf{r}\cdot\mathbf{r}'}{r^2}\right) = r - \hat{r}\cdot\mathbf{r}'$$

and

$$k|\mathbf{r} - \mathbf{r}'| \approx kr - (k\hat{r})\cdot\mathbf{r}'. \qquad (20.53)$$

At this point we introduce the wave number

$$\mathbf{k}' = k\hat{r} \qquad (20.54)$$

and we can write

$$k|\mathbf{r} - \mathbf{r}'| \approx kr - \mathbf{k}'\cdot\mathbf{r}'. \qquad (20.55)$$

Going back to the integral equation, we have

$$\psi_{\mathbf{k}}^{(+)}(\mathbf{r}) \approx \frac{e^{i\mathbf{k}\cdot\mathbf{r}}}{(2\pi)^{3/2}} - \frac{e^{ikr}}{4\pi r} \int d^3r'\, e^{-i\mathbf{k}'\cdot\mathbf{r}'}\, U(\mathbf{r}')\, \psi_{\mathbf{k}}^{(+)}(\mathbf{r}'). \qquad (20.56)$$

Note that the last equation has exactly the form of the asymptotic scattering solution (20.13) with the identification of the integral

$$f_{\mathbf{k}}(\hat{r}) = -\sqrt{\frac{\pi}{2}} \int d^3r'\, e^{-i\mathbf{k}'\cdot\mathbf{r}'}\, U(\mathbf{r}')\, \psi_{\mathbf{k}}^{(+)}(\mathbf{r}') \qquad (20.57)$$

as the scattering amplitude.

As we have stated in the beginning of this approach, although it does not provide a solution to the scattering problem, it is suitable for the development of systematic approximations. For example, the most drastic approximation to (20.56) would be to approximate the scattering wave function inside the integral with a plane wave, namely,

$$\psi_{\mathbf{k}}^{(1)}(\mathbf{r}) = \frac{e^{i\mathbf{k}\cdot\mathbf{r}}}{(2\pi)^{3/2}} - \frac{e^{ikr}}{4\pi r} \int \frac{d^3r'}{(2\pi)^{3/2}}\, e^{i(\mathbf{k}-\mathbf{k}')\cdot\mathbf{r}'}\, U(\mathbf{r}'). \qquad (20.58)$$

This is the so-called *Born approximation*. The second-order approximation would be to approximate the scattering wave function inside the integral with the Born approximation just obtained, namely,

$$\psi_{\mathbf{k}}^{(2)}(\mathbf{r}) = \frac{e^{i\mathbf{k}\cdot\mathbf{r}}}{(2\pi)^{3/2}} - \frac{e^{ikr}}{4\pi r} \int d^3r' \, e^{-i\mathbf{k}'\cdot\mathbf{r}'} \, U(\mathbf{r}')\psi_{\mathbf{k}}^{(1)}(\mathbf{r}') \,. \tag{20.59}$$

This iterative procedure can be continued until the desired level of accuracy is achieved. The series generated this way is called the *Born Series*.

Considering now the scattering amplitude in the Born approximation, we have

$$f_{\mathbf{k}}^{(B)}(\hat{r}) = -\frac{2m}{\hbar^2}\sqrt{\frac{\pi}{2}} \int \frac{d^3r'}{(2\pi)^{3/2}} \, e^{i(\mathbf{k}-\mathbf{k}')\cdot\mathbf{r}'} \, V(\mathbf{r}') \,. \tag{20.60}$$

Note that this integral is just the Fourier transform of the potential

$$\tilde{V}(\mathbf{q}) = \int \frac{d^3r}{(2\pi)^{3/2}} e^{i\mathbf{q}\cdot\mathbf{r}} \, V(\mathbf{r}) \tag{20.61}$$

and

$$f_{\mathbf{k}}^{(B)}(\hat{r}) = -\frac{2m}{\hbar^2}\sqrt{\frac{\pi}{2}} \, \tilde{V}(\mathbf{q}) \,. \tag{20.62}$$

The wave number $\mathbf{q}$, often referred to as *momentum transfer*, is

$$\mathbf{q} = \mathbf{k} - \mathbf{k}' \implies q^2 = 2k^2\left(1 - \cos(\hat{r}\cdot\hat{k})\right) = 4k^2\sin^2(\hat{r}\cdot\hat{k}/2) \,. \tag{20.63}$$

Before we close this section on the scattering amplitude, we shall prove the relation used in the previous section on wave packet scattering

$$\int d^3k \, \tilde{\psi}(\mathbf{k}) \, e^{-i\mathbf{k}\cdot\mathbf{r}_0} \, f_{\mathbf{k}}(\hat{r}) \, e^{ikr} = 0 \,, \tag{20.64}$$

where $\tilde{\psi}(\mathbf{k})$ is the momentum wave function of the incident wave packet. This relation is equivalent to

$$\int \frac{d^3k}{(2\pi)^{3/2}} \, \tilde{\psi}(\mathbf{k}) \, e^{i\mathbf{k}\cdot(\mathbf{r}-\mathbf{r}_0)} = \int d^3k \, \tilde{\psi}(\mathbf{k}) \, e^{-i\mathbf{k}\cdot\mathbf{r}_0} \, \psi_{\mathbf{k}}^{(+)}(\mathbf{r}) \,. \tag{20.65}$$

Substituting in (20.64) the integral expression for the scattering amplitude (20.57), we obtain

$$\int d^3k \, \tilde{\psi}(\mathbf{k}) \, e^{-i\mathbf{k}\cdot\mathbf{r}_0} \int d^3r' \frac{e^{ik|\mathbf{r}-\mathbf{r}'|}}{|\mathbf{r}-\mathbf{r}'|} U(\mathbf{r}') \, \psi_{\mathbf{k}}^{(+)}(\mathbf{r}') \,. \tag{20.66}$$

Note that **r** is in the asymptotic region where the short-range potential is very small, while the integration variable **r'** is restricted in the range of the potential. The wave number integral in this expression is

$$\int d^3k \, \tilde{\psi}(\mathbf{k}) \, e^{-i\mathbf{k}\cdot\mathbf{r}_0 + ik|\mathbf{r}-\mathbf{r}'|} \psi_{\mathbf{k}}^{(+)}(\mathbf{r}') \approx \psi_{\mathbf{k}_0}^{(+)}(\mathbf{r}') \int d^3k \, \tilde{\psi}(\mathbf{k}) \, e^{i\mathbf{k}\cdot\left(-\mathbf{r}_0 + \hat{k}_0|\mathbf{r}-\mathbf{r}'|\right)},$$

where in the last step we made the approximations $\psi_{\mathbf{k}}^{(+)}(\mathbf{r}') \approx \psi_{\mathbf{k}_0}^{(+)}(\mathbf{r}')$ and $k = \mathbf{k}\cdot\hat{k} \approx \mathbf{k}\cdot\hat{k}_0$. These approximations are justified by the localization of the wave packet around $\mathbf{k}_0$. Note, however, that the last integral is $\psi\left(\hat{k}_0|\mathbf{r} - \mathbf{r}'|, 0\right)$, which is just the initial wave packet at the location $\hat{k}_0|\mathbf{r} - \mathbf{r}'| \approx \hat{k}_0\, r$. However, these points are at the direction of incidence and *behind the target*, where the incident wave packet $\psi(\mathbf{r}, 0)$ is practically zero. This justifies in neglecting the mother expression (20.64) in our analysis of the previous section.

Example 20.1 Consider the asymptotic scattering wave function

$$\psi_{\mathbf{k}}^{(+)}(\mathbf{r}) \sim \frac{e^{i\mathbf{k}\cdot\mathbf{r}}}{(2\pi)^{3/2}} + \frac{f_{\mathbf{k}}(\theta)}{(2\pi)^{3/2}} \frac{e^{ikr}}{r}. \tag{20.67}$$

Note that the probability current density $\mathcal{J}$ is a sum of the incident current density $\mathcal{J}_{inc}$ due to the incident plane wave $\frac{e^{i\mathbf{k}\cdot\mathbf{r}}}{(2\pi)^{3/2}}$, the scattered current density $\mathcal{J}_{sc}$ due to the scattered spherical wave $\frac{f_{\mathbf{k}}(\theta)}{(2\pi)^{3/2}} \frac{e^{ikr}}{r}$, and an interference term $\mathcal{J}_{int}$, where both waves contribute.

(a) Calculate all these current densities $\mathcal{J}$, $\mathcal{J}_{inc}$, $\mathcal{J}_{sc}$, and $\mathcal{J}_{int}$ in the asymptotic region.

(b) Calculate the surface integrals $\oint d\mathbf{S} \cdot \mathcal{J}$ and $\oint d\mathbf{S} \cdot \mathcal{J}_{inc}$ on a spherical surface centered at the target.

(c) Use the expression $\oint d\mathbf{S} \cdot (\mathcal{J}_{sc} + \mathcal{J}_{int})$ in order to prove the so-called *"Optical Theorem"* for the total scattering cross section

$$\sigma = \frac{4\pi}{k} Im\left(f_{\mathbf{k}}(0)\right). \tag{20.68}$$

The following approximation can be used in the asymptotic region:

$$\int_{-1}^{1} d(\cos\theta)\, F(\theta)\, e^{-ikr\cos\theta} \approx \frac{i}{kr}\left(F(0)e^{-ikr} - F(\pi)e^{ikr}\right) + O(1/r^2). \tag{20.69}$$

You may also assume that, due to cylindrical symmetry, there is no dependence of the scattering amplitude on the azimuthal angle ϕ.

From the standard expression of the current density, we have

$$\mathcal{J} = \frac{\hbar}{2mi}\left(\psi_{\mathbf{k}}^* \nabla \psi_{\mathbf{k}} - c.c.\right) = \mathcal{J}_{inc} + \mathcal{J}_{sc} + \mathcal{J}_{int}, \tag{20.70}$$

where

$$\mathcal{J}_{inc} = \frac{\hbar \mathbf{k}}{m(2\pi)^3}, \quad \mathcal{J}_{sc} = \frac{\hbar k \hat{r}}{m(2\pi)^3}\frac{|f_{\mathbf{k}}(\theta)|^2}{r^2} \tag{20.71}$$

and

$$\mathcal{J}_{int} = \frac{\hbar}{2mi(2\pi)^3}\left\{ \hat{r}\left(\frac{ik}{r}\left(f_{\mathbf{k}}(\theta)e^{ikr(1-\cos\theta)} + c.c.\right) - \frac{1}{r^2}\left(f_{\mathbf{k}}(\theta)e^{ikr(1-\cos\theta)} - c.c.\right)\right)\right.$$

$$\left. + \mathbf{k}\frac{i}{r}\left(f_{\mathbf{k}}(\theta)e^{ikr(1-\cos\theta)} + f_{\mathbf{k}}^*(\theta)e^{-ikr(1-\cos\theta)}\right) + \hat{\theta}(\ldots)\right\}. \tag{20.72}$$

Probability conservation for the stationary state $\psi_{\mathbf{k}}^{(+)}(\mathbf{r})$ implies $\oint d\mathbf{S} \cdot \mathcal{J} = 0$. In addition to this, we have also

$$\oint d\mathbf{S} \cdot \mathcal{J}_{inc} = \frac{\hbar}{m(2\pi)^3}r^2 \int d\Omega\, \hat{r} \cdot \mathbf{k} = \frac{\hbar k r^2}{m(2\pi)^2}\int_{-1}^{1} d(\cos\theta)\cos\theta = 0. \tag{20.73}$$

Therefore, we have $\oint d\mathbf{S} \cdot (\mathcal{J}_{sc} + \mathcal{J}_{int}) = 0$ or

$$\oint d\mathbf{S} \cdot \mathcal{J}_{sc} = \frac{\hbar k}{m(2\pi)^3}\int d\Omega |f_{\mathbf{k}}(\theta)|^2 = \frac{\hbar k}{m(2\pi)^3}\sigma = -\oint d\mathbf{S} \cdot \mathcal{J}_{int} = -r^2 \oint d\Omega\, \hat{r} \cdot \mathcal{J}_{int}$$

$$= \frac{-\hbar k r}{2m(2\pi)^2}\int_{-1}^{1} d(\cos\theta)\left(\left(f_{\mathbf{k}}(\theta)e^{ikr(1-\cos\theta)} + c.c.\right) + \cos\theta\left(f_{\mathbf{k}}(\theta)e^{ikr(1-\cos\theta)} + c.c.\right)\right).$$

Terms of $O(1/r^2)$ are omitted. We proceed with the angle integration using the suggested approximation and we obtain $\oint d\mathbf{S} \cdot \mathcal{J}_{int} = -\frac{\hbar}{im(2\pi)^2}\left(f_{\mathbf{k}}(0) - f_{\mathbf{k}}^*(0)\right)$ or

$$\sigma = \frac{4\pi}{k}Im(f_{\mathbf{k}}(0)). \tag{20.74}$$

This is the *Optical Theorem* telling us that the total scattering cross section is proportional to the imaginary part of the scattering amplitude *behind* the target and, thus, demonstrating the extreme wave-like nature of the scattering process of particle collisions in Quantum Mechanics.

Example 20.2 Verify that the integral equation

$$\psi_k(x) = \frac{e^{ikx}}{\sqrt{2\pi}} - \frac{im}{\hbar^2 k}\int_{-\infty}^{+\infty} dx'\, e^{ik|x-x'|}\, V(x')\,\psi_k(x') \tag{20.75}$$

is equivalent to the Schroedinger equation describing one-dimensional scattering of a particle of mass m and energy $E = \hbar^2 k^2/2m$ in the potential $V(x)$.

(a) Consider the case of a finite range potential and use this equation for $x \gg x'$ in order to derive an expression involving the scattering amplitude f_k.

(b) Consider the special case of the potential $V(x) = -\frac{\hbar^2 g^2}{2m}\delta(x - a)$ and calculate exactly the scattering amplitude.

Acting on both sides of the integral equation with $-\frac{\hbar^2 d^2}{2m dx^2}$ we obtain[7]

$$-\frac{\hbar^2}{2m}\psi_k''(x) = \frac{\hbar^2 k^2}{2m}\frac{e^{ikx}}{\sqrt{2\pi}} + \frac{i}{2k}\int_{-\infty}^{+\infty} dx' \frac{d^2}{dx^2}\left(e^{ik|x-x'|}\right) V(x')\psi_k(x') =$$

$$\frac{\hbar^2 k^2}{2m}\frac{e^{ikx}}{\sqrt{2\pi}} + \frac{i}{2k}\int_{-\infty}^{+\infty} dx' \left(ik\frac{d^2|x-x'|}{dx^2} - k^2\left(\frac{d|x-x'|}{dx}\right)^2\right) e^{ik|x-x'|} V(x')\psi_k(x')$$

$$= \frac{\hbar^2 k^2}{2m}\frac{e^{ikx}}{\sqrt{2\pi}} + \frac{i}{2k}\int_{-\infty}^{+\infty} dx' \left(2ik\delta(x-x') - k^2\right) e^{ik|x-x'|} V(x')\psi_k(x')$$

or

$$-\frac{\hbar^2}{2m}\psi_k''(x) = \frac{\hbar^2 k^2}{2m}\frac{e^{ikx}}{\sqrt{2\pi}} - V(x)\psi_k(x) - \frac{im}{k}\int_{-\infty}^{+\infty} dx' e^{ik|x-x'|} V(x')\psi_k(x')$$

or

$$-\frac{\hbar^2}{2m}\psi_k''(x) = -V(x)\psi_k(x) + E\psi_k(x),$$

which is just the Schroedinger equation.

(b) For the delta function potential, we have $\psi_k(x) = \frac{e^{ikx}}{\sqrt{2\pi}} - \frac{ig^2}{2k}e^{ik|x-a|}\psi_k(a)$, from which we obtain $\psi_k(a) = e^{ika}/\sqrt{2\pi}\left(1 + ig^2/2k\right)$ and substituting it into the above expression we have

$$\psi_k(x) = \frac{e^{ikx}}{\sqrt{2\pi}} + \left(\frac{-ig^2/2k}{1+ig^2/2k}\right)\frac{e^{ik|x-a|}}{\sqrt{2\pi}} \tag{20.76}$$

from which we can read off the scattering amplitude as

$$f_k = \frac{-ig^2/2k}{1+ig^2/2k}. \tag{20.77}$$

Example 20.3 Consider the scattering of particles of mass m and energy E by the potential

$$V(r) = \begin{cases} 0 & (0 \le r < a) \\ V_0 & (a < r < b) \\ 0 & (r > b) \end{cases} \tag{20.78}$$

(a) Calculate the scattering amplitude in the Born approximation.

(b) Consider the case of low energies ($kb << 1$) and using (a) calculate the total scattering cross section.

---

[7]Note that

$$\frac{d|x|}{dx} = \Theta(x) - \Theta(-x) \quad and \quad \frac{d^2|x|}{dx^2} = \Theta'(x) - \Theta'(-x) = 2\delta(x).$$

(a) We have

$$f_{\mathbf{k}}(\hat{r}) = -\frac{m}{2\hbar^2\pi} \int d^3r' \, e^{i\mathbf{q}\cdot\mathbf{r}'} \, V(\mathbf{r}')$$

$$= -\frac{m}{2\hbar^2\pi} \int_0^{+\infty} dr' \, r'^2 \, V(r') \int_0^{2\pi} d\phi' \int_{-1}^{+1} d(\cos\theta') \, e^{iqr'\cos\theta'}, \qquad (20.79)$$

where

$$\mathbf{q} = \mathbf{k} - k\hat{r} \implies q^2 = 2k^2 - 2k^2 \left(\hat{k}\cdot\hat{r}\right) = 2k^2 - 2k^2\cos\theta = 4k^2\sin^2(\theta/2). \qquad (20.80)$$

Proceeding further, we have

$$f_{\mathbf{k}}(\theta) = -\frac{mV_0}{\hbar^2} \int_a^b dr' \, r'^2 \int_{-1}^{+1} d(\cos\theta') \, e^{iqr'\cos\theta'} =$$

$$-\frac{2mV_0}{\hbar^2 q} \int_a^b dr' r' \sin(qr') = \frac{2mV_0}{\hbar^2 q} \frac{d}{dq} \int_a^b dr' \cos(qr') = \frac{2mV_0}{\hbar^2 q} \frac{d}{dq} \left(\frac{1}{q}\left(\sin(qb) - \sin(qa)\right)\right)$$

or

$$f_{\mathbf{k}}(\theta) = \frac{2mV_0}{\hbar^2 q} \left(-\frac{1}{q^2}\left(\sin(qb) - \sin(qa)\right) + \frac{1}{q}\left(b\cos(bq) - a\cos(qa)\right)\right). \qquad (20.81)$$

(b) At low energies

$$(ka) < (kb) << 1 \quad or \quad (qa) < (qb) << 1,$$

we have ($\sin\epsilon \sim \epsilon$ and $\cos\epsilon \sim 1 - \epsilon^2/2$)

$$f_k \approx -\frac{mV_0}{\hbar^2}\left(b^3 - a^3\right) \qquad (20.82)$$

and

$$\sigma = \int d\Omega \, |f_k|^2 = \frac{4\pi m^2 V_0^2}{\hbar^4}\left(b^3 - a^3\right)^2. \qquad (20.83)$$

20.6 Scattering in Central Potentials

For a central potential $V(r)$, the solutions $\psi_{\mathbf{k}}^{(+)}(\mathbf{r})$ of the integral scattering equation

$$\psi_{\mathbf{k}}^{(+)}(\mathbf{r}) = \frac{e^{i\mathbf{k}\cdot\mathbf{r}}}{(2\pi)^{3/2}} - \frac{1}{4\pi} \int d^3r' \, \frac{e^{ik|\mathbf{r}-\mathbf{r}'|}}{|\mathbf{r}-\mathbf{r}'|} U(r') \, \psi_{\mathbf{k}}^{(+)}(\mathbf{r}') \qquad (20.84)$$

exhibit *cylindrical symmetry*, depending only on the energy, i.e., k, on the radius r and on the angle $\arccos(\hat{k}\cdot\hat{r})$. Taking the direction of incidence $\hat{k}$ as the polar axis $\hat{z}$ of the reference frame, the latter is the polar angle θ and $\psi_{\mathbf{k}}^{(+)}(\mathbf{r}) = \psi_k^{(+)}(r,\theta)$. Since $\psi_k^{(+)}$ does not depend on the azimuthal angle ϕ, it will be an eigenfunction of $\hat{L}_z$ of zero eigenvalue.

Since the Hamiltonian commutes with the orbital angular momentum, they can have common eigenstates

$$R_{k,\ell}(r) Y_{\ell 0}(\theta) \propto R_{k,\ell}(r) P_\ell(\cos\theta), \qquad (20.85)$$

where $P_\ell(\cos\theta)$ are the *Legendre Polynomials*.[8] Thus, the solutions of (20.84) can be expanded in terms of these angular momentum and energy eigenfunctions (*spherical waves*) as[9]

$$\psi_k^{(+)}(r,\theta) = \sum_{\ell=0}^{\infty} R_{k,\ell}(r)\, P_\ell(\cos\theta). \tag{20.86}$$

This is an analysis of $\psi_k^{(+)}$ in terms of *spherical waves*, in contrast to a Fourier transformation which is an analysis in plane waves, i.e., eigenfunctions of the momentum. We can actually go from one basis to the other by expanding a plane wave in terms of spherical waves. At this point we shall use the expansion (13.36) which we discussed earlier in Chap. 13, expressing a plane wave as an infinite superposition of spherical waves

$$e^{ikz} = \sum_{\ell=0}^{\infty} i^\ell (2\ell+1)\, P_\ell(\cos\theta)\, j_\ell(kr), \tag{20.87}$$

where $\theta = \arccos(\hat{k}\cdot\hat{r})$. We can also expand the scattering amplitude $f_k(\theta)$ in terms of Legendre polynomials as

$$f_k(\theta) = \sum_{\ell=0}^{\infty} f_\ell\, P_\ell(\cos\theta). \tag{20.88}$$

Substituting the expansions (20.87) and (20.88) in the asymptotic wave function, we obtain

$$\psi_k^{(+)}(r,\theta) \sim \frac{1}{(2\pi)^{3/2}} \sum_{\ell=0}^{\infty} \left(i^\ell (2\ell+1) j_\ell(kr) + \frac{e^{ikr}}{r} f_\ell \right) P_\ell(\cos\theta). \tag{20.89}$$

The radial part of the asymptotic wave function will be a combination of spherical Bessel and spherical Neumann functions

$$\psi_k^{(+)}(r,\theta) = \sum_{\ell=0}^{\infty} (A_\ell j_\ell(kr) + B_\ell n_\ell(kr))\, P_\ell(\cos\theta). \tag{20.90}$$

Nevertheless, since these expressions are valid in the asymptotic region of $r \to \infty$, we may use the asymptotic expression for the spherical Bessel and Neumann func-

---

[8] $P_\ell(\cos\theta) = \sqrt{\frac{4\pi}{2\ell+1}}\, Y_{\ell 0}(\theta)$. From the orthonormality relation of spherical harmonics, we can obtain an orthonormality relation for Legendre polynomials

$$\int d\Omega\, Y_{\ell 0}^*(\theta,0)\, Y_{\ell' 0}(\theta,0) = \delta_{\ell\ell'} \implies \int_{-1}^{1} d\xi\, P_\ell(\xi)\, P_{\ell'}(\xi) = \frac{2\delta_{\ell\ell'}}{(2\ell+1)}.$$

[9] See also [1–4].

tions

$$j_\ell(kr) \sim \frac{\sin(kr - \ell\pi/2)}{kr}, \quad n_\ell(kr) \sim -\frac{\cos(kr - \ell\pi/2)}{kr}.$$

Then, $\psi_k^{(+)}(r, \theta)$ can always be written as[10]

$$\psi_k^{(+)}(r, \theta) \sim \sum_{\ell=0}^{\infty} \frac{\alpha_\ell}{kr} \sin(kr - \ell\pi/2 + \delta_\ell) P_\ell(\cos\theta), \quad (20.91)$$

i.e., in terms of the so-called *phase shift* δ_ℓ. Comparing (20.91)–(20.89), we obtain

$$\alpha_\ell \sin(kr - \ell\pi/2 + \delta_\ell) = i^\ell(2\ell + 1)\sin(kr - \ell\pi/2) + k f_\ell e^{ikr} \quad (20.92)$$

leading to

$$\alpha_\ell = i^\ell(2\ell + 1) e^{i\delta_\ell}$$

$$f_\ell = \frac{(2\ell+1)}{2ik}\left(e^{2i\delta_\ell} - 1\right). \quad (20.93)$$

Thus, finally, the scattering amplitude can be written as

$$f_k(\theta) = \sum_{\ell=0}^{\infty} \frac{(2\ell + 1)}{2ik}\left(e^{2i\delta_\ell} - 1\right) P_\ell(\cos\theta). \quad (20.94)$$

All the scattering properties of the potential are incorporated in the phase shift δ_ℓ. We can also obtain a very elegant expression for the total scattering cross section as a series of partial wave cross sections. Starting from

$$|f_k(\theta)|^2 = \frac{1}{4k^2} \sum_{\ell=0}^{\infty} \sum_{\ell'=0}^{\infty} (2\ell + 1)(2\ell' + 1)\left(e^{-2i\delta_\ell} - 1\right)\left(e^{2i\delta_{\ell'}} - 1\right) P_\ell(\cos\theta) P_{\ell'}(\cos\theta)$$

and integrating over all angles

$$\frac{1}{4k^2} \sum_{\ell=0}^{\infty} \sum_{\ell'=0}^{\infty} (2\ell + 1)(2\ell' + 1)\left(e^{-2i\delta_\ell} - 1\right)\left(e^{2i\delta_{\ell'}} - 1\right) \int d\Omega\, P_\ell(\cos\theta)) P_{\ell'}(\cos\theta),$$

thanks to the orthogonality of Legendre polynomials, we obtain

$$\sigma = \frac{4\pi}{k^2} \sum_{\ell=0}^{\infty} (2\ell + 1)\sin^2\delta_\ell. \quad (20.95)$$

---

[10] $A_\ell = \alpha_\ell \cos\delta_\ell,\ B_\ell = -\alpha_\ell \sin\delta_\ell.$

A surprising relation between the total cross section and the scattering amplitude emerges if we consider the scattering amplitude *behind the target*, i.e., at $\theta = 0$. Noting that $P_\ell(1) = 1$, we have

$$f_k(0) = \frac{1}{2ik} \sum_{\ell=0}^{\infty} (2\ell + 1) \left(e^{2i\delta_\ell} - 1 \right) = \frac{1}{2k} \sum_{\ell=0}^{\infty} (2\ell + 1) \sin(2\delta_\ell) + \frac{i}{k} \sum_{\ell=0}^{\infty} (2\ell + 1) \sin^2 \delta_\ell$$

or

$$Im(f_k(0)) = \frac{1}{k} \sum_{\ell=0}^{\infty} (2\ell + 1) \sin^2 \delta_\ell .$$

Comparing with the expression for the total cross section we obtain

$$\sigma = \frac{4\pi}{k} Im(f_k(0)) . \tag{20.96}$$

This identity is the so-called *Optical Theorem*. It is generally valid and it is not particular for central potentials (for a general proof see Example 20.1 in this chapter). The Optical Theorem is an embodiment of the extreme wave-like nature of the quantum scattering process, relating the total scattering probability to what happens right behind the target, a region that would not be accessible in a classical particle collision.

Scattering on a hard sphere. As an application of the above we shall consider the scattering of particles of a given mass and energy by a spherical impenetrable region. Such a situation is described by a infinitely repulsive central potential

$$V(r) = \begin{cases} +\infty & (0 \le r \le r_0) \\ \\ 0 & (r > r_0) \end{cases} . \tag{20.97}$$

The radial wave function in the outside region $r > r_0$ can be parametrized in terms of the phase shift δ_ℓ as

$$R_{k,\ell}(r) = \alpha_\ell \left(\cos \delta_\ell \, j_\ell(kr) - \sin \delta_\ell \, n_\ell(kr) \right) . \tag{20.98}$$

It has to vanish in the interior of the sphere. By continuity, we have

$$R_{k,\ell}(r_0) = 0 \implies \tan \delta_\ell = \frac{j_\ell(kr_0)}{n_\ell(kr_0)} . \tag{20.99}$$

The total cross section is

$$\sigma = \frac{4\pi}{k^2} \sum_{\ell=0}^{\infty} (2\ell + 1) \sin^2 \delta_\ell = \frac{4\pi}{k^2} \sum_{\ell=0}^{\infty} \frac{(2\ell + 1) \, j_\ell^2(kr_0)}{\left(j_\ell^2(kr_0) + n_\ell^2(kr_0) \right)} . \tag{20.100}$$

In the case of *low energies* $kr_0 \to 0$ we may approximate the spherical Bessel and Neumann functions by

$$j_\ell(kr) \approx (kr)^\ell/(2\ell+1)!!, \qquad n_\ell(kr) \approx (kr_0)^{-(\ell+1)}(2\ell+1)!!/(2\ell+1)$$

and obtain an approximate expression for the total cross section, namely,

$$\sigma \approx 4\pi r_0^2 \sum_{\ell=0}^{\infty} \frac{(kr_0)^{4\ell}(2\ell+1)^3}{((2\ell+1)!!)^4} \approx 4\pi r_0^2 + \cdots \qquad (20.101)$$

The dominant term in this case is $4\pi r_0^2$, being just the area of the sphere. Thus, we may conclude that at low energies the total scattering cross section of a hard sphere equals its area. In contrast to the classical case, where the cross section would be expected to by πr_0^2, we see that in the quantum scattering case the whole spherical area contributes to the scattering process, reflecting its wave-like nature.

In the case of *high energies* or $kr_0 \gg \ell$, we may use the approximate asymptotic expressions of the spherical functions $j_\ell(x) \sim \sin(x-\ell\pi/2)/x$ and $n_\ell(x) \sim -\cos(x-\ell\pi/2)/x$ and have

$$\sigma \approx \frac{4\pi}{k^2} \sum_{\ell=0}^{kr_0}(2\ell+1)\sin^2(kr_0 - \ell\pi/2). \qquad (20.102)$$

The upper limit of the sum has been set to kr_0, since our high energies assumption is $kr_0 \gg \ell$. We may further approximate this sum with an integral according to

$$\sum_{\ell=0}^{kr_0}(2\ell+1) \ldots \to 2\int_0^{kr_0} d\ell\, \ell \ldots$$

and obtain

$$\sigma \approx 2\pi r_0^2. \qquad (20.103)$$

Again, in contrast to the classical case, where the cross section would be just πr_0^2, here, in the high energy limit $kr_0 \gg 1$ the wave nature of the process cannot be ignored.

Before we close this first section on scattering in central potentials, we shall derive a useful integral formula for the *phase shift*. Note that the scattering amplitude has been expressed as a sum of terms, each corresponding to a different angular momentum quantum number and depending, on the phase shift δ_ℓ, which embodies all dependence on the scattering potential. Let's write down the radial Schroedinger equation for the scattering problem at hand using the *one-dimensional wave function* $u_{k,\ell}(r) = r\, R_{k,\ell}(r)$. Remember that $u_{k,\ell}(0) = 0$. We have

$$\left\{ -\frac{\hbar^2}{2m}\frac{d^2}{dr^2} + \frac{\hbar^2\ell(\ell+1)}{2mr^2} + V(r) \right\} u_{k,\ell}(r) = E u_{k,\ell}(r).\qquad(20.104)$$

If the potential were absent, we would have the free radial Schroedinger equation

$$\left\{ -\frac{\hbar^2}{2m}\frac{d^2}{dr^2} + \frac{\hbar^2\ell(\ell+1)}{2mr^2} \right\} u_{k,\ell}^{(0)}(r) = E u_{k,\ell}^{(0)}(r).\qquad(20.105)$$

Of course, $u_{k,\ell}^{(0)}(r)$ is just $r\, j_\ell(kr)$. The energy values E (spectrum) would be the same for both problems. Multiplying (20.104) by $u_{k,\ell}^{(0)}(r)$ and (20.105) by $u_{k,\ell}(r)$ and subtracting them, we obtain

$$-\frac{\hbar^2}{2m}\left(u_{k,\ell}^{(0)}u_{k,\ell}'' - u_{k,\ell}u_{k,\ell}^{(0)\,''} \right) + V(r) u_{k,\ell}^{(0)}u_{k,\ell} = 0$$

or

$$\left(u_{k,\ell}^{(0)}u_{k,\ell}' - u_{k,\ell}u_{k,\ell}^{(0)\,'} \right)' = \frac{2m}{\hbar^2} V(r)u_{k,\ell}u_{k,\ell}^{(0)}.$$

Integrating, we obtain

$$u_{k,\ell}^{(0)}(r)\, u_{k,\ell}'(r) - u_{k,\ell}(r)u_{k,\ell}^{(0)\,'}(r) = \frac{2m}{\hbar^2}\int_0^r dr'\, V(r')\, u_{k,\ell}^{(0)}(r')u_{k,\ell}(r').\qquad(20.106)$$

In the asymptotic region ($r \to \infty$), we may replace the wave functions in the left-hand side with their asymptotic forms

$$u_{k,\ell} \sim \frac{1}{k}\sin(kr - \ell\pi/2 + \delta_\ell) \quad\text{and}\quad u_{k,\ell}^{(0)} \sim \frac{1}{k}\sin(kr - \ell\pi/2).$$

Then, we have

$$\sin\delta_\ell \approx -\frac{2mk^2}{\hbar^2}\int_0^\infty dr'\, r'\, V(r')\, j_\ell(kr')\, u_{k,\ell}(r').\qquad(20.107)$$

For a relatively weak potential, this can be further approximated by

$$\sin\delta_\ell \approx -\frac{2mk^2}{\hbar^2}\int_0^\infty dr\, r^2\, V(r)\, j_\ell^2(kr).\qquad(20.108)$$

Example 20.4 Consider the scattering of particles of mass m and energy E by the potential

$$V(r) = \begin{cases} 0 & (0 \leq r < a) \\ \frac{U_0}{r} & (a < r < 2a) \\ 0 & (r > 2a). \end{cases} \qquad (20.109)$$

Calculate the scattering amplitude in the Born approximation. Are there special values of the energy for which the Born approximate scattering amplitude vanishes? We have

$$f_k(\Omega) = -\frac{mU_0}{2\hbar^2\pi} \int_{r'\in[a,2a]} dr'\, r'^2 \frac{e^{i\mathbf{q}\cdot\mathbf{r}'}}{r'} = -\frac{mU_0}{\hbar^2} \int_a^{2a} dr'\, r' \int_{-1}^{1} d(\cos\theta')\, e^{iqr'\cos\theta'},$$

where $\mathbf{q} = \mathbf{k} - \hat{r}k \implies q^2 = 4k^2 \sin^2(\theta/2)$ or

$$f_k(\theta) = \frac{imU_0}{\hbar^2 q} \int_a^{2a} dr' \left(e^{iqr'} - e^{-iqr'} \right) = \frac{2mU_0}{\hbar^2 q^2} \left(2\cos^2(qa) - 1 - \cos(qa) \right). \quad (20.110)$$

This becomes zero for $\cos(qa) = 1$, $-1/2$, e.g., $qa = 2\pi/3$ or $\sin(\theta/2) = \frac{\pi}{3(ka)}$. This can only happen for low values of $(ka) < \pi/3$.

20.7 Bound States and Resonances

In this section, we shall simplify our discussion on short-range central potentials by considering potentials that are exactly zero beyond a radius a. As a matter of fact any short-range potential could be approximated with such a potential that vanishes beyond a certain *effective range a*. Since $V(r \geq a) = 0$, the radial wave function in the region $r \geq a$ is just

$$R_{k,\ell}(r) = (j_\ell(kr)\cos\delta_\ell - n_\ell(kr)\sin\delta_\ell) \qquad (20.111)$$

with the asymptotic behavior

$$R_{k,\ell} \sim \frac{1}{kr} \sin(kr - \ell\pi/2 + \delta_\ell).$$

The wave function in the internal $r < a$ region depends on the details of the potential. In any case, it will be related to (20.111) through continuity. The same is true for its derivative. We may express this continuity requirement in terms of the *logarithmic derivative* at the cutoff point a

$$\lambda_\ell \equiv \frac{1}{R_{k,\ell}(a)} \left(\frac{dR_{k,\ell}}{dr} \right)_{r=a} . \tag{20.112}$$

Substituting (20.111) in (20.112), we obtain

$$\lambda_\ell = \frac{j'_\ell(ka) \cos \delta_\ell - n'_\ell(ka) \sin \delta_\ell}{j_\ell(ka) \cos \delta_\ell - n_\ell(ka) \sin \delta_\ell}, \tag{20.113}$$

where the primes denote derivative with respect to the radius r. We may use this relation to express the phase shift in terms of the parameter λ_ℓ as

$$\cot \delta_\ell = \frac{n'_\ell(ka) - \lambda_\ell n_\ell(ka)}{j'_\ell(ka) - \lambda_\ell j_\ell(ka)} . \tag{20.114}$$

Note that in the case of the hard sphere ($\lambda_\ell = \infty$) this formula gives the right result for the phase shift obtained in the previous section, namely, $\tan \delta_\ell = j_\ell/n_\ell$.

The partial scattering amplitude and bound states. The expression (20.94) shows the scattering amplitude as a sum over angular momenta (partial spherical waves) with each term proportional to the quantity

$$S_\ell(k) = e^{2i\delta_\ell} - 1 \tag{20.115}$$

which we may call *partial scattering amplitude*. This is an important quantity the usefulness of which will be demonstrated shortly. To do this we consider a particularly simple example of central potential that belongs to the class of potentials that vanish beyond a point, namely, the case of the *spherical well* $V(r) = -V_0 \Theta(a - r)$ that we considered in Chap. 13. In order to make contact with our previous analysis, we rephrase formula (20.114) in terms of Hankel functions[11] and have

$$S_\ell(k) + 1 = e^{2i\delta_\ell} = \frac{h_\ell^{(+)'}(ka) - \lambda_\ell h_\ell^{(+)}(ka)}{h_\ell^{(-)'}(ka) - \lambda_\ell h_\ell^{(-)}(ka)} . \tag{20.116}$$

Let's consider now $S_\ell(k)$ as a function of k in the *full complex plane* [1, 3, 4]. This would include purely imaginary values of k which would correspond to negative energies, i.e., energies below the value of the potential at infinity and, therefore, bound states. It turns out that *bound states correspond to poles of the scattering amplitude located at the imaginary k-axis*. A heuristic argument supporting this statement is the following: Since the scattering amplitude S_ℓ is proportional to the probability amplitude to find the particle in the scattering range of the potential for a given incident flux, in the limit that the incident flux tends to zero, an infinite S_ℓ is required in order to have a nonzero probability amplitude to find a particle in the vicinity of the potential. This limiting situation of the presence of a particle without

---

[11]$h_\ell^{(\pm)} = n_\ell \pm i j_\ell$.

any incident flux corresponds to the existence of bound states. Therefore, bound states are related to the infinities of the scattering amplitude or, equivalently, to its poles.

Going back to the expression for the scattering amplitude of the spherical well, we see that its poles occur at wave numbers for which

$$\frac{h_\ell^{(-)'}(ka)}{h_\ell^{(-)}(ka)} = \lambda_\ell. \tag{20.117}$$

On the other hand in our analysis of the spherical well in Chap. 13, we calculated from the internal wave function that the logarithmic derivative at a is

$$\lambda_\ell = \frac{j_\ell'(qa)}{j_\ell(qa)} \quad \left(where \ \frac{\hbar^2 q^2}{2m} = E + V_0\right)$$

and that the condition for bound states of energy $E = -\frac{\hbar^2\kappa^2}{2m}$ is

$$\frac{h_\ell^{(-)'}(i\kappa a)}{h_\ell^{(-)}(i\kappa a)} = \frac{j_\ell'(qa)}{j_\ell(qa)}. \tag{20.118}$$

This condition coincides with the above pole condition (20.117) for $k = i\kappa$, i.e.,

$$S_\ell^{-1}(i\kappa a) = 0. \tag{20.119}$$

The fact that the poles of the scattering amplitude at the imaginary wave number axis correspond to bound states is not special to the spherical well but is a general property of all attractive potentials. Note also that the opposite is true, i.e., if there are no poles on the imaginary axis, there are no bound states.

The partial scattering amplitude and low-energy resonances. We know that the total cross section is a sum of terms

$$\sigma_\ell = \frac{4\pi}{k^2}(2\ell + 1)\sin^2\delta_\ell.$$

It is clear that each partial cross section approaches a maximum whenever

$$\delta_\ell(E) = \frac{\pi}{2} \tag{20.120}$$

or, equivalently

$$S_\ell(E) + 1 = e^{2i\delta_\ell(E)} = -1. \tag{20.121}$$

Let's see if such a situation can occur in the case of central potentials that vanish beyond a point. We start from the formula (20.114) or its equivalent

$$e^{2i\delta_\ell} = \frac{h_\ell^{(+)'}(ka) - \lambda_\ell\, h_\ell^{(+)}(ka)}{h_\ell^{(-)'}(ka) - \lambda_\ell\, h_\ell^{(-)}(ka)} \tag{20.122}$$

and apply it at *low-energies* $ka \ll 1$, where we may use the approximate expressions

$$h_\ell^{(\pm)}(ka) \approx (ka)^{-(\ell+1)}\frac{(2\ell+1)!!}{(2\ell+1)} \pm i\,\frac{(ka)^\ell}{(2\ell+1)!!} \tag{20.123}$$

and

$$h_\ell^{(\pm)'}(ka) \approx -k(\ell+1)(ka)^{-(\ell+2)}\frac{(2\ell+1)!!}{(2\ell+1)} \pm ik\ell\frac{(ka)^{\ell-1}}{(2\ell+1)!!} \tag{20.124}$$

and obtain

$$e^{2i\delta_\ell} \approx \frac{\ell+1+a\lambda_\ell - i(ka)^{2\ell+1}(2\ell+1)\,(\ell-a\lambda_\ell)}{\ell+1+a\lambda_\ell + i(ka)^{2\ell+1}(2\ell+1)\,(\ell-a\lambda_\ell)}\,. \tag{20.125}$$

Now it is clear from this expression that, if for some energy E_R the following is true

$$a\,\lambda_\ell(E_R) + \ell + 1 = 0, \tag{20.126}$$

the phase shift will be $\pi/2$ and the partial cross section will be maximum. This value of the energy E_R defines a so-called *resonance*. Expanding $\lambda_\ell(E)$ around a resonance, we may write

$$\lambda_\ell(E) \approx \lambda_\ell(E_R) + (E - E_R)\,\lambda_\ell'(E_R) + \cdots \tag{20.127}$$

and obtain

$$e^{2i\delta_\ell(E)} \approx \frac{E - E_R - i\Gamma}{E - E_R + i\Gamma}\,, \tag{20.128}$$

where

$$\Gamma \equiv (k_R a)^{2\ell+1}\frac{(2\ell+1)^2}{a\lambda_\ell'(E_R)}\,. \tag{20.129}$$

Near a resonance the partial cross section is

$$\sigma_\ell = \frac{4\pi(2\ell+1)}{k^2}\sin^2\delta_\ell \approx \frac{4\pi(2\ell+1)}{k^2}\left(\frac{\Gamma^2}{(E-E_R)^2+\Gamma^2}\right)\,. \tag{20.130}$$

Thus, it is clear that at $E = E_R$ the cross section attains a maximum value of $4\pi(2\ell+1)/k^2$. The quantity Γ signifies the *"width"* of the resonance and depends on the details of the potential. The partial scattering amplitude near a resonance is

$$S_\ell(E) = \frac{-2i\Gamma}{E - E_R + i\Gamma}, \tag{20.131}$$

implying that the resonances can be associated with poles at $E = E_R - i\Gamma$ in the complex energy plane.

Closing this section, we should stress again that the analytic behavior of the scattering amplitude in the complex plane and the association of bound states and resonances with its poles is not special to central potentials or to the simple examples of the square well or the class of potentials that vanish beyond a point but is a very general property of all potentials.

20.8 Coulomb Scattering

Up to now we have considered potentials of finite range, i.e., potentials that tend to zero at infinity faster than $1/r$. The *Coulomb potential*, responsible for the interactions of electrically charged particles, being ee'/r is of *infinite range* and was left out of our discussion, since a number of assumptions made are not valid for it. The difference from finite range potentials was already evident in the case of the discrete spectrum. In contrast to the spherical well in which the asymptotic behavior of the radial wave functions is proportional to $e^{-\kappa r}/r$, in the case of the hydrogen atom the wave functions go as $r^{n-1} e^{-\kappa r} = \exp[-\kappa r + (a_0\kappa)^{-1} \ln r]/r$, exhibiting an additional phase $\gamma \ln r$ that grows logarithmically.

Like all central potential scattering problems, the Coulomb scattering exhibits cylindrical symmetry. Nevertheless, the ideal type of coordinates to solve the time-independent Schroedinger equation is not cylindrical coordinates but the so-called *parabolic coordinates*, defined as [1]

$$\begin{cases} \xi = r + z \\[2mm] \eta = r - z \\[2mm] \varphi = \arctan(y/x) \end{cases} \tag{20.132}$$

The scattering of charged particles of mass m, electric charge e and energy $E = \hbar^2 k^2/2m$ by an electrostatic potential e'/r results in a Coulomb potential (energy)

$$V(r) = \frac{ee'}{r}. \tag{20.133}$$

Introducing the energy-dependent parameter

$$\gamma = \frac{mee'}{\hbar^2 k}, \tag{20.134}$$

we are led to the following form of the Schroedinger equation in terms of the above
parabolic coordinates

$$\left\{ \frac{4}{(\xi + \eta)} \left[\frac{\partial}{\partial \xi} \left(\xi \frac{\partial}{\partial \xi} \right) + \frac{\partial}{\partial \eta} \left(\eta \frac{\partial}{\partial \eta} \right) \right] + \frac{1}{\xi \eta} \frac{\partial^2}{\partial \varphi^2} \right\} \psi = \left(\frac{4\gamma k}{(\xi + \eta)} - k^2 \right) \psi.$$
(20.135)

Expecting, as we remarked, an axial symmetry, we search for solutions that will be
independent of φ. Then, we consider a trial solution in the form of a product

$$\psi(\xi, \eta) = f_1(\xi) \, f_2(\eta).$$
(20.136)

Inserting it into the Schroedinger equation, we obtain

$$\left(\xi \frac{f_1''}{f_1} + \frac{f_1'}{f_1} + \frac{k^2}{4} \xi \right) + \left(\eta \frac{f_2''}{f_2} + \frac{f_2'}{f_2} + \frac{k^2}{4} \eta \right) = \gamma k.$$
(20.137)

Since the first parenthesis depends only on ξ while the second parenthesis depends
only on η, the only way that their sum could be a constant is if both of them are
constants

$$\begin{cases} \xi \frac{f_1''}{f_1} + \frac{f_1'}{f_1} + \frac{k^2}{4} \xi = c_1 \\[2mm] \eta \frac{f_2''}{f_2} + \frac{f_2'}{f_2} + \frac{k^2}{4} \eta = c_2 \end{cases}$$
(20.138)

and there is a condition

$$c_1 + c_2 = \gamma k.$$
(20.139)

The scattering solution we are looking for is expected to contain an *incident wave*

$$e^{ikz} = e^{i \frac{k}{2} (\xi - \eta)}$$

and a *outgoing wave*

$$e^{ikr} = e^{i \frac{k}{2} (\xi + \eta)}.$$

Thus, a factor $e^{ik\xi/2}$ should be present in $f_1(\xi)$. Taking $f_1(\xi)$ to be exactly this factor

$$f_1(\xi) = e^{ik\xi/2},$$
(20.140)

we determine the constant c_1 from the first equation to be

$$c_1 = ik/2.$$
(20.141)

The second equation becomes

$$\eta \frac{f_2''}{f_2} + \frac{f_2'}{f_2} + \frac{k^2}{4}\eta = \gamma k - ik/2. \tag{20.142}$$

This equation can be transformed through the change of variable

$$f_2(\eta) = g(\eta)\, e^{-ik\eta/2} \tag{20.143}$$

into a mathematically familiar Hypergeometric equation

$$\eta\, g''(\eta) + (1 - ik\eta)\, g'(\eta) - \gamma k\, g(\eta) = 0 \tag{20.144}$$

with a known tabulated solution

$$g(\eta) = {}_1F_1(-i\gamma;\ 1;\ ik\eta). \tag{20.145}$$

The asymptotic ($r \to \infty$ or $\eta \to \infty$) behavior of ${}_1F_1(a;\ c;\ w)$ for purely imaginary argument w is known to be

$$_1F_1(a;\ 1;\ w) \approx \frac{|w|^{-a}}{\Gamma(1-a)} e^{-ia\pi/2} - i\,\frac{|w|^{a-1}\, e^{w}}{\Gamma(a)}\, e^{ia\pi/2}. \tag{20.146}$$

From this, we deduce that

$$g(\eta) \approx \frac{(k\eta)^{i\gamma}}{\Gamma(1+i\gamma)} e^{\gamma\pi/2} - i\,\frac{(k\eta)^{-1-i\gamma} e^{ik\eta}}{\Gamma(-i\gamma)} e^{\gamma\pi/2}. \tag{20.147}$$

We may use the property of gamma functions $\Gamma(n+1) = n\Gamma(n)$ for complex arguments $\Gamma(1 - i\gamma) = (-i\gamma)\Gamma(-i\gamma)$ and also set $\Gamma(1 \pm i\gamma) = |\Gamma(1 \pm i\gamma)|e^{\pm i\sigma}$, where σ is a phase. Then, we have

$$g(\eta) = \frac{e^{\gamma\pi/2}}{|\Gamma(1-i\gamma)|}\left(e^{-i\sigma}\, e^{i\gamma \ln(k\eta)} - \frac{\gamma}{k\eta}\, e^{-i\gamma \ln(k\eta)} e^{ik\eta} e^{i\sigma}\right). \tag{20.148}$$

The full asymptotic wave function will be

$$\psi(\xi, \eta) = e^{ik(\xi-\eta)/2}\, g(\eta)$$

$$\sim \frac{e^{\gamma\pi/2}}{|\Gamma(1-i\gamma)|}\left(e^{-i\sigma}\, e^{ik(\xi-\eta)/2 + i\gamma \ln(k\eta)} - \frac{\gamma}{k\eta}\, e^{ik(\xi+\eta)/2 - i\gamma \ln(k\eta)} e^{i\sigma}\right)$$

or

$$\psi(r, z) \sim \frac{1}{(2\pi)^{3/2}}\left(e^{ikz + i\gamma \ln(k(r-z))} - \frac{\gamma e^{2i\sigma}}{k(r-z)}\, e^{ikr - i\gamma \ln(k(r-z))}\right). \tag{20.149}$$

The scattered part of the wave function, written in terms of the radius r and the polar angle θ, is

$$\psi_{sc}(r, \theta) \sim -\frac{\gamma e^{2i\sigma}}{k(2\pi)^{3/2}} (1 - \cos\theta)^{-1+i\gamma} \frac{e^{ikr - i\gamma \ln(kr)}}{r}. \tag{20.150}$$

From this, we can read off the scattering amplitude

$$f_k(\theta) = -\frac{\gamma e^{2i\sigma}}{k} (1 - \cos\theta)^{-1+i\gamma} \tag{20.151}$$

and have an almost-standard expression for the scattered spherical wave

$$\psi_{sc}(r, \theta) \sim \frac{f_k(\theta)}{(2\pi)^{3/2}} \frac{e^{ikr - i\gamma \ln(kr)}}{r}. \tag{20.152}$$

The Coulomb differential cross section is

$$\sigma(\theta) = \frac{\gamma^2}{k^2(1 - \cos\theta)^2} = \frac{\gamma^2}{4k^2 \sin^4(\theta/2)}. \tag{20.153}$$

According to our findings in previous sections, the poles of the scattering amplitude at purely imaginary wave numbers in the case of the attractive Coulomb potential should correspond to the hydrogen atom energy levels. In this case

$$E \rightarrow \frac{\hbar^2(i\kappa)^2}{2m} = -\frac{\hbar^2\kappa^2}{2m} \quad and \quad \gamma \rightarrow i\frac{me^2}{\hbar^2\kappa}. \tag{20.154}$$

The scattering amplitude is proportional to

$$e^{2i\sigma} = \frac{\Gamma(1 + i\gamma)}{\Gamma(1 - i\gamma)} = i\gamma\frac{\Gamma(i\gamma)}{\Gamma(1 - i\gamma)} \rightarrow -\frac{me^2}{\hbar^2\kappa} \frac{\Gamma(-me^2/\hbar^2\kappa)}{\Gamma(1 + me^2/\hbar^2\kappa)}. \tag{20.155}$$

The poles of gamma functions arise at negative integer values

$$\Gamma^{-1}(-n) = 0 \quad (n = 1, 2, \ldots).$$

Therefore, the scattering amplitude will have poles at

$$me^2/\hbar^2\kappa = n \implies E_n = -\frac{me^4}{2\hbar^2 n^2}, \tag{20.156}$$

which are just the hydrogen energy levels.

20.9 Scattering of Identical Particles

Up to now in our discussion of scattering, we have ignored the spin of particles. It is straightforward to generalize most of what has been said to the case of spinning particles and spin-dependent forces. However, there is a particular case where spin plays a very important role with dramatic consequences. This is the case of scattering of identical particles. Actually, the scattering of identical particles is the best way to identify their bosonic or fermionic nature [1–3].

Identical boson scattering. Consider a pair of identical particles of integer spins. As a working example we may consider a pair of π-mesons (spin zero bosons) and assume for simplicity that their interaction potential depends only on the relative distance. As we know the wave function in center of mass coordinates will be

$$\Psi(\mathbf{r}_1, \mathbf{r}_2) = e^{i\mathbf{K}\cdot\mathbf{R}}\,\psi(\mathbf{r})\,, \tag{20.157}$$

where $\mathbf{R} = \frac{1}{2}(\mathbf{r}_1 + \mathbf{r}_2)$ is their center of mass coordinate and $\mathbf{r} = \mathbf{r}_1 - \mathbf{r}_2$ their relative position coordinate. Since the wave function of two identical bosons has to be symmetric in their interchange, the relative wave function will have to be even

$$\Psi(\mathbf{r}_1, \mathbf{r}_2) = \Psi(\mathbf{r}_2, \mathbf{r}_1) \implies \psi(\mathbf{r}) = \psi(-\mathbf{r})\,. \tag{20.158}$$

If the system is in an eigenstate of angular momentum, it is clear that only states of even ℓ will be allowed, since $Y_{\ell m}(-\hat{r}) = (-1)^{\ell}\,Y_{\ell m}(\hat{r})$. The asymptotic form $e^{ikz} + f_k(\theta)e^{ikr}/r$ will have to be replaced with a symmetrized expression[12]

$$\psi(\mathbf{r}) \sim \frac{1}{(2\pi)^{3/2}} \left(e^{ikz} + e^{-ikz} + (f_k(\theta) + f_k(\pi - \theta))\frac{e^{ikr}}{r} \right)\,. \tag{20.159}$$

Thus, the scattering amplitude is

$$f_k^{(S)}(\theta) = f_k(\theta) + f_k(\pi - \theta)\,. \tag{20.160}$$

The physical interpretation of the two terms is shown in Fig. 20.5. The two processes shown are indistinguishable. The scattering cross section will be

$$\sigma(\theta) = |f_k(\theta) + f_k(\pi - \theta)|^2 = |f_k(\theta)|^2 + |f_k(\pi - \theta)|^2 + 2Re\left\{f_k^*(\theta)f_k(\pi - \theta)\right\}\,. \tag{20.161}$$

If the two particles were distinguishable, the cross section would be just the sum of the first two terms. The existence of interference term is a measurable proof of the indistinguishability of the two identical particles. For instance, if we consider scattering at right angles, i.e., $\theta = \pi/2$, we obtain $\sigma(\pi/2) = 4|f_k(\pi/2)|^2$, which is twice what we would obtain if the two particles were distinguishable.

---

[12]Note that $\mathbf{r} \to -\mathbf{r}$ corresponds to $\theta, \phi \to \pi - \theta, \pi + \phi$.

Fig. 20.5 Identical particle scattering

Identical fermion scattering. If the two identical particles are fermions, we should take into consideration their spin. For the sake of simplicity, we may assume here that their interaction potential does not depend on spin but only on their relative distance. Then, their overall relative wave function can be a product of a spatial and a spinorial part, the latter being an eigenfunction of the total spin of the pair $\mathbf{S} = \mathbf{s}_1 + \mathbf{s}_2$, namely, $\psi(\mathbf{r}) \, |S, \, S_z\rangle$. For two spin–1/2 identical fermions, $|S, \, S_z\rangle$ is one of the following:

$$
triplet \implies
\begin{cases}
|1, 1\rangle = |\uparrow\rangle^{(1)} |\uparrow\rangle^{(2)} \\[6pt]
|1, 0\rangle = \frac{1}{\sqrt{2}} \left(|\uparrow\rangle^{(1)} |\downarrow\rangle^{(2)} + |\downarrow\rangle^{(1)} |\uparrow\rangle^{(2)} \right) \\[6pt]
|1, -1\rangle = |\downarrow\rangle^{(1)} |\downarrow\rangle^{(2)}
\end{cases}
\tag{20.162}
$$

and

$$
singlet \implies |0, 0\rangle = \frac{1}{\sqrt{2}} \left(|\uparrow\rangle^{(1)} |\downarrow\rangle^{(2)} - |\downarrow\rangle^{(1)} |\uparrow\rangle^{(2)} \right) .
\tag{20.163}
$$

The *Symmetrization Postulate* dictates that the total wave function of two identical fermions should be antisymmetric under their interchange. That leads to two possibilities, namely, *a symmetric spatial wave function with an antisymmetric spinorial one (i.e., the singlet)* and *an antisymmetric spatial wave function with a symmetric spinorial one (i.e., the triplet)*. If the system is in an eigenstate of the orbital angular momentum, the first possibility requires even values of $\ell = 0, 2, \ldots$, while the second possibility requires odd values $\ell = 1, 3, \ldots$. The corresponding cross sections are

$$
\sigma_{S=0}(\theta) = |f_k(\theta) + f_k(\pi - \theta)|^2
$$
$$
\sigma_{S=1}(\theta) = |f_k(\theta) - f_k(\pi - \theta)|^2.
\tag{20.164}
$$

Note that there is a vanishing cross section for triplet-scattering at right angles

$$\sigma_{S=1}(\pi/2) = 0. \tag{20.165}$$

In contrast, singlet scattering of identical fermions gives the same cross sections as in the case of identical bosons.

Example 20.5 Consider the scattering of two α-particles, i.e., ^4He nuclei, interacting through Coulomb forces due to their $-2e$ positive charge. These particles, composed of four nucleons (two protons and two neutrons), are bosons. Calculate the corresponding differential cross section.

We have

$$\sigma(\theta) = |f_k(\theta) + f_k(\pi - \theta)|^2, \tag{20.166}$$

where the scattering amplitude is just the Coulomb scattering amplitude with $\gamma = 4m_\alpha e^2/\hbar^2 k$, i.e.,

$$f_k(\theta) = -\frac{\gamma}{k} e^{2i\sigma} (1 - \cos\theta)^{-1+i\gamma}. \tag{20.167}$$

The phase σ is $e^{2i\sigma} = \Gamma(1 + i\gamma)/\Gamma(1 - i\gamma)$. Therefore, we have

$$\sigma(\theta) = \frac{\gamma^2}{k^2} \left((1 - \cos\theta)^{-2} + (1 + \cos\theta)^{-2} + (1 - \cos^2\theta)^{-1} \left(e^{i\gamma \ln\left(\frac{1+\cos\theta}{1-\cos\theta}\right)} + c.c. \right) \right)$$

$$= \frac{\gamma^2}{4k^2} \left(\frac{1}{\sin^4(\theta/2)} + \frac{1}{\cos^4(\theta/2)} + \frac{2\cos\left(2\gamma \ln(\tan(\theta/2))\right)}{\sin^2(\theta/2)\cos^2(\theta/2)} \right). \tag{20.168}$$

Example 20.6 Consider the scattering of two ^3He nuclei, interacting through Coulomb forces due to their $-2e$ positive charge. These particles, composed of three nucleons (two protons and one neutron), are fermions. Calculate the corresponding differential cross section assuming all spin states to be equally probable.

We know that the possible spin states are the triplet and the singlet correspondingly. The cross section will be

$$\sigma(\theta) = \frac{3}{4}|f_k(\theta) - f_k(\pi - \theta)|^2 + \frac{1}{4}|f_k(\theta) + f_k(\pi - \theta)|^2. \tag{20.169}$$

The factor $3/4$ refers to the three spin states of the triplet out of the four possible spin states, while the factor $1/4$ refers to the singlet. Due to the antisymmetry of the overall wave function, the triplet states have an antisymmetric spatial counterpart, while the singlet corresponds to a symmetrical spatial wave function. Substituting the Coulomb scattering amplitude $f_k(\theta)$, we end up with

$$\sigma(\theta) = \frac{\gamma^2}{4k^2} \left(\frac{1}{\sin^4(\theta/2)} + \frac{1}{\cos^4(\theta/2)} - \frac{\cos\left(2\gamma \ln(\tan(\theta/2))\right)}{\sin^2(\theta/2)\cos^2(\theta/2)} \right). \tag{20.170}$$

20.10 Scattering as a Transition Process

The scattering of particles by a finite range potential can be viewed as a transition process [1, 2] from one unperturbed state to another, these states being possibly plane wave eigenstates of the free Hamiltonian $\hat{H}_0 = \mathbf{p}^2/2m$. In this point of view, the scattering potential acts as a perturbation inducing transitions from an initial state $|\mathbf{k}\rangle$ to a final state $|\mathbf{k}'\rangle$.

Consider a particle described by the Hamilton operator $\hat{H} = \hat{H}_0 + \hat{V}$. We shall denote the eigenstates of the "free" part $\hat{H}_0$ as $|\psi_n^{(0)}\rangle$ and the corresponding eigenvalues with $E_n^{(0)}$, labeling them in terms of a discrete index n purely in a symbolic fashion. We can always expand the state of the system—satisfying the Schroedinger equation with the full Hamiltonian $\hat{H}$—in terms of the complete set of eigenstates $|\psi_n^{(0)}\rangle$ as

$$|\psi(t)\rangle = \sum_n c_n(t)\, e^{-\frac{i}{\hbar}E_n^{(0)}t}|\psi_n^{(0)}\rangle . \qquad (20.171)$$

Substituting the above in the Schroedinger equation, we obtain a system of differential equations for the coefficients $c_n(t)$, namely,

$$i\hbar\frac{dc_n}{dt} = \sum_{n'} V_{nn'}\, c_{n'}(t)\, e^{i\omega_{nn'}t} , \qquad (20.172)$$

where we have introduced the matrix elements of the perturbation with respect to the unperturbed states

$$V_{nn'} = \langle \psi_n^{(0)}|\hat{V}|\psi_{n'}^{(0)}\rangle \qquad (20.173)$$

and the frequencies

$$\omega_{nn'} = \frac{1}{\hbar}\left(E_n^{(0)} - E_{n'}^{(0)} \right) . \qquad (20.174)$$

We shall assume that initially $(t = -\infty)$ the system occupies an initial state $|\psi_i^{(0)}\rangle$. This is equivalent to assuming that

$$c_{ni}(-\infty) = \delta_{ni} . \qquad (20.175)$$

Next, we may transform the differential equation (20.172) into an integral equation as

$$c_n(t) = \delta_{ni} - \frac{i}{\hbar}\sum_{n'} V_{nn'} \int_{t_0}^t dt'\, c_{n'}(t')\, e^{i\omega_{nn'}t'+\alpha t'} . \qquad (20.176)$$

For technical reasons, we have replaced the lower integration limit with t_0, although we shall ultimately take $t_0 \to -\infty$. We have also introduced a *convergence factor* $\alpha \to 0$ to ensure convergence while we approach the lower integration limit.

First-order perturbation theory would give us a solution $(c_{n'}(t') \to \delta_{n'i})$

$$c_n^{(1)}(t) \approx \delta_{ni} - \frac{i}{\hbar}V_{ni} \int_{t_0}^t dt'\, e^{i\omega_{ni}t'+\alpha t'} . \qquad (20.177)$$

We may replace this approximate equation with an *exact* one of analogous form, namely,

$$c_n(t) = \delta_{ni} - \frac{i}{\hbar} T_{ni} \int_{t_0}^{t} dt' \, e^{i\omega_{ni}t' + \alpha t'}, \tag{20.178}$$

introducing the unknown quantity $T_{nn'}$. This equation is essentially the *definition* of the matrix $T_{nn'}$. Equivalently, it can be considered as an *Ansatz* for the solution of the original differential equation. Performing the time integration, we obtain, in the limit $\alpha t \to 0$, $\alpha t_0 \to -\infty$,

$$\int_{t_0}^{t} dt' \, e^{i\omega_{ni}t' + \alpha t'} = \frac{e^{i\omega_{ni}t}}{\alpha + i\omega_{ni}}.$$

Thus, we have

$$c_n(t) = \delta_{ni} - \frac{i}{\hbar} \frac{e^{i\omega_{ni}t} \, T_{ni}}{(\alpha + i\omega_{ni})}. \tag{20.179}$$

This equation enables us to replace the time-dependent coefficients $C_n(t)$ with the time-independent matrix elements $T_{nn'}$. An equation depending exclusively on $T_{nn'}$, which could in principle be solved, can be obtained by substituting the latter into the original differential equation. Doing that, we get

$$T_{in} = V_{in} - \frac{i}{\hbar} \sum_{n'} \frac{V_{in'} T_{n'n}}{(\alpha + i\omega_{n'n})}. \tag{20.180}$$

The matrix $T_{nn'}$ is called *transition matrix* for reasons that will become evident shortly.

We are mostly interest in transitions from an initial state $|\psi_i^{(0)}\rangle$ at $t_0 \to -\infty$ to a final state $|\psi_f^{(0)}\rangle \neq |\psi_i^{(0)}\rangle$. Thus, the *transition probability* will be

$$|c_f(t)|^2 = \frac{1}{\hbar^2} \frac{|T_{fi}|^2}{(\omega_{if}^2 + \alpha^2)} e^{2\alpha t}. \tag{20.181}$$

The corresponding *transition rate* will be

$$\frac{d|c_f|^2}{dt} = \frac{1}{\hbar^2} \left(\frac{2\alpha}{\omega_{if}^2 + \alpha^2} \right) |T_{fi}|^2 e^{2\alpha t}. \tag{20.182}$$

At this point we may remember that

$$\lim_{\alpha \to 0} \left\{ \frac{2\alpha}{\omega_{if}^2 + \alpha^2} \right\} = 2\pi \, \delta(\omega_{if}). \tag{20.183}$$

Therefore, the transition rate is

$$\Gamma_{i \to f} = \frac{2\pi}{\hbar} \delta(E_i^{(0)} - E_f^{(0)}) |T_{fi}|^2 . \tag{20.184}$$

This exact formula for the transition rate has the form of the *Golden Rule* encountered in first-order time-dependent perturbation theory, however with the potential matrix element V_{fi} having been replaced by the transition matrix element T_{fi}. It is actually the exact statement of the Golden Rule.

How is all this related to the cross section? In order to eventually arrive to that let's consider the total transition rate from an initial state $|\mathbf{k}\rangle$ to a set of analogous states $\{|\mathbf{k}'\rangle\}$, going at the same time to the *continuum limit*, since the wave number labeling these states takes up continuous values

$$\Gamma = \sum_{\mathbf{k}'} \frac{d|c_{\mathbf{k}'}|^2}{dt} \implies \Gamma = \frac{V}{(2\pi)^3} \int d^3k' \, \frac{2\pi}{\hbar} \delta(E - E') |T_{\mathbf{k}'\mathbf{k}}|^2 . \tag{20.185}$$

The continuum limit is taken by replacing the sum with an integral according to the rule

$$\frac{1}{V} \sum_n \to \int \frac{d^3k'}{(2\pi)^3} ,$$

where V is the total space volume. Using the fact that $\delta(E - E') = (2m/\hbar^2 k) \, \delta(k - k')$, we obtain

$$\frac{d\Gamma}{d\Omega} = \frac{Vmk}{\hbar^3 (2\pi)^2} |T_{\mathbf{k}'\mathbf{k}}|^2 \big|_{k'=k} , \tag{20.186}$$

where Ω corresponds to the $\hat{k}'$ direction, i.e., the direction of the final state. This rate is the total number of scattered particles per unit time. In order to connect this rate to the differential cross section, we must divide it by the incident flux $(\hbar k/m) \times (1/V)$. Thus, we may write

$$\frac{d\sigma}{d\Omega} = \frac{Vm}{\hbar k} \frac{d\Gamma}{d\Omega} = \left(\frac{Vm}{2\pi \hbar^2} \right)^2 |T_{\mathbf{k}'\mathbf{k}}|^2 \big|_{k'=k} . \tag{20.187}$$

This expression should coincide with the standard differential cross section expression $d\sigma/d\Omega = |f_{\mathbf{k}}(\hat{k}')|^2$. Therefore, the following relation should be true

$$f_{\mathbf{k}}(\hat{k}') = -\frac{Vm}{2\pi \hbar^2} T_{\mathbf{k}\mathbf{k}'} \tag{20.188}$$

with $k' = k$. Let's see if this is true, at least in the Born approximation. In the Born approximation

$$T_{\mathbf{kk'}}^{(B)} = \langle \mathbf{k'}|\hat{V}|\mathbf{k}\rangle = \int d^3 r \, \frac{e^{i\mathbf{k'}\cdot\mathbf{r}}}{\sqrt{V}} \, V(\mathbf{r}) \, \frac{e^{-i\mathbf{k}\cdot\mathbf{r}}}{\sqrt{V}} = \tilde{V}(\mathbf{q})$$

(for $\mathbf{q} = \mathbf{k} - \mathbf{k'}$), which, according to (20.62) is related to the Born-scattering amplitude as

$$f_{\mathbf{k}}^{(B)}(\hat{k}') = -\frac{2m}{\hbar^2}\sqrt{\frac{\pi}{2}} \left(\frac{V}{(2\pi)^{3/2}}\right) \tilde{V}(\mathbf{q}) = -\frac{mV}{2\pi\hbar^2} T_{\mathbf{kk'}}^{(B)},$$

which is exactly the relation (20.188) above. We have inserted in (20.61) and (20.62) an extra $V/(2\pi)^{3/2}$ factor to account for keeping the volume finite. Actually, this relation is not special to the Born approximation but can be proven to hold exactly.

20.11 The Lippmann–Schwinger Equation

Consider the transition matrix introduced in the previous section. Suppose that there exist states $|\psi_n^{(+)}\rangle$ such that

$$T_{n'n} = \langle \psi_{n'}^{(0)}|\hat{V}|\psi_n^{(+)}\rangle, \tag{20.189}$$

where $|\psi_n^{(0)}\rangle$ are the unperturbed states. Consider now the equation satisfied by the transition matrix, derived in the previous section

$$T_{n'n} = V_{n'n} - \frac{i}{\hbar}\sum_m \frac{V_{n'm}T_{mn}}{i\omega_{mn}+\alpha}$$

and insert the above expression for the transition matrix. We have

$$\langle \psi_{n'}^{(0)}|\hat{V}|\psi_n^{(+)}\rangle = \langle \psi_{n'}^{(0)}|\hat{V}|\psi_n^{(0)}\rangle - \frac{i}{\hbar}\sum_m \frac{1}{i\omega_{mn}+\alpha}\langle \psi_{n'}^{(0)}|\hat{V}|\psi_m^{(0)}\rangle\langle \psi_m^{(0)}|\hat{V}|\psi_n^{(+)}\rangle$$

or, dropping the $\langle \psi_{n'}^{(0)}|\hat{V}$ on the left, we obtain

$$|\psi_n^{(+)}\rangle = |\psi_n^{(0)}\rangle - \frac{i}{\hbar}\sum_m |\psi_m^{(0)}\rangle \frac{\langle \psi_m^{(0)}|\hat{V}|\psi_n^{(+)}\rangle}{i\omega_{mn}+\alpha}. \tag{20.190}$$

Next, we may rewrite

$$\frac{i}{\hbar}\sum_m |\psi_m^{(0)}\rangle \frac{1}{i\omega_{mn}+\alpha}\langle \psi_m^{(0)}| = \frac{1}{\hat{H}_0 - E_n - i\hbar\alpha}.$$

Note that we make no distinction between the energy eigenvalues $E_n^{(0)}$ of $\hat{H}_0$ and the energy eigenvalues of E_n of the complete Hamiltonian, since the scattering spectrum is exactly the same for both Hamiltonians. Returning to the equation, we write it in the form

$$|\psi_E^{(+)}\rangle = |\psi_E^{(0)}\rangle + \frac{1}{E - \hat{H}_0 + i\hbar\alpha} \hat{V} |\psi_E^{(+)}\rangle . \qquad (20.191)$$

This is the so-called *Schwinger–Lippmann equation* [1, 3]. Multiplying it with $(\hat{H}_0 - E)$, we obtain, in the limit $\alpha \to 0$,

$$\hat{H} |\psi_E^{(+)}\rangle = E |\psi_E^{(+)}\rangle . \qquad (20.192)$$

Thus, the states $|\psi_E^{(+)}\rangle$, originally defined through the transition matrix, are really the scattering eigenstates of the full Hamiltonian.

The Lippmann–Schwinger equation can also be written in terms of *Green's operator*, defined as a solution of the operator equation

$$\left(\hat{H}_0 - E\right) \hat{G}(E) = -\mathbf{I} . \qquad (20.193)$$

There are two operators satisfying this equation, namely,

$$\hat{G}^{(\pm)}(E) = \frac{1}{E - \hat{H}_0 \pm i\hbar\alpha} . \qquad (20.194)$$

It is $\hat{G}^{(+)}(E)$ the one that is suitable for the scattering process,[13] since its matrix elements give the Green's function with outgoing spherical wave behavior that we have employed

$$\langle \mathbf{r} | \hat{G}^{(+)}(E) | \mathbf{r}' \rangle = -\frac{m}{2\pi\hbar^2} G_k^{(+)}(R) . \qquad (20.195)$$

In terms of the operator $\hat{G}^{(+)}(E)$, the Lippmann–Schwinger equation is

$$|\psi_E^{(+)}\rangle = |\psi_E^{(0)}\rangle + \hat{G}^{(+)}(E) \hat{V} |\psi_E^{(+)}\rangle . \qquad (20.196)$$

Note that this is exactly the same scattering integral equation

$$\psi_{\mathbf{k}}^{(+)}(\mathbf{r}) = \psi_{\mathbf{k}}^{(0)}(\mathbf{r}) - \frac{m}{2\pi\hbar^2} \int d^3r' \, G_k^{(+)}(\mathbf{r}, \mathbf{r}') V(\mathbf{r}') \psi_{\mathbf{k}}^{(+)}(\mathbf{r}')$$

that we derived in a previous section written in a less formal way. Note also that it is straightforward to translate the integral expression for the scattering amplitude

---

[13] There exist also states $|\psi_E^{(-)}\rangle$ satisfying the Schwinger–Lippmann equation with the Green's operator $\hat{G}^{(-)}(E)$, namely, $|\psi_E^{(-)}\rangle = |\psi_E^{(0)}\rangle + \frac{1}{E - \hat{H}_0 - i\hbar\alpha} \hat{V} |\psi_E^{(-)}\rangle$. These states correspond to unphysical ingoing spherical waves.

(20.57) into

$$f_{\mathbf{k}}(\hat{k}') = -\frac{4\pi^2 m}{\hbar^2} \langle \psi_{\mathbf{k}'}^{(0)}| \hat{V} |\psi_{\mathbf{k}}^{(+)}\rangle = -\frac{4\pi^2 m}{\hbar^2} T_{\mathbf{k}'\mathbf{k}} \,. \tag{20.197}$$

This relation is exact. For a finite volume this has to be multiplied with an extra $V/(2\pi)^3$ factor and give

$$f_{\mathbf{k}}(\hat{k}') = -\frac{mV}{2\pi\hbar^2} T_{\mathbf{k}'\mathbf{k}} \,. \tag{20.198}$$

We can formally *"solve"* the Lippmann–Schwinger equation as

$$|\psi_E^{(+)}\rangle = \left(1 - \hat{G}^{(+)}(E)\,\hat{V}\right)^{-1} |\psi_E^{(0)}\rangle \,. \tag{20.199}$$

The *"solution"* rests on the knowledge of the inverse operator appearing above. We have

$$\left(1 - \hat{G}^{(+)}(E)\,\hat{V}\right)^{-1} = \left(\hat{G}^{(+)}(E)\left(E - \hat{H}_0 - \hat{V}\right)\right)^{-1} = \left(\hat{G}^{(+)}(E)\left(E - \hat{H}\right)\right)^{-1}$$

$$(E - \hat{H})^{-1}(E - \hat{H}_0) = (E - \hat{H})^{-1}(E - \hat{H} + \hat{V}) = 1 + \frac{1}{E - \hat{H}}\hat{V} \,.$$

Thus, the formal solution is

$$|\psi_E^{(+)}\rangle = |\psi_E^{(0)}\rangle + \frac{1}{E - \hat{H} + i\hbar\alpha}\hat{V}|\psi_E^{(0)}\rangle \,. \tag{20.200}$$

We have to admit that this expression is not very useful. Although it contains only the unperturbed states in the right-hand side, it has the full Hamiltonian acting on them. Despite that, this expression could prove to be useful for a perturbative expansion. Let's try to write the above inverse operator in a more suitable form. We have

$$\left(E - \hat{H} + i\hbar\alpha\right)^{-1} = \left(\left(\hat{G}^{(+)}(E)\right)^{-1} - \hat{V}\right)^{-1} =$$

$$\left(\left(\hat{G}^{(+)}(E)\right)^{-1}\left(1 - \hat{G}^{(+)}(E)\hat{V}\right)\right)^{-1} = \frac{1}{1 - \hat{G}^{(+)}(E)\hat{V}}\hat{G}^{(+)}(E) \,.$$

Then, our equation becomes

$$|\psi_E^{(+)}\rangle = |\psi_E^{(0)}\rangle + \frac{1}{1 - \hat{G}^{(+)}(E)\hat{V}}\hat{G}^{(+)}(E)\hat{V}|\psi_E^{(0)}\rangle$$

or

$$|\psi_E^{(+)}\rangle = \sum_{n=0}^{\infty} \left(\hat{G}^{(+)}(E)\hat{V} \right)^n |\psi_E^{(0)}\rangle . \tag{20.201}$$

This is the so-called *Born Series*

$$|\psi_E^{(+)}\rangle = |\psi_E^{(0)}\rangle + \hat{G}^{(+)}(E)\hat{V}|\psi_E^{(0)}\rangle + \hat{G}^{(+)}(E)\hat{V}\hat{G}^{(+)}(E)\hat{V}|\psi_E^{(0)}\rangle + \cdots \tag{20.202}$$

Example 20.7 Show that the scattering states $|\psi_E^{(+)}\rangle$ are orthonormal. Do the same for the states $|\psi_E^{(-)}\rangle$. Are these sets complete?

We have

$$\langle \psi_E^{(+)}|\psi_{E'}^{(+)}\rangle = \langle \psi_E^{(0)}| \left(1 + \hat{V}\frac{1}{E - \hat{H}} \right) |\psi_{E'}^{(+)}\rangle \tag{20.203}$$

$$= \langle \psi_E^{(0)}|\psi_{E'}^{(+)}\rangle + \frac{1}{E - E'}\langle \psi_E^{(0)}|\hat{V}|\psi_{E'}^{(+)}\rangle = \langle \psi_E^{(0)}|\psi_{E'}^{(+)}\rangle + \langle \psi_E^{(0)}|\frac{1}{\hat{H}_0 - E'}\hat{V}|\psi_{E'}^{(+)}\rangle$$

$$= \langle \psi_E^{(0)}|\psi_{E'}^{(+)}\rangle - \langle \psi_E^{(0)}| \left(|\psi_{E'}^{(+)}\rangle - |\psi_{E'}^{(0)}\rangle \right) = \langle \psi_E^{(0)}|\psi_{E'}^{(0)}\rangle = \delta(E - E'). \tag{20.204}$$

In an analogous fashion we can show that $\langle \psi_E^{(-)}|\psi_{E'}^{(-)}\rangle = \delta(E - E')$. Each of these sets is complete, provided that we include the bound states if they exist

$$\int_0^{\infty} dE \, |\psi_{E'}^{(\pm)}\rangle\langle \psi_{E'}^{(\pm)}| + \sum_{\Delta} |\Delta\rangle\langle\Delta| = \mathbf{I}. \tag{20.205}$$

20.12   The Scattering Matrix

Let's return to the definition of the transition matrix

$$c_f(t) = \delta_{fi} - \frac{i}{\hbar}T_{fi}\int_{t_0}^t dt' e^{i\omega_{fi}t' + \alpha t'} . \tag{20.206}$$

The coefficient $c_f(t)$ is the probability amplitude to start from a state $|\psi_i^{(0)}\rangle$ at the remote past $t_0 \to -\infty$ and make a transition to a state $|\psi_f^{(0)}\rangle$ at the remote future $t \to +\infty$. This amplitude can be written in terms of the time-evolution operator as

$$c_f(+\infty) = \langle \psi_f^{(0)}|\hat{U}(+\infty, -\infty)|\psi_i^{(0)}\rangle . \tag{20.207}$$

This quantity, i.e., the matrix elements of the time-evolution operator from the remote past to the remote future, between any two states,

$$S_{fi} = \langle \psi_f^{(0)}|\hat{U}(+\infty, -\infty)|\psi_i^{(0)}\rangle \tag{20.208}$$

is called the *Scattering Matrix* [1] or, simply the *S-matrix* between these states. The *S*-matrix is expressed in terms of the transition matrix as

$$S_{fi} = \delta_{fi} - \frac{i}{\hbar} T_{fi} \lim_{t, -t_0 \to \infty} \left\{ \frac{e^{i\omega_{fi}t + \alpha t}}{i\omega_{fi} + \alpha} \right\} , \tag{20.209}$$

having performed the time integration. The quantity in brackets in the limit $t \to \infty$, $\alpha \to 0$ and $\alpha t \to 0$ is just[14] $2\pi\delta(\omega_{fi})$. Thus, we have

$$S_{fi} = \delta_{fi} - 2\pi i \, \delta(E_f - E_i) \, T_{fi} . \tag{20.210}$$

Let us now recall the equations

$$|\psi_f^{(\pm)}\rangle = |\psi_f^{(0)}\rangle + \frac{1}{E_f - \hat{H} \pm i\hbar\alpha} \hat{V} |\psi_f^{(0)}\rangle$$

$$|\psi_i^{(\pm)}\rangle = |\psi_i^{(0)}\rangle + \frac{1}{E_i - \hat{H} \pm i\hbar\alpha} \hat{V} |\psi_i^{(0)}\rangle . \tag{20.211}$$

We may form the inner product

$$\langle \psi_f^{(-)} | \psi_i^{(+)} \rangle = \langle \psi_f^{(0)} | \left(1 - \hat{V} \frac{1}{E_f - \hat{H} + i\hbar\alpha} \right) | \psi_i^{(+)} \rangle \tag{20.212}$$

$$= \langle \psi_f^{(0)} | \psi_i^{(0)} \rangle + \langle \psi_f^{(0)} | \frac{1}{E_i - \hat{H}_0 + i\hbar\alpha} \hat{V} | \psi_i^{(+)} \rangle - \langle \psi_f^{(0)} | \hat{V} \frac{1}{E_f - \hat{H} + i\hbar\alpha} | \psi_i^{(+)} \rangle$$

$$= \delta_{if} + \left(\frac{1}{E_i - E_f + i\hbar\alpha} - \frac{1}{E_f - E_i + i\hbar\alpha} \right) \langle \psi_f^{(0)} | \hat{V} | \psi_i^{(+)} \rangle . \tag{20.213}$$

The parenthesis is

$$\frac{1}{E_i - E_f + i\hbar\alpha} + \frac{1}{E_i - E_f - i\hbar\alpha} = \frac{-2i\hbar\alpha}{(E_f - E_i)^2 + (\hbar\alpha)^2} = -2\pi i \, \delta(E_i - E_f). \tag{20.214}$$

Thus, finally we have

$$\langle \psi_f^{(-)} | \psi_i^{(+)} \rangle = \delta_{fi} - 2\pi i \, \delta(E_f - E_i) \, T_{fi} . \tag{20.215}$$

Comparing this to our identical earlier expression for the scattering matrix, we conclude that the S-matrix can also be defined as

$$S_{fi} = \langle \psi_f^{(-)} | \psi_i^{(+)} \rangle . \tag{20.216}$$

Since the S-matrix is defined as the matrix elements of the unitary time-evolution operator, it is automatically unitary, i.e.,

---

[14]For $t \to \infty$, $\alpha \to 0$ and $\alpha t \to 0$, we have $e^{(i\omega+\alpha)t}/(i\omega + \alpha) \to 2\pi\delta(\omega)$.

$$\sum_n S_{fn} S_{in}^* = \delta_{if} . \tag{20.217}$$

Problems and Exercises

20.1 Calculate the differential cross section for the central potentials $V(r) = V_0 e^{-r/a}$ and $V(r) = V_0 e^{-r^2/a^2}$ in the Born approximation.

20.2 Consider the scattering of a particle of mass m by a repulsive potential $V(r) = \Theta(a - r)V_0$.

(a) Calculate the phase shift at high energies $(\hbar k)^2 \gg 2mV_0$.

(b) In the opposite limit $(ka) \ll 1$, show that the differential cross section has the form $\sigma(\theta) = A + B \cos\theta$ with $B \ll A$.

(c) Suppose that $\xi \equiv \hbar^2/mV_0 a^2 \ll 1$. What is the meaning of that in the case $ka \ll 1$? Calculate A and B in the limit $\xi \to 0$, $(ka) \to 0$.

(d) Calculate the phase shift for $\ell = 0$ *exactly*.

20.3 Consider the scattering of particles from a periodic potential $V(\mathbf{r}) = V(\mathbf{r} + \mathbf{a})$. Write down the expression for the scattering amplitude in the Born approximation and show that it is nonzero for special values of the momentum transfer $\mathbf{q}$.

20.4 Consider a particle of mass m and energy $E > 0$ moving in the short-range central potential $V(r)$. Show that the radial wave function satisfies the integral equation

$$R_{E,\ell}(r) = j_\ell(kr) - \frac{m}{2\hbar^2\pi} \int_0^\infty dr' \, r'^2 \, \mathcal{G}_{k,\ell}(r, r') \, V(r') \, R_{E,\ell}(r'),$$

where the *radial Green's function* is defined through the equation

$$\left(\frac{1}{r^2} \frac{d}{dr} r^2 \frac{d}{dr} - \frac{\ell(\ell+1)}{r^2} + k^2 \right) \mathcal{G}_{k,\ell}(r, r') = -\frac{4\pi}{r^2} \delta(r - r').$$

Prove that the solution to this equation is

$$\mathcal{G}_{k,\ell}(r, r') = \begin{cases} 4i\pi k \, j_\ell(kr) \, h_\ell(kr') & (r' > r) \\ 4i\pi k \, j_\ell(kr') \, h_\ell(kr) & (r > r') \end{cases}$$

where $h_\ell(x) = j_\ell(x) + i n_\ell(x)$.

20.5 Consider the system of two electrons interacting through their mutual Coulomb repulsion $e^2/|\mathbf{r}_1 - \mathbf{r}_2|$. Classify all the possible cases of total spin of the system and calculate the differential cross section for each case.

20.6 Two spin-0 bosons of masses m_1 and m_2 interact through the potential

$$V(\mathbf{r}_1, \mathbf{r}_2) = \begin{cases} +\infty & |\mathbf{r}_1 - \mathbf{r}_2| < a \\ \\ 0 & |\mathbf{r}_1 - \mathbf{r}_2| > a \end{cases}$$

(a) Calculate the total scattering cross section at low energies.

(b) Do the same in the case that the two bosons are identical ($m_1 = m_2$).

20.7 The potential

$$V = V_{ab}|a\rangle\langle b| + V_{ba}|b\rangle\langle a|.$$

Consider the *Schwinger–Lippmann equation* $|\psi\rangle = |\psi_0\rangle + G(E)V|\psi\rangle$ and determine the state $|\psi\rangle$.

20.8 Consider the potential

$$V(r) = -\frac{g^2}{(r^2 + a^2)^2}.$$

(a) Calculate the scattering amplitude $f_k(\theta)$ in the Born approximation.

(b) Using (a) calculate the total cross section.

20.9 Consider a double slit cut in a very thin material on the (x, y)-plane and a beam of particles incident along the $\hat{z}$-axis. The situation can be modeled with a potential

$$V(x, y, z) = \begin{cases} V_0\delta(z) \; for \begin{cases} x \in [-b-a, -b+a] & -c < y < c \\ \\ x \in [b-a, b+a] & -c < y < c \end{cases} \\ \\ 0 & otherwise \end{cases}$$

where the centers of the slits are at the points $x = \pm b$ and their widths $2a << b$. Calculate the differential scattering cross section in the Born approximation for incidence along the z-axis.

20.10 Consider scattering in a hard sphere ($V(r) = +\infty$ for $r \le a$) in the case $\ell = 0$. Investigate the existence of resonances, defined by $\delta_0(E) = \pi/2$.

References

1. E. Merzbacher, *Quantum Mechanics*, 3rd edn. (Wiley, 1998)
2. A. Messiah, *Quantum Mechanics* (Dover publications, 1958). Single-volume reprint of the John Wiley & Sons, New York, two-volume 1958 edition
3. G. Baym, *Lectures in Quantum Mechanics*, Lecture Notes and Supplements in Physics (ABP, 1969)
4. K. Gottfried, T.-M. Yan, *Quantum Mechanics: Fundamentals* (Springer, Berlin, 2004)

Chapter 21
Quantum Behavior

21.1 Quantum Measurements

Consider a quantum system in an arbitrary state $|\psi\rangle$ and an apparatus designed to measure a particular observable $\mathcal{Q}$ of the system. Whatever is the state of the system at the instant prior to the measurement the measured value of $\mathcal{Q}$ is bound to be one of its eigenvalues q. This means that after the measurement the system will necessarily occupy one of the eigenstates $|q\rangle$ of $\mathcal{Q}$.[1] This is consistent with the fact that the eigenstates are the only states with vanishing uncertainty $\Delta\mathcal{Q}$. Any subsequent measurements of $\mathcal{Q}$ will yield the same eigenvalue q, meaning that the system continues to occupy the same eigenstate $|q\rangle$. It is said that the system has been *prepared* in the state $|q\rangle$.

Consider now that instead of a single measurement of $\mathcal{Q}$ that yielded the eigenvalue q we performed a series of measurements *under the exact same conditions* or equivalently measurements on a statistical *ensemble* of identical copies of the system. The result would be a series of, different in general, eigenvalues of $\mathcal{Q}$

$$q', q'', \ldots \tag{21.1}$$

For a large number of measurements, the frequency (probability) of each particular eigenvalue $\mathcal{P}(q)$ can be used to determine the state $|\psi\rangle$ of the system prior to measurement through the rule

$$\mathcal{P}(q) = |\langle q|\psi\rangle|^2 = |\psi(q)|^2. \tag{21.2}$$

---

[1] The measurement of an observable $\mathcal{Q}$, while the system is in a state $|\psi\rangle$, should not be confused with the action of the corresponding operator in the Hilbert space, since the latter yields another state $|\psi'\rangle$ (in general, a superposition of eigenstates of $\mathcal{Q}$), while as a result of a measurement the system will necessarily occupy one eigenstate of $\mathcal{Q}$.

© Springer Nature Switzerland AG 2019
K. Tamvakis, *Basic Quantum Mechanics*, Undergraduate Texts in Physics,
https://doi.org/10.1007/978-3-030-22777-7_21

441

The fact that as a result of the measurement the system has made a transition from a superposition of eigenstates to a single eigenstate

$$|\psi\rangle = \sum_{q'} \psi(q')|q'\rangle \implies |q\rangle \tag{21.3}$$

has been termed *collapse of the wave function*.[2] This transition is in sharp contrast to the unitary time evolution of the state that takes place when no measurement is performed and it is central to the acausal and probabilistic behavior associated with Quantum Mechanics.

Let us be more specific and consider the system of a particle characterized only by its spin $s = 1/2$. Our spin measuring apparatus, based on the Stern–Gerlach effect, can be oriented to measure any of the components S_x, S_y, or S_z. The result will always be one of the eigenvalues $\pm\hbar/2$ of these operators. Let's start with the apparatus oriented so that it can measure S_z. As a result, after the measurement the system will be prepared in an S_z eigenstate. Let's assume that this is the one corresponding to the eigenvalue $+\hbar/2$, namely

$$|\psi\rangle = |\uparrow\rangle \implies \begin{bmatrix} 1 \\ 0 \end{bmatrix}. \tag{21.4}$$

Then, we reorient the apparatus so that we measure S_x. After the measurement of S_x, the system will occupy one of the states

$$|\psi_+\rangle = \frac{1}{\sqrt{2}}(|\uparrow\rangle + |\downarrow\rangle) \implies \frac{1}{\sqrt{2}} \begin{bmatrix} 1 \\ 1 \end{bmatrix} \tag{21.5}$$

or

$$|\psi_-\rangle = \frac{1}{\sqrt{2}}(|\uparrow\rangle - |\downarrow\rangle) \implies \frac{1}{\sqrt{2}} \begin{bmatrix} 1 \\ -1 \end{bmatrix}. \tag{21.6}$$

A series of measurements will yield $\pm\hbar/2$ with equal probability 50%

$$|\langle\psi_\pm|\uparrow\rangle|^2 = |\langle\psi_\pm|\downarrow\rangle|^2 = \frac{1}{2}. \tag{21.7}$$

Note that although prior to the measurement the system was prepared in the $|\uparrow\rangle$ state, it subsequently made a transition to a statistical mixture of "spin up" and "spin down". The state of the *ensemble* after the measurement cannot be represented in terms a state vector. An appropriate representation for the "mixed state" of the system

---

[2]We leave aside the more ambiguous case of a degenerate eigenvalue which corresponds to more than one eigenstates $|q^{(\alpha)}\rangle$. A possible way around this problem is by replacing (21.2) with $|\langle q|\Pi_q|\psi\rangle|^2$ where $\Pi_q = \sum_\alpha |q^{(\alpha)}\rangle\langle q^{(\alpha)}|$ is the projection operator to the subspace spanned by the set of eigenstates $|q^{(\alpha)}\rangle$ that correspond to the degenerate eigenvalue q.

can be achieved in terms of the *density matrix* or *density operator* to be defined in the next section.

There is an interesting way to view the measurement process as the action of a *projection operator*. Following this line of thinking we may write any observable Λ in the form

$$\Lambda = \sum_j \lambda_j |j\rangle\langle j| = \sum_j \lambda_j \Pi_j \tag{21.8}$$

in terms of its eigenvalues and eigenstates and the corresponding projectors $\Pi_j = |j\rangle\langle j|$. The projection operators satisfy $\sum_j \Pi_j = I$ and $\Pi_i \Pi_j = \delta_{ij} \Pi_i$. The basic axioms of Quantum Mechanics dictate that when the system is subject to a measurement of the observable Λ and the system is in a state $|\psi\rangle$, as a result of the measurement, we shall obtain an eigenvalue λ_k with probability

$$\mathcal{P}_k = |\langle\psi|k\rangle|^2 = \langle\psi|\Pi_k|\psi\rangle \tag{21.9}$$

and, furthermore, the system will make a transition to the eigenstate $|k\rangle$

$$|\psi\rangle \rightarrow |k\rangle = \frac{1}{\sqrt{\mathcal{P}_k}} \Pi_k |\psi\rangle . \tag{21.10}$$

21.2 The Density Matrix

Up to now, whenever we have considered a quantum system we have assumed that it is described by a state vector $|\psi\rangle$ in Hilbert space which evolves according to the Schroedinger equation. The example discussed above of a system subject to a series of measurements has shown that a different situation is also possible in which the representation of the system in terms of a ket is not adequate. Nevertheless, whenever the state of the system corresponds to just one ket in Hilbert space we say that the system is in a *"pure state"*. The expectation value of any observable in a pure state is $\langle \mathcal{Q} \rangle = \langle\psi|\mathcal{Q}|\psi\rangle$. We can always introduce the operator

$$\rho = |\psi\rangle\langle\psi| \tag{21.11}$$

and write the above expectation value in terms of it as

$$\langle \mathcal{Q} \rangle = Tr(\mathcal{Q}\rho) . \tag{21.12}$$

This can be shown by expanding $|\psi\rangle$ in an orthonormal basis as $|\psi\rangle = \sum_j \psi_j |j\rangle$ and writing

$$\langle \mathcal{Q} \rangle = \sum_{i,j} \psi_i^* \mathcal{Q}_{ij} \psi_j = \sum_{ij} \mathcal{Q}_{ij} \rho_{ji} = Tr(\mathcal{Q}\rho) \tag{21.13}$$

since $\rho_{ij} = \psi_i \psi_j^*$. The operator ρ is called a *Density Matrix* [1]. Notice that the above-defined density matrix (21.11) for a pure state is a projection operator, having the standard properties of projection operators, namely[3]

$$\rho^\dagger = \rho, \quad \rho^2 = \rho \quad and \quad Tr(\rho) = 1. \tag{21.14}$$

Although the introduction of the density matrix for a system in a state $|\psi\rangle$ seems rather academic, it turns out that in more general situations the concept of a density matrix is very useful. In order to illustrate such a more general situation, we shall consider again the particularly simple system of a particle characterized by its spin alone. The system is prepared in an S_z eigenstate and subsequently is subject to an S_x measurement and ends up as a mixture of S_x eigenstates

$$|\psi_\pm\rangle = \frac{1}{\sqrt{2}} (|\uparrow\rangle \pm |\downarrow\rangle), \tag{21.15}$$

corresponding to $S_x = \pm\hbar/2$, each with a 50% probability. It is said that as a result of the measurement the system is in a *"mixed state"*. A usefull way to represent the mixed state is through a density matrix

$$\rho = \frac{1}{2}|\psi_+\rangle\langle\psi_+| + \frac{1}{2}|\psi_-\rangle\langle\psi_-|. \tag{21.16}$$

Each term projects into the $|\psi_+\rangle$ or $|\psi_-\rangle$ state, while the number 1/2 multiplying each projector is the probability of each state. Note that this is quite different than being in a pure state that is a superposition $|\psi\rangle = \frac{1}{\sqrt{2}} (|\psi_+\rangle + e^{i\phi}|\psi_-\rangle)$. For instance, the expectation value of an observable in this state, say S_x, depends on the phase difference of the superposed wave vectors (*coherent superposition*), namely, $\langle\psi|S_x|\psi\rangle = \frac{\hbar}{2} \cos\phi$. In contrast, a mixed state, defined by the density matrix (21.16) does not depend on the possible phases of the superposed terms. In this sense, it represents an *incoherent superposition*. Note that in the above mixed state the expectation value of S_x, reasonably defined as $\langle \cdots \rangle = \sum_a \mathcal{P}_a \langle\psi_a| \cdots |\psi_a\rangle$, is

$$\langle S_x \rangle = \frac{1}{2}\langle\psi_+|S_x|\psi_+\rangle + \frac{1}{2}\langle\psi_-|S_x|\psi_-\rangle = 0. \tag{21.17}$$

Thus, for a general mixed state, we define the corresponding density matrix as

$$\rho = \sum_a \mathcal{P}_a |\psi_a\rangle\langle\psi_a|, \tag{21.18}$$

where $\mathcal{P}_a$ is the probability for each pure state participating in the mixture. Expanding the participating states in an orthonormal basis, we get

---

[3] We have assumed $|\psi\rangle$ to be a normalized state.

$$\rho = \sum_a \mathcal{P}_a \sum_{i,j} \psi_{ai}^* \psi_{aj} |i\rangle\langle j| = \sum_{i,j} |i\rangle \rho_{ji} \langle j| \tag{21.19}$$

with

$$\rho_{ij} \equiv \sum_a \mathcal{P}_a \psi_{ai} \psi_{aj}^* . \tag{21.20}$$

The expectation value of an observable in such a mixed state is defined as

$$\langle \mathcal{Q} \rangle = \sum_a \mathcal{P}_a \langle \psi_a | \mathcal{Q} | \psi_a \rangle = \sum_{i,j} \sum_a \mathcal{P}_a \psi_{ai}^* \psi_{aj} \mathcal{Q}_{ij} = \sum_{i,j} \rho_{ji} \mathcal{Q}_{ij} \tag{21.21}$$

and can be put in the form of the trace formula

$$\langle \mathcal{Q} \rangle = Tr(\mathcal{Q}\rho) . \tag{21.22}$$

Note, however, that in contrast to the case of the density matrix for a pure state

$$\rho^2 \neq \rho . \tag{21.23}$$

Nevertheless, still $Tr(\rho) = 1$, since

$$Tr(\rho) = \sum_\alpha \mathcal{P}_\alpha Tr(|\psi_\alpha\rangle\langle\psi_\alpha|) = \sum_\alpha \mathcal{P}_\alpha \langle\psi_\alpha|\psi_\alpha\rangle = \sum_\alpha \mathcal{P}_\alpha = 1, \tag{21.24}$$

assuming that these states are normalized. We can also prove that for a mixed state

$$Tr(\rho^2) < 1 . \tag{21.25}$$

The proof goes as follows. For $Tr(\rho^2)$, we have

$$Tr\left(\sum_{\alpha,\beta} \mathcal{P}_\alpha \mathcal{P}_\beta |\psi_\alpha\rangle\langle\psi_\alpha|\psi_\beta\rangle\langle\psi_\beta|\right) = \sum_{\alpha\beta} \mathcal{P}_\alpha \mathcal{P}_\beta |\langle\psi_\alpha|\psi_\beta\rangle|^2 \leq \sum_{\alpha,\beta} \mathcal{P}_\alpha \mathcal{P}_\beta < 1,$$

in view of the Schwarz inequality $\langle\alpha|\alpha\rangle \langle\beta|\beta\rangle \geq |\langle\alpha|\beta\rangle|^2$ and the fact that the probabilities are numbers less than one.

The general properties of any density matrix, regardless of whether it represents a pure or a mixed state, are

$$\rho^\dagger = \rho, \quad Tr(\rho) = 1, \quad and \quad \langle\psi|\rho|\psi\rangle \geq 0 \ \forall|\psi\rangle . \tag{21.26}$$

Any operator that satisfies these properties can be viewed as the density matrix of some system. As a Hermitian operator, it has a set of orthogonal eigenstates and real eigenvalues

$$\rho|\tilde{\psi}_n\rangle = \varpi_n|\tilde{\psi}_n\rangle . \tag{21.27}$$

Then, it can be written as

$$\rho = \sum_n \varpi_n|\tilde{\psi}_n\rangle\langle\tilde{\psi}_n| . \tag{21.28}$$

Note that the last expression differs from (21.18) in that the states $|\psi_\alpha\rangle$ in (21.18) are not necessarily orthogonal, while the eigenstates $|\tilde{\psi}_n\rangle$ in (21.28) are. The real numbers ϖ_n are the probabilities associated with each eigenstate. The important thing is that any density matrix can be written in this form.

The state vectors $|\psi_\alpha\rangle$ that define the mixed state evolve with time in the standard way through the Schroedinger equation. At any instant of time t, we have

$$\rho(t) = \sum_\alpha P_\alpha|\psi_\alpha(t)\rangle\langle\psi_\alpha(t)| . \tag{21.29}$$

Differentiating with respect to time, we have

$$\frac{d\rho}{dt} = \sum_\alpha P_\alpha \left(\frac{d}{dt}|\psi_\alpha(t)\rangle\right)\langle\psi_\alpha(t)| + \sum_\alpha P_\alpha|\psi_\alpha(t)\rangle\left(\frac{d}{dt}\langle\psi_\alpha(t)|\right) \tag{21.30}$$

or

$$\frac{d\rho}{dt} = \sum_\alpha P_\alpha\left(-\frac{i}{\hbar}H|\psi_\alpha(t)\rangle\right)\langle\psi_\alpha(t)| + \sum_\alpha P_\alpha|\psi_\alpha(t)\rangle\left(\frac{i}{\hbar}H\langle\psi_\alpha(t)|\right) \tag{21.31}$$

or

$$\frac{d\rho}{dt} = -\frac{i}{\hbar}[H, \rho] . \tag{21.32}$$

This equation describes the time evolution of a mixed state. It is called the *von Neumann equation* (or the quantum analogue of the Liouville equation). Notice the difference in sign from the similar looking Heisenberg's equation. The obvious solution to (21.32) is

$$\rho(t) = U(t)\rho(0)U^\dagger(t) \tag{21.33}$$

with $U(t) = e^{-\frac{i}{\hbar}Ht}$ the standard time-evolution operator.

The evolution rates of the expectation values of various observables in mixed states can be obtained from (21.32) through the trace formulae as

$$\frac{d\langle Q\rangle}{dt} = Tr\left(Q\frac{d\rho}{dt}\right) = -\frac{i}{\hbar}Tr\left(Q[H, \rho]\right) = -\frac{i}{\hbar}Tr\left(\rho Q H\right) + \frac{i}{\hbar}Tr\left(\rho H Q\right)$$

or

$$\frac{d\langle Q \rangle}{dt} = \frac{i}{\hbar} Tr\left(\rho[H, Q]\right) = \frac{i}{\hbar} \langle [H, Q] \rangle. \tag{21.34}$$

This is the standard formula for the time evolution of the expectation values of pure states as well.

It should be noted that there is not an one-to-one correspondence between a mixed state and a density matrix. Two different mixed states can correspond to the same density matrix. We can illustrate this in the system of one spin. Consider a mixed state corresponding to spin "up" and spin "down" with equal probability. The density matrix will be

$$\rho = \frac{1}{2} |\uparrow\rangle\langle\uparrow| + \frac{1}{2} |\downarrow\rangle\langle\downarrow| \implies \begin{pmatrix} 1/2 & 0 \\ & \\ 0 & 1/2 \end{pmatrix} = \frac{1}{2}\mathbf{I}. \tag{21.35}$$

Nevertheless, a different mixed state with spin in the $S_x = +\frac{\hbar}{2}$ state and in the $S_x = -\frac{\hbar}{2}$ state with equal probability corresponds to the same density matrix. We have

$$\rho = \frac{1}{2} |\psi_+\rangle\langle\psi_+| + \frac{1}{2} |\psi_-\rangle\langle\psi_-| = \ldots = \frac{1}{2}\mathbf{I} \tag{21.36}$$

with $|\psi_\pm\rangle = \frac{1}{\sqrt{2}}(|\uparrow\rangle \pm |\downarrow\rangle)$.

In a composite system, it is possible to be in a pure state and yet have the density matrix corresponding to a subsystem to represent a mixed state. Consider any composite system $\mathcal{E}_1 \otimes \mathcal{E}_2$ (See the section on tensor product spaces in the Appendix). We may define a density matrix for the subsystem $\mathcal{E}_1$ as

$$\rho_1 = Tr_2(\rho), \tag{21.37}$$

where ρ is the density matrix of the composite system and the trace refers to summation over states of the subsystem $\mathcal{E}_2$. To be specific (but very general), we consider a pure state of the full system

$$|\Psi\rangle = \sum_{i_1 i_2} \Psi_{i_1 i_2} |E_{i_1 i_2}\rangle, \tag{21.38}$$

where

$$|E_{i_1 i_2}\rangle = |e_{i_1}^{(1)}\rangle \otimes |e_{i_2}^{(2)}\rangle \tag{21.39}$$

is an orthonormal basis. The corresponding density matrix will be

$$\rho = |\Psi\rangle\langle\Psi| = \sum_{i_1 i_2 j_1 j_2} \Psi_{i_1 i_2} \Psi_{j_1 j_2}^* \left(|e_{i_1}^{(1)}\rangle \otimes |e_{i_2}^{(2)}\rangle\right) \left(\langle e_{j_1}^{(1)}| \otimes \langle e_{j_2}^{(2)}|\right). \tag{21.40}$$

Taking the partial trace over $\mathcal{E}_2$, we obtain

$$Tr_2(\rho) = \left(\sum_{i_2} \Psi_{i_1 i_2} \Psi^*_{j_1 i_2} \right) |e^{(1)}_{i_1}\rangle \langle e^{(1)}_{j_1}| . \tag{21.41}$$

Thus, we have

$$\rho_1 = \sum_{i_1 j_1} (\rho_1)_{i_1 j_1} |e^{(1)}_{i_1}\rangle \langle e^{(1)}_{j_1}| \tag{21.42}$$

with

$$(\rho_1)_{i_1 j_1} = \sum_{i_2} \Psi_{i_1 i_2} \Psi^*_{j_1 i_2} . \tag{21.43}$$

Note that although we started with a pure state for the full system, the subsystem density matrix corresponds to a mixed state. This is related to a property of quantum systems under the name *Entanglement*, which will be the subject of the next section.

Example 21.1 Consider the state of the combined system of two spins to be $|\psi\rangle = \frac{1}{\sqrt{2}} (|\uparrow\rangle|\downarrow\rangle + |\downarrow\rangle|\uparrow\rangle)$. Construct the density matrix $\hat{\rho}_1$ of the subsystem of one spin and study its properties.

The components of $|\psi\rangle$ in a tensor product basis $|e^{(1)}_{i_1}\rangle \otimes |e^{(2)}_{i_2}\rangle$ are $\psi_{i_1 i_2}$ are

$$\psi_{\uparrow\uparrow} = 0, \ \psi_{\uparrow\downarrow} = \frac{1}{\sqrt{2}}, \ \psi_{\downarrow\uparrow} = \frac{1}{\sqrt{2}}, \ \psi_{\downarrow\downarrow} = 0. \tag{21.44}$$

The $\mathcal{E}_1$-density matrix will be

$$(\rho_1)_{j_1 i_1} = \sum_{i_2} \psi^*_{i_1 i_2} \psi_{j_1 i_2} , \tag{21.45}$$

its elements being

$$(\rho_1)_{\uparrow\uparrow} = \psi^*_{\uparrow\uparrow}\psi_{\uparrow\uparrow} + \psi^*_{\uparrow\downarrow}\psi_{\uparrow\downarrow} = \tfrac{1}{2}$$

$$(\rho_1)_{\uparrow\downarrow} = \psi^*_{\uparrow\uparrow}\psi_{\uparrow\downarrow} + \psi^*_{\uparrow\downarrow}\psi_{\downarrow\downarrow} = 0$$

$$\implies \rho_1 = \begin{pmatrix} \tfrac{1}{2} & 0 \\ 0 & \tfrac{1}{2}. \end{pmatrix} \tag{21.46}$$

$$(\rho_1)_{\downarrow\uparrow} = \psi^*_{\downarrow\uparrow}\psi_{\uparrow\uparrow} + \psi_{\downarrow\downarrow}\psi_{\downarrow\uparrow} = 0$$

$$(\rho_1)_{\downarrow\downarrow} = \psi^*_{\downarrow\uparrow}\psi_{\uparrow\downarrow} + \psi_{\downarrow\downarrow}\psi_{\downarrow\downarrow} = \tfrac{1}{2}$$

In accordance with the previously stated general properties, we see that

$$Tr(\rho_1) = 1 \quad and \quad Tr(\rho_1^2) = 1/2 < 1. \tag{21.47}$$

21.3   Entanglement

In classical physics, if we happen to possess complete knowledge of the state of the system, we will necessarily know also everything that is to know about its parts. This is not true anymore for a quantum system, i.e., although we may know the state of a composite system, we shall not necessarily know the states of its subsystems. This property of quantum systems is called *Entanglement*[4] and we shall illustrate it in the framework of the typical composite system of two particles of spin $1/2$. Its simplest version is when the spatial degrees of freedom are ignored and it reduces to a system of two spins S_1 and S_2. Each one-particle subsystem can be described in terms of the states $|\uparrow\rangle$ and $|\downarrow\rangle$. Thus, a general state of the *particle-"1"* subsystem is

$$|\psi_a\rangle = a_+|\uparrow\rangle + a_-|\downarrow\rangle.$$

Analogously for *particle-"2"* subsystem

$$|\psi_b\rangle = b_+|\uparrow\rangle + b_-|\downarrow\rangle.$$

Out of these one-particle states, we may obtain states corresponding to the combined system through their *tensor product* such as $|\downarrow\rangle \otimes |\downarrow\rangle$ or $|\uparrow\rangle \otimes |\downarrow\rangle$. By convention subsystem "1" is put to the left while subsystem "2" is put to the right. A simplified notation is also $|\downarrow\rangle|\downarrow\rangle$ and $|\uparrow\rangle|\downarrow\rangle$. These product states $|i\rangle \otimes |j\rangle$ represent all the possible outcomes for the individual spins and constitute an orthonormal basis spanning the combined system Hilbert space $\mathcal{E}_{1\cup2} = \mathcal{E}_1 \otimes \mathcal{E}_2$. Nevertheless, product states are not the most general states of the combined system. A general state may be refered to as $|\Psi_{ab}\rangle$ and can be analyzed in the above basis as

$$|\Psi_{ab}\rangle = c_{++}|\uparrow\rangle|\uparrow\rangle + c_{+-}|\uparrow\rangle|\downarrow\rangle + c_{-+}|\downarrow\rangle|\uparrow\rangle + c_{--}|\downarrow\rangle|\downarrow\rangle. \quad (21.48)$$

Normalization corresponds to the condition

$$|c_{++}|^2 + |c_{+-}|^2 + |c_{-+}|^2 + |c_{--}|^2 = 1. \quad (21.49)$$

Consider now the expectation values of the z-component of each spin in the above general state

$$\langle \hat{S}_{1z} \rangle = \frac{\hbar}{2}\left(|c_{++}|^2 + |c_{+-}|^2 - |c_{-+}|^2 - |c_{--}|^2\right)$$
$$\langle \hat{S}_{2z} \rangle = \frac{\hbar}{2}\left(|c_{++}|^2 - |c_{+-}|^2 + |c_{-+}|^2 - |c_{--}|^2\right) \quad (21.50)$$

On the other hand, the expectation value of the product of these operators is

$$\langle \hat{S}_{1z}\hat{S}_{2z} \rangle = \frac{\hbar^2}{4}\left(|c_{++}|^2 - |c_{+-}|^2 - |c_{-+}|^2 + |c_{--}|^2\right). \quad (21.51)$$

From these expressions after some algebra and using the normalization condition, we may obtain

---

[4]See [2].

$$\langle \hat{S}_{1z}\hat{S}_{2z}\rangle - \langle \hat{S}_{1z}\rangle \langle \hat{S}_{2z}\rangle = \hbar^2 \left(|c_{+-}|^2|c_{-+}|^2 - |c_{++}|^2|c_{--}|^2\right). \tag{21.52}$$

Consider now the particular case of a state with $c_{++} = c_{--} = 0$ and $c_{+-} = c_{-+}$

$$|\Psi\rangle = \frac{1}{\sqrt{2}}(|\uparrow\rangle|\downarrow\rangle + |\downarrow\rangle|\uparrow\rangle). \tag{21.53}$$

In this state

$$\langle \hat{S}_{1z}\rangle = \langle \hat{S}_{2z}\rangle = 0, \tag{21.54}$$

while

$$\langle \hat{S}_{1z}\hat{S}_{2z}\rangle = -\hbar^2. \tag{21.55}$$

This means that in this state, although each of the parts of the system has a vanishing spin expectation value, this is not the case for the expectation value of the product of spins. In sharp contrast, if the system is in a state that all outcomes are equally probable (e.g. $c_{+-} = c_{-+} = c_{++} = c_{--}$) the quantity (21.52) will vanish and the expectation value of the product factorizes into the product of the expectation values, i.e.

$$\langle \hat{S}_{1z}\hat{S}_{2z}\rangle^2 - \langle \hat{S}_{1z}\rangle^2 \langle \hat{S}_{2z}\rangle^2 = 0 \tag{21.56}$$

and, since the square of an amplitude corresponds to a probability, we may conclude that the vanishing of this expression corresponds to that fact that the two single particle observables are *uncorrelated*. This is in line with the fact that if the probability distribution for two general observables factorizes

$$\mathcal{P}(A, B) = \mathcal{P}(A)\mathcal{P}(B), \tag{21.57}$$

these are considered statistically independent and uncorrelated. Thus, the quantity

$$\mathcal{C}(A, B) \equiv \langle A\,B\rangle - \langle A\rangle \langle B\rangle \tag{21.58}$$

can be referred to as the *correlation* of the two observables A and B. If the *correlation* is nonzero their measured values will be correlated.

Consider again the two spin state (21.48) in the particular case that it were a product state, namely, if

$$c_{++}c_{--} = c_{+-}c_{-+}. \tag{21.59}$$

Then, it is clear that in this case the state can be written as

$$|\Psi\rangle = \frac{1}{c_{++}}(c_{-+}|\downarrow\rangle + c_{++}|\uparrow\rangle)(c_{+-}|\downarrow\rangle + c_{++}|\uparrow\rangle), \tag{21.60}$$

which is a tensor product. It is clear also that (21.59) and (21.60) lead to a vanishing spin correlation (21.52)

$$C(S_{1z}, S_{2z}) = 0. \tag{21.61}$$

This is general. *Tensor product states have a vanishing correlation.* Note that a general state (21.48) is characterized by four complex numbers or eight real numbers which are reduced to six due to the normalization condition and an irrelevant overall phase. In contrast, a tensor product state is characterized by four real numbers (two real numbers for each normalized factor), as one would expect, since a one-spin state is characterized by two numbers and a state of two independent spins by four.

The property that general states like (21.48) embody nonzero correlations, while product states do not, is called *Entanglement*. Entanglement is a quantitative property in the sense that there are various degrees of entanglement. For example the state[5]

$$|0, 0\rangle = \frac{1}{\sqrt{2}} (|\uparrow\rangle|\downarrow\rangle - |\downarrow\rangle|\uparrow\rangle), \tag{21.62}$$

as it will become clear, is a state of maximum entanglement. We may compute the spin expectation value in this state[6] and find

$$\langle \mathbf{S}_1 \rangle = \langle \mathbf{S}_2 \rangle = 0. \tag{21.63}$$

Note that this is in sharp contrast to the fact that for any normalized *single spin state* $\alpha|\uparrow\rangle + \beta|\downarrow\rangle$ we have $\langle \mathbf{S} \rangle^2 = \hbar^2/4$. This means that in single spin states not all spin expectation values can be zero. This continues to be true for product states. Nevertheless in the two spin state $|0, 0\rangle$ all spin expectation values vanish according to (21.63), something that can be interpreted as ignorance about individual spins. In view of the fact that

$$\langle \hat{S}_{1x}^2 \rangle = \langle \hat{S}_{1y}^2 \rangle = \langle \hat{S}_{1z}^2 \rangle = \frac{\hbar^2}{4}$$

$$\langle \hat{S}_{2x}^2 \rangle = \langle \hat{S}_{2y}^2 \rangle = \langle \hat{S}_{2z}^2 \rangle = \frac{\hbar^2}{4}$$

we are completely uncertain about the value of each individual spin. On the other hand we obtain

$$\langle S_{1x} S_{2x} \rangle = \langle S_{1y} S_{2y} \rangle = \langle S_{1z} S_{2z} \rangle = -\frac{\hbar^2}{4},$$

which corresponds to a maximal correlation C for each of these spin operator products.

As we showed in the previous section, the subsystem density matrix will in general correspond to a mixed state even if the composite system is in a pure state.

---

[5]This is the *singlet* eigenstate of total spin zero.
[6]We have

$$\hat{S}_x|\uparrow\rangle = \frac{\hbar}{2}|\downarrow\rangle, \ \hat{S}_x|\downarrow\rangle = \frac{\hbar}{2}|\uparrow\rangle,$$

$$\hat{S}_y|\uparrow\rangle = i\frac{\hbar}{2}|\downarrow\rangle, \ \hat{S}_y|\downarrow\rangle = -i\frac{\hbar}{2}|\uparrow\rangle.$$

Nevertheless, when the composite system is in a tensor product state

$$|\Psi\rangle = |\psi^{(1)}\rangle|\psi^{(2)}\rangle = \sum_{i_1 i_2} \psi_{i_1}^{(1)} \psi_{i_2}^{(2)} \left(|e_{i_1}^{(1)}\rangle \otimes |e_{i_2}^{(2)}\rangle \right) \qquad (21.64)$$

the subsystems $\mathcal{E}_1$ and $\mathcal{E}_2$ are in pure states. For example, the $\mathcal{E}_1$ density matrix

$$\rho^{(1)} = Tr_2(\rho) \implies \rho_{i_1 j_1}^{(1)} = \sum_{i_2} \rho_{i_1 i_2, j_1 i_2} = \sum_{i_2} \Psi_{j_1 i_2} \Psi_{i_1 i_2}^*$$

$$= \sum_{i_2} \psi_{j_1}^{(1)} \psi_{i_2}^{(2)} \left\{ \psi_{i_1}^{(1)} \right\}^* \left\{ \psi_{i_2}^{(2)} \right\}^* = \psi_{j_1}^{(1)} \left\{ \psi_{i_1}^{(1)} \right\}^* \sum_{i_2} \left| \psi_{i_2}^{(2)} \right|^2 = \psi_{j_1}^{(1)} \left\{ \psi_{i_1}^{(1)} \right\}^*$$
$$(21.65)$$

or

$$\rho^{(1)} = |\psi^{(1)}\rangle\langle\psi^{(1)}|, \qquad (21.66)$$

meaning that the subsystem $\mathcal{E}_1$ is in the pure one-particle state $|\psi^{(1)}\rangle$ and there is no entanglement. Note that (21.66), as a projection operator has only one non-zero eigenvalue equal to one, while all the other eigenvalues vanish.[7]

The case of a tensor product state is a very special one and in the general case the $\mathcal{E}_1$-subsystem will occupy a mixed state. An extreme situation of *maximal entanglement* is when all eigenvalues of the density matrix $\rho^{(1)}$ are equal. Given that $Tr_1(\rho^{(1)}) = 1$, this means that the eigenvalues are equal to $1/N$, with N being the dimension of $\mathcal{E}_1$. An example of such a maximally entangled state is the two spin state $|\Psi\rangle = \frac{1}{\sqrt{2}} (|\uparrow\rangle|\downarrow\rangle - |\downarrow\rangle|\uparrow\rangle)$. Notice that in this case of maximal entanglement whatever is the result of measurement of S_{1z} in $\mathcal{E}_1$, the state of the $\mathcal{E}_2$ subsystem is completely predictable. If our S_{iz} outcome is the state $|\psi^{(1)}\rangle = |\uparrow\rangle$, we know with certainty that the $\mathcal{E}_2$ will be in the state $|\psi^{(2)}\rangle = |\downarrow\rangle$. These and analogous properties have given rise to the so-called *EPR paradox*.

21.4   Bell's Theorem, etc.

The measurement process and the collapse of the wave function are intimately connected with a nonlocal aspect of Quantum Mechanics that has puzzled physicists and has led to various *"paradoxes"*. Characteristic is the so-called EPR "paradox", named after Einstein, Podolsky, and Rosen, who posed it as a *gedanken experiment* [3]. A simplified variant of the EPR setup may consist of two spin 1/2 particles prepared in a total spin zero state

---

[7]Equivalently, any matrix of the form $\alpha_i \alpha_j^*$ like (21.65) has only one non-zero eigenvalue equal to one, while all the other eigenvalues vanish.

$$|0, 0\rangle \;=\; \frac{1}{\sqrt{2}}\,(|\uparrow\rangle_\ell|\downarrow\rangle_r \,-\, |\downarrow\rangle_\ell|\uparrow\rangle_r)$$

and emitted at opposite directions (designated as "left" and "right"). When the particles are far apart an S_z measurement is performed at either of them. After a number of trials, what is found is that whenever the spin of the particle on the left is found to be "up"—and this occurs 50% of the time—the spin of the particle on the right is found to be "down". Analogously, whenever the spin of the particle on the left is found to be "down", the spin of the particle on the right is found to be "up". This means that measuring the spin of the particle on the right is enough to determine the spin of the particle on the left. Note that the particle on the left was not in a spin "down" or spin "up" state initially. So if for example the particle on the right was measured to have spin "up", the particle on the left must have "jumped" into a spin "down" state. Note again that the measurements are performed when the particles are far enough so that no light signal can communicate the result of each measurement. Although this is not really a paradox, since no propagation of information faster than light takes place, but can be accommodated in the entanglement properties of Quantum Mechanics, it has prompted a class of physicists to view Quantum Theory as incomplete and pursue alternative theories of so-called *"hidden variables"* that would circumvent the concepts of probabilities at a fundamental level or the above nonlocal aspects of entanglement. Nevertheless, a quantitative test of the predictions of any local hidden variable theory versus the predictions of Quantum Mechanics has been developed by J. Bell and it has been experimentally verified that *"no local[8] hidden variable theory can agree with all the predictions of Quantum Mechanics"*. This is *Bell's Theorem* [4] and it has ruled out all local hidden variable theories. The related experiment was performed in the early 80s by Aspect [5] and was based on the violation of Bell's inequality derived in 1964. Instead of deriving this inequality and giving a proof of Bell's theorem, we shall consider a simplified version of a thought experiment developed by D. Mermin that suffices to demonstrate its essential assertion that quantum mechanical behavior cannot be reproduced by local hidden variable theories [6].

Consider again the previous EPR setup with the additional specifications that each of the detectors used for the above measurements has a switch with two positions (say "1" and "2") and two lights (red (R) and green (G)). When a switch is in position "1" it is set to measure S_z, while, when it is in position "2" it is set to measure S_n (i.e., spin in some other direction $\hat{n}$ making an angle θ with the $\hat{z}$ axis). When "spin up" is measured, the red light flashes, while, when "spin down" is measured, the flashing light is the green one. After a number of trials, it is observed that[9]

(1) $12GG$ or $21GG$ never occurs;

(2) $22GG$ comes out $\alpha\%$ of the time;

(3) $11RR$ never occurs.

---

[8]"Local" means that the theory does not contain any faster than light propagation.

[9]$12GG$ means that the left switch is on "1" and the left light is green, while the right switch is on "2" and the right light is green and so on.

In the framework of a hidden variable theory, each particle would be assumed to be emitted always having a set of definite values of its variables. Thus, from (1) we would draw the conclusion that if one of the particles can make a detector switched to 2 to flash green, then the other particle would necessarily make the other detector switched to 1 to flash red. Then, since $22GG$ comes out $\alpha\%$ of the time, so would $11RR$. However, according to the results this never occurs. Therefore, the above reasoning, based on admitting the existence of definite values for all the particle variables (manifest and hidden) at all times, must be wrong. The alternative is to treat the situation with the rules of Quantum Mechanics. We have assumed that

$$|S_z = \hbar/2\rangle = |1, R\rangle, \quad |S_z = -\hbar/2\rangle = |1, G\rangle$$

$$|S_n = \hbar/2\rangle = |2, R\rangle, \quad |S_n = -\hbar/2\rangle = |2, G\rangle.$$

$$(21.67)$$

These kets are related by

$$|2, G\rangle = \cos(\theta/2)|1, R\rangle + \sin(\theta/2)|1, G\rangle$$

$$|2, R\rangle = -\sin(\theta/2)|1, R\rangle + \cos(\theta/2)|1, G\rangle.$$

$$(21.68)$$

We may choose now to prepare the particles leaving the source to be in a normalized state[10]

$$|\Psi\rangle = \beta|1, R\rangle_\ell|1, G\rangle_r + \beta|1, G\rangle_\ell|1, R\rangle_r + \gamma|1, G\rangle_\ell|1, G\rangle_r. \quad (21.69)$$

Our observations are

$$P_{12GG} = P_{21GG} = 0$$

$$P_{22GG} = \alpha\%$$

$$(21.70)$$

$$P_{11RR} = 0.$$

Let's use now the rules of Quantum Mechanics to compute these probabilities. Setting $\cos(\theta/2) = c$, $\sin(\theta/2) = s$ for shorthand, we have

$$\langle\Psi|1, G\rangle_\ell|2, G\rangle_r = \langle\Psi|1, G\rangle_\ell (c|1, R\rangle_r + s|1, G\rangle_r) =$$

$$= (\beta\langle1, R|_\ell\langle1, G|_r + \beta\langle1, G|_\ell\langle1, R|_r + \gamma\langle1, G|_\ell\langle1, G|_r)|1, G\rangle_\ell (c|1, R\rangle_r + s|1, G\rangle_r)$$

$$= \beta c + \gamma s \implies P_{12GG} = |\beta\cos(\theta/2) + \gamma\sin(\theta/2)|^2 = 0. \quad (21.71)$$

$$\langle\Psi|2, G\rangle_\ell|2, G\rangle_r =$$

---

[10]$2\beta^2 + \gamma^2 = 1.$

$$\langle \Psi | \left(c^2 |1, R\rangle_\ell |1, R\rangle_r + cs |1, R\rangle_\ell |1, G\rangle_r + cs |1, G\rangle_\ell |1, R\rangle_r + s^2 |1, G\rangle_\ell |1, G\rangle_r \right)$$
$$(21.72)$$

$$= 2\beta cs + \gamma s^2 \implies P_{22GG} = |\beta 2\beta \cos(\theta/2) \sin(\theta/2) + \gamma \sin^2(\theta/2)|^2 = \alpha\%.$$
$$(21.73)$$

Thus, we obtain

$$\tan(\theta/2) = -\beta/\gamma$$
$$\gamma \sin^2(\theta/2) + 2\beta \cos(\theta/2) \sin(\theta/2) = \sqrt{\alpha}/10$$
$$(21.74)$$

or

$$\theta = 2\arctan\left(-\beta/\sqrt{1 - 2\beta^2}\right) \quad and \quad \beta^2 \sqrt{1 - 2\beta^2}/(1 - \beta^2) = \sqrt{\alpha}/10. \quad (21.75)$$

For the choice $\beta^2 = 3/8$ (Mermin), we obtain $\theta \approx 101.5^o$ and $\alpha = 9$. Thus, Quantum Mechanics can accommodate the results of the (gedanken) experiment while no local hidden variable theory can. This suffices to rule out the whole class of these theories. Note however that this rules out the classical notion of locality. Quantum Mechanics is still local in the sence that no faster than light signals exist. Actually the modern theoretical physics framework is that of *local* quantum field theory.

Problems and Exercises

21.1 A beam of particles with spin $s = 1$ are subject to measurement of S_x. The resulting beam of spin $+\hbar$ is subject to a second measurement. What are the possible results and the corresponding probabilities? Write down a density matrix to describe the mixed state resulting from the final measurement.

21.2 Consider the system of a spin-1/2 particle. Ignoring spatial degrees of freedom consider the state $|\psi\rangle = a| \uparrow\rangle + b| \downarrow\rangle$. Calculate the density matrix $\hat{\rho}$ corresponding to this state and verify that

$$Tr(\hat{\rho}) = 1, \quad and \quad \hat{\rho}^2 = \hat{\rho}.$$

21.3 Consider the system of two spins in the state $|\Psi\rangle = \frac{1}{\sqrt{2}} (| \uparrow\rangle| \uparrow\rangle + | \downarrow\rangle| \downarrow\rangle)$. Calculate the expectation values $\langle S_{1z} S_{2z}\rangle$, $\langle S_{1z}\rangle$, and $\langle S_{2z}\rangle$ in this state. Do the same for the state $|\Psi'\rangle = \frac{1}{\sqrt{2}} (| \uparrow\rangle| \downarrow\rangle - | \downarrow\rangle| \uparrow\rangle)$. Give an interpretation of the results in terms of entanglement.

21.4 Consider the system of two spins in a general state

$$|\Psi\rangle = C_{++}| \uparrow\rangle| \uparrow\rangle + C_{+-}| \uparrow\rangle| \downarrow\rangle + C_{-+}| \downarrow\rangle| \uparrow\rangle + C_{--}| \downarrow\rangle| \downarrow\rangle.$$

Construct the density matrix corresponding to this state. Verify the standard properties $Tr(\rho) = 1$ and $\rho^2 = \rho$. Construct the density matrix of the *spin-1* subsystem. Verify again the property $Tr(\rho_1) = 1$ and prove that $Tr(\rho_1^2) < 1$.

21.5 Consider a system with a density matrix

$$\rho = p_1|\psi_1\rangle\langle\psi_1| + p_2|\psi_2\rangle\langle\psi_2|$$

where $|\psi_1\rangle$, $|\psi_2\rangle$ is a pair of two *nonorthogonal* states. Find the eigenvalues of the density matrix.

References

1. E. Merzbacher, *Quantum Mechanics*, 3rd edn. (Wiley, Hoboken, 1998)
2. L. Susskind, A. Friedman, *Quantum Mechanics: The Theoretical Minimum* (Basic Books, New York, 2014)
3. A. Einstein, B. Podolsky, N. Rosen, Phys. Rev. **47**(10), 777 (1935)
4. J. Bell, *Speakable and Unspeakable in Quantum Mechanics* (Cambridge University Press, Cambridge, 2004)
5. A. Aspect et al., Phys. Rev. Lett. **47**, 460 (1981)
6. D. Mermin, Physics today (1985)

Chapter 22
Quantization of EM Fields

22.1 Classical Electrodynamics

The physics of classical electric $\mathbf{E}(\mathbf{r}, t)$ and magnetic fields $\mathbf{B}(\mathbf{r}, t)$ is described by *Maxwell's equations*[1]

$$\nabla \cdot \mathbf{E} = \rho$$

$$\nabla \cdot \mathbf{B} = 0$$

$$\nabla \times \mathbf{E} = -\frac{\partial \mathbf{B}}{\partial t} \qquad (22.1)$$

$$\nabla \times \mathbf{E} = \mathbf{J} + \frac{\partial \mathbf{E}}{\partial t}$$

in terms of the sources of *electric charge density* $\rho(\mathbf{r}, t)$ and *electric current density* $\mathbf{J}(\mathbf{r}, t)$. The pair of homogeneous Maxwell's equations $\nabla \cdot \mathbf{B} = 0$ and $\nabla \times \mathbf{E} + \dot{\mathbf{B}} = 0$ can be solved in terms of the *scalar* and *vector potential* $\phi(\mathbf{r}, t)$ and $\mathbf{A}(\mathbf{r}, t)$ defined by

$$\mathbf{E} = -\nabla \phi - \frac{1}{c}\frac{\partial \mathbf{A}}{\partial t}$$

$$\mathbf{B} = \nabla \times \mathbf{A} \qquad (22.2)$$

The remaining two Maxwell's equations, in terms of the potentials, can be set in the form

---

[1] In the present and the next chapter, we adopt the Heaviside–Lorentz system of units for reasons of notational simplicity. The only difference with previous chapters in the replacement of the electric charge in Coulomb's law according to $e^2 \rightarrow e^2/4\pi$. The electromagnetic quantities q, E, B in the Heaviside–Lorentz system are related to the analogous quantities q', E', B' in the MKSA system by the rule

$$q' = q\sqrt{\epsilon_0}, \ E' = E/\sqrt{\epsilon_0}, \ B' = B\sqrt{\mu_0}$$

where ϵ_0 and μ_0 are the *electric* and *magnetic susceptibility of the vacuum* ($\epsilon_0\mu_0 = c^{-2}$).

© Springer Nature Switzerland AG 2019
K. Tamvakis, *Basic Quantum Mechanics*, Undergraduate Texts in Physics,
https://doi.org/10.1007/978-3-030-22777-7_22

$$-\nabla^2\phi + \frac{1}{c^2}\frac{\partial^2\phi}{\partial t^2} - \frac{1}{c}\frac{\partial}{\partial t}\left(\nabla\cdot\mathbf{A} + \frac{1}{c}\frac{\partial\phi}{\partial t}\right) = \rho$$

$$-\nabla^2\mathbf{A} + \frac{1}{c^2}\frac{\partial^2\mathbf{A}}{\partial t^2} + \nabla\left(\nabla\cdot\mathbf{A} + \frac{1}{c}\frac{\partial\phi}{\partial t}\right) = \frac{1}{c}\mathbf{J}$$

(22.3)

It is straightforward to see that acting on the first of these equations with $\frac{1}{c}\frac{\partial}{\partial t}$, taking the divergence of the second and adding them, we obtain the *continuity equation*

$$\nabla\cdot\mathbf{J} + \frac{\partial\rho}{\partial t} = 0,\qquad(22.4)$$

expressing locally the conservation of electric charge.

Gauge invariance. The correspondence between electric and magnetic fields and their potentials is not one to one. Different potentials can correspond to the same fields. In fact, we may transform the potentials according to the rule

$$\phi \rightarrow \phi' = \phi + \frac{\partial\Lambda}{\partial t}$$

$$\mathbf{A} \rightarrow \mathbf{A}' = \mathbf{A} - \nabla\Lambda$$

(22.5)

in terms of the arbitrary function $\Lambda(\mathbf{r}, t)$, while the electric and magnetic field stay the same

$$\mathbf{E}' = -\nabla\phi' - \frac{1}{c}\frac{\partial\mathbf{A}'}{\partial t} = -\nabla\phi - \frac{1}{c}\nabla\dot{\Lambda} - \frac{1}{c}\frac{\partial\mathbf{A}}{\partial t} + \frac{1}{c}\nabla\dot{\Lambda} = \mathbf{E}$$

$$\mathbf{B}' = \nabla\times\mathbf{A}' = \nabla\times\mathbf{A} - \nabla\times(\nabla\Lambda) = \mathbf{B}.$$

These are the *gauge transformations*, encountered before. Although we have discussed gauge transformations already in Chap. 17 the next few lines serve as a reminder of the basic points.

Gauge transformations can be used to our advantage in order to facilitate calculations. For example, starting with a pair of potentials ϕ, $\mathbf{A}$ we may choose a gauge transformation function Λ such that

$$\nabla^2\Lambda = \nabla\cdot\mathbf{A}.$$

Then, the transformed vector potential will satisfy

$$\nabla\cdot\mathbf{A}' = 0.$$

The choice of Λ is called a *gauge choice*. Two very popular gauge choices are the *Coulomb or "transverse" gauge* ($\nabla\cdot\mathbf{A} = 0$) and the *Lorentz gauge* ($\nabla\cdot\mathbf{A} + \frac{1}{c}\frac{\partial\phi}{\partial t} = 0$). Maxwel's equations in the Coulomb gauge take the form

$$-\nabla^2 \phi = \rho$$

$$-\nabla^2 \mathbf{A} + \frac{1}{c^2}\frac{\partial^2 \mathbf{A}}{\partial t^2} + \frac{1}{c}\nabla \dot{\phi} = \frac{\mathbf{J}}{c}$$

$$(22.6)$$

where the scalar potential equation does not involve any time derivatives, being the same as in the electrostatic case. In contrast, Maxwell's equations in the Lorentz gauge retain a more symmetrical form between scalar and vector potential, namely

$$-\nabla^2 \phi + \frac{1}{c^2}\frac{\partial^2 \phi}{\partial t^2} = \rho$$

$$-\nabla^2 \mathbf{A} + \frac{1}{c^2}\frac{\partial^2 \mathbf{A}}{\partial t^2} = \frac{\mathbf{J}}{c}.$$

$$(22.7)$$

The equation for the scalar potential in the Coulomb gauge, not involving any time derivatives, is not a dynamical equation of motion but a constraint equation that can be solved as in the static case. Its solution is

$$Coulomb\ gauge \implies \phi(\mathbf{r}, t) = \frac{1}{4\pi} \int d^3 r' \frac{\rho(\mathbf{r}', t)}{|\mathbf{r} - \mathbf{r}'|}. \qquad (22.8)$$

Electromagnetic sources. At the microscopic level electric charges manifest themselves on point-like particles. A way to represent such a charge distribution in terms of the densities ρ, $\mathbf{J}$ is as

$$\rho(\mathbf{r}, t) = \sum_a q_a \delta(\mathbf{r} - \mathbf{r}^{(a)}(t))$$

$$\mathbf{J}(\mathbf{r}, t) = \sum_a q_a \dot{\mathbf{r}}^{(a)}(t) \delta(\mathbf{r} - \mathbf{r}^{(a)}(t)),$$

$$(22.9)$$

where $\mathbf{r}_a(t)$ is the instantaneous position (trajectory) of the "a"-particle. Note that these densities satisfy the continuity equation

$$\dot{\rho} = \sum_a q_a \dot{x}_i^{(a)} \frac{\partial}{\partial x_i^{(a)}} \delta(\mathbf{r} - \mathbf{r}^{(a)}(t)) = -\sum_a q_a \dot{x}_i^{(a)} \frac{\partial}{\partial x_i} \delta(\mathbf{r} - \mathbf{r}^{(a)}(t))$$

$$= -\nabla_i \left(\sum_a q_a \dot{x}_i^{(a)} \delta(\mathbf{r} - \mathbf{r}^{(a)}(t)) \right) = -\nabla \cdot \mathbf{J}.$$

For a distribution of point charges the Coulomb gauge scalar potential takes the familiar Coulomb law form

$$\phi(\mathbf{r}, t) = \sum_a \frac{q_a}{|\mathbf{r} - \mathbf{r}^{(a)}(t)|}. \qquad (22.10)$$

Pure electromagnetism. In the absence of sources, Maxwell's equations in the Coulomb gauge reduce to

$$-\nabla^2 \phi = 0$$

$$-\nabla^2 \mathbf{A} + \tfrac{1}{c^2}\tfrac{\partial^2 \mathbf{A}}{\partial t^2} + \tfrac{1}{c}\nabla\dot\phi = 0 \tag{22.11}$$

and are satisfied with a vanishing scalar potential $\phi = 0$. Thus, the electromagnetic field is represented only by $\mathbf{A}$ and the equation

$$-\nabla^2 \mathbf{A} + \frac{1}{c^2}\frac{\partial^2 \mathbf{A}}{\partial t^2} = 0. \tag{22.12}$$

Not all three components of $\mathbf{A}$ are dynamical variables, since they have to satisfy the Coulomb gauge condition $\nabla \cdot \mathbf{A} = 0$.

It is now straightforward to find solutions of the equation of motion (22.12). Partial solutions in the form of plane waves

$$\mathbf{A}_k \propto e^{\pm i k \cdot x},$$

where

$$k \cdot x = \omega t - \mathbf{k} \cdot \mathbf{r},$$

exist, provided

$$\omega(k) = c\,k. \tag{22.13}$$

A general superposition of such solutions would be

$$\mathbf{A}(\mathbf{r}, t) = \int \frac{d^3 k}{(2\pi)^{3/2}\sqrt{2\omega}}\, \mathbf{e}(k)\left(a_k\, e^{-ik\cdot x} + a_k^*\, e^{ik\cdot x} \right). \tag{22.14}$$

The factor $(2\omega)^{-1/2}$ is just a convention and could have been absorbed in the functions a_k which are a shorthand for $a(\mathbf{k})$. Nevertheless, the Coulomb transversality condition $\nabla \cdot \mathbf{A} = 0$ implies that the vectors $\mathbf{e}$ are orthogonal to the propagation direction of each mode $\hat{k}$, namely $\mathbf{e} \cdot \mathbf{k} = 0$. Since there are two linearly independent such vectors, the most general solution reads

$$\mathbf{A}(\mathbf{r}, t) = \sum_{\lambda=1,2} \int \frac{d^3 k}{(2\pi)^{3/2}\sqrt{2\omega}}\, \mathbf{e}^{(\lambda)}(k)\left(a_k^{(\lambda)}\, e^{-ik\cdot x} + a_k^{(\lambda)*}\, e^{ik\cdot x} \right). \tag{22.15}$$

The vectors $\mathbf{e}^{(\lambda)}(k)$ are taken to be mutually orthogonal unit vectors and are called *polarization vectors* and here are taken to be real. The triplet $\mathbf{e}^{(1)}(k)$, $\mathbf{e}^{(2)}(k)$, $\hat{k}$ constitute a Cartesian orthonormal basis

$$\mathbf{e}^{(\lambda)}(k) \cdot \mathbf{e}^{(\lambda')}(k) = \delta_{\lambda\lambda'}, \quad \mathbf{k} \cdot \mathbf{e}^{(\lambda)}(k) = 0. \tag{22.16}$$

Note that the wave number and frequency of each mode contributing to the superposition of plane waves (22.15) satisfy the relativistic energy–momentum relation for a particle of energy $\hbar\omega$, momentum $\hbar\mathbf{k}$, and vanishing mass

$$(\hbar\omega(\mathbf{k}))^2 - (\hbar c\mathbf{k})^2 = 0. \qquad (22.17)$$

Of course, there is no particle at the classical level but only electromagnetic waves. It is at the quantum level that relativistic particles of vanishing mass (photons) will emerge out of the Maxwell theory.

The *electromagnetic energy density* is

$$\mathcal{U}_{EM} = \frac{1}{2}\mathbf{E}^2 + \frac{1}{2}\mathbf{B}^2 = \frac{1}{2c^2}\left(\frac{\partial\mathbf{A}}{\partial t}\right)^2 + (\nabla\times\mathbf{A})^2. \qquad (22.18)$$

The *electromagnetic energy* or, equivalently, the Hamiltonian of the electromagnetic field, will be the integral of $\mathcal{U}_{EM}$ over all space

$$H_{EM} = \frac{1}{2}\int d^3r\left(\frac{1}{2c^2}\left(\frac{\partial\mathbf{A}}{\partial t}\right)^2 + (\nabla\times\mathbf{A})^2\right). \qquad (22.19)$$

From the point of view of the canonical formalism the fundamental canonical variable of the system is the vector potential $\mathbf{A}(\mathbf{r}, t)$, while the corresponding *canonical momentum* Π can be read off from Hamilton's equation[2]

$$\dot{\Pi}_j = -\frac{\partial H_{EM}}{\partial A_j} \implies \dot{\Pi}_j = \nabla^2 A_j = \frac{1}{c^2}\ddot{A}_j$$

or

$$\Pi_j = \frac{1}{c^2}\dot{A}_j = -\frac{E_j}{c}. \qquad (22.20)$$

Thus, the canonical momentum is essentially the electric field. This is going to be important to setup the quantization procedure of the system. Not all canonical momenta are independent, since

$$\nabla_j\Pi_j = -\frac{1}{c}(\nabla\cdot\mathbf{E}) = 0.$$

---

[2] The magnetic term in the energy density can be written as $-\frac{1}{2}A_j\nabla^2 A_j$ up to total divergences, which do not contribute to the energy since they reduce to surface integrals at infinity where the fields vanish.

22.2 Quantization of the EM Field

The starting point for the quantization of the system is to introduce the Hilbert space operators that correspond to the canonical variables of the system and write down their equal time commutation relations. In our case, we have

$$A_j(\mathbf{r}, t) \;\to\; \hat{A}_j(\mathbf{r}, t)$$

$$\Pi_j(\mathbf{r}, t) = \tfrac{1}{c^2}\dot{A}_j(\mathbf{r}, t) \;\to\; \hat{\Pi}_j(\mathbf{r}, t) = \tfrac{1}{c^2}\dot{\hat{A}}_j(\mathbf{r}, t). \tag{22.21}$$

Note that the position $\mathbf{r}$ is just a real parameter labeling space points and doesn't correspond to the position of any physical object. By analogy with the simpler quantum systems that we have encountered up to now, we write for their commutation relation

$$\left[\hat{A}_i(\mathbf{r}, t),\, \dot{\hat{A}}_j(\mathbf{r}', t) \right] = i\hbar c^2\, \delta_{ij}\, \delta(\mathbf{r} - \mathbf{r}').$$

Nevertheless, this commutation relation is obviously wrong, since by acting from the left with ∇_i, we obtain, due to the transversality condition $\nabla \cdot \mathbf{A} = 0$, that $0 = i\hbar\nabla_i\delta(\mathbf{r} - \mathbf{r}')$, which is not true. Therefore, the correct commutation relation is [1, 2]

$$\left[\hat{A}_i(\mathbf{r}, t),\, \dot{\hat{A}}_j(\mathbf{r}', t) \right] = i\hbar c^2\, \delta_{ij}^{(\perp)}(\mathbf{r} - \mathbf{r}'), \tag{22.22}$$

where the delta function in the right-hand side has the property

$$\nabla_i \delta_{ij}^{(\perp)}(\mathbf{r} - \mathbf{r}') = 0. \tag{22.23}$$

An obvious integral representation for the so-called *transverse delta function* $\delta^{(\perp)}$ is

$$\delta_{ij}^{(\perp)}(\mathbf{r} - \mathbf{r}') = \int \frac{d^3k}{(2\pi)^3} \left(\delta_{ij} - \frac{k_i k_j}{k^2} \right) e^{i\mathbf{k}\cdot(\mathbf{r}-\mathbf{r}')}. \tag{22.24}$$

The rest of the equal time commutation relations are trivial, namely

$$\left[\hat{A}_i(\mathbf{r}, t),\, \hat{A}_j(\mathbf{r}', t) \right] = \left[\hat{\Pi}_i(\mathbf{r}, t),\, \hat{\Pi}_j(\mathbf{r}', t) \right] = 0. \tag{22.25}$$

The Coulomb gauge equation of motion retains its classical form in terms of the operator $\hat{A}_i(\mathbf{r}, t)$

$$\left(\frac{1}{c^2}\frac{\partial^2}{\partial t^2} - \nabla^2 \right) \hat{A}_i(\mathbf{r}, t) = 0 \tag{22.26}$$

and has the same type of plane-waves expansion

$$\hat{A}_i(\mathbf{r}, t) = c\hbar^{1/2} \sum_{\lambda=1,2} \int \frac{d^3k}{(2\pi)^{3/2}\sqrt{2\omega}} e_i^{(\lambda)}(k) \left(\hat{a}_k^{(\lambda)} e^{-ik \cdot x} + \hat{a}_k^{(\lambda)\,\dagger} e^{ik \cdot x} \right) \quad (22.27)$$

with the crucial difference that the coefficients $\hat{a}_k^{(\lambda)}$ are now operators. We have also fixed the coefficient $c\hbar^{1/2}$ in order to simplify the commutation relations to be derived shortly. Differentiation by time gives us the corresponding plane wave expansion for the canonical momentum $\Pi_i = \frac{1}{c^2}\dot{\hat{A}}_i$

$$\Pi_i = -\frac{i\sqrt{\hbar}}{c} \sum_{\lambda} \int \frac{d^3k}{(2\pi)^{3/2}} \sqrt{\frac{\omega}{2}} e_i^{(\lambda)}(k) \left(e^{-ik \cdot x} \hat{a}_k^{(\lambda)} - e^{ik \cdot x} \hat{a}_k^{(\lambda)\,\dagger} \right). \quad (22.28)$$

In order to find out what kind of commutation relations the operators $a_k^{(\lambda)}$ satisfy, we proceed by substituting these expressions in the commutation relation (22.22). To simplify matters, we assume from the beginning that the commutators $[\hat{a}_k, \hat{a}_{k'}]$ vanish, while the commutators $[\hat{a}_k, \hat{a}_{k'}^{\dagger}]$ satisfy harmonic oscillator-type commutation relations, namely,

$$\left[\hat{a}_k^{(\lambda)}, \hat{a}_{k'}^{(\lambda')\,\dagger} \right] = \hbar \delta_{\lambda\lambda'} \delta(\mathbf{k} - \mathbf{k}'). \quad (22.29)$$

We obtain

$$\left[\hat{A}_i(\mathbf{r}, t), \hat{\Pi}_j(\mathbf{r}', t) \right] = -\frac{i\hbar}{2} \sum_{\lambda} \int \frac{d^3k}{(2\pi)^3} e_i^{(\lambda)}(k) e_j^{(\lambda)}(k) \left(e^{i\mathbf{k} \cdot (\mathbf{r}-\mathbf{r}')} + e^{-i\mathbf{k} \cdot (\mathbf{r}-\mathbf{r}')} \right).$$

Next we see that the symmetric matrix $\sum_{\lambda} e_i^{(\lambda)} e_j^{(\lambda)}$ cannot by anything else but a combination of the only existing symmetric matrices δ_{ij} and $k_i k_j$. Therefore, we write

$$\sum_{\lambda} e_i^{(\lambda)}(k) e_j^{(\lambda)}(k) = C_1 \delta_{ij} + C_2 k_i k_j.$$

Taking the trace, we obtain

$$\sum_{\lambda} \left(\mathbf{e}^{(\lambda)} \right)^2 = 3C_1 + C_2 k^2 \implies 2 = 3C_1 + C_2 k^2.$$

Multiplying by k_i, we obtain

$$\sum_{\lambda} \left(\mathbf{k} \cdot \mathbf{e}^{(\lambda)}(k) \right) e_j^{(\lambda)}(k) = C_1 k_j + C_2 k^2 k_j \implies 0 = C_1 + k^2 C_2.$$

From these we get

$$C_1 = 1, \quad C_2 = -1/k^2.$$

Thus, we have the identity

$$\sum_{\lambda} e_i^{(\lambda)}(k) e_j^{(\lambda)}(k) = \delta_{ij} - \frac{k_i k_j}{k^2}. \tag{22.30}$$

Substituting above, we verify the correct commutators established earlier

$$\left[\hat{A}_i(\mathbf{r}, t), \hat{\Pi}_j(\mathbf{r}', t) \right] = i\hbar \int \frac{d^3 k}{(2\pi)^3} \left(\delta_{ij} - \frac{k_i k_j}{k^2} \right) e^{i\mathbf{k}\cdot(\mathbf{r}-\mathbf{r}')} = i\hbar \delta_{ij}^{(\perp)}(\mathbf{r} - \mathbf{r}'). \tag{22.31}$$

Summarizing, we conclude that the correct commutation relations for the operator coefficients are

$$\left[\hat{a}_k^{(\lambda)}, \hat{a}_{k'}^{(\lambda')\,\dagger} \right] = \delta_{\lambda\lambda'}\, \delta(\mathbf{k} - \mathbf{k}')$$

$$\left[\hat{a}_k^{(\lambda)}, \hat{a}_{k'}^{(\lambda')} \right] = 0 \tag{22.32}$$

$$\left[\hat{a}_k^{(\lambda)\,\dagger}, \hat{a}_{k'}^{(\lambda')\,\dagger} \right] = 0.$$

Photons. We may now proceed and obtain an expression of the Hamiltonian in terms of the mode operators $\hat{a}_k^{(\lambda)}$ and $\hat{a}_k^{(\lambda)\,\dagger}$. After quite a bit of calculation, we may arrive at the expression

$$\hat{H} = \frac{\hbar}{2} \sum_{\lambda=1,2} \int d^3 k \, \omega(k) \left(\hat{a}_k^{(\lambda)\,\dagger} \hat{a}_k^{(\lambda)} + \hat{a}_k^{(\lambda)} \hat{a}_k^{(\lambda)\,\dagger} \right). \tag{22.33}$$

An equivalent expression for the Hamiltonian is

$$\frac{\hbar}{2} \sum_{\lambda=1,2} \int d^3 k \, \omega(k) \left(2\hat{a}_k^{(\lambda)\,\dagger} \hat{a}_k^{(\lambda)} + \delta(\mathbf{k} - \mathbf{k}) \right) = \hbar \sum_{\lambda=1,2} \int d^3 k \, \omega(k) \hat{a}_k^{(\lambda)\,\dagger} \hat{a}_k^{(\lambda)} + \mathcal{E}_0$$

where $\mathcal{E}_0$ is the infinite constant $4\pi c\hbar^2 \delta(0) \int_0^\infty dk\, k^3$. This constant is unobservable[3] and can be ignored. Thus, our final expression for the Hamiltonian is

$$\hat{H} = \sum_{\lambda=1,2} \int d^3 k \, \hbar\omega(k) \, \hat{a}_k^{(\lambda)\,\dagger} \hat{a}_k^{(\lambda)}. \tag{22.34}$$

Taken together with the commutation relations (22.32), this expression for the Hamiltonian seems to tell us that the *quantized electromagnetic field is a sixfold infinity of independent harmonic oscillators*, each characterized by a wave vector $\mathbf{k}$, a polarization $\mathbf{e}^{(\lambda)}(\mathbf{k})$ and a frequency $\omega(k) = c|\mathbf{k}|$.

The Hamiltonian (22.34) is positive definite and its eigenvalues are nonnegative. The lowest energy of the system (*ground state energy*) is zero and the *ground state*, symbolized by $|0\rangle$ is defined as the state annihilated by $\hat{a}_k^{(\lambda)}$, for all $\mathbf{k}$, and λ

---

[3] At least in the framework of flat spacetime.

$$\hat{a}_k^{(\lambda)}|0\rangle = 0.$$ (22.35)

Next, it is straightforward to prove that

$$\left[\hat{H}, \hat{a}_k^{(\lambda)\,\dagger} \right] = \hbar\omega(k)\,\hat{a}_k^{(\lambda)\,\dagger}$$

$$\left[\hat{H}, \hat{a}_k^{(\lambda)} \right] = -\hbar\omega(k)\,\hat{a}_k^{(\lambda)}.$$ (22.36)

Acting on the ground state with the first relation, we obtain

$$\hat{H}\left(\hat{a}_k^{(\lambda)}\right)^\dagger |0\rangle - \left(\hat{a}_k^{(\lambda)}\right)^\dagger \hat{H}|0\rangle = \hbar\omega(k)\left(\hat{a}_k^{(\lambda)}\right)^\dagger |0\rangle$$

or

$$\hat{H}\left(\hat{a}_k^{(\lambda)}\right)^\dagger |0\rangle = \hbar\omega(k)\left(\hat{a}_k^{(\lambda)}\right)^\dagger |0\rangle,$$ (22.37)

meaning that the state

$$|1_{\mathbf{k}}^{(\lambda)}\rangle \equiv \left(\hat{a}_k^{(\lambda)}\right)^\dagger |0\rangle$$ (22.38)

is an energy eigenstate of energy $\hbar\omega(k)$. The obvious interpretation of this state is *"a state of one massless particle of momentum $\hbar\mathbf{k}$ and energy $\hbar\omega(k) = \hbar c|\mathbf{k}|$, with two possible polarizations λ."* These particles satisfy the relativistic energy–momentum relation $E^2 = c^2 p^2 + m^2 c^4$ with vanishing mass

$$(\hbar\omega(k))^2 = c^2\,(\hbar\mathbf{k})^2.$$ (22.39)

It is straightforward to construct multiparticle states as

$$|(N_1)_{\mathbf{k}_1}^{(\lambda_1)}, (N_2)_{\mathbf{k}_2}^{(\lambda_2)}, \ldots\rangle = \frac{1}{\sqrt{N_1!N_2!\ldots}} \left(\hat{a}_{k_1}^{(\lambda_1)\,\dagger}\right)^{N_1} \left(\hat{a}_{k_2}^{(\lambda_2)\,\dagger}\right)^{N_2} \ldots |0\rangle$$ (22.40)

this being a state with N_1 particles of $\mathbf{k}_1$, λ_1, N_2 particles of $\mathbf{k}_2$, λ_2, and so on. It is clear that these particles are bosons, since their states are symmetric by construction, namely,

$$|1_{\mathbf{k}_1}^{(\lambda_1)}, 1_{\mathbf{k}_2}^{(\lambda_2)}\rangle = \left(\hat{a}_{k_2}^{(\lambda_2)}\right)^\dagger \left(\hat{a}_{k_1}^{(\lambda_1)}\right)^\dagger |0\rangle = \left(\hat{a}_{k_1}^{(\lambda_1)}\right)^\dagger \left(\hat{a}_{k_2}^{(\lambda_2)}\right)^\dagger |0\rangle = |1_{\mathbf{k}_2}^{(\lambda_2)}, 1_{\mathbf{k}_1}^{(\lambda_1)}\rangle.$$

These particles, being the quantum excitations of the electromagnetic field are the *photons*, massless vector particles having two polarization states and analogous to their classical counterparts, i.e., the transverse electromagnetic waves [1, 2].

Angular momentum and circularly polarized photons. It is known from Classical Electrodynamics that the classical electromagnetic field carries momentum. The quantity $\varpi = \mathbf{S}/c = (\mathbf{E} \times \mathbf{B})/c$ is interpreted as the *momentum density* of the electromagnetic field. Similarly, the quantity

$$\mathbf{J} = \frac{1}{c} \int d^3 r \, \mathbf{r} \times (\mathbf{E}(\mathbf{r}, t) \times \mathbf{B}(\mathbf{r}, t)) \tag{22.41}$$

corresponds to the *total angular momentum of the electromagnetic field* [2]. This expression can be directly carried over to the quantum case, as long as $\mathbf{E}$ and $\mathbf{B}$ refer to the corresponding operators. There is no need to specify the ordering of operators due to the form of the commutation relation. This formula can be modified as follows:

$$(\mathbf{r} \times (\mathbf{E} \times \mathbf{B}))_i = \epsilon_{ijk}\epsilon_{k\ell m}\epsilon_{mpq} x_j E_\ell \nabla_p A_q = \epsilon_{ijk} \left(\delta_{kp}\delta_{\ell q} - \delta_{kq}\delta_{p\ell} \right) x_j E_\ell \nabla_p A_q$$

$$= \epsilon_{ijk} x_j E_\ell \nabla_k A_\ell - \epsilon_{ijk} x_j E_\ell \nabla_\ell A_k = \epsilon_{ijk} x_j E_\ell \nabla_k A_\ell - \epsilon_{ijk} \nabla_\ell \left(x_j E_\ell A_k \right) + \epsilon_{ijk}\delta_{j\ell} E_\ell A_k$$

$$+ \epsilon_{ijk} x_j \left(\nabla \cdot \mathbf{E} \right) A_k = E_\ell \left(\epsilon_{ijk} x_j \nabla_k \right) A_\ell + \epsilon_{ijk} E_j A_k \,,$$

where we have dropped a total divergence that will contribute to the integral over all space only through a vanishing surface term at infinity. Finally, we have

$$\hat{J}_i = \frac{1}{c} \int d^3 r \, \mathbf{E} \left(\mathbf{r} \times \nabla \right)_i \mathbf{A} + \frac{1}{c} \int d^3 r \, \left(\mathbf{E} \times \mathbf{A} \right)_i \,. \tag{22.42}$$

The interpretation of the two terms is transparent, the first referring to the orbital angular momentum, while the second referring to the photon spin. We may proceed substituting into this expression the quantized electromagnetic field expansion in terms of the operators $\hat{a}_k^{(\lambda)}$. However, the orbital term will not give any contribution since

$$\int d^3 r \, \mathbf{r} \times \nabla e^{\pm i k \cdot \mathbf{r}} e^{\pm i k' \cdot \mathbf{r}} \propto \int d^3 r \, \mathbf{r} \times \left(\pm \mathbf{k} \pm \mathbf{k}' \right) e^{\pm i k \cdot \mathbf{r}} e^{\pm i k' \cdot \mathbf{r}}$$

$$\propto \left(\pm \mathbf{k} \pm \mathbf{k}' \right) \times \nabla_k \delta(\pm \mathbf{k} \pm \mathbf{k}') = 0 \,.$$

The intrinsic angular momentum term gives

$$\frac{i\hbar}{2} \sum_{\lambda, \lambda'} \int d^3 k \int d^3 k' \sqrt{\frac{\omega_k}{\omega_{k'}}} \left(\mathbf{e}_k^{(\lambda)} \times \mathbf{e}_{k'}^{(\lambda')} \right) \left(\hat{a}_k^{(\lambda)} \hat{a}_{k'}^{(\lambda')} \int \frac{d^3 r}{(2\pi)^3} e^{i(\mathbf{k}+\mathbf{k}') \cdot \mathbf{r}} e^{-i(\omega_k+\omega_{k'})t} \right.$$

$$+ \hat{a}_k^{(\lambda)} \hat{a}_{k'}^{(\lambda') \, \dagger} \int \frac{d^3 r}{(2\pi)^3} e^{i(\mathbf{k}-\mathbf{k}') \cdot \mathbf{r}} e^{-i(\omega_k-\omega_{k'})t} - \hat{a}_k^{(\lambda) \, \dagger} \hat{a}_{k'}^{(\lambda')} \int \frac{d^3 r}{(2\pi)^3} e^{-i(\mathbf{k}-\mathbf{k}') \cdot \mathbf{r}} e^{i(\omega_k-\omega_{k'})t}$$

$$\left. - \hat{a}_k^{(\lambda) \, \dagger} \hat{a}_{k'}^{(\lambda') \, \dagger} \int \frac{d^3 r}{(2\pi)^3} e^{-i(\mathbf{k}+\mathbf{k}') \cdot \mathbf{r}} e^{i(\omega_k+\omega_{k'})t} \right)$$

$$= \frac{i\hbar}{2} \sum_{\lambda\lambda'} \int d^3k \left(\left(\mathbf{e}_k^{(\lambda)} \times \mathbf{e}_{-k}^{(\lambda')} \right) \hat{a}_k^{(\lambda)} \hat{a}_{-k}^{(\lambda')} e^{-2i\omega_k t} + \left(\mathbf{e}_k^{(\lambda)} \times \mathbf{e}_k^{(\lambda')} \right) \hat{a}_k^{(\lambda)} \hat{a}_k^{(\lambda')\,\dagger} \right.$$

$$\left. - \left(\mathbf{e}_k^{(\lambda)} \times \mathbf{e}_k^{(\lambda')} \right) \hat{a}_k^{(\lambda)\,\dagger} \hat{a}_k^{(\lambda')} - \left(\mathbf{e}_k^{(\lambda)} \times \mathbf{e}_{-k}^{(\lambda')} \right) \hat{a}_k^{(\lambda)\,\dagger} \hat{a}_{-k}^{(\lambda')\,\dagger} e^{2i\omega_k t} \right).$$

Each of the time-dependent terms vanishes according to

$$\sum_{\lambda,\lambda'} \int d^3k \left(\mathbf{e}_k^{(\lambda)} \times \mathbf{e}_{-k}^{(\lambda')} \right) \hat{a}_k^{(\lambda)} \hat{a}_{-k}^{(\lambda')} e^{-2i\omega_k t} = \sum_{\lambda,\lambda'} \int d^3k \left(\mathbf{e}_{-k}^{(\lambda)} \times \mathbf{e}_k^{(\lambda')} \right) \hat{a}_{-k}^{(\lambda)} \hat{a}_k^{(\lambda')} e^{-2i\omega_k t}$$

$$= - \sum_{\lambda\lambda'} \int d^3k \left(\mathbf{e}_k^{(\lambda')} \times \mathbf{e}_{-k}^{(\lambda)} \right) \hat{a}_k^{(\lambda')} \hat{a}_{-k}^{(\lambda)} e^{-2i\omega_k t} = - \sum_{\lambda'\lambda} \int d^3k \left(\mathbf{e}_k^{(\lambda)} \times \mathbf{e}_{-k}^{(\lambda')} \right) \hat{a}_k^{(\lambda)} \hat{a}_{-k}^{(\lambda')} e^{-2i\omega_k t}.$$

Therefore, we finally obtain

$$\mathbf{J} = \frac{i\hbar}{2} \sum_{\lambda,\lambda'} \int d^3k \left(\left(\mathbf{e}_k^{(\lambda)} \times \mathbf{e}_k^{(\lambda')} \right) \hat{a}_k^{(\lambda)} \hat{a}_k^{(\lambda')\,\dagger} - \left(\mathbf{e}_k^{(\lambda)} \times \mathbf{e}_k^{(\lambda')} \right) \hat{a}_k^{(\lambda)\,\dagger} \hat{a}_k^{(\lambda')} \right).$$

$$(22.43)$$

Since

$$\left(\mathbf{e}_k^{(\lambda)} \times \mathbf{e}_k^{(\lambda')} \right) \hat{a}_k^{(\lambda)} \hat{a}_k^{(\lambda')\,\dagger} = \left(\mathbf{e}_k^{(\lambda)} \times \mathbf{e}_k^{(\lambda')} \right) \left(\hat{a}_k^{(\lambda)\,\dagger} \hat{a}_k^{(\lambda')} + \delta_{\lambda\lambda'} \delta(0) \right)$$

$$= \left(\mathbf{e}_k^{(\lambda)} \times \mathbf{e}_k^{(\lambda')} \right) \hat{a}_k^{(\lambda)\,\dagger} \hat{a}_k^{(\lambda')},$$

we may simplify $\mathbf{J}$ to

$$\mathbf{J} = \frac{i\hbar}{2} \sum_{\lambda,\lambda'} \int d^3k \left(\mathbf{e}_k^{(\lambda)} \times \mathbf{e}_k^{(\lambda')} \right) \left(\hat{a}_k^{(\lambda')\,\dagger} \hat{a}_k^{(\lambda)} - \hat{a}_k^{(\lambda)\,\dagger} \hat{a}_k^{(\lambda')} \right) \qquad (22.44)$$

or

$$\mathbf{J} = i\hbar \int d^3k \left(\mathbf{e}_k^{(1)} \times \mathbf{e}_k^{(2)} \right) \left(\hat{a}_k^{(2)\,\dagger} \hat{a}_k^{(1)} - \hat{a}_k^{(1)\,\dagger} \hat{a}_k^{(2)} \right) \qquad (22.45)$$

or

$$\mathbf{J} = i\hbar \int d^3k \, \hat{k} \left(\hat{a}_k^{(2)\,\dagger} \hat{a}_k^{(1)} - \hat{a}_k^{(1)\,\dagger} \hat{a}_k^{(2)} \right). \qquad (22.46)$$

The one-photon states $|1_k^{(\lambda)}\rangle$, defined as eigenstates of momentum $\hbar\mathbf{k}$ are not angular momentum eigenstates. Acting with $\mathbf{J}$ on them changes their polarization

$$\left(\hat{q} \cdot \hat{J} \right) |1_q^{(1)}\rangle = i\hbar \int d^3k \, \hat{q} \cdot \hat{k} \left(\hat{a}_k^{(2)\,\dagger} \hat{a}_k^{(1)} - \hat{a}_k^{(1)\,\dagger} \hat{a}_k^{(2)} \right) \hat{a}_q^{(1)\,\dagger} |0\rangle$$

$$= i\hbar \int d^3k\, \hat{a}_k^{(2)\,\dagger} \delta(\mathbf{k} - \mathbf{q})\, |0\rangle = i\hbar \hat{a}_q^{(2)\,\dagger} |0\rangle = i\hbar |1_\mathbf{q}^{(2)}\rangle$$

and similarly

$$\left(\hat{q} \cdot \hat{J}\right) |1_\mathbf{q}^{(2)}\rangle = -i\hbar |1_\mathbf{q}^{(1)}\rangle .$$

At this point we may introduce the polarization vectors

$$\mathbf{e}_k^{(\pm)} = \frac{1}{\sqrt{2}} \left(\mathbf{e}_k^{(1)} \pm i\mathbf{e}_k^{(2)}\right) . \tag{22.47}$$

The combination appearing in the plane wave expansion becomes

$$\sum_\lambda \mathbf{e}_k^{(\lambda)} \hat{a}_k^{(\lambda)} = \frac{1}{\sqrt{2}} \left(\left(\mathbf{e}_k^{(+)} + \mathbf{e}_k^{(-)}\right) \hat{a}_k^{(1)} - i\left(\mathbf{e}_k^{(+)} - \mathbf{e}_k^{(-)}\right) \hat{a}_k^{(2)} \right)$$

$$= \mathbf{e}_k^{(+)} \frac{1}{\sqrt{2}} \left(\hat{a}_k^{(1)} - i\hat{a}_k^{(2)}\right) + \mathbf{e}^{(-)} \frac{1}{\sqrt{2}} \left(\hat{a}_k^{(1)} + i\hat{a}_k^{(2)}\right) .$$

Therefore, introducing

$$\hat{a}_k^{(\pm)} = \frac{1}{\sqrt{2}} \left(\hat{a}_k^{(1)} \mp i\hat{a}_k^{(2)}\right) \tag{22.48}$$

we may write the expansion of **A** as

$$\mathbf{A}(\mathbf{r}, t) = \hbar^{1/2} c \sum_{a=\pm} \int \frac{d^3k}{(2\pi)^{3/2}\sqrt{2\omega_k}} \mathbf{e}_k^{(a)} \left(\hat{a}_k^{(a)} e^{-ik\cdot x} + \hat{a}_k^{(a)\,\dagger} e^{ik\cdot x}\right) . \tag{22.49}$$

The states

$$|1_\mathbf{k}^{(\pm)}\rangle = \hat{a}_k^{(\pm)\,\dagger} |0\rangle \tag{22.50}$$

are angular momentum eigenstates. Indeed, we have

$$\left(\hat{k} \cdot \hat{J}\right) |1_\mathbf{k}^{(\pm)}\rangle = \pm\hbar |1_\mathbf{k}^{(\pm)}\rangle . \tag{22.51}$$

This is enough to convince us that the spin of the photon is $j = 1$. The fact that no eigenstate of spin zero has arisen is entirely attributable to the transverse gauge condition imposed. In fact, the standard angular momentum commutation relation $[\hat{J}_i, \hat{J}_j] = i\hbar\epsilon_{ijk}\hat{J}_k$ cannot be recovered unless the Coulomb gauge transversality condition is put aside and a conventional commutation relation $\left[\hat{A}_i(\mathbf{r}, t), \hat{E}_j(\mathbf{r}', t)\right] = -i\hbar\delta_{ij}\delta(\mathbf{r} - \mathbf{r}')$ is adopted. The angular momentum algebra is a direct reflection of the rotational symmetry, which has been broken by the gauge condition.

Problems and Exercises

22.1 Consider the quantized electromagnetic field in the absence of sources. Calculate the commutators

$$\left[\hat{E}_i(\mathbf{r}, t), \hat{E}_j(\mathbf{r}', t)\right], \quad \left[\hat{B}_i(\mathbf{r}, t), \hat{B}_j(\mathbf{r}', t)\right], \quad and \quad \left[\hat{E}_i(\mathbf{r}, t), \hat{B}_j(\mathbf{r}', t)\right].$$

22.2 Consider regular functions[4] $f(r)$, $g(r)$, ... that quickly approach zero outside a region of volume (ΔV) and in terms of them define the Coulomb gauge operators

$$\mathbf{A}[f] = \frac{1}{(\Delta V)} \int d^3 r \, f(r) \, \mathbf{A}(\mathbf{r}, t), \quad \mathbf{E}[g] = \frac{1}{(\Delta V)} \int d^3 r \, g(r) \, \mathbf{E}(\mathbf{r}, t).$$

(a) Calculate the commutator $\left[\hat{A}_i[f], \hat{E}_j[g]\right]$.
(b) Evaluate the uncertainties

$$(\Delta A)^2 = \langle 0|\mathbf{A}^2[f]|0\rangle - (\langle 0|\mathbf{A}[f]|0\rangle)^2, \quad (\Delta E)^2 = \langle 0|\mathbf{E}^2[g]|0\rangle - (\langle 0|\mathbf{E}[g]|0\rangle)^2$$

in the lowest energy state $|0\rangle$.
(c) Verify the Heisenberg inequality.

22.3 Starting from the quantized Coulomb gauge EM field

$$\hat{A}_i(\mathbf{r}, t) = c\hbar^{1/2} \sum_{\lambda=1,2} \int \frac{d^3 k}{(2\pi)^{3/2}\sqrt{2\omega}} \, e_i^{(\lambda)}(k) \left(\hat{a}_k^{(\lambda)} e^{-ik\cdot x} + \hat{a}_k^{(\lambda)\,\dagger} e^{ik\cdot x}\right)$$

derive the expression of the Hamiltonian in terms of creation–annihilation operators

$$\hat{H} = \sum_{\lambda=1,2} \int d^3 k \, \hbar\omega(k) \, \hat{a}_k^{(\lambda)\,\dagger} \hat{a}_k^{(\lambda)}.$$

22.4 Starting[5] from the spin definition $\mathbf{S} = \int d^3 r \, \mathbf{E} \times \mathbf{A}$ and using the commutation relation $\left[A_i(\mathbf{r}, t), E_j(\mathbf{r}', t)\right] = -i\delta_{ij}\delta(\mathbf{r} - \mathbf{r}')$, *without using the Coulomb gauge transversality condition*, derive the spin angular momentum algebra $[S_i, S_j] = i\epsilon_{ijk}S_k$.

22.5 *Coherent states* for the photon field can also be defined as in the case of the simple harmonic oscillator, namely as

---

[4]Without loss of generality, we could take these functions to be Gaussians $f(r) \propto \exp[-r^2/(\Delta V)^{2/3}]$.
[5]$c = \hbar = 1$ units are assumed.

$$|z, n_{\mathbf{k}}^{(\lambda)}\rangle = e^{-|z|^2/2} \sum_{n=0}^{\infty} \frac{z^n}{\sqrt{n!}} |n_{\mathbf{k}}^{(\lambda)}\rangle = e^{-|z|^2/2} \sum_{n=0}^{\infty} \frac{z^n}{n!} \left(\left(a_{\mathbf{k}}^{(\lambda)} \right)^{\dagger} \right)^n |0\rangle \,.$$

(a) Show that the following relation is true:

$$\frac{1}{\pi} \int dz \int dz^* \, |z\rangle\langle z| = 1 \,.$$

(b) Show that any coherent state can be expressed linearly in terms of other coherent states as

$$|z'\rangle = \frac{1}{\pi} \int dz \int dz^* \, e^{-|z-z'|^2} \, e^{(z^*z'-zz'^*)/2} |z\rangle \,.$$

References

1. A. Messiah, *Quantum Mechanics* (Dover publications, Mineola, 1958). Single-volume reprint of the John Wiley & Sons, New York, two-volume 1958 edition
2. J.J. Sakurai, *Advanced Quantum Mechanics*, 1st edn. (1967)

Chapter 23
Matter–Radiation Interaction

23.1 The Interaction Hamiltonian

A nonrelativistic system composed of N electrically charged particles interacting with an electromagnetic field is described by the Hamiltonian

$$\hat{H}_m = \sum_{a=1}^{N} \frac{1}{2m_a} \left(\mathbf{p}_a - \frac{q_a}{c} \mathbf{A}(\mathbf{r}_a, t) \right)^2 + e\phi(\mathbf{r}_a, t) + V(\mathbf{r}_1, \mathbf{r}_2, \ldots), \quad (23.1)$$

where $\mathbf{A}$, ϕ are the electromagnetic potentials and V the potential corresponding to any additional interaction. The parameters m_a and q_a are the masses and electric charges of the N particles. The set of the N charged particles may refer to the protons and electrons composing atoms and molecules, however, due to their smaller mass only electrons have a dominant role in the interactions with the electromagnetic field. Furthermore, since in equilibrium conditions only a few of the bound electrons of each atom or molecule can play an active role, the effective number N of active charged particles is small. The above Hamiltonian is also written as

$$\hat{H}_m = -\sum_{a=1}^{N} \frac{\hbar^2}{2m_a} \nabla_a^2 + \sum_{a=1}^{N} \frac{i\hbar q_a}{2m_a c} \left(\nabla_a \cdot \mathbf{A} + \mathbf{A} \cdot \nabla_a \right) + \sum_{a=1}^{N} \frac{q_a^2}{2m_a c^2} \mathbf{A}^2$$

$$+ \sum_{a=1}^{N} q_a \phi(\mathbf{r}_a, t) + V(\mathbf{r}_1, \mathbf{r}_2, \ldots). \quad (23.2)$$

In the Coulomb gauge, we impose the transversality condition, $\nabla \cdot \mathbf{A} = 0$ and make the replacement $\nabla \cdot \mathbf{A} + \mathbf{A} \cdot \nabla \rightarrow 2\mathbf{A} \cdot \nabla$. We may also solve for the scalar potential from Laplace's equation and obtain

© Springer Nature Switzerland AG 2019
K. Tamvakis, *Basic Quantum Mechanics*, Undergraduate Texts in Physics,
https://doi.org/10.1007/978-3-030-22777-7_23

$$\phi(\mathbf{r}) = \frac{1}{4\pi} \int d^3 r' \frac{\rho(\mathbf{r})}{|\mathbf{r}' - \mathbf{r}|} = \sum_{b=1}^{N} \frac{1}{4\pi} \frac{q_b}{|\mathbf{r}_b - \mathbf{r}_a|}. \tag{23.3}$$

As a result, the Hamiltonian becomes

$$\hat{H}_m = -\sum_{a=1}^{N} \frac{\hbar^2}{2m_a} \nabla_a^2 + \sum_{a=1}^{N} \frac{i\hbar q_a}{m_a c} \mathbf{A}(\mathbf{r}_a, t) \cdot \nabla_a + \sum_{a=1}^{N} \frac{q_a^2}{2m_a c^2} \mathbf{A}^2(\mathbf{r}_a, t)$$

$$+ \frac{1}{4\pi} \sum_{a>b}^{N} \frac{q_a q_b}{|\mathbf{r}_b - \mathbf{r}_a|} + V(\mathbf{r}_1, \mathbf{r}_2, \ldots) \tag{23.4}$$

or, equivalently

$$\hat{H}_m = \sum_{a=1}^{N} \frac{\mathbf{p}_a^2}{2m_a} - \sum_{a=1}^{N} \frac{q_a}{m_a c} \mathbf{A}_a \cdot \mathbf{p}_a + \sum_{a=1}^{N} \frac{q_a^2 \mathbf{A}_a^2}{2m_a c^2} + \frac{1}{4\pi} \sum_{a>b}^{N} \frac{q_a q_b}{|\mathbf{r}_b - \mathbf{r}_a|} + V(\mathbf{r}_1, \mathbf{r}_2, \ldots)$$

$$\tag{23.5}$$

with the shorthand $\mathbf{A}_a = \mathbf{A}(\mathbf{r}_a, t)$. Note however that there is an additional term due to the interaction of the magnetic field and the *spin magnetic dipole moment* of the particles $-\sum_a \frac{g_a q_a}{2m_a c} \mathbf{s}_a \cdot \mathbf{B}$, where g_a is the Lande factor of each particle. Thus, the final expression for the matter-Hamiltonian is

$$\hat{H}_m = \sum_{a=1}^{N} \frac{\mathbf{p}_a^2}{2m_a} - \sum_{a=1}^{N} \frac{q_a}{m_a c} \mathbf{A}_a \cdot \mathbf{p}_a + \sum_{a=1}^{N} \frac{q_a^2 \mathbf{A}_a^2}{2m_a c^2} - \sum_{a=1}^{N} \frac{g_a q_a}{2m_a c} \mathbf{s}_a \cdot \mathbf{B}_a$$

$$+ \frac{1}{4\pi} \sum_{a>b}^{N} \frac{q_a q_b}{|\mathbf{r}_b - \mathbf{r}_a|} + V(\mathbf{r}_1, \mathbf{r}_2, \ldots) \tag{23.6}$$

with $\mathbf{B}_a = \nabla_a \times \mathbf{A}_a$.

The full Hamiltonian of the combined system of particles and electromagnetic field includes also the Hamiltonian of the electromagnetic field itself in the absence of matter

$$\hat{H}_{EM} = \frac{1}{2} \int d^3 r \left(\mathbf{E}^2 + \mathbf{B}^2 \right) = \frac{1}{2} \int d^3 r \left(\frac{1}{c^2} (\dot{\mathbf{A}})^2 + (\nabla \times \mathbf{A})^2 \right). \tag{23.7}$$

The Hamiltonian can be written as a sum

$$\hat{H} = \hat{H}_m + \hat{H}_{EM} = \hat{H}_m^{(0)} + \hat{H}_{EM} + \hat{H}_{int}, \tag{23.8}$$

where

$$\hat{H}_m^{(0)} = \sum_{a=1}^N \frac{\mathbf{p}_a^2}{2m_a} + \frac{1}{4\pi} \sum_{a>b}^N \frac{q_a q_b}{|\mathbf{r}_b - \mathbf{r}_a|} + V(\mathbf{r}_1, \mathbf{r}_2, \ldots)$$

$$\hat{H}_{EM} = \frac{1}{2} \int d^3r \left(\frac{1}{c^2} \left(\dot{\mathbf{A}} \right)^2 + (\nabla \times \mathbf{A})^2 \right) \tag{23.9}$$

$$\hat{H}_{int} = -\sum_{a=1}^N \frac{q_a}{m_a c} \mathbf{A}_a \cdot \mathbf{p}_a + \sum_{a=1}^N \frac{q_a^2 A_a^2}{2m_a c^2} - \sum_{a=1}^N \frac{q_a q_a}{2m_a c} \mathbf{s}_a \cdot \mathbf{B}_a$$

For a system corresponding to an atom or a molecule the eigenstates of $\hat{H}_m^{(0)}$ are the standard atomic or molecular states depending on the potential terms. On the other hand, the eigenstates of $\hat{H}_{EM}$ are the photon states $|n_k^{(\alpha)}, n_q^{(\beta)}, \ldots\rangle$ derived in the previous chapter. The interaction part $\hat{H}_{int}$ can safely be treated as a perturbation since the particle charges correspond to a small parameter ($e^2/4\pi\hbar c = 1/137$). Thus, the problem of charged matter interacting with electromagnetic radiation can be approached via perturbation theory [1, 2].

23.2 Emission and Absorption

Consider the combined system of an atom and electromagnetic radiation. To a very good approximation, we may ignore the nuclei and restrict ourselves in the system of radiation and (a few) electrons. Then, the interaction Hamiltonian is

$$\hat{H}_{int} = -\frac{e}{m_e c} \sum_{a=1}^N \mathbf{A}_a \cdot \mathbf{p}_a + \frac{e^2}{2m_e c^2} \sum_{a=1}^N \mathbf{A}_a^2 - \frac{e}{m_e c} \sum_{a=1}^N \mathbf{s}_a \cdot \mathbf{B}_a . \tag{23.10}$$

The energy eigenstates of the unperturbed system of photons and electrons are[1]

$$|n_k^{(\lambda)}\rangle |\psi\rangle , \tag{23.11}$$

where $|n_k^{(\lambda)}\rangle$ is an n-photon eigenstate of $\hat{H}_{EM}$ and $|\psi\rangle$ is an eigenstate of the unperturbed atom Hamiltonian $\hat{H}_0$ (e.g., Hydrogen atom energy eigenstate). If the interaction $\hat{H}_{int}$ were absent and the system occupied such an eigenstate it would continue undisturbed in this state. Nevertheless, under the influence of the perturbation (23.10) the system will make transitions between these and other eigenstates

$$|n_k^{(\lambda)}\rangle |\psi\rangle \rightarrow |n_{k'}^{(\lambda')}\rangle |\psi'\rangle .$$

Examples of such processes are the *photon emission*

$$|n_k^{(\lambda)}\rangle |\psi_B\rangle \rightarrow |(n+1)_k\rangle |\psi_A\rangle \tag{23.12}$$

---

[1] As usually, we use the shorthand $|n_k^{(\lambda)}\rangle \otimes |\psi\rangle = |n_k^{(\lambda)}\rangle |\psi\rangle$.

or the *photon absorption*

$$|n_{\mathbf{k}}^{(\lambda)}\rangle\,|\psi_A\rangle \;\rightarrow\; |(n-1)_{\mathbf{k}}\rangle|\psi_B\rangle\,. \tag{23.13}$$

Processes like these have been instrumental in the developement of Quantum Theory since the early time of the Bohr model, but their quantitative description requires the quantum treatment of the electromagnetic field. Before we elaborate on each of these elementary processes, let's remind ourselves of the basic points of time-dependent perturbation theory in the interaction picture.

Given a Hamiltonian $\hat{H} = \hat{H}_0 + \hat{H}_{int}$ (in our case $\hat{H}_0 = \hat{H}_m^{(0)} + \hat{H}_{EM}$) in the interaction picture, observables evolve in time with $\hat{H}_0$, while the interaction-picture state vectors obey a Schroedinger equation with a Hamiltonian $\hat{H}_I(t) = \hat{U}_0^\dagger(t, t_0)\hat{H}_{int}(t)\hat{U}_0(t, t_0)$. The equivalent integral equation is

$$|\psi_I(t)\rangle = |\psi_I(t_0)\rangle - \frac{i}{\hbar}\int_{t_0}^t dt'\,\hat{H}_I(t')|\psi_I(t')\rangle\,.$$

The first-order probability of a transition $|0\rangle \rightarrow |n\rangle \neq |0\rangle$ is

$$\mathcal{P}_{0\rightarrow}(t) = \frac{1}{\hbar^2}\left|\int_{t_0}^t dt'\,e^{\frac{i}{\hbar}(E_n-E_0)(t'-t_0)}\langle n|\hat{H}_{int}(t')|0\rangle\right|^2\,.$$

Spontaneous photon emission in the dipole approximation [2]. The term *spontaneous emission* refers to a process in which a photon is emitted while the atom makes a transition from an excited state to a lower energy state, even if there are no photons in the initial state. The transition probability for such a process is

$$\mathcal{P}(t)|_{B\rightarrow A+\gamma} = \frac{1}{\hbar^2}\left|\int_{t_0}^t dt'\,e^{\frac{i}{\hbar}(E_A-E_B)(t'-t_0)}\langle A|\langle 1_{\mathbf{k}}^{(\lambda)}|\hat{H}_{int}(t')|B\rangle|0\rangle\right|^2\,. \tag{23.14}$$

In the above matrix element it is clear that only the linear in $\mathbf{A}$ terms will contribute, since the quadratic term $\mathbf{A}^2$ will give $\hat{a}^2$, $\hat{a}^{\dagger\,2}$ and $\hat{a}\hat{a}^\dagger$ terms, which, together with the $\hat{a}_{\mathbf{k}}$ term from the one photon state have an overall odd annihilation signature and give zero. Assuming that we have only one active electron in the atom, this matrix element will be

$$\langle A|\langle 1_{\mathbf{k}}^{(\lambda)}|\left(-\frac{e}{m_e c}\mathbf{A}\cdot\mathbf{p} - \frac{e}{m_e c}\mathbf{s}\cdot(\nabla\times\mathbf{A})\right)|0\rangle|B\rangle\,.$$

The relevant photonic part is

$$\langle 1_{\mathbf{k}}^{(\lambda)}|\mathbf{A}|0\rangle = \hbar^{1/2}c\sum_{\lambda'}\int\frac{d^3k'}{(2\pi)^{3/2}\sqrt{2\omega'}}\mathbf{e}_{k'}^{(\lambda')}\langle 0|\hat{a}_{\mathbf{k}}^{(\lambda)}\hat{a}_{k'}^{(\lambda')\,\dagger}|0\rangle e^{i\omega't}e^{-i\mathbf{k}'\cdot\mathbf{r}}$$

$$= \hbar^{1/2} c \sum_{\lambda'} \int \frac{d^3 k'}{(2\pi)^{3/2}\sqrt{2\omega'}} \mathbf{e}_{k'}^{(\lambda')} \delta_{\lambda\lambda'} \delta(\mathbf{k} - \mathbf{k}') e^{i\omega' t} e^{-i\mathbf{k}'\cdot\mathbf{r}} = \frac{\hbar^{1/2} c}{(2\pi)^{3/2}\sqrt{2\omega}} \mathbf{e}_k^{(\lambda)} e^{i\omega t} e^{-i\mathbf{k}\cdot\mathbf{r}}.$$

Therefore, the matrix element will be

$$\langle A|\langle 1_{\mathbf{k}}^{(\lambda)}|\hat{H}_{int}(t)|B\rangle|0\rangle = -\frac{e\hbar^{1/2}}{m_e(2\pi)^{3/2}\sqrt{2\omega}} \mathbf{e}_k^{(\lambda)} \cdot \left(\langle A|e^{-i\mathbf{k}\cdot\mathbf{r}}\mathbf{p}|B\rangle - i\langle A|e^{-i\mathbf{k}\cdot\mathbf{r}}\mathbf{s}|B\rangle \times \mathbf{k} \right) e^{i\omega t}.$$
(23.15)

Next, we may assume that the radiation wavelength is considerably larger that the effective size of the atom, i.e.,

$$\lambda >> a_0 \implies k\, a_0 << 1,$$

something that allows for the approximation

$$e^{-i\mathbf{k}\cdot\mathbf{r}} \approx 1.$$
(23.16)

This approximation is called the *dipole approximation* for reasons that will be clear shortly. A process that gives a nonzero transition amplitude within this approximation is said to correspond to a $E1$ transition. Thus, the matrix element takes the form

$$\langle A|\langle 1_{\mathbf{k}}^{(\lambda)}|\hat{H}_{int}(t)|B\rangle|0\rangle = -\frac{e\hbar^{1/2}}{m_e(2\pi)^{3/2}\sqrt{2\omega}} \left(\mathbf{e}_k^{(\lambda)} \cdot \mathbf{p}_{AB} - i\mathbf{e}_k^{(\lambda)} \cdot (\mathbf{s}_{AB} \times \mathbf{k}) \right) e^{i\omega t},$$
(23.17)

where $\mathbf{p}_{AB} = \langle A|\mathbf{p}|B\rangle$ and $\mathbf{s}_{AB} = \langle A|\mathbf{s}|B\rangle$. However, since the atomic states were supposed to be different, their spatial parts are orthogonal and the spin matrix element $\mathbf{s}_{AB}$ will not contribute. Note also that the momentum matrix element can be related to the position matrix element in the following way:

$$\left[\hat{H}_0, x_i \right] = \frac{1}{2m_e} \left[p^2, x_i \right] = -\frac{i\hbar}{m_e} p_i \implies \langle A| \left[\hat{H}_0, \mathbf{r} \right] |B\rangle = -\frac{i\hbar}{m_e}\mathbf{p}_{AB}$$

or

$$\mathbf{p}_{AB} = \frac{im_e}{\hbar}(E_A - E_B)\, \mathbf{r}_{AB}.$$
(23.18)

Then, the matrix element becomes

$$\langle A|\langle 1_{\mathbf{k}}^{(\lambda)}|\hat{H}_{int}(t)|B\rangle|0\rangle = -\frac{i(E_A - E_B)}{(2\pi)^{3/2}\hbar^{1/2}\sqrt{2\omega}} \left(\mathbf{e}_k^{(\lambda)} \cdot \mathbf{d}_{AB} \right) e^{i\omega t},$$
(23.19)

where $\mathbf{d} = e\mathbf{r}$ is the *electric dipole* operator. This justifies the terminology (*electric dipole transition* or E1).

Proceeding to the corresponding probability, we have

$$\mathcal{P}(t)|_{B \to A+\gamma} = \frac{(E_A - E_A)^2}{2\omega(2\pi)^3\hbar^3} \left|\mathbf{e}_k^{(\lambda)} \cdot \mathbf{d}_{AB}\right|^2 \left|\int_{t_0}^{t} dt' \, e^{\frac{i}{\hbar}(E_A - E_B + \hbar\omega)t'}\right|^2$$

$$\approx \frac{(E_A - E_A)^2}{2\omega(2\pi)^3\hbar^3} \left|\mathbf{e}_k^{(\lambda)} \cdot \mathbf{d}_{AB}\right|^2 (2\pi\hbar)\delta(E_B - E_A - \hbar\omega)\, t \qquad (23.20)$$

for $t \gg t_0$. The emission rate per unit solid angle for photons of any energy $\hbar\omega(k)$ and direction Ω will be

$$d\Gamma = d^3k \, (\mathcal{P}(t)/t) = d\Omega \int_0^\infty dk \, k^2 \frac{(E_A - E_A)^2}{2\omega(2\pi\hbar)^2} \left|\mathbf{e}_k^{(\lambda)} \cdot \mathbf{d}_{AB}\right|^2 \delta(E_B - E_A - \hbar\omega)$$

or

$$\left.\frac{d\Gamma}{d\Omega}\right|_{B \to A+\gamma} = \frac{\omega^3}{8\pi^2\hbar c^3} |\mathbf{e}_k^{(\lambda)} \cdot \mathbf{d}_{AB}|^2\,, \qquad (23.21)$$

where $\omega = ck = (E_B - E_A)/\hbar$.

Selection rules. The atomic states $|A\rangle$ are usually eigenstates of angular momentum and parity. However, the fact that the position operator (and, of course, the electric dipole moment operator) is a vector operator and anti-commutes with parity implies a set of specific rules (*Selection Rules*) which have to be obeyed in order to obtain a nonzero transition amplitude to this order of approximation.

Let's write the position operator in terms of *spherical components*

$$\hat{V}_{+1} = -\frac{1}{\sqrt{2}}(x - iy)\,, \quad \hat{V} = \frac{1}{\sqrt{2}}(x + iy) \quad \hat{V}_0 = z\,.$$

The *Wigner–Eckart Theorem* states that

$$\langle j_A, m_A|\hat{V}_q|j_B, m_B\rangle = \langle j_B, 1, m_B, q|j_B, 1, j_A, m_A\rangle \frac{\langle A||\hat{V}_q||B\rangle}{\sqrt{2j_A + 1}}\,. \qquad (23.22)$$

The matrix element in the right-hand side vanishes unless

$$|j_a - j_B| = 1 \quad (E1 \; selection \; rule). \qquad (23.23)$$

Similarly, under parity we have

$$\hat{P}\,\mathbf{r}\,\hat{P} = -\mathbf{r}$$

and, therefore,

$$\langle A|\mathbf{r}|B\rangle = -\varpi_A\varpi_B\langle A|\mathbf{r}|B\rangle\,,$$

where ϖ are the parity eigenvalues of the atomic states, or

$$\varpi_A \varpi_B = -1 \quad (E1 \ selection \ rule) . \tag{23.24}$$

Summation over polarizations. The rate of spontaneous emission (23.21) depends on the polarization of the emitted photon. In the case that this polarization is not measured, we are interested on the rate regardless of the photon polarization and we should add the overall possible polarizations. Therefore, we have

$$\left.\frac{d\Gamma}{d\Omega}\right|_{B \to A+\gamma} = \frac{\omega^3}{8\pi^2\hbar c^3} \left(\sum_{\lambda=1,2} \left(e_k^{(\lambda)}\right)_i \left(e_k^{(\lambda)}\right)_j \right) (d_{AB})_i \, (d_{AB})_j$$

$$= \frac{\omega^3}{8\pi^2\hbar c^3} \left(\delta_{ij} - \frac{k_i k_j}{k^2} \right) (d_{AB})_i \, (d_{AB})_j . \tag{23.25}$$

Taking $\mathbf{d}_{AB}$ to be the $\hat{z}$-axis of the Ω_k cartesian system of coordinates, we obtain

$$\left.\frac{d\Gamma}{d\Omega}\right|_{B \to A+\gamma} = \frac{\omega^3}{8\pi^2\hbar c^3} (\mathbf{d}_{AB})^2 \sin^2\theta . \tag{23.26}$$

This is the rate per solid angle. The total rate is

$$\left.\Gamma^{(tot)}\right|_{B \to A+\gamma} = \frac{\omega^3}{3\pi\hbar c^3} (\mathbf{d}_{AB})^2 = \frac{4\alpha\omega^3}{3c^2} (\mathbf{r}_{AB})^2 . \tag{23.27}$$

Electric quadrupole and magnetic dipole spontaneous emission [2]. In the cases that the given states $|A\rangle$ and $|B\rangle$ are such that no electric dipole radiation is permitted, the transition may still be feasible if we go beyond the dipole approximation. Indeed, keeping the linear term in the expansion of the exponential

$$e^{-i\mathbf{k}\cdot\mathbf{r}} \approx 1 - i\mathbf{k}\cdot\mathbf{r} + \cdots , \tag{23.28}$$

we obtain the following contribution to the matrix element:

$$\langle A|\langle 1_{\mathbf{k}}^{(\lambda)}|\hat{H}_{int}(t)|B\rangle|0\rangle = \frac{ie\hbar^{1/2}}{m_e(2\pi)^{3/2}\sqrt{2\omega}} \langle A|(\mathbf{k}\cdot\mathbf{r})\left(e_k^{(\lambda)}\cdot\mathbf{p}\right)|B\rangle e^{i\omega t} . \tag{23.29}$$

The spin part does not contribute since by assumption the spatial part of $\langle A|\mathbf{r}|B\rangle$ vanishes. Note that the order of operators does not matter, since $\mathbf{k}\cdot\mathbf{e} = 0$, namely,

$$(\mathbf{k}\cdot\mathbf{r})\left(e_k^{(\lambda)}\cdot\mathbf{p}\right) = \left(e_k^{(\lambda)}\cdot\mathbf{p}\right)(\mathbf{k}\cdot\mathbf{r}) + i\hbar\left(\mathbf{k}\cdot e_k^{(\lambda)}\right) = \left(e_k^{(\lambda)}\cdot\mathbf{p}\right)(\mathbf{k}\cdot\mathbf{r}) .$$

Thus, we have

$$\langle A|\langle 1_{\mathbf{k}}^{(\lambda)}|\hat{H}_{int}(t)|B\rangle|0\rangle = \frac{ie\hbar^{1/2}}{m_e(2\pi)^{3/2}\sqrt{2\omega}} e^{i\omega t} k_i \left(e_k^{(\lambda)}\right)_j \langle A|x_i p_j|B\rangle. \quad (23.30)$$

We may write

$$x_i p_j = \frac{1}{2}\left(x_i p_j + p_i x_j\right) + \frac{1}{2}\left(x_i p_j - p_i x_j\right). \quad (23.31)$$

The first term leads to the following contribution to the matrix element

$$\frac{ie\hbar^{1/2}}{2m_e(2\pi)^{3/2}\sqrt{2\omega}} e^{i\omega t} k_i \left(e_k^{(\lambda)}\right)_j \langle A|\left(x_i p_j + p_i x_j\right)|B\rangle \quad (23.32)$$

and transitions due to this terms are called *electric quadrupole transitions* or just *E*2. The second term gives rise to the contribution

$$\frac{ie\hbar^{1/2}}{2m_e(2\pi)^{3/2}\sqrt{2\omega}} e^{i\omega t} k_i \left(e_k^{(\lambda)}\right)_j \langle A|\left(x_i p_j - p_i x_j\right)|B\rangle \quad (23.33)$$

and transitions due to this terms are called *magnetic dipole transitions* or just *M*1. We shall see shortly that the selection rules for the two cases are mutually exclusive and only one of the two contribution arises depending on the particular states $|A\rangle$, $|B\rangle$.

Making use of the identity

$$\left[\hat{H}_0, x_i x_j\right] = \frac{1}{m_e}\left(x_i p_j + p_i x_j\right) \quad (23.34)$$

we obtain

$$\langle A|\left(x_i p_j + p_i x_j\right)|B\rangle = m_e(E_A - E_B)\langle A|x_i x_j|B\rangle \quad (23.35)$$

or

$$m_e(E_A - E_B)k_i \left(e_k^{(\lambda)}\right)_j \langle A|x_i x_j|B\rangle = m_e(E_A - E_B)k_i \left(e_k^{(\lambda)}\right)_j \langle A|\left(x_i x_j - \frac{1}{3}\delta_{ij}r^2\right)|B\rangle, \quad (23.36)$$

where the last term gives just $k_i \left(e_k^{(\lambda)}\right)_j \delta_{ij} = \mathbf{k}\cdot\mathbf{e}_k^{(\lambda)} = 0$. Then, the quadrupole term is written in terms of the *quadrupole moment* operator

$$\hat{T}_{ij} = x_i x_j - \frac{1}{3}\delta_{ij}r^2 \quad (23.37)$$

as

$$\langle A|\langle 1_{\mathbf{k}}^{(\lambda)}|\hat{H}_{int}(t)|B\rangle|0\rangle\Big|_{E2} = \frac{ie\hbar^{1/2}(E_A - E_B)}{2(2\pi)^{3/2}\sqrt{2\omega}} e^{i\omega t} k_i \left(e_k^{(\lambda)}\right)_j \langle A|\hat{T}_{ij}|B\rangle. \quad (23.38)$$

The angular momentum selection rule corresponding to quadrupole transitions can be easily deduced from the Wigner–Eckart theorem to be

$$|j_A - j_B| \leq 2 \leq j_A + j_B \quad (E2 \; selection \; rule) . \tag{23.39}$$

Similarly, for the parity selection rule we have

$$\varpi_A \varpi_B = +1 \quad (E2 \; selection \; rule) . \tag{23.40}$$

Let us now concentrate on the magnetic dipole term. We have the identity

$$(\mathbf{k} \cdot \mathbf{r})(\mathbf{e} \cdot \mathbf{p}) - (\mathbf{k} \cdot \mathbf{p})(\mathbf{e} \cdot \mathbf{r}) = (\mathbf{k} \times \mathbf{e}) \cdot (\mathbf{r} \times \mathbf{p}) = (\mathbf{k} \times \mathbf{e}) \cdot \mathbf{L} .$$

Therefore, we may write the $M1$ contribution to the matrix element as

$$\langle A | \langle 1_{\mathbf{k}}^{(\lambda)} | \hat{H}_{int}(t) | B \rangle | 0 \rangle \Big|_{M1} = \frac{ie\hbar^{1/2}}{2m_e (2\pi)^{3/2} \sqrt{2\omega}} e^{i\omega t} (\mathbf{k} \times \mathbf{e}) \cdot \langle A | \mathbf{L} | B \rangle . \tag{23.41}$$

The corresponding selection rules are

$$|j_B - j_A| \leq 1 \quad and \quad \varpi_A \varpi_B = +1 \quad (M1 \; selection \; rules) . \tag{23.42}$$

Induced absorption and emission [2]. An *absorption* process is one that an atom in a state $|A\rangle$ and radiation in a n-photon state make a transition to a state $|B\rangle$ and radiation in a $(n-1)$-photon state

$$A + n\gamma \rightarrow B + (n-1)\gamma \tag{23.43}$$

The first-order transition probability for such a process is

$$\mathcal{P}(t) = \frac{1}{\hbar^2} \left| \int_{t_0}^{t} dt' e^{\frac{i}{\hbar}(E_B - E_A)(t' - t_0)} \langle B | \langle (n-1)_k^{(\lambda)} | \hat{H}_{int}(t') | A \rangle | n_k^{(\lambda)} \rangle \right|^2 . \tag{23.44}$$

The relevant matrix element is

$$- \frac{e}{m_e c} \langle B | \langle (n-1)_k^{(\lambda)} | \mathbf{A} \cdot \mathbf{p} | A \rangle | n_k^{(\lambda)} \rangle \tag{23.45}$$

with the photonic part

$$\sum_{\lambda'} \int \frac{d^3 k'}{(2\pi)^{3/2} \sqrt{2\omega'}} \mathbf{e}_{k'}^{(\lambda')} \langle (n-1)_k^{(\lambda)} | \hat{a}_{k'}^{(\lambda')} | n_k^{(\lambda)} \rangle e^{-i\omega' t} e^{i \mathbf{k'} \cdot \mathbf{r}} = \frac{\sqrt{n_k^{(\lambda)}}}{(2\pi)^{3/2} \sqrt{2\omega}} \mathbf{e}_k^{(\lambda)} e^{-i\omega t} e^{i \mathbf{k} \cdot \mathbf{r}} .$$

Thus, we have

$$\mathcal{P}(t) = \frac{n_k^{(\lambda)} e^2}{m_e^2 \hbar^2 (2\pi)^3 2\omega} \left| \int_{t_0}^{t} dt' e^{\frac{i}{\hbar}(E_B - E_A)(t'-t_0)} e^{-i\omega t'} \langle B | \mathbf{e}_k^{(\lambda)} \cdot \mathbf{p} e^{i\mathbf{k}\cdot\mathbf{r}} | A \rangle \right|^2 . \quad (23.46)$$

For $t \gg t_0$, we have

$$\Gamma_{abs} = \lim_{t \to \infty} \{\mathcal{P}(t)/t\} \approx \frac{n_k^{(\lambda)} e^2}{8 m_e^2 \pi^2 \hbar\omega} \delta(E_B - E_A - \hbar\omega) \left| \langle B | \mathbf{e}_k^{(\lambda)} \cdot \mathbf{p} e^{-i\mathbf{k}\cdot\mathbf{r}} | A \rangle \right|^2 .$$
$$(23.47)$$

Using the expression obtainted in the case of spontaneous emission in the dipole approximation ($e^{-i\mathbf{k}\cdot\mathbf{r}} \approx 1$)

$$\frac{e^2}{m_e^2} |\langle B | \mathbf{e}_k^{(\lambda)} \cdot \mathbf{p} | A \rangle|^2 = \frac{(E_B - E_A)^2}{\hbar^2} |\mathbf{e}_k^{(\lambda)} \cdot \mathbf{d}_{BA}|^2, \quad (23.48)$$

we get Γ_{abs} in the form

$$\Gamma_{abs} = \frac{n_k^{(\lambda)} (E_B - E_A)^2}{(2\pi\hbar)^2 2\omega} \delta(E_B - E_A - \hbar\omega) |\mathbf{e}_k^{(\lambda)} \cdot \mathbf{d}_{BA}|^2. \quad (23.49)$$

The absorption rate per solid angle will be

$$\frac{d\Gamma_{abs}}{d\Omega} = \frac{n_k^{(\lambda)} \omega^3}{8\pi^2 \hbar c^3} |\mathbf{e}_k^{(\lambda)} \cdot \mathbf{d}_{BA}|^2, \quad (23.50)$$

where $\hbar\omega = E_B - E_A$. The selection rules are the $E1$ selection rules encountered in the case of spontaneous emission.

Entirely analogous is the case of induced emission

$$B + n\gamma \to A + (n+1)\gamma \quad (23.51)$$

After similar steps we end up with the rate

$$\frac{d\Gamma_{em}}{d\Omega} = \frac{(n_k^{(\lambda)} + 1)\omega^3}{8\pi^2 \hbar c^3} |\mathbf{e}_k^{(\lambda)} \cdot \mathbf{d}_{AB}|^2 . \quad (23.52)$$

Application: black body radiation [2]. Consider a collection of atoms with two active atomic energy levels $E_A < E_B$ and the emission and absorption processes

$$B \to A + \gamma \qquad A + \gamma \to B \quad (23.53)$$

the energy of the exchanged photon being $E_B - E_A = \hbar\omega$. We may approximately treat the atoms as distinguishable particles obeying Boltzman statistics. Then, for temperatures $T \gg E_B/k_B$ their energy level occupation numbers will be

$$\mathcal{N}_B/\mathcal{N}_A \approx e^{-(E_B-E_A)/k_BT} = e^{-\frac{\hbar\omega}{k_BT}}. \tag{23.54}$$

The rate of photon absorption by level A will be $\mathcal{N}_A \Gamma_{abs}$, while the rate of photon emission by level B will be $\mathcal{N}_B \Gamma_{em}$. *Thermodynamic Equilibrium* is achieved when these two rates are equal, namely,

$$\mathcal{N}_A \Gamma_{abs} = \mathcal{N}_B \Gamma_{em}. \tag{23.55}$$

Going back to the expressions for the absorption and emission rates, we have

$$\begin{aligned}
\frac{\Gamma_{abs}}{\Gamma_{em}} &= \frac{|\langle n_k^{(\lambda)} - 1|\langle B|\mathbf{e}_k^{(\lambda)} \cdot \mathbf{p}e^{i\mathbf{k}\cdot\mathbf{r}}|A\rangle|n_k^{(\lambda)}\rangle|^2}{|\langle n_k^{(\lambda)} + 1|\langle A|\mathbf{e}_k^{(\lambda)} \cdot \mathbf{p}e^{-i\mathbf{k}\cdot\mathbf{r}}|B\rangle|n_k^{(\lambda)}\rangle|^2} \\[2mm]
&= \frac{n_k^{(\lambda)}}{\left(n_k^{(\lambda)} + 1\right)} \frac{|\langle B|\mathbf{e}_k^{(\lambda)} \cdot \mathbf{p}e^{i\mathbf{k}\cdot\mathbf{r}}|A\rangle|^2}{|\langle A|\mathbf{e}_k^{(\lambda)} \cdot \mathbf{p}e^{-i\mathbf{k}\cdot\mathbf{r}}|B\rangle|^2}.
\end{aligned} \tag{23.56}$$

However, since[2]

$$\langle B|(\mathbf{e}\cdot\mathbf{p})e^{i\mathbf{k}\cdot\mathbf{r}}|A\rangle = \langle A|e^{-i\mathbf{k}\cdot\mathbf{r}}(\mathbf{e}\cdot\mathbf{p})|B\rangle^* = \langle A|(\mathbf{e}\cdot\mathbf{p})e^{-i\mathbf{k}\cdot\mathbf{r}}|B\rangle^*,$$

we have

$$\frac{\Gamma_{abs}}{\Gamma_{em}} = \frac{n_k^{(\lambda)}}{n_k^{(\lambda)} + 1}. \tag{23.57}$$

Inserting this into the equilibrium condition, we obtain

$$\frac{n_k^{(\lambda)}}{n_k^{(\lambda)} + 1} = e^{-\hbar\omega/k_BT}$$

or

$$n_k^{(\lambda)} = \frac{1}{e^{\hbar\omega/k_BT} - 1}. \tag{23.58}$$

This formula relates the number of photons of a given frequency to temperature and can be used to derive *Planck's Law of Black Body Radiation*. The energy density of the gas of photons of frequencies in the interval $(\omega, \omega + d\omega)$ is (integrating over all directions)

$$U(\omega)d\omega = 2 \int \frac{d^3k}{(2\pi)^3} \frac{\hbar\omega}{e^{\hbar\omega/k_BT} - 1}$$

or

---

[2]Note that $(\mathbf{e}\cdot\mathbf{p})e^{\pm i\mathbf{k}\cdot\mathbf{r}} = e^{\pm i\mathbf{k}\cdot\mathbf{r}}(\mathbf{e}\cdot\mathbf{p}) \pm \hbar(\mathbf{k}\cdot\mathbf{e})e^{\pm i\mathbf{k}\cdot\mathbf{r}} = e^{\pm i\mathbf{k}\cdot\mathbf{r}}(\mathbf{e}\cdot\mathbf{p}).$

$$.U(\omega)d\omega = \frac{\hbar^2}{c^3\pi^2}\left(\frac{d\omega\,\omega^3}{e^{\hbar\omega/k_B T}-1}\right). \tag{23.59}$$

The factor of 2 corresponds to the two possible polarization states. This is Planck's Law giving us the frequency distribution of the black-body radiation spectrum. The equilibrium condition for the two level system we analyzed above correspond to those of a cavity that acts as a black body.

Problems and Exercises

23.1 Consider a charged particle bound by an isotropic harmonic oscillator potential in its ground state interacting with quantized electromagnetic radiation. Calculate the induced absorption rate per solid angle for the dipole transition

$$|A\rangle + n\gamma \rightarrow |B\rangle + (n-1)\gamma,$$

where $|A\rangle$ is the ground state and $|B\rangle$ is one of the first excited states.

23.2 Calculate the spontaneous emission rate for the dipole transition of a Hydrogen atom from the $2p$ state to the ground state.

23.3 In order to describe two-photon processes $A + \gamma \rightarrow B + \gamma'$ one has to employ higher orders in perturbation theory involving matrix elements

$$C^{(2)}(t) = \int_{t_0}^{t} dt' \int_{t_0}^{t} dt'' e^{iE_B t'/\hbar} e^{-iE_A t''/\hbar} \langle B|\langle n_{\mathbf{k}'}^{(\lambda')}| T\left(\hat{H}_{int}(t')\hat{H}_{int}(t'')\right)|A\rangle |n_{\mathbf{k}}^{(\lambda)}\rangle.$$

Insert a complete set of atomic states and evaluate $C^{(2)}$ for arbitrary atomic states $|A\rangle$, $|B\rangle$.

References

1. A. Messiah, *Quantum Mechanics* (Dover publications, 1958). Single-volume reprint of the John Wiley & Sons, New York, two-volume 1958 edition
2. J.J. Sakurai, *Advanced Quantum Mechanics*, 1st edn. (1967)

Chapter 24
The Path Integral Formulation of QM

The basic objects of interest in the study of microscopic quantum systems are the probability amplitudes. State vectors and hermitian operators are tools leading to matrix elements and inner products that correspond to measurable probability amplitudes. The central physical issue, of course, is the time evolution of these probability amplitudes. This is formulated in terms of a differential equation, namely Schroedinger's equation. Nevertheless, it has been realized quite a few years ago that an alternative formulation of Quantum Mechanics is possible, not based on operators and differential equations but on the concept of a generalized summation over possible paths of the system in configuration space. In this formulation of Quantum Mechanics, employing the mathematical concept of functional integration or *path integrals*, probability amplitudes are expressible as path integrals [1]. This formulation is in some ways closer to classical thinking, having, of course, made the drastic replacement of the classical trajectory with the concept of summation over all possible paths (path integration). Although the path integral formulation seems more complicated than the conventional operator approach in elementary problems, it turns out that in advanced problems like the quantization of systems of an infinite number of degrees of freedom, this method has been proven quite fruitful and powerful in comparison to the standard methods of canonical quantization.

24.1 Propagators

Consider a system that is in a state $|\psi(t_0)\rangle$ at an initial time t_0. The state of the system at a later time $t > t_0$ will be

$$|\psi(t)\rangle = e^{-\frac{i}{\hbar}\hat{H}(t-t_0)}|\psi(t_0)\rangle, \tag{24.1}$$

© Springer Nature Switzerland AG 2019
K. Tamvakis, *Basic Quantum Mechanics*, Undergraduate Texts in Physics,
https://doi.org/10.1007/978-3-030-22777-7_24

where $\hat{H}$ is the Hamiltonian of the system. In the $\{x\}$ representation, (24.1) is translated as follows:

$$\langle x|\psi(t)\rangle = \langle x|e^{-\frac{i}{\hbar}\hat{H}(t-t_0)}|\psi(t_0)\rangle = \langle x|e^{-\frac{i}{\hbar}\hat{H}(t-t_0)}\left(\int_{-\infty}^{+\infty} dx'\,|x'\rangle\langle x'|\right)|\psi(t_0)\rangle\,,$$

or

$$\psi(x,t) = \int_{-\infty}^{+\infty} dx'\,\mathcal{K}(x,x';t-t_0)\,\psi(x',t_0)\,, \tag{24.2}$$

where

$$\mathcal{K}(x,x';t-t_0) \equiv \langle x|e^{-\frac{i}{\hbar}\hat{H}(t-t_0)}|x'\rangle\,. \tag{24.3}$$

The function $\mathcal{K}(x,x';t-t_0)$ is called *the propagator* [2, 3] of the system and it corresponds to the matrix elements of the evolution operator in the basis of position eigenvectors. Its physical meaning is that of *the probability amplitude to make a transition from a position x' at time t_0 to a position x at a later time t.*

Let us calculate the propagator for a free particle moving in one dimension. We have

$$\mathcal{K}_0(x,x';t-t_0) = \langle x|e^{-i\frac{\hat{p}^2}{2m\hbar}(t-t_0)}|x'\rangle = \langle x|e^{-i\frac{\hat{p}^2}{2m\hbar}(t-t_0)}\left(\int_{-\infty}^{+\infty} dp|p\rangle\langle p|\right)|x'\rangle\,, \tag{24.4}$$

where we have inserted a complete set of energy eigenstates, which for the free particle coincide with the momentum eigenstates. Next, we have

$$\mathcal{K}_0(x,x';t-t_0) = \int_{-\infty}^{+\infty} dp\,e^{-i\frac{p^2}{2m\hbar}(t-t_0)}\langle x|p\rangle\langle p|x'\rangle = \int_{-\infty}^{+\infty} \frac{dp}{2\pi\hbar}\,e^{-i\frac{p^2}{2m\hbar}(t-t_0)}\,e^{\frac{i}{\hbar}p(x-x')}\,. \tag{24.5}$$

Performing a Gaussian integration,[1] we obtain

$$\mathcal{K}_0(x,x';t-t_0) = \sqrt{\frac{m}{2\pi\hbar i(t-t_0)}}\,e^{i\frac{m(x-x')^2}{2\hbar(t-t_0)}}\,. \tag{24.6}$$

This is generalized very easily to three dimensions. The three-dimensional free propagator is

$$\mathcal{K}_0(\mathbf{r},\mathbf{r}';t-t_0) = \left(\frac{m}{2\pi i\hbar(t-t_0)}\right)^{3/2} e^{i\frac{m|\mathbf{r}-\mathbf{r}'|^2}{2\hbar(t-t_0)}}\,. \tag{24.7}$$

Note that the propagator, even in the interacting case, will always be a function of the time difference $t - t'$ and not of the initial and final times separately, due to the fact that *time translation $t \to t + \alpha$* is an exact symmetry of all our equations. Time translation invariance is also referred to as *homogeneity of time.*

---

[1] $\int_{-\infty}^{+\infty} dx\,e^{-ax^2+bx} = \sqrt{\frac{\pi}{a}}e^{\frac{b^2}{4a}}$. See the section on Gaussian integrals in the Appendix.

An examination of the expressions (24.6) and (24.7) of the free propagator reveals an interesting connection with the *Classical Action* of the system. For a classical particle, its *Classical Action* $S[x]$ is defined as the time integral of its *Lagrangian* $\mathcal{L}$, given in most cases as the difference of its kinetic energy minus its potential energy $T - V$. The Action of a particle is a function of the initial and final positions of the particle $x_i = x(t_0)$ and $x_f = x(t)$ and its trajectory between those two points. Thus, it is a function of the trajectory as a whole, having different values for different paths followed. In mathematical terms, it is a *functional* (i.e., a function of a function) of the path followed by the system. For a free particle, the Lagrangian is just the kinetic energy and the action for one-dimensional motion is

$$S = \int_{t_0}^{t} dt' \frac{1}{2} m \dot{x}^2(t') . \tag{24.8}$$

Since there is no force, we have $\dot{x} = const. = (x' - x)/(t - t_0)$ and

$$S = (t - t_0) \frac{1}{2} m \left(\frac{x' - x}{t - t_0} \right)^2 = \frac{m(x' - x)^2}{(t - t_0)} . \tag{24.9}$$

Thus, the free propagator is proportional to the exponential of the Action

$$\mathcal{K}_0(x, x'; t - t_0) = \left(\frac{m}{2\pi i \hbar} \right)^{1/2} e^{\frac{i}{\hbar} S} . \tag{24.10}$$

In Classical Mechanics, the equations of motion of a system, according to *Hamilton's Principle* or, equivalently, the *Least Action Principle*, correspond to an extremum (minimum) of the Action. In other words, the physical trajectory of the system is the one giving minimal Action. Classical Mechanics should be recovered in the limit $\hbar \to 0$. This is the case indeed, since in that limit, the exponential of the propagator $e^{\frac{i}{\hbar} S}$ will give appreciable values only for that path that corresponds to a minimum of S, namely the classical trajectory.

For the general case of an interacting system, we repeat the same steps inserting a complete set of energy eigenstates

$$\mathcal{K}(x, x'; t - t_0) = \langle x | e^{-\frac{i}{\hbar} \hat{H}(t - t_0)} | x' \rangle = \langle x | e^{-\frac{i}{\hbar} \hat{H}(t - t_0)} \left(\sum_E | E \rangle \langle E | \right) | x' \rangle$$

$$= \sum_E \langle x | E \rangle \, e^{-\frac{i}{\hbar} E(t - t_0)} \langle E | x' \rangle \tag{24.11}$$

or

$$\mathcal{K}(x, x'; t - t_0) = \sum_E \psi_E(x) \, e^{-\frac{i}{\hbar} E(t - t_0)} \psi_E^*(x') , \tag{24.12}$$

where $\psi_E(x) = \langle x|E \rangle$ are the energy eigenfunctions. The exact expression of the propagator, of course, depends on the particular system and requires quite a bit of mathematical machinery to evaluate. As an example, we may consider the harmonic oscillator. Substituting in (24.12) the expressions for the wavefunctions in terms of Hermite polynomials (7.31), we obtain

$$\mathcal{K}(x, x'; t) = \sqrt{\frac{m\omega}{\hbar\pi}} e^{-\frac{1}{2}(\xi^2 + \xi'^2)} e^{-i\omega t/2} \sum_{n=0}^{\infty} \frac{e^{in\omega t}}{2^n n!} H_n(\xi) H_n(\xi'), \qquad (24.13)$$

where $\xi = x\sqrt{\frac{m\omega}{\hbar}}$. Now, using the completeness relation of Hermite polynomials (7.35) with $z = e^{i\omega t}/2$, and after some manipulation, we get the propagator in the form

$$\mathcal{K}(x, x'; t) = \left(\frac{m\omega}{2\hbar\pi i \sin(\omega t)} \right)^{1/2} e^{i\frac{m\omega}{2\hbar}\left[(x^2 + x'^2)\cot(\omega t) - \frac{2xx'}{\sin(\omega t)}\right]}. \qquad (24.14)$$

Note that in the free-particle limit $\omega \to 0$, the free propagator is recovered. Note also that, in contrast to the free propagator, (24.14) is not a function of the difference $x - x'$. This is because only the free-particle system is invariant under spatial translations $x \to x + \alpha$ (homogeneity of space). This is not a symmetry anymore in the presence of forces.

24.2 A Property of Propagators

Consider the system of a particle, in general subject to forces, and its propagator for a transition between points x at time t_0 and x' at time t. We can always divide the time interval $t - t_0$ in N equal segments considering the intermediate times

$$t_0 < t_1 < t_2 < \ldots\ldots < t_{N-1} < t_N = t \quad \Longrightarrow \quad t_n = t_0 + \frac{n}{N}(t - t_0) \qquad (24.15)$$

with $n = 0, 1, \ldots, N$. Each equal segment is $(t - t_0)/N$. The propagator is

$$\mathcal{K}(x, x'; t - t_0) = \langle x| e^{-\frac{i}{\hbar}\hat{H}(t_1 - t_0)} e^{-\frac{i}{\hbar}\hat{H}(t_2 - t_1)} \ldots e^{-\frac{i}{\hbar}\hat{H}(t_N - t_{N-1})} |x'\rangle. \qquad (24.16)$$

Inserting complete sets of position eigenstates between the successive evolution operators, we obtain

$$= \langle x| e^{-\frac{i}{\hbar}\hat{H}(t_1 - t_0)} \left(\int dx_1 |x_1\rangle\langle x_1| \right) e^{-\frac{i}{\hbar}\hat{H}(t_2 - t_1)} \left(\int dx_2 |x_2\rangle\langle x_2| \right) \ldots e^{-\frac{i}{\hbar}\hat{H}(t_N - t_{N-1})} |x'\rangle$$

$$= \int dx_1 \int dx_2 \ldots \int dx_{N-1} \mathcal{K}(x, x_1; t_1 - t_0) \ldots \ldots \mathcal{K}(x_{N-1}, x'; t - t_{N-1}).$$

$$(24.17)$$

Thus, the propagator has been written as a convoluted product of N propagators, each corresponding to an intermediate time interval of equal length, with $N - 1$ integrations on intermediate positions between the start and end points x and x'

$$\mathcal{K}(x, x'; T) = \int dx_1 \ldots \int dx_{N-1} \mathcal{K}(x, x_1; T/N) \ldots \mathcal{K}(x_{N-1}, x'; T/N).$$

$$(24.18)$$

This expression means that the amplitude to make a transition between the points x, x' is composed of the amplitudes for all possible paths traversed at intermediate times.

24.3   The Feynman Path Integral

Let us now take the expression (24.18) and consider the limit of $N \to \infty$ or an infinitesimal time interval $T/N = \epsilon$. The intermediate propagators are (to first order in ϵ)

$$\mathcal{K}(x_{j-1}, x_j; \epsilon) = \langle x_{j-1} | e^{-\frac{i}{\hbar} \epsilon (\hat{T} + V(\hat{x}))} | x_j \rangle \approx \langle x_{j-1} | e^{-\frac{i}{\hbar} \epsilon \hat{T}} \left(1 - \frac{i}{\hbar} \epsilon V(\hat{x}) \right) | x_j \rangle$$

$$= \left(1 - \frac{i}{\hbar} \epsilon V(x_j) \right) \langle x_{j-1} | e^{-\frac{i}{\hbar} \epsilon \hat{T}} | x_j \rangle = \mathcal{K}_0(x_{j-1}, x_j; \epsilon) \left(1 - \frac{i}{\hbar} \epsilon V(x_j) \right)$$

$$= \sqrt{\frac{m}{2\pi \hbar i \epsilon}} e^{i \frac{m}{2\hbar \epsilon} (x_{j-1} - x_j)^2} \left(1 - \frac{i}{\hbar} \epsilon V(x_j) \right) \qquad (24.19)$$

or

$$\mathcal{K}(x_{j-1}, x_j; \epsilon) \approx \sqrt{\frac{m}{2\pi \hbar i \epsilon}} e^{\frac{i}{\hbar} \epsilon \left(\frac{1}{2} m \left(\frac{x_{j-1} - x_j}{\epsilon} \right)^2 - V(x_j) \right)}. \qquad (24.20)$$

Note now that, although *a priori* two intermediate points x_{j-1}, x_j could have any values between $(-\infty, +\infty)$, since the integrations are Gaussian

$$\ldots \int_{-\infty}^{+\infty} dx_{j-1} \int_{-\infty}^{+\infty} dx_j \, e^{i \frac{m}{2\hbar \epsilon} (x_{j-1} - x_j)^2}$$

in the $\epsilon \to 0$ limit they will be dominated by neighboring points $x_{j-1} \approx x_j$. This enables us to define an approximately continuous function $x(t')$, such that $t_0 \le t' \le t$ and

$$x(t_j) = x_j, \quad x(t_0) = x, \quad x(t) = x'. \qquad (24.21)$$

Then, going back to the full expression (24.18), we have

$$\mathcal{K}(x, x'; T) = \left(\frac{m}{2\pi\hbar i \epsilon}\right)^{(N-1)/2} \Pi_{j=1}^{N-1} \int dx_j e^{\frac{i}{\hbar}\epsilon \sum_{j=1}^{N}\left(\frac{1}{2}m\left(\frac{x_{j-1}-x_j}{\epsilon}\right)^2 - V(x_j)\right)}. \quad (24.22)$$

In the exponent, we can now make the replacements

$$\frac{x_j - x_{j-1}}{\epsilon} \rightarrow \frac{dx(t')}{dt'} \quad and \quad \epsilon \sum_j \cdots \rightarrow \int_{t_0}^{t} dt' \dots . \quad (24.23)$$

Thus, the exponential becomes

$$e^{\frac{i}{\hbar}\int_{t_0}^{t} dt'\left(\frac{1}{2}m\dot{x}^2(t') - V(x(t'))\right)} = e^{\frac{i}{\hbar}S[x]}. \quad (24.24)$$

The collection of integrations over intermediate paths is defined to be the (functional) *integration measure*

$$[Dx] = \lim_{\substack{\epsilon \to 0 \\ N \to \infty}} \left\{\left(\frac{m}{2\pi\hbar i \epsilon}\right)^{(N-1)/2} \underbrace{\int dx_1 \dots \int dx_{N-1}}\right\}. \quad (24.25)$$

Thus, we have arrived at the expression of the propagator as the *Feynman Path Integral* which expresses the time evolution of the transition probability amplitude as an integral over all possible paths [1]

$$\mathcal{K}(x, x'; t - t_0) = \int_{x(t_0)=x}^{x(t)=x'} [Dx]\, e^{\frac{i}{\hbar}S[x]}, \quad (24.26)$$

the exponent being the Classical Action

$$S[x] = \int_{t_0}^{t} dt' \left(\frac{1}{2}m\dot{x}^2(t') - V(x(t'))\right). \quad (24.27)$$

Note that the integration measure $[Dx]$ signifies integration over a full space of functions $x(t)$, satisfying the given boundary conditions at the end points. A basic property of the functional integration measure is that it is invariant under translation by a given function

$$x(t) \rightarrow x'(t) = x(t) + f(t) \implies [Dx'] = [Dx], \quad (24.28)$$

where $f(t)$ is a *known function* (e.g., $f(t) = \cos(\omega t)$). This is essentially not different than the property of an ordinary differential to be invariant under translation by a constant ($d(x + a) = dx$).

Example 24.1 Rewrite the Feynman path integral for the propagator of a particle moving in one dimension subject to a potential $V(x)$ as a functional integral over *phase space*.

We start with

$$\mathcal{K}(x, x'; T) = \int_{x(0)=x}^{x(T)=x'} [Dx] \, e^{\frac{i}{\hbar} S[x]} \tag{24.29}$$

with

$$S = \int_0^T dt \left(\frac{1}{2} m \dot{x}^2 - V(x) \right). \tag{24.30}$$

This path integral is determined up to a multiplicative normalization constant, which we can take it to be equal to the indefinite path integral

$$\int [Dp] \, e^{-\frac{i}{\hbar} \int_0^T dt \frac{p^2}{2m}}, \tag{24.31}$$

which, thanks to the shift invariance of the integration measure, can be also written as

$$\int [Dp] \exp\left[-\frac{i}{2m\hbar} \int_0^T dt \, (p - m\dot{x})^2 \right]. \tag{24.32}$$

We insert this constant to our starting path integral and we obtain

$$\mathcal{K}(x, x'; T) = \int_{x(0)=x}^{x(T)=x'} [Dx] \int [Dp] \, e^{\frac{i}{\hbar} \int_0^T dt \left(p\dot{x} - \frac{p^2}{2m} - V(x) \right)} \tag{24.33}$$

or

$$\mathcal{K}(x, x'; T) = \int_{x(0)=x}^{x(T)=x'} [Dx] \int [Dp] \, e^{\frac{i}{\hbar} \int_0^T dt \, (p\dot{x} - H(p,x))}. \tag{24.34}$$

This is the propagator written as a path integral over the phase space x, p. Note that the momentum path integral is indefinite.

Example 24.2 Evaluate the path integral for a simple harmonic oscillator

$$\mathcal{K}(x, x'; T) = \int_{x(0)=x}^{x(T)=x'} [Dx] \, e^{\frac{i}{\hbar} \left(\frac{1}{2} m\dot{x}^2 - \frac{1}{2} m\omega^2 x^2 \right)}. \tag{24.35}$$

We introduce the eigenfunctions of the differential operator $-\frac{d^2}{dt^2} - \omega^2$

$$-\left(\frac{d^2}{dt^2} + \omega^2 \right) u_n(t) = \lambda_n u_n(t) \tag{24.36}$$

with homogeneous boundary conditions $u_n(T) = u_n(0) = 0$. These are the orthonormal set

$$u_n(t) = \sqrt{\frac{2}{T}} \sin(\pi n t / T) \quad (n = 1, 2, \ldots). \tag{24.37}$$

The corresponding eigenvalues are

$$\lambda_n = -\omega^2 + \left(\frac{\pi n}{T} \right)^2. \tag{24.38}$$

Next we write the integration function as

$$x(t) = x_0(t) + \sum_{n=1}^{\infty} \tilde{x}_n u_n(t), \tag{24.39}$$

where $x_0(t)$ is a solution of

$$\ddot{x}_0 + \omega^2 x_0 = 0 \tag{24.40}$$

with boundary conditions $x_0(T) = x'$, $x_0(0) = x$. This can be determined exactly to be

$$x_0(t) = x \left[\cos(\omega t) - \sin(\omega t)\cot(\omega T)\right] + x' \frac{\sin(\omega t)}{\sin(\omega T)}. \tag{24.41}$$

Substituting $x(t)$ in the action and using the boundary conditions, we obtain

$$S[x] = S[x_0] + \frac{1}{2} m \sum_{n=1}^{\infty} \lambda_n \tilde{x}_n^2. \tag{24.42}$$

Actually, $S[x_0]$ can be computed to be

$$S[x_0] = \frac{1}{2} m \omega \left[\left(x^2 + x'^2\right) \cot(\omega T) - 2xx'(\sin(\omega T))^{-1}\right]. \tag{24.43}$$

Therefore, the propagator is

$$\mathcal{K}(x, x'; T) = e^{\frac{i}{\hbar} S[x_0]} \int_{\tilde{x}(0)=0}^{\tilde{x}(T)=0} [D\tilde{x}] e^{\frac{im}{2\hbar} \sum_{n=1}^{\infty} \lambda_n \tilde{x}_n^2}. \tag{24.44}$$

The integration measure is

$$[D\tilde{x}] \propto \Pi_j \tilde{x}(t_j) \propto \Pi_j \sum_n \tilde{x}_n \sin(\pi n t_j / T). \tag{24.45}$$

Note, however, that the *"matrix"* $S_{nj} = \sqrt{\frac{2}{T}} \sin(n\pi t_j/T)$, thanks to the orthonormality of these eigenfunctions, is unitary and, therefore, the integration measure is just

$$[D\tilde{x}] \propto \Pi_n \tilde{x}_n = d\tilde{x}_1 \ldots d\tilde{x}_n \ldots. \tag{24.46}$$

Thus, the integration reduces to a product of ordinary gaussian integrals with the simple result

$$\mathcal{K}(x, x'; T) \propto e^{\frac{i}{\hbar} S[x_0]} \Pi_n \left(\frac{2\pi i \hbar}{m \lambda_n}\right)^{1/2} = e^{\frac{i}{\hbar} S[x_0]} \Pi_n \left(\frac{2\pi i \hbar}{m \left(-\omega^2 + \left(\frac{n\pi}{T}\right)^2\right)}\right)^{1/2}. \tag{24.47}$$

The multiplicative constant can be determined by the demand that the propagator reduces to the known free one for $\omega = 0$. Then, we obtain

$$\mathcal{K}(x, x'; T) \propto e^{\frac{i}{\hbar} S[x_0]} \Pi_n \left(\frac{1}{\left(1 - \left(\frac{\omega T}{n\pi}\right)^2\right)}\right)^{1/2}. \tag{24.48}$$

The appearing infinite product is known to be $\sqrt{\omega T / \sin(\omega T)}$. Finally, we obtain

$$\mathcal{K}(x, x'; T) = \sqrt{\frac{m\omega}{2\pi i \hbar \sin(\omega T)}} e^{\frac{im\omega}{2\hbar} \left[\left(x^2 + x'^2\right)\cot(\omega T) - 2xx'(\sin(\omega T))^{-1}\right]}. \tag{24.49}$$

Problems and Exercises

24.1 Prove that the expression for the free propagator in the limit $t_f \to t_i$ reduces to a delta function $\delta(\mathbf{r}' - \mathbf{r})$.

24.2 Evaluate the Feynman path integral for the propagator $\mathcal{K}(x', x; t_f, t_i)$ $(x' = x(t_f),\ x = x(t_i))$ of the *forced harmonic oscillator*

$$L = \frac{1}{2}m\dot{x}^2 - \frac{1}{2}m\omega^2 x^2 + F(t)x \,.$$

24.3 Evaluate the Feynman path integral for the propagator $\mathcal{K}(x', x; t_f, t_i)$ $(x' = x(+\infty),\ x = x(-\infty))$ of the *forced harmonic oscillator*

$$L = \frac{1}{2}m\dot{x}^2 - \frac{1}{2}m\omega^2 x^2 + F(t)x \,.$$

References

1. R.P. Feynman, A.R. Hibbs, in *Quantum Mechanics and Path Integrals, Emended*, ed. by E. Styer (Dover, United States, 2010)
2. G. Baym, *Lectures in Quantum Mechanics*, Lecture Notes and Supplements in Physics (ABP, 1969)
3. E. Merzbacher, *Quantum Mechanics*, 3rd edn. (Wiley, Hoboken, 1998)

Appendix
Mathematical Supplement

A.1 Probabilities

A.1.1 Basic Definitions

Consider a physical system on which we perform N measurements[1] of a given quantity A (e.g., the temperature of a liquid in an insulated container), under exactly the same conditions.[2] Let the set of distinct values measured be

$$\alpha_1, \ \alpha_2, \ \alpha_3, \ \ldots \tag{A.1}$$

and let $N_j = N(\alpha_j)$ be the number of times a particular value α_j turns up. The relative *frequency* of the value α_j will be

$$f_j \equiv \frac{N_j}{N}. \tag{A.2}$$

Obviously, we must have

$$f_j \leq 1 \tag{A.3}$$

and

$$\sum_j f_j = 1. \tag{A.4}$$

The *Probability* of a particular value is defined as the frequency of this value *in the limit of a large number of measurments $N \to \infty$*

---

[1] The term *measurement* here is used loosely in the classical sense and is not related to the nontrivial issue of measurement on a quantum system.

[2] Equivalently, we may consider a set of N identical copies of the system (a so-called *statistical ensemble*) and perform a measurement of A in each of them.

© Springer Nature Switzerland AG 2019
K. Tamvakis, *Basic Quantum Mechanics*, Undergraduate Texts in Physics,
https://doi.org/10.1007/978-3-030-22777-7

$$P_j = \frac{N_j}{N}. \tag{A.5}$$

The two defining properties of probability are

$$0 \le P_j \le 1,$$
$$\sum_j P_j = 1. \tag{A.6}$$

Implicit in the above definition of probability is the fact that the measurements α_j and α_k are *mutually exclusive* and that the probability of obtaining a value *either* α_j *or* α_k is

$$P_{j\cup k} = P_j + P_k. \tag{A.7}$$

Note that the above definition of probability refers to single independent measurements of the given quantity. It is possible to consider *joint probabilities* corresponding to two simultaneous measurments (or, alternatively, single measurements in two copies of the statistical ensemble) and ask the question *"what is the probability of a value α_j and a value α_k*. In the case that the measurements are *"statistically independent"* the probability for the pair (α_j, α_k) is the product

$$P_{j\cap k} = P_j P_k. \tag{A.8}$$

According to the *Law of Large numbers* of Probability Theory the *most probable value* (e.g., the value for which the probability has a maximum), in the limit $N \to \infty$, approaches the so called *expectation value* (also *mean or average value*), defined as

$$\langle A \rangle = \sum_j \alpha_j P_j. \tag{A.9}$$

Similarly, the expectation value of any function $F(A)$ is defined as

$$\langle F(A) \rangle = \sum_j F(\alpha_j) P_j. \tag{A.10}$$

Nevertheless, it is possible that two different probability distributions to yield the same expectation value while they differ in their spreading of individual values around it. A good measure of their difference is the so-called *Standard Deviation* (ΔA), defined as

$$(\Delta A)^2 = \langle (A - \langle A \rangle)^2 \rangle. \tag{A.11}$$

An equivalent expression for (A.11) is

$$(\Delta A)^2 = \langle A^2 \rangle - \langle A \rangle^2. \tag{A.12}$$

Notice that (A.11) ensures that always $\langle A^2 \rangle \ge \langle A \rangle^2$.

A.1.2 Continuous Distributions

It is often the case that the measured values of a given quantity cover a continuous range. In that case, it is meaningful to talk about the probability of measured values in a small, i.e., infinitesimal region $[\alpha, \alpha + d\alpha]$. This probability must also be small, i.e., infinitesimal as well, and, therefore, proportional to $d\alpha$

$$dP(\alpha, \alpha + d\alpha) \propto d\alpha. \tag{A.13}$$

Then, it is useful to intoduce the concept of *probability density*

$$\mathcal{P}(\alpha) \equiv \frac{dP(\alpha)}{d\alpha}. \tag{A.14}$$

The defining property of the probability density is

$$\int_{\alpha_1}^{\alpha_2} d\alpha \, \mathcal{P}(\alpha) = 1, \tag{A.15}$$

where $[\alpha_1, \alpha_2]$ is the full range of values taken by α. Such a continuous probability density function is referred to as a *probability distribution*.

The definitions of the expectation value and the standard deviation follow in a straightforward fashion from (A.9) and (A.11) as

$$\langle A \rangle = \int_{\alpha_1}^{\alpha_2} d\alpha \, \alpha \, \mathcal{P}(\alpha)$$

$$(\Delta A)^2 = \int_{\alpha_1}^{\alpha_2} d\alpha \, \alpha^2 \, \mathcal{P}(\alpha) - \left(\int_{\alpha_1}^{\alpha_2} d\alpha \, \alpha \, \mathcal{P}(\alpha) \right)^2. \tag{A.16}$$

As an example, we may consider a Gaussian probability distribution for a position variable x taking values in the entire real line. It is given by

$$\mathcal{P}(x) = \left(\frac{\alpha}{\pi} \right)^{1/2} e^{-\alpha x^2}, \tag{A.17}$$

where $\alpha > 0$ is a parameter. The coefficient $\sqrt{\alpha/\pi}$ in front of the exponential is just what is necessary to meet the requirement that $\int dx \, \mathcal{P} = 1$. The expectation value vanishes independently of the parameter α as a result of the $x \to -x$ reflection symmetry of the distribution

$$\langle x \rangle = \int_{-\infty}^{+\infty} dx \, x \, \mathcal{P}(x) = 0. \tag{A.18}$$

The standard deviation is

$$(\Delta x)^2 = \langle x^2 \rangle = \int_{-\infty}^{+\infty} dx\, x^2 \mathcal{P}(x) = \frac{1}{2\alpha} \tag{A.19}$$

and depends strongly on the parameter α. Thus, for $\alpha \ll 1$ we have a very large spread in the values of x, while, for $\alpha \gg 1$, we have a more or less localized distribution of values around $x = 0$.

A.1.3 Probability Amplitudes

Probabilities or the probability densities are always real positive functions. It is possible, however, to introduce a *probability amplitude* or a *probability density amplitude* defined so that its absolute square gives the corresponding probability or probability density

$$P_j = |\Psi_j|^2 \quad or \quad \mathcal{P}(\alpha) = |\psi(\alpha)|^2 . \tag{A.20}$$

The probability amplitude is a complex quantity. In terms of the probability amplitude, the statement that the *probability of obtaining any value is equal to certainty* reads

$$\sum_j |\Psi_j|^2 = 1 \quad or \quad \int_{-\infty}^{+\infty} d\alpha\, |\psi(\alpha)|^2 = 1 . \tag{A.21}$$

A.2 Fourier Transforms

Consider a square-integrable function $f(x)$ that takes up complex values. By definition,

$$\int_{-\infty}^{+\infty} dx\, |f(x)|^2 < \infty . \tag{A.22}$$

Its *Fourier transform* is defined as

$$g(k) = \int_{-\infty}^{+\infty} \frac{dx}{\sqrt{2\pi}} f(x)\, e^{ikx} . \tag{A.23}$$

It can be shown that $g(k)$ is also a square-integrable function, i.e.,

$$\int_{-\infty}^{+\infty} dk\, |g(k)|^2 < \infty \tag{A.24}$$

and that the *inverse Fourier transform* exists, namely

$$f(x) = \int_{-\infty}^{+\infty} \frac{dk}{\sqrt{2\pi}} \, g(k) \, e^{-ikx} \,. \tag{A.25}$$

Furthermore, the following relation is true (Plancherel's Theorem):

$$\int_{-\infty}^{+\infty} dx \, |f(x)|^2 = \int_{-\infty}^{+\infty} dk \, |g(k)|^2 \,. \tag{A.26}$$

Fourier transformations can also be defined in three dimensions as

$$g(\mathbf{k}) = \int \frac{d^3 r}{(2\pi)^{3/2}} \, f(\mathbf{r}) \, e^{i\mathbf{k}\cdot\mathbf{r}} \quad and \quad f(\mathbf{r}) = \int \frac{d^3 k}{(2\pi)^{3/2}} \, g(\mathbf{k}) \, e^{-i\mathbf{k}\cdot\mathbf{r}} \,, \tag{A.27}$$

while *Plancherel's Theorem* reads

$$\int d^3 r \, |f(\mathbf{r})|^2 = \int d^3 k \, |g(\mathbf{k})|^2 \,. \tag{A.28}$$

A.3 Generalized Functions

The *Theta function* or *step function* $\Theta(x)$ is defined as

$$\Theta(x) = \begin{cases} 0 & (x < 0) \\ 1 & (x > 0). \end{cases} \tag{A.29}$$

Obviously, it is not continuous at $x = 0$ and $\Theta(0)$ is not defined. The derivative of $\Theta(x)$ vanishes everywhere except at $x = 0$, where it does not exist. Despite that we may write formally

$$\int_{-\infty}^{+\infty} dx \, \frac{d\Theta}{dx} = \Theta(+\infty) - \Theta(-\infty) = 1 \,. \tag{A.30}$$

Therefore, we may introduce another generalized function that vanishes everywhere except at $x = 0$, where it is infinite, but, despite that, it has an integral over the entire real line equal to 1. This is the *Dirac Delta function*, defined as

$$\delta(x) = \begin{cases} 0 & (x \neq 0) \\ \infty & (x = 0). \end{cases} \tag{A.31}$$

and

$$\int_{-\infty}^{+\infty} dx \, \delta(x) = 1 . \tag{A.32}$$

It is obvious that for any regular function $f(x)$, we shall have

$$f(x) \, \delta(x) = f(0) \, \delta(x) \tag{A.33}$$

or

$$\int_{-\infty}^{+\infty} dx \, f(x) \, \delta(x) = f(0) . \tag{A.34}$$

All the above definitions and identities are generalized to an arbitrary point x_0 as

$$\delta(x - x_0) = \begin{cases} 0 & (x \neq x_0) \\ \infty & (x = x_0) \end{cases} \quad \left(\int_{-\infty}^{+\infty} dx \delta(x - x_0) f(x) = f(x_0) \right) . \tag{A.35}$$

Note also that the delta function is an even function

$$\delta(x - x_0) = \delta(x_0 - x) . \tag{A.36}$$

There are many representations of the delta function as a limit of ordinary functions. For example,

$$\delta(x) = \lim_{\epsilon \to 0} \frac{e^{-x^2/\epsilon}}{\sqrt{\pi\epsilon}} . \tag{A.37}$$

An alternative very useful integral representation, stated without proof for the sake of economy, is

$$\delta(x) = \int_{-\infty}^{+\infty} \frac{dk}{2\pi} e^{ikx} . \tag{A.38}$$

As a demonstration of the usefulness of the delta function we may use it to prove the existence of the inverse Fourier transform and Plancherel's theorem. We have

$$f(x) = \int_{-\infty}^{+\infty} \frac{dk}{\sqrt{2\pi}} g(k) e^{ikx} = \int_{-\infty}^{+\infty} \frac{dk}{\sqrt{2\pi}} \left(\int_{-\infty}^{+\infty} \frac{dx'}{\sqrt{2\pi}} f(x') e^{-ikx'} \right) e^{ikx}$$

$$= \int_{-\infty}^{+\infty} dx' f(x') \int_{-\infty}^{+\infty} \frac{dk}{2\pi} e^{ik(x-x')} =, \int_{-\infty}^{+\infty} dx' f(x') \delta(x - x') = f(x) .$$

Similarly

$$\int_{-\infty}^{+\infty} dx \, |f(x)|^2 = \int_{-\infty}^{+\infty} dx \left(\int_{-\infty}^{+\infty} \frac{dk}{\sqrt{2\pi}} g^*(k) e^{ikx} \right) \left(\int_{-\infty}^{+\infty} \frac{dk'}{\sqrt{2\pi}} g(k') e^{-ik'x} \right)$$

$$= \int_{-\infty}^{+\infty} dk \int_{-\infty}^{+\infty} dk' \, g^*(k) g(k') \int_{-\infty}^{+\infty} \frac{dx}{2\pi} e^{ix(k-k')} = \int_{-\infty}^{+\infty} dk \, g^*(k) \int_{-\infty}^{+\infty} dk' \, g(k') \, \delta(k - k')$$

$$= \int_{-\infty}^{+\infty} dk |g(k)|^2 .$$

When the argument of the delta function is a function itself it is easy to prove

$$\delta(f(x)) = \sum_j \frac{\delta(x - x_j)}{|f'(x_j)|} , \tag{A.39}$$

where x_j are the roots of the equation $f(x) = 0$. For example,

$$\delta(x^2 - x_0^2) = \frac{1}{2|x_0|} \left(\delta(x - x_0) + \delta(x + x_0) \right) . \tag{A.40}$$

A *three-dimensional delta function* can be defined in a straightforward way as a product

$$\delta(\mathbf{r} - \mathbf{r}_0) = \delta(x - x_0)\delta(y - y_0)\delta(z - z_0) . \tag{A.41}$$

In terms of it the defining property of the delta function takes the form

$$\int d^3 r' \, f(\mathbf{r}') \, \delta(\mathbf{r}' - \mathbf{r}) = f(\mathbf{r}) . \tag{A.42}$$

The corresponding integral representation reads

$$\delta(\mathbf{r} - \mathbf{r}_0) = \int \frac{d^3 k}{(2\pi)^3} e^{i\mathbf{k} \cdot \mathbf{r}} . \tag{A.43}$$

A.4 Gaussian and Other Integrals

A.4.1 One-Dimensional Integrals

Consider the integral

$$I = \int_{-\infty}^{+\infty} dx \, e^{-x^2} , \tag{A.44}$$

often referred to as *Euler's integral* or *Gaussian integral*, since the integrated function corresponds to the *Gaussian distribution*. It can easily be evaluated with elementary methods considering its square and employing polar coordinates as

$$I^2 = \int_{-\infty}^{+\infty} dx \int_{-\infty}^{+\infty} dy\, e^{-(x^2+y^2)} = \int_0^\infty dr \int_0^{2\pi} d\theta\, r\, e^{-r^2}$$

$$= 2\pi \int_0^\infty dr\, r\, e^{-r^2} = \pi \int_0^\infty d\xi\, e^{-\xi} = \pi \implies \int_{-\infty}^{+\infty} dx\, e^{-x^2} = \sqrt{\pi}.$$

Starting from this integral, we can deduce the more general one

$$I(\alpha) = \int_{-\infty}^{+\infty} dx\, e^{-\alpha x^2} = \sqrt{\frac{\pi}{\alpha}} \quad with \quad \mathrm{Re}(\alpha) \geq 0. \tag{A.45}$$

Integrals involving a Gaussian exponential times a power can be easily computed from $I(\alpha)$ by diferentation. For instance,

$$\int_{-\infty}^{+\infty} dx\, x^2\, e^{-\alpha x^2} = -\frac{dI}{d\alpha} = \frac{\sqrt{\pi}}{2\alpha\sqrt{\alpha}} \tag{A.46}$$

or for general even powers

$$I_{2n}(\alpha) = \int_{-\infty}^{+\infty} dx\, x^{2n}\, e^{-\alpha x^2} = (-1)^n \frac{d^n I}{d\alpha^n} = (-1)^n \frac{\partial^n}{\partial \alpha^n}\left(\sqrt{\frac{\pi}{\alpha}}\right). \tag{A.47}$$

Integrals with odd powers are automatically zero due to oddness

$$I_{2n+1}(\alpha) = \int_{-\infty}^{+\infty} dx\, x^{2n+1}\, e^{-\alpha x^2} = 0. \tag{A.48}$$

Integrals of the type

$$I(\alpha, \beta) = \int_{-\infty}^{+\infty} dx\, e^{-\alpha x^2 + \beta x} \tag{A.49}$$

can also be computed in the following fashion:

$$I = \int_{-\infty}^{+\infty} dx\, e^{-\alpha\left(x^2 - 2\frac{\beta}{\alpha}x + \frac{\beta^2}{4\alpha^2}\right)} e^{\frac{\beta^2}{4\alpha}} = e^{\frac{\beta^2}{4\alpha}} \int_{-\infty}^{+\infty} dx'\, e^{-\alpha x'^2}$$

or

$$I(\alpha, \beta) = \int_{-\infty}^{+\infty} dx\, e^{-\alpha x^2 + \beta x} = \sqrt{\frac{\pi}{\alpha}}\, e^{\frac{\beta^2}{4\alpha}}. \tag{A.50}$$

As a result, we may conclude that the Fourier transform of a Gaussian is also a Gaussian, namely

$$\int_{-\infty}^{+\infty} dx\, e^{-\alpha x^2 + ikx} = \sqrt{\frac{\pi}{\alpha}}\, e^{-\frac{k^2}{4\alpha}}. \tag{A.51}$$

Table A.1 Usefull Integrals

$\int_{-\infty}^{+\infty} dx\, e^{-\alpha x^2 + \beta x}$	$= \sqrt{\frac{\pi}{\alpha}}\, e^{\frac{\beta^2}{4\alpha}}$
$\int_{-\infty}^{+\infty} dx\, x^{2n}\, e^{-\alpha x^2}$	$= (-1)^n \frac{\partial^n}{\partial \alpha^n} \left(\sqrt{\frac{\pi}{\alpha}} \right)$
$\int_{-\infty}^{+\infty} dx\, x^{2n+1}\, e^{-\alpha x^2}$	$= 0$
$\int_{-\infty}^{+\infty} dx\, x^n\, e^{-\alpha x^2 + \beta x}$	$= \sqrt{\frac{\pi}{\alpha}} \frac{\partial^n}{\partial \beta^n} \exp\left(\beta^2 / 4\alpha \right)$
$\int_0^{+\infty} dx\, x^n\, e^{-\alpha x}$	$= \frac{n!}{\alpha^{n+1}}$

From the above integral, we can also compute

$$I_n(\alpha, \beta) = \int_{-\infty}^{+\infty} dx\, x^n\, e^{-\alpha x^2 + \beta x} = \frac{\partial^n I}{\partial \beta^n} = \sqrt{\frac{\pi}{\alpha}} \frac{\partial^n}{\partial \beta^n} \exp\left(\beta^2 / 4\alpha \right), \quad (A.52)$$

for odd and even values of the integer n.

Another set of useful integrals that can be calculated with elementary methods are the integrals

$$J_n = \int_0^{+\infty} dx\, x^n\, e^{-x}. \qquad (A.53)$$

Introducing a parameter α we have

$$J_n(\alpha) = \int_0^{+\infty} dx\, x^n\, e^{-\alpha x} = (-1)^n \frac{\partial^n}{\partial \alpha^n} \int_0^{+\infty} dx\, e^{-\alpha x}$$

$$= (-1)^n \frac{\partial^n \alpha^{-1}}{\partial \alpha^n} = (-1)^n (-1)2(-1)3 \ldots (-1)n\, \alpha^{-(n+1)} = \frac{n!}{\alpha^{n+1}}. \quad (A.54)$$

Thus, we obtain

$$J_n = \int_0^{+\infty} dx\, x^n\, e^{-x} = n!. \qquad (A.55)$$

We have summarized our results in the Table A.1.

A.4.2 Three-Dimensional Integrals

Three-dimensional Gaussian integrals can also be derived in an analogous fashion. Starting from the basic integral

$$I(\alpha) = \int d^3 r\, e^{-\alpha r^2} = \left(\frac{\pi}{\alpha} \right)^{3/2}, \qquad (A.56)$$

thanks to spherical symmetry, we have

$$I_{ij}(\alpha) = \int d^3r \, x_i x_j \, e^{-\alpha r^2} = \frac{\delta_{ij}}{3} \int d^3r \, r^2 \, e^{-\alpha r^2}$$

$$= -\frac{\delta_{ij}}{3} \frac{\partial}{\partial \alpha} \int d^3r \, e^{-\alpha r^2} = \frac{\delta_{ij}}{3} \frac{3}{2} \frac{\pi^{3/2}}{\alpha^{5/2}}$$

or

$$I_{ij}(\alpha) = \int d^3r \, x_i x_j \, e^{-\alpha r^2} = \frac{\delta_{ij}}{2\alpha} \left(\frac{\pi}{\alpha}\right)^{3/2} . \tag{A.57}$$

Linear integrals $I_j(\alpha) = \int d^3r \, x_j \, e^{-\alpha r^2}$ vanish due to oddness.

We can also compute

$$I(\alpha, \mathbf{b}) = \int d^3r \, e^{-\alpha r^2 + \mathbf{b} \cdot \mathbf{r}} = e^{\frac{b^2}{4\alpha}} \left(\frac{\pi}{\alpha}\right)^{3/2} . \tag{A.58}$$

Next, we may compute the vector integrals

$$J_i = \int d^3r \, x_i \, e^{-\alpha r^2 + \mathbf{b} \cdot \mathbf{r}} = \frac{\partial}{\partial b_i} I(\alpha, \mathbf{b})$$

or

$$\mathbf{J} = \int d^3r \, \mathbf{r} \, e^{-\alpha r^2 + \mathbf{b} \cdot \mathbf{r}} = \frac{\mathbf{b}}{2\alpha} e^{\frac{b^2}{4\alpha}} \left(\frac{\pi}{\alpha}\right)^{3/2} . \tag{A.59}$$

General integrals of the type

$$I_{i_1 i_2 \dots i_n}(\alpha, \mathbf{b}) = \int d^3r \, x_{i_1} x_{i_2} \dots x_{i_n} \, e^{-\alpha r^2 + \mathbf{b} \cdot \mathbf{r}} \tag{A.60}$$

can be calculated from the formula

$$I_{i_1 i_2 \dots i_n}(\alpha, \mathbf{b}) = \frac{\partial^n I(\alpha, \mathbf{b})}{\partial b_{i_1} \partial b_{i_2} \dots} = \left(\frac{\pi}{\alpha}\right)^{3/2} \frac{\partial^n e^{\frac{b^2}{4\alpha}}}{\partial b_{i_1} \partial b_{i_2} \dots} . \tag{A.61}$$

A.5   Operator Identities

Addition and multiplication of operators are defined through the following rules:

$$A + B = B + A$$

$$A + 0 = 0 + A = A$$

$$A + (B + C) = (A + B) + C$$

$$A(BA) = (AB)C$$

$$AI = IA = A$$ (A.62)

$$AA^{-1} = A^{-1}A = I.$$

Linear operators, satisfy

$$Aa = aA, \quad (a \in C).$$ (A.63)

Only the case of *Time Reversal* corresponds to an *antilinear* operator satisfying

$$Ta = a^*T.$$ (A.64)

The multiplication of operators is not commutative. For a string of operators, we have

$$(ABC\ldots)^{-1} = \ldots C^{-1}B^{-1}A^{-1}$$

$$(ABC\ldots)^\dagger = \ldots C^\dagger B^\dagger A^\dagger,$$ (A.65)

where † is the Hermitian conjugation operation.[3] In contrast,

$$(ABC\ldots)^* = A^*B^*C^*\ldots.$$ (A.66)

A useful quantity is the *Commutator* of two operators, defined as

$$[A, B] = AB - BA.$$ (A.67)

The following set of commutator identities are useful:

$$[aA + bB, C] = a[A, C] + b[B, C],$$

$$[AB, C] = A[B, C] + [A, C]B,$$ (A.68)

$$[[A, B], C] + [[C, A], B] + [[B, C], A] = 0.$$

Functions of operators can also be defined for functions that are expandable in power series. For example, for a function expandable around zero $f(x) = \sum_{n=0}^{\infty} \frac{f^{(n)}(0)}{n!} x^n$, we may define for an operator A, the operator

---

[3]The definition of Hermitian conjugation for operators acting on vectors $|\psi_1\rangle, |\psi_2\rangle, \ldots$ is $\langle\psi_1|\hat{A}^\dagger|\psi_2\rangle = \langle\psi_2|\hat{A}|\psi_1\rangle^*$.

$$f(A) = \sum_{n=0}^{\infty} \frac{f^{(n)}(0)}{n!} A^n = \mathbf{I} + f'(0) A + \frac{1}{2} f''(0) A^2 + \dots . \tag{A.69}$$

Typical specific cases are

$$e^A = \sum_{n=0}^{\infty} \frac{A^n}{n!}, \quad \ln(1+A) = \sum_{n=1}^{\infty} (-1)^{n+1} \frac{A^n}{n} . \tag{A.70}$$

The exponentials of operators are of particular importance, since for any Hermitian operator A, the exponential operator $U(\alpha) = e^{i\alpha A}$, where α is a real number, is *unitary*

$$U^\dagger(\alpha) = e^{-i\alpha A^\dagger} = e^{-i\alpha A} = U^{-1}(\alpha) \implies U U^\dagger = U^\dagger U = \mathbf{I}. \tag{A.71}$$

For two non-commuting operators A and B, the product of their exponentials is a complicated function that cannot be written in a closed form. There is, however, a notable exception when their commutator is a complex number

$$[A, B] = c \in \mathbf{C} \implies [A, [A, B]] = [B, [A, B]] = 0$$

and

$$e^A e^B = e^{A+B} e^{\frac{1}{2}[A, B]} . \tag{A.72}$$

Note also the following version of this relation:

$$e^A e^B = e^B e^A e^{[A, B]} . \tag{A.73}$$

A.6 Coordinate Systems

Cartesian Coordinates and Differential Operators. In the standard *Cartesian Coordinates* a position vector is analyzed in terms of the three unit vectors $\hat{\mathbf{x}}$, $\hat{\mathbf{y}}$, $\hat{\mathbf{z}}$ (the "hat" here does not denote an operator but a unit c-number vector) as

$$\mathbf{r} = x \hat{\mathbf{x}} + y \hat{\mathbf{y}} + z \hat{\mathbf{z}} \tag{A.74}$$

or

$$\mathbf{r} = \sum_{i=1,2,3} x_i \hat{\mathbf{x}}_i , \tag{A.75}$$

where $x_1 = x$, $x_2 = y$, $x_3 = z$. An arbitrary vector function will be

$$\mathbf{A} = A_x\,\hat{\mathbf{x}} + A_y\,\hat{\mathbf{y}} + A_z\,\hat{\mathbf{z}} = \sum_i A_i\hat{\mathbf{x}}_i, \tag{A.76}$$

where $A_i = \mathbf{A} \cdot \hat{\mathbf{x}}_i$.

The *Gradient* operator is defined as

$$\nabla = \hat{\mathbf{x}}\frac{\partial}{\partial x} + \hat{\mathbf{y}}\frac{\partial}{\partial y} + \hat{\mathbf{z}}\frac{\partial}{\partial z} = \sum_i \hat{\mathbf{x}}_i\frac{\partial}{\partial x_i}. \tag{A.77}$$

The *Divergence* of a vector function is

$$\nabla \cdot \mathbf{A} = \sum_i \frac{\partial A_i}{\partial x_i} = \frac{\partial A_x}{\partial x} + \frac{\partial A_y}{\partial y} + \frac{\partial A_z}{\partial z}. \tag{A.78}$$

The *Curl* of a vector function is defined as

$$\nabla \times \mathbf{B} = \begin{vmatrix} \hat{\mathbf{x}} & \hat{\mathbf{y}} & \hat{\mathbf{z}} \\ \frac{\partial}{\partial x} & \frac{\partial}{\partial y} & \frac{\partial}{\partial z} \\ B_x & B_y & B_z \end{vmatrix} \tag{A.79}$$

or

$$(\nabla \times \mathbf{B})_i = \sum_{j,k} \epsilon_{ijk}\frac{\partial B_k}{\partial x_j}. \tag{A.80}$$

The *Laplacian* ∇^2 of a scalar or vector function is the operator

$$\nabla^2 = \sum_i \frac{\partial^2}{\partial x_i^2} = \frac{\partial^2}{\partial x^2} + \frac{\partial^2}{\partial y^2} + \frac{\partial^2}{\partial z^2}. \tag{A.81}$$

It is straightforward to show that the curl of a gradient and the divergence of a curl vanish identically

$$\nabla \times \nabla f = 0, \quad \nabla \cdot (\nabla \times \mathbf{g}) = 0. \tag{A.82}$$

Cylindrical Coordinates. These are defined in terms of the standard Cartesian ones as

$$\begin{cases} \rho = \sqrt{x^2 + y^2} \\ \phi = \arctan(y/x) \quad and \\ z = z \end{cases} \quad \begin{cases} x = \rho \cos\phi \\ y = \rho \sin\phi \\ z = z \end{cases} \tag{A.83}$$

with

$$0 \leq \rho < \infty, \quad -\infty < z < +\infty, \quad 0 \leq \phi \leq 2\pi. \tag{A.84}$$

A system of *cylindrical unit vectors* can be introduced as

$$\begin{cases} \hat{\rho} = \hat{\mathbf{x}}\cos\phi + \hat{\mathbf{y}}\sin\phi \\[2mm] \hat{\phi} = -\hat{\mathbf{x}}\sin\phi + \hat{\mathbf{y}}\cos\phi \\[2mm] \hat{\mathbf{z}} = \hat{\mathbf{z}} \end{cases} \textbf{with} \begin{cases} \hat{\rho}\cdot\hat{\phi} = \hat{\rho}\cdot\hat{\mathbf{z}} = \hat{\phi}\cdot\hat{\mathbf{z}} = 0 \\[2mm] \hat{\rho}\times\hat{\phi} = \hat{\mathbf{z}} \\[2mm] \hat{\phi}\times\hat{\mathbf{z}} = \hat{\rho}. \end{cases} \tag{A.85}$$

The inverse relations are also usefull[4]

$$\begin{cases} \hat{x} = \hat{\rho}\cos\phi - \hat{\phi}\sin\phi \\[2mm] \hat{y} = \hat{\rho}\sin\phi + \hat{\phi}\cos\phi. \end{cases} \tag{A.86}$$

In contrast to the Cartesian unit vectors $\hat{\rho}$ and $\hat{\phi}$ are not constant but depend on the angle ϕ. Thus, we have

$$\frac{\partial\hat{\rho}}{\partial\phi} = \hat{\phi}, \quad \frac{\partial\hat{\phi}}{\partial\phi} = -\hat{\rho}. \tag{A.87}$$

The position vector in terms of the cylindrical ones is

$$\mathbf{r} = \rho\hat{\rho} + z\hat{\mathbf{z}}. \tag{A.88}$$

An arbitrary vector function can also be decomposed as

$$\mathbf{A} = \hat{\rho}A_\rho + \hat{\phi}A_\phi + \hat{\mathbf{z}}A_z. \tag{A.89}$$

The gradient in cylindrical coordinates takes the form

$$\nabla = \hat{\rho}\frac{\partial}{\partial\rho} + \frac{\hat{\phi}}{\rho}\frac{\partial}{\partial\phi} + \hat{\mathbf{z}}\frac{\partial}{\partial z}. \tag{A.90}$$

From this, we get easily the Laplacian

$$\nabla^2 = \frac{\partial^2}{\partial\rho^2} + \frac{1}{\rho}\frac{\partial}{\partial\rho} + \frac{1}{\rho^2}\frac{\partial^2}{\partial\phi^2} + \frac{\partial^2}{\partial z^2}. \tag{A.91}$$

---

[4]Usefull relations are also

$$\frac{\partial}{\partial x} = \cos\phi\frac{\partial}{\partial\rho} - \frac{\sin\phi}{\rho}\frac{\partial}{\partial\phi}, \quad \frac{\partial}{\partial y} = \sin\phi\frac{\partial}{\partial\rho} + \frac{\cos\phi}{\rho}\frac{\partial}{\partial\phi}.$$

The divergence of (A.89) in cylindrical coordinates is

$$\nabla \cdot \mathbf{A} = \frac{\partial A_\rho}{\partial \rho} + \frac{1}{\rho} A_\rho + \frac{1}{\rho}\frac{\partial A_\phi}{\partial \phi} + \frac{\partial A_z}{\partial z}. \tag{A.92}$$

Similarly, we can obtain the curl of (A.89)

$$\nabla \times \mathbf{A} = \hat{\rho}\left(\frac{1}{\rho}\frac{\partial A_z}{\partial \phi} - \frac{\partial A_\phi}{\partial z}\right) + \hat{\phi}\left(\frac{\partial A_\rho}{\partial z} - \frac{\partial A_z}{\partial \rho}\right) + \hat{z}\left(\frac{\partial A_\phi}{\partial \rho} + \frac{1}{\rho}A_\phi - \frac{1}{\rho}\frac{\partial A_\rho}{\partial \phi}\right). \tag{A.93}$$

Spherical Coordinates. These are defined in terms of the Cartesian ones as

$$\begin{cases} r = \sqrt{x^2 + y^2 + z^2} \\ \theta = \arctan(\sqrt{x^2 + y^2}/z) \quad \textbf{and} \\ \phi = \arctan(y/x) \end{cases} \begin{cases} x = r\sin\theta\cos\phi \\ y = r\sin\theta\sin\phi \\ z = r\cos\theta \end{cases} \tag{A.94}$$

with $0 \le r < \infty$, $0 \le \theta \le \pi$ and $0 \le \phi \le 2\pi$. Unit vectors can be introduced as

$$\begin{cases} \hat{\mathbf{r}} = \hat{\mathbf{x}}\sin\theta\cos\phi + \hat{\mathbf{y}}\sin\theta\sin\phi + \hat{\mathbf{z}}\cos\theta \\ \hat{\theta} = \hat{\mathbf{x}}\cos\theta\cos\phi + \hat{\mathbf{y}}\cos\theta\sin\phi - \hat{\mathbf{z}}\sin\theta \quad \textbf{and} \\ \hat{\phi} = -\hat{\mathbf{x}}\sin\phi + \hat{\mathbf{y}}\cos\phi \end{cases} \begin{cases} \hat{\mathbf{r}}\cdot\hat{\theta} = \hat{\mathbf{r}}\cdot\hat{\phi} = \hat{\theta}\cdot\hat{\phi} = 0 \\ \hat{\mathbf{r}}\times\hat{\theta} = \hat{\phi} \\ \hat{\theta}\times\hat{\phi} = \hat{\mathbf{r}}. \end{cases} \tag{A.95}$$

We also have

$$\begin{cases} \frac{\partial\hat{\mathbf{r}}}{\partial\theta} = \hat{\theta} & \frac{\partial\hat{\mathbf{r}}}{\partial\phi} = \sin\theta\,\hat{\phi} \\ \frac{\partial\hat{\theta}}{\partial\theta} = -\hat{\mathbf{r}} & \frac{\partial\hat{\theta}}{\partial\phi} = \cos\theta\,\hat{\phi}. \end{cases} \tag{A.96}$$

The gradient in spherical coordinates takes the form

$$\nabla = \hat{\mathbf{r}}\frac{\partial}{\partial r} + \frac{\hat{\theta}}{r}\frac{\partial}{\partial\theta} + \frac{\hat{\phi}}{r\sin\theta}\frac{\partial}{\partial\phi} \tag{A.97}$$

and the Laplacian is

$$\nabla^2 = \frac{\partial^2}{\partial r^2} + \frac{2}{r}\frac{\partial}{\partial r} + \frac{\cot\theta}{r^2}\frac{\partial}{\partial\theta} + \frac{1}{r^2}\frac{\partial^2}{\partial\theta^2} + \frac{1}{r^2\sin^2\theta}\frac{\partial^2}{\partial\phi^2}. \tag{A.98}$$

Consider now a vector function expanded in spherical coordinates

$$\mathbf{A} = \hat{\mathbf{r}}\,A_r + \hat{\theta}\,A_\theta + \hat{\phi}\,A_\phi. \tag{A.99}$$

Its divergence is

$$\nabla \cdot \mathbf{A} = \frac{\partial A_r}{\partial r} + \frac{2}{r} A_r + \frac{\cot\theta}{r} A_\theta + \frac{1}{r} \frac{\partial A_\theta}{\partial \theta} + \frac{1}{r\sin\theta} \frac{\partial A_\phi}{\partial \phi} . \tag{A.100}$$

Similarly, its curl is

$$\nabla \times \mathbf{A} = \frac{\hat{\mathbf{r}}}{r\sin\theta} \left(\cos\theta A_\phi + \frac{\partial A_\phi}{\partial \theta} - \frac{\partial A_\theta}{\partial \phi} \right) + \hat{\theta} \left(\frac{1}{r\sin\theta} \frac{\partial A_r}{\partial \phi} - \frac{1}{r} A_\phi - \frac{\partial A_\phi}{\partial r} \right)$$
$$+ \hat{\phi} \left(\frac{1}{r} A_\theta + \frac{\partial A_\theta}{\partial r} - \frac{1}{r} \frac{\partial A_r}{\partial \theta} \right) . \tag{A.101}$$

A.7 Tensor Products

Tensor Products. Consider two vector spaces $\mathcal{E}_1$ and $\mathcal{E}_2$. Their dimensions N_1 and N_2 are in general different. A *tensor product* is an operation that maps any pair of vectors that belong to $\mathcal{E}_1$ and $\mathcal{E}_2$ to a vector of a new vector space of dimension $N = N_1 N_2$ denoted by $\mathcal{E}_1 \otimes \mathcal{E}_2$

$$|a\rangle \in \mathcal{E}_1, \quad |b\rangle \in \mathcal{E}_2 \implies |a\rangle \otimes |b\rangle \in \mathcal{E}_1 \otimes \mathcal{E}_2. \tag{A.102}$$

By convention the vector of space "1" is put on the left while the vector of space "2" on the right. The operation is linear in the sense

$$\left(\alpha|a\rangle + \alpha'|a'\rangle \right) \otimes |b\rangle = \alpha|a\rangle \otimes |b\rangle + \alpha'|a'\rangle \otimes |b\rangle \tag{A.103}$$

and

$$|a\rangle \otimes \left(\beta|b\rangle + \beta'|b'\rangle \right) = \beta|a\rangle \otimes |b\rangle + \beta'|a\rangle \otimes |b'\rangle \tag{A.104}$$

for any complex numbers α, α', β, β'. In addition to that, the null element of each vector space maps any vector to the null element of the tensor product space

$$0 \otimes |b\rangle = |a\rangle \otimes 0 = 0. \tag{A.105}$$

For Euclidean vector spaces with an inner product, the rules for the inner product in the tensor product space is

$$(\langle a| \otimes \langle b|) \cdot (|c\rangle \otimes |d\rangle) = \langle a|c\rangle \langle b|d\rangle . \tag{A.106}$$

For an orthonormal basis $|e_{i_1}\rangle$ in $\mathcal{E}_1$ and an orthonormal basis $|\varepsilon_{i_2}\rangle$ in $\mathcal{E}_2$, we can obtain an orthonormal basis in $\mathcal{E}_1 \otimes \mathcal{E}_2$ as

$$|E_{i_1 i_2}\rangle = |e_{i_1}\rangle \otimes |\varepsilon_{i_2}\rangle \implies \langle E_{i_1 i_2}|E_{i'_1 i'_2}\rangle = \delta_{i_1 i'_1}\delta_{i_2 i'_2}. \qquad \text{(A.107)}$$

Note, however, that not all vectors belonging to the tensor product space $\mathcal{E}$ are tensor products of vectors of $\mathcal{E}_1$ and $\mathcal{E}_2$. For example, the vector

$$|\Psi\rangle = |a\rangle \otimes |b\rangle + |c\rangle \otimes |d\rangle \qquad \text{(A.108)}$$

clearly belongs to $\mathcal{E}$ and can be expanded in the basis $|E_{i_1 i_2}\rangle$ as

$$|\Psi\rangle = \sum_{i_1 i_2} \left(a_{i_1} b_{i_2} + c_{i_1} d_{i_2}\right) |e_{i_1}\rangle \otimes |\varepsilon_{i_2}\rangle = \sum_{i_1 i_2} \left(a_{i_1} b_{i_2} + c_{i_1} d_{i_2}\right) |E_{i_1 i_2}\rangle. \qquad \text{(A.109)}$$

In terms of the given bases, vectors can be represented as collumns and rows. To be concrete let's take $N_1 = 2$ and $N_2 = 3$. Then, we have

$$|a\rangle \in \mathcal{E}_1 \implies \begin{pmatrix} a_1 \\ a_2 \end{pmatrix} \qquad |b\rangle \in \mathcal{E}_2 \implies \begin{pmatrix} b_1 \\ b_2 \\ b_3 \end{pmatrix},$$

$$|a\rangle \otimes |b\rangle \implies \begin{pmatrix} a_1 \\ a_2 \end{pmatrix} \otimes \begin{pmatrix} b_1 \\ b_2 \\ b_3 \end{pmatrix} = \begin{pmatrix} a_1 b_1 \\ a_1 b_2 \\ a_1 b_3 \\ a_2 b_1 \\ a_2 b_2 \\ a_2 b_3 \end{pmatrix}.$$

Operators that act in the spaces $\mathcal{E}_1$ and $\mathcal{E}_2$ acting on the states $|a\rangle \in \mathcal{E}_1$ and $|b\rangle \in \mathcal{E}_2$ have a straightforward generalization in product space. For example, an operator

$$\hat{A} \in \mathcal{E}_1 \quad \rightarrow \quad \hat{A}|a\rangle = |a'\rangle \qquad \text{(A.110)}$$

is generalized to

$$\hat{A} \otimes \mathbf{I} \quad \rightarrow \quad \left(\hat{A} \otimes \mathbf{I}\right)|a\rangle \otimes |b\rangle = |a'\rangle \otimes |b\rangle. \qquad \text{(A.111)}$$

Similarly for an operator acting in $\mathcal{E}_2$ as $\hat{B}|b\rangle = |b'\rangle$

$$\mathbf{I} \otimes \hat{B} \;\rightarrow\; \left(\mathbf{I} \otimes \hat{B}\right)|a\rangle \otimes |b\rangle = |a\rangle \otimes |b'\rangle . \tag{A.112}$$

Thus, a *tensor product of operators* $\hat{A} \otimes \hat{B}$ is defined to act according to

$$\left(\hat{A} \otimes \hat{B}\right)|a\rangle \otimes |b\rangle = |a'\rangle \otimes |b'\rangle . \tag{A.113}$$

Action on more general states is obvious

$$\left(\hat{A} \otimes \hat{B}\right)\left(\,\lambda|c\rangle \otimes |d\rangle + \mu|e\rangle \otimes |f\rangle\,\right) = \lambda|c'\rangle \otimes |d'\rangle + \mu|e'\rangle \otimes |f'\rangle . \tag{A.114}$$

A matrix representation of the tensor product of operators can be obtained in a straightforward fashion as follows:

$$\hat{A} \otimes \hat{B} = \left(\sum_{i_1 j_1} |e_{i_1}\rangle A_{i_1 j_1} \langle e_{j_1}|\right) \otimes \left(\sum_{i_2 j_2} |\varepsilon_{i_2}\rangle B_{i_2 j_2} \langle \varepsilon_{j_2}|\right) =$$

$$\sum_{i_1 i_2 j_1 j_2} \left(|e_{i_1}\rangle \otimes |\varepsilon_{i_2}\rangle\right) A_{i_1 j_1} B_{i_2 j_2} \left(\langle e_{j_1}| \otimes \langle \varepsilon_{j_2}|\right) = \sum_{i_1 i_2 j_1 j_2} |E_{i_1 i_2}\rangle A_{i_1 j_1} B_{i_2 j_2} \langle E_{j_1 j_2}|$$

or

$$\hat{C} = \hat{A} \otimes \hat{B} \implies C_{i_1 i_2, j_1 j_2} = A_{i_1 j_1} B_{i_2 j_2} . \tag{A.115}$$

As an example, we may consider the tensor product of two spin operators

$$s_{1x} \otimes s_{2x} = \frac{\hbar^2}{2} \begin{pmatrix} 0 & 1 \\ 1 & 0 \end{pmatrix} \otimes \begin{pmatrix} 0 & 1 \\ 1 & 0 \end{pmatrix} = \frac{\hbar^2}{2} \begin{pmatrix} 0\begin{pmatrix} 0 & 1 \\ 1 & 0 \end{pmatrix} & 1\begin{pmatrix} 0 & 1 \\ 1 & 0 \end{pmatrix} \\ 1\begin{pmatrix} 0 & 1 \\ 1 & 0 \end{pmatrix} & 0\begin{pmatrix} 0 & 1 \\ 1 & 0 \end{pmatrix} \end{pmatrix}$$

$$= \frac{\hbar^2}{2} \begin{pmatrix} 0 & 0 & 0 & 1 \\ 0 & 0 & 1 & 0 \\ 0 & 1 & 0 & 0 \\ 1 & 0 & 0 & 0 \end{pmatrix}$$

Note that in a shorthand notation, the $\otimes$ symbol will be often ommited and the tensor product will be simply denoted as $|a\rangle|b\rangle$ or AB for the operators, provided that it is understood that $|a\rangle$ and A refer to $\mathcal{E}_1$ while $|b\rangle$ and B refer to $\mathcal{E}_2$.

Index

© Springer Nature Switzerland AG 2019
K. Tamvakis, *Basic Quantum Mechanics*, Undergraduate Texts in Physics,
https://doi.org/10.1007/978-3-030-22777-7

Printed in the United States
By Bookmasters